New Frontiers in Astronomy

New Frontiers in Astronomy

Edited by Adam Powell

SYRAWOOD
PUBLISHING HOUSE
New York

Published by Syrawood Publishing House,
750 Third Avenue, 9th Floor,
New York, NY 10017, USA
www.syrawoodpublishinghouse.com

New Frontiers in Astronomy
Edited by Adam Powell

International Standard Book Number: 978-1-68286-854-6 (Hardback)

Cataloging-in-Publication Data

New frontiers in astronomy / edited by Adam Powell.
p. cm.
Includes bibliographical references and index.
ISBN 978-1-68286-854-6
1. Astronomy. 2. Astronomical instruments. 3. Space sciences. I. Powell, Adam.
QB43.3 .N49 2020
520--dc23

TABLE OF CONTENTS

PREFACE

Astronomy is the natural science that involves the observation and explanation of various events occurring outside Earth and its atmosphere. It applies chemistry, mathematics and physics to study the origin, phenomena and evolution of celestial bodies. The celestial objects studied under this discipline include nebulae, planets, stars, moons, galaxies and comets. It also studies phenomena like quasars, cosmic microwave background radiation, supernova, pulsars and gamma-ray bursts. Astronomy can be further classified into solar astronomy, astrophysics, stellar astronomy, astrochemistry, galactic astronomy, astrobiology and cosmology. This book attempts to understand the multiple branches that fall under the discipline of astronomy and how such concepts have practical applications. It strives to provide a fair idea about this field and to help develop a better understanding of the latest advances within this area of study. Scientists and students actively engaged in this field will find it full of crucial and unexplored concepts.

The researches compiled throughout the book are authentic and of high quality, combining several disciplines and from very diverse regions from around the world. Drawing on the contributions of many researchers from diverse countries, the book's objective is to provide the readers with the latest achievements in the area of research. This book will surely be a source of knowledge to all interested and researching the field.

In the end, I would like to express my deep sense of gratitude to all the authors for meeting the set deadlines in completing and submitting their research chapters. I would also like to thank the publisher for the support offered to us throughout the course of the book. Finally, I extend my sincere thanks to my family for being a constant source of inspiration and encouragement.

Editor

1

High-Precision Heading Determination based on the Sun for Mars Rover

Yinhu Zhan, [1,2] **Shaojie Chen**, [2,3] **and Donghan He** [2]

[1] *State Key Laboratory of Geo-Information Engineering, Xian, China*
[2] *Zhengzhou Institute of Surveying and Mapping, Zhengzhou, China*
[3] *National Time Service Center, Chinese Academy of Sciences, Xian, China*

Correspondence should be addressed to Shaojie Chen; 867347382@qq.com

Academic Editor: Jean-Pierre Barriot

Since the American Mars Exploration Rover Opportunity landed on Mars in 2004, it has travelled more than 40 km, and heading-determination technology based on its sun sensor has played an important role in safe driving of the rover. A high-precision heading-determination method will always play a significant role in the rover's autonomous navigation system, and the precision of the measured heading strongly affects the navigation results. In order to improve the heading precision to the 1-arcminute level, this paper puts forward a novel calibration algorithm for solving the comparable distortion of large-field sun sensor by introducing an antisymmetric matrix. The sun sensor and inclinometer alignment model are then described in detail to maintain a high-precision horizon datum, and a strict sun image centroid-extraction algorithm combining subpixel edge detection with circle or ellipse fitting is presented. A prototype comprising a sun sensor, electronic inclinometer, and chip-scale atomic clock is developed for testing the algorithms, models, and methods presented in this paper. Three field tests were conducted in different months during 2017. The results show that the precision of the heading determination reaches 0.28–0.97′ (1σ) and the centroid error of the sun image and the sun elevation are major factors that affect the heading precision.

1. Introduction

Spirit and Opportunity, the twin rovers of NASA's Mars Exploration Mission, landed on Mars in 2004. Some discoveries of the rovers, such as the evidence of liquid water on the surface of Mars, have been a source of excitement. Opportunity has been kept in service for 13 years and has travelled over 42.195 km, becoming a veritable marathon champion. High-precision attitude-determination technology of Mars rovers, especially the heading-determination technology, has played an important role in safe driving and the realization of scientific goals. Because there is no satellite navigation system like GPS on Mars, the rovers utilize Pancam cameras as sun sensors for heading, which can restrict the error growth of the IMU and improve the dead-reckoning accuracy [1]. When the rover remains static, the sun sensor makes a 10-minute tracking observation of the sun, and the quaternion estimator (QUEST) method is used to calculate the attitude and heading, the precision of which reaches 1.5°. Because the field view of sun sensor on Spirit and Opportunity is only 16°, a rotation platform with a dial is needed. The rover first rotates the sun sensor to the predicted elevation of the sun and then horizontally scans the sky to find the sun [2]. Long-term searching and observation of the sun do not fulfill the real-time navigation requirement for Mars rovers, and the low-precision heading greatly affects the dead reckoning.

Several Jet Propulsion Laboratory (JPL) studies have also used the sun sensor for heading determination for Rocky 7 and FIDO field rover, but the method is similar to that of Spirit and Opportunity, and the precision in the field test is also to within a few degrees [3, 4]. Deans et al. bundled a fish-eye camera and an inclinometer together, and the precision of the heading determination is superior to 1° [5]. Most published heading-determination methods require long-duration observations only when inclinometer data is unavailable. Enright et al. and Furgale et al. use a digital sun sensor with 140° field of view and an inclinometer called HMR-3000 for heading, and practical tests on Earth indicate that the precision reaches 1° [6, 7]. Yang et al. use a large-field-of-view sun sensor for heading determination, and the precision reaches 0.1123° when observing the sun for 30 minutes, which is the best reported precision so far [8]. Illyas et al. present a novel algorithm for micro-planetary rover-heading determination using a low-cost sun sensor, and a large number of experiments show that the heading precision reaches 0.09° (1σ), which plays an important role in reducing the accumulated heading error of MEMS sensors [9]. In a GPS-denied environment, visual navigation can provide accurate localization [10], but the error grows sharply with the distance travelled. Lambert et al. develop a novel approach incorporating the sun sensor and inclinometer measurements directly into the visual odometry pipeline to reduce the error growth of path estimation, and the resulting localization error is only 1.1% of a 10-km distance travelled [11].

The Mars rover-heading determination method via the inertial navigation system and star sensor has also been thoroughly researched. A novel method based on star sensors and the strap-down inertial navigation system (SINS) is put forward to accurately determine the rover's position and attitude, and the initial position and attitude determination for planetary rovers by INS/Star Sensor integration is researched [12]. A high-precision SINS/star sensor deeply integrated navigation scheme is effected by He et al. [13]. In recent years, the inertial navigation system, star sensor, and visual navigation system have been integrated to improve the navigation accuracy and reliability [14]. However, Mars has some atmosphere, which makes stars invisible in the daytime. Furthermore, the rovers always move during the day and rest at night, which makes the star sensor difficult to apply.

It is evident that heading determination by the sun will always play a significant role in Mars exploration rovers, and the precision of the heading strongly affects the navigation results. Because the rovers move with a maximum speed of 5 cm/s on the surface of Mars and are stationary most of time, this paper only considers the problem of static heading determination. The main focus of this paper is using only one image of the sun to achieve a heading-determination precision on the order of 1 arcminute. This paper is organized as follows. First, we introduce the basic theories of heading determination using the sun, and major factors that affect the precision of the heading are analyzed. Next, the sun sensor calibration model, sun sensor, inclinometer alignment model, and sun image centroid-extraction algorithm are described in detail. After that, our prototype is introduced and three field tests are conducted to verify our models, methods, and algorithms. Finally, we give the basic conclusion according to the results of the field tests.

2. Basic Theories of Heading Determination Using the Sun

In this section, we introduce the main concepts related to this paper and provide basic theories of heading determination using sun sensor, inclinometer, and local clock. Table 1 describes the main frames in this paper and gives their names, notations, and definitions.

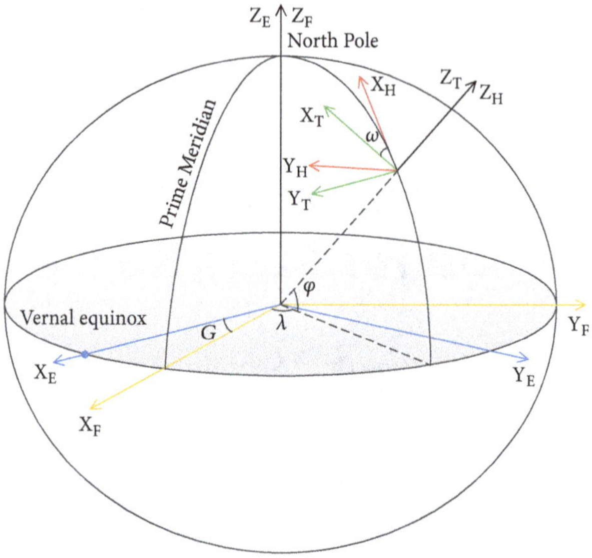

FIGURE 1: Space relation of the important frames.

Due to the objective conditions, all our field tests are conducted on Earth, so we only discuss the heading-determination problem in the Earth reference frame. Figure 1 provides the definitions and space relation of the important frames, and ω is the rover's heading to be estimated.

We assume that the sun vectors in the E and H frames are $\boldsymbol{S}_{\mathrm{E}}$ and $\boldsymbol{S}_{\mathrm{H}}$, respectively. $\boldsymbol{R}_{\mathrm{E}}^{\mathrm{H}}$ represents the rotation matrix from frame E to frame H, which can be written by

$$\boldsymbol{S}_{\mathrm{H}} = \boldsymbol{R}_{\mathrm{E}}^{\mathrm{H}} \cdot \boldsymbol{S}_{\mathrm{E}}. \tag{1}$$

According to Figure 1, $\boldsymbol{R}_{\mathrm{EH}}$ can be derived by double rotations:

$$\boldsymbol{R}_{\mathrm{E}}^{\mathrm{H}} = \boldsymbol{R}_{\mathrm{Y}}(\varphi - 90^\circ) \cdot \boldsymbol{R}_{\mathrm{Z}}(G + \lambda + 180^\circ), \tag{2}$$

where G is the Greenwich sidereal time calculated by the existing formula according to observation epoch, and λ and φ are the longitude and latitude of the rover provided by dead reckoning on Mars. Because $\boldsymbol{S}_{\mathrm{E}}$ can be calculated by ephemeris such as DE405, we can obtain the sun vector $\boldsymbol{S}_{\mathrm{H}}$ relative to the local horizontal frame by (1). Furthermore, we can determine the sun's predicted azimuth A_{H} relative to local north according to $\boldsymbol{S}_{\mathrm{H}}$.

Figure 2 gives the schematic of sun azimuth measurement using a sun sensor. Suppose the sun sensor is adjusted to be horizontal and the sun sensor frame C coincides with the transition frame T. Once a sun image is captured, we can calculate A_{C} or A_{T} according to sun image centroid coordinates, projection model, and distortion model. Then, the rover's heading ω can be calculated by

$$\omega = A_{\mathrm{H}} - A_{\mathrm{T}}. \tag{3}$$

It is evident that the precision of A_{T} determines the heading results. In order to improve the measurement precision of the sun azimuth, we must strictly solve the following three problems:

TABLE 1: Definition of main frames.

Name	Notation	X axis	Y axis	Z axis
Horizontal frame	H	Local north	Local west	Opposite gravity
Earth-centered inertial frame	E	Vernal equinox	Right-handed	North pole
Earth-centered fixed frame	F	Prime meridian	Right-handed	North pole
Transition frame	T	Rover heading	Right-handed	Opposite gravity
Inclinometer frame	I	X axis of itself	Y axis of itself	Right-handed
Sun sensor frame	C	Horizontal pixels	Right-handed	Optical axis

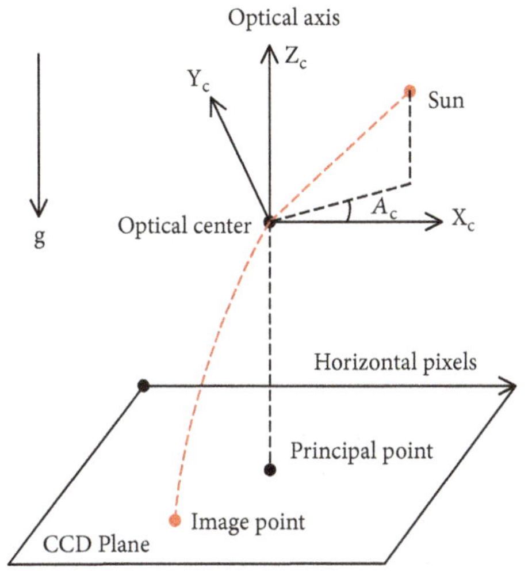

FIGURE 2: Schematic of sun azimuth measurement.

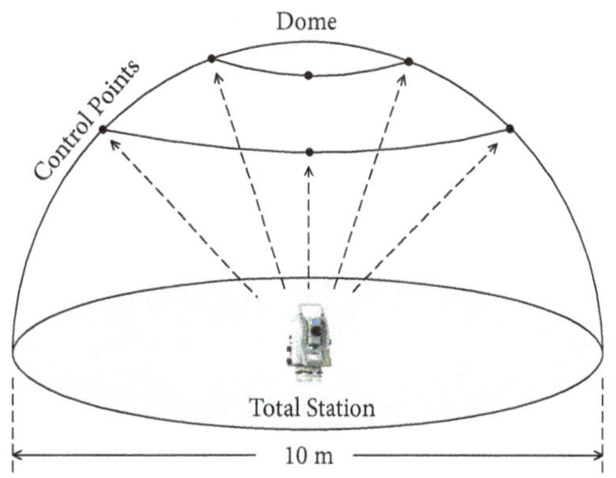

FIGURE 3: 10-m diameter dome and measurement sketch of control points.

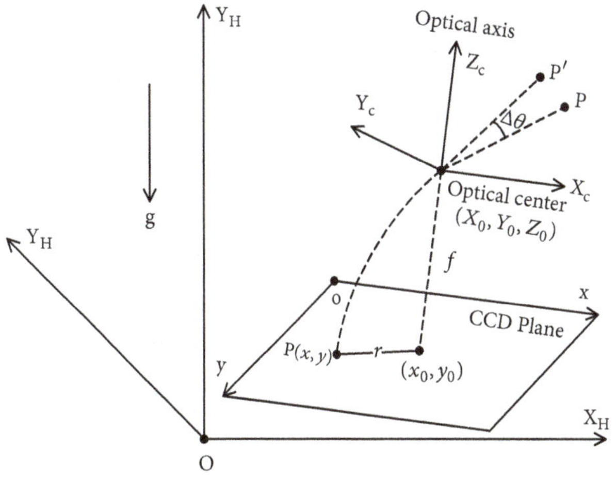

FIGURE 4: Geometric or physical meaning of the 12 sun sensor parameters.

(1) The sun sensor always utilizes large-field-of-view lens to avoid platform rotation, but comparable imaging distortion is introduced. An accurate distortion calibration model is needed.

(2) In practice, the sun sensor is difficult to keep absolutely horizontal, so a suitable and accurate algorithm for inclination correction must be seriously considered.

(3) Because the sun is a disk object, it always occupies a regular area on the image plane, which means a proper image-processing algorithm is needed to calculate the centroid of the sun image. Due to the large field of view, the sun sensor always suffers poor resolution. We must optimize the existing centroid-extraction algorithm to improve the centroid-extraction precision.

3. Algorithms

3.1. Sun Sensor Calibration Model

3.1.1. Error Equation. Because there is no regular control point on the surface of Mars, the sun sensor is usually calibrated in the laboratory before launch. We build a 10-m diameter dome, on the surface of which we uniformly install 37 super-bright optical fibers as control points. The 3D coordinates of the control points are surveyed by high-precision total station Leica TDRA6000, which has an orientation measurement precision of $0.5''$. Figure 3 shows the dome model and measurement sketch of the control points.

The purpose of sun sensor calibration is to obtain the parameters of the sun sensor, including the image principal point (x_0, y_0), focal length f, radial distortion parameters (k_1, k_2, k_3), rotation parameters (a, b, c), and translation parameters (X_0, Y_0, Z_0). Figure 4 depicts the geometric or physical meaning of the 12 parameters.

(x_0, y_0), f, and (k_1, k_2, k_3) are interior elements relative to the sun sensor; (a, b, c) and (X_0, Y_0, Z_0) are exterior orientation elements relative to local horizontal frame; and $\Delta\theta$ is the error of the half angle of view caused by radial distortion. Because the sun sensor utilizes a solid-angle projection, the real half angle of view is represented as

$$\theta = 2\arcsin\left(\frac{r}{2f}\right) + \Delta\theta. \quad (4)$$

Suppose (x, y) are coordinates of the image centroid of a control point. The polar distance r is represented as

$$r = \sqrt{(x - x_0)^2 + (y - y_0)^2}, \quad (5)$$

and we adopt the polynomial model to describe the radial distortion [15]:

$$\Delta\theta = k_1\left(\arcsin\left(\frac{r}{2f}\right)\right)^2 + k_2\left(\arcsin\left(\frac{r}{2f}\right)\right)^3 + k_3\left(\arcsin\left(\frac{r}{2f}\right)\right)^4. \quad (6)$$

The azimuth of the control point in the sun sensor frame is written as

$$A_C = \begin{cases} \operatorname{atan}\dfrac{y_p}{x_p} & x_p > 0,\ y_p > 0 \\ \operatorname{atan}\dfrac{y_p}{x_p} + \pi & x_p < 0 \\ \operatorname{atan}\dfrac{y_p}{x_p} + 2\pi & x_p > 0,\ y_p < 0, \end{cases} \quad (7)$$

where $x_p = x - x_0$ and $y_p = y - y_0$. Hence, the vector of the control point in the sun sensor frame is represented as

$$\mathbf{S}_C = \begin{bmatrix} X_C \\ Y_C \\ Z_C \end{bmatrix} = \begin{bmatrix} \sin\theta\cos A_C \\ \sin\theta\sin A_C \\ \cos\theta \end{bmatrix} \quad (8)$$

and the vector of the control point in horizontal frame can be written as

$$\mathbf{S}_H = [X_H\ \ Y_H\ \ Z_H]^T. \quad (9)$$

We move the horizontal frame to make its origin coincide with that of the sun sensor frame, and the vector of the control point in the new frame becomes

$$\mathbf{S} = [X_H + X_0\ \ Y_H + Y_0\ \ Z_H + Z_0]^T. \quad (10)$$

$\mathbf{S}$ is normalized as

$$\mathbf{S}_0 = \frac{\mathbf{S}}{|\mathbf{S}|} = [X\ \ Y\ \ Z]^T. \quad (11)$$

The relationship between $\mathbf{S}_C$ and $\mathbf{S}_0$ can be described by a rotation matrix $\boldsymbol{R}$, as follows:

$$\mathbf{S}_0 = \boldsymbol{R} \cdot \mathbf{S}_C. \quad (12)$$

Equation (12) is our basic observation equation of sun sensor calibration. Unlike previous studies, we first employ the antisymmetric matrix $\mathbf{Q}$ instead of Euler angles to express $\boldsymbol{R}$, which is beneficial for reducing the amount of calculation through reduced use of trigonometric sines and cosines. It has been proven that all the rotation matrices can be expressed by an antisymmetric matrix as follows [16]:

$$\boldsymbol{R} = (\mathbf{I} - \mathbf{Q})^{-1}(\mathbf{I} + \mathbf{Q}), \quad (13)$$

where $\mathbf{I}$ and $\mathbf{Q}$ are written as

$$\mathbf{I} = \begin{bmatrix} 1 & 0 & 0 \\ 0 & 1 & 0 \\ 0 & 0 & 1 \end{bmatrix}$$
$$\mathbf{Q} = \begin{bmatrix} 0 & -c & -b \\ c & 0 & -a \\ b & a & 0 \end{bmatrix}. \quad (14)$$

Then, (12) can be written as

$$(\mathbf{I} - \mathbf{Q}) \cdot \mathbf{S}_0 = (\mathbf{I} + \mathbf{Q}) \cdot \mathbf{S}_C. \quad (15)$$

$[v_1\ \ v_2\ \ v_3]^T$ represents the error vector of (15), and the error equation is expressed by

$$\begin{bmatrix} V_1 \\ V_2 \\ V_3 \end{bmatrix} = \begin{bmatrix} 0 & -Z_C - Z & -Y_C - Y \\ -Z_C - Z & 0 & X_C + X \\ Y_C + Y & X_C + X & 0 \end{bmatrix} \begin{bmatrix} a \\ b \\ c \end{bmatrix} - \begin{bmatrix} X - X_C \\ Y - Y_C \\ Z - Z_C \end{bmatrix}. \quad (16)$$

Equation (16) needs to be linearized before being solved. Appendix A gives the partial derivatives of V_i with respect to the 12 parameters, which can be used to linearize (16). Because there are 37 control points, 37×3 linearized error equations can be obtained. We express the error equations in the form of a vector:

$$\boldsymbol{v} = \boldsymbol{A} \cdot \boldsymbol{X} - \boldsymbol{l}, \quad (17)$$

where $\boldsymbol{v}$ is the residual vector; $\boldsymbol{A}$ is coefficient matrix; $\boldsymbol{X} = [a\ \ b\ \ c]^T$, which is the vector of unknown parameters; and $\boldsymbol{l}$ is the free term vector. The least-square method is used to estimate the 12 parameters:

$$\boldsymbol{X} = (\boldsymbol{A}^T\boldsymbol{A})^{-1}\boldsymbol{A}^T\boldsymbol{l}. \quad (18)$$

3.1.2. Calibration Results. Figure 5 is an image of the 37 control points from our sun sensor introduced in Section 4.1. After threshold operation, we use the gray centroid method to detect the image centroids of the control points [17]. Figure 6 depicts the residuals of the 37 image points after calibration.

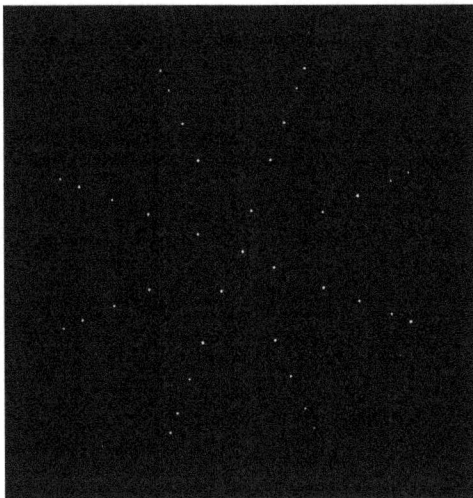

FIGURE 5: Image of the 37 control points.

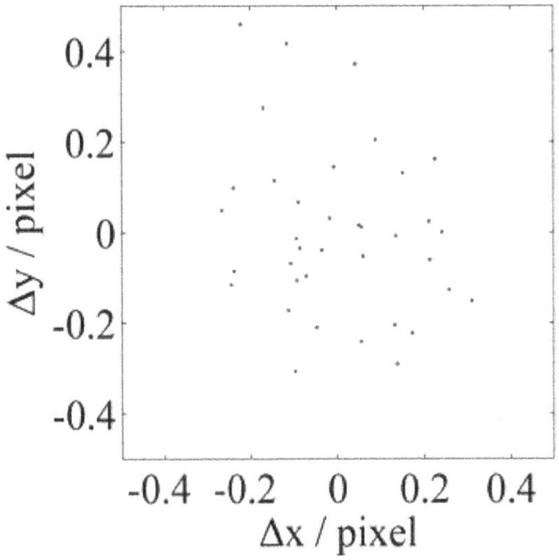

FIGURE 6: Residuals of the fish-eye sun sensor calibration.

The standard deviations of residuals along x and y axes are ±0.169 pixels and ±0.185 pixels, respectively. The calibration precision is similar to the method based on the 2D or 3D calibration board [4, 18]. As the minimum altitude of control point is about 15°, the effective calibration field of the sun sensor reaches 150°. Later, we put the sun sensor near the center of the dome and rotate it in 8 directions in order. In each direction, the sun sensor captures an image of the 37 control points for calibration. The results of the 6 interior parameters of the sun sensor in each direction are listed in Table 2.

In Table 2, the standard deviations of x_0, y_0, and f are ±0.081 pixels, ±0.097 pixels, and ±0.068 pixels, respectively, which indicate high consistency of the calibration results in 8 directions. However, the mean values of (x_0, y_0), f, and (k_1, k_2, k_3) for the 8 directions cannot be adopted as the final results because of the strong correlations between f and (k_1, k_2, k_3) and between (x_0, y_0) and (a, b, c). The standard deviation of the residuals, which should be as small as possible, is a good criterion to choose the best set of parameters. Direction 4 achieves the smallest standard deviation of the residuals, so the fourth group of interior parameters should be the final result.

3.2. Sun Sensor and Inclinometer Alignment Model. Prior studies generally use the sun as a control point to determine the relationship between the sun sensor and inclinometer, which must wait for the sun to move across several long traces and obtain low-precision results due to having only one control point in the sky [6, 7, 19]. However, the 10-m diameter dome with 37 control points provides ideal conditions for the sun sensor and inclinometer calibration. Because the coordinates of the 37 control points are surveyed by the high-precision total station, they contain local gravity information, which can be used to determine the relationship between the sun sensor and inclinometer.

In Section 3.1, (x_0, y_0), f, and (k_1, k_2, k_3) are determined; hence, we use the least-square method to calculate (a, b, c) and (X_0, Y_0, Z_0), which describe the position and attitude of the sun sensor frame relative to the horizontal frame. Then, the rotation matrix $\boldsymbol{R}_C^H$ from the sun sensor frame to the horizontal frame is calculated by Appendix C, and $\boldsymbol{R}_C^H$ can be described by 3-step rotations around the X_C, Y_C, and Z_C axes in order:

(1) Rotate an angle of γ around the X_C axis.

(2) Rotate an angle of ψ around the Y_C axis.

(3) Rotate an angle of κ around the Z_C axis.

$\boldsymbol{R}_C^H$ is written as

$$\boldsymbol{R}_C^H = \boldsymbol{R}_Z(\kappa)\,\boldsymbol{R}_Y(\psi)\,\boldsymbol{R}_X(\gamma). \tag{19}$$

Appendix C gives the expressions for $\boldsymbol{R}_Z(\kappa)$, $\boldsymbol{R}_Y(\psi)$, and $\boldsymbol{R}_X(\gamma)$. Then, (19) is expanded as

$$\boldsymbol{R}_C^H = \begin{bmatrix} \cos\psi\cos\kappa & \sin\gamma\sin\psi\cos\kappa + \cos\gamma\sin\kappa & -\cos\gamma\sin\psi\cos\kappa + \sin\gamma\sin\kappa \\ -\cos\psi\sin\kappa & -\sin\gamma\sin\psi\sin\kappa + \cos\gamma\cos\kappa & \cos\gamma\sin\psi\sin\kappa + \sin\gamma\cos\kappa \\ \sin\psi & -\sin\gamma\cos\psi & \cos\gamma\cos\psi \end{bmatrix}. \tag{20}$$

Table 2: Results of the 6 interior parameters in 8 directions.

Raw azimuth	x_0 (pixel)	y_0 (pixel)	f (pixel)	k_1	k_2	k_3	Std (Δx)	Std (Δy)
0°	1502.292	1585.044	855.242	0.212471	-0.740583	0.877803	0.1621	0.1728
45°	1502.209	1584.992	855.142	0.221231	-0.764446	0.897239	0.1604	0.1852
90°	1502.280	1585.034	855.156	0.211819	-0.734883	0.870884	0.1788	0.1553
135°	1502.365	1585.170	855.201	0.181307	-0.653738	0.807112	0.1640	0.1481
180°	1502.293	1585.219	855.047	0.161114	-0.594148	0.755274	0.1454	0.1674
225°	1502.209	1585.248	855.101	0.160012	-0.594534	0.758023	0.1547	0.1603
270°	1502.133	1585.201	855.086	0.191343	-0.678503	0.825247	0.1763	0.1502
315°	1502.141	1585.089	855.214	0.203631	-0.713076	0.853851	0.1809	0.1521

Table 3: Values of γ and ψ in 8 directions.

Direction number	Approximate azimuth	γ ($''$)	ψ ($''$)
1	0°	1123.9	121.7
2	45°	1120.6	142.1
3	90°	1124.1	139.5
4	135°	1126.6	122.5
5	180°	1129.3	129.2
6	225°	1124.7	135.1
7	270°	1122.1	144.0
8	315°	1125.0	138.4

Here, we provide the expressions of γ and ψ:

$$\gamma = -\operatorname{atan}\frac{\boldsymbol{R}_{\mathrm{C}}^{\mathrm{H}}(3,2)}{\boldsymbol{R}_{\mathrm{H}}^{\mathrm{C}}(3,3)}$$
$$\psi = \operatorname{asin}\boldsymbol{R}_{\mathrm{C}}^{\mathrm{H}}(3,1). \quad (21)$$

When the outputs of the inclinometer are adjusted to zero, we rotate the sun sensor frame by steps (1) and (2); then the Z axis of the new frame opposes local gravity. In other words, if the outputs of the inclinometer are adjusted to zero and the angles γ and ψ are known, the relationship between sun sensor and inclinometer can be determined.

In the experiment of Section 3.1.2, we always adjust the legs of the prototype introduced in Section 4.1 to keep the outputs of the inclinometer close to zero (smaller than $1''$ in practice). Table 3 lists the results of γ and ψ in each direction.

Standard deviations of γ and ψ are $\pm 2.6''$ and $\pm 8.6''$, respectively. The mean values ($\gamma = 1124.5'', \psi = 134.1''$) are adopted as the final parameters. Table 3 indicates the high-precision relationship determination between the sun sensor and inclinometer, which is the basis of high-precision heading determination.

3.3. Sun Image Centroid-Extraction Algorithm. Because large-field sun sensors suffer from poor angular resolution, we must improve the precision of the sun image centroid extraction as much as possible. The gray centroid method is widely used for the sun image centroid extraction, but the precision is not high when the sun image is irregular [17, 20]. Cui et al. consider the sun image as a circle and use the Sobel operator to detect the pixel-level edge; thus, the centroid is achieved by circle fitting of the edge points [21]. Yang et al. adopt the Zernike moment to obtain the subpixel edge points and make the precision of the circular sun image centroid reach 0.07 pixel, which is obviously better than that of the gray centroid algorithm [22]. However, for the general projection models, such as the pinhole and solid-angle models, when the sun is away from the boresight direction, the shape of the sun image is more similar to an ellipse than a circle. Our algorithm for sun image centroid extraction is similar to [22], but the difference is that both circle and ellipse sun image are considered, and a reasonable criterion for shape judgment is put forward. Our algorithm is realized by 6 steps:

(1) The Sobel operator is used to detect the pixel-level edge points (x, y) of the sun image.

(2) For each pixel-level edge point, the Zernike moment is used to detect the subpixel edge point (x', y'). Appendix C gives the computation method in detail. Figures 7(a) and 7(b) show a practical circular and elliptical sun image, respectively, from our fish-eye sun sensor and its edge-detection results. Obviously, the Zernike moment produces a smoother edge than the Sobel operator, which is beneficial for improving the centroid fitting precision.

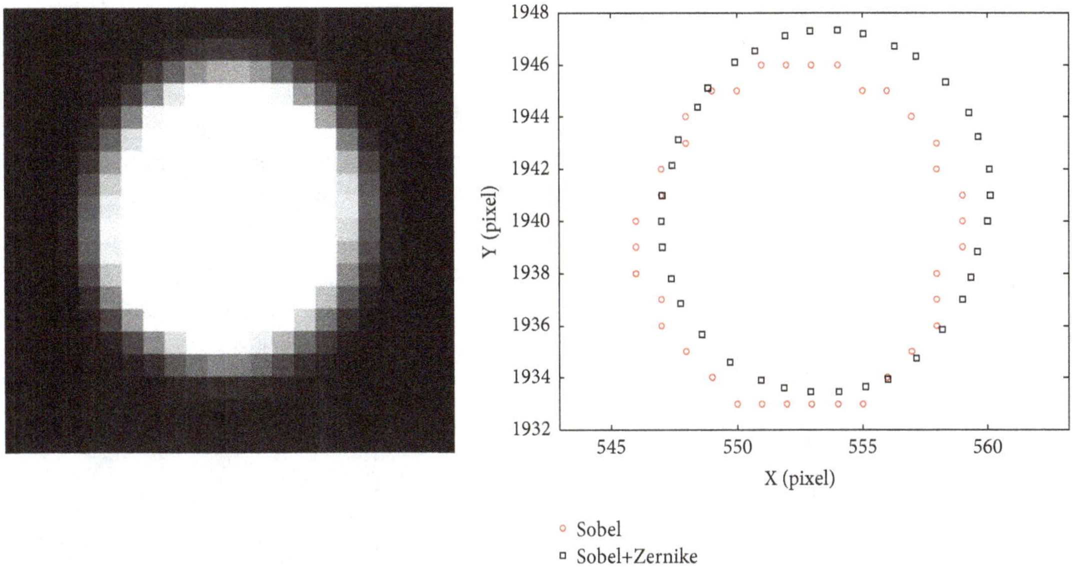

(a) A circular sun image and its edge-detection results

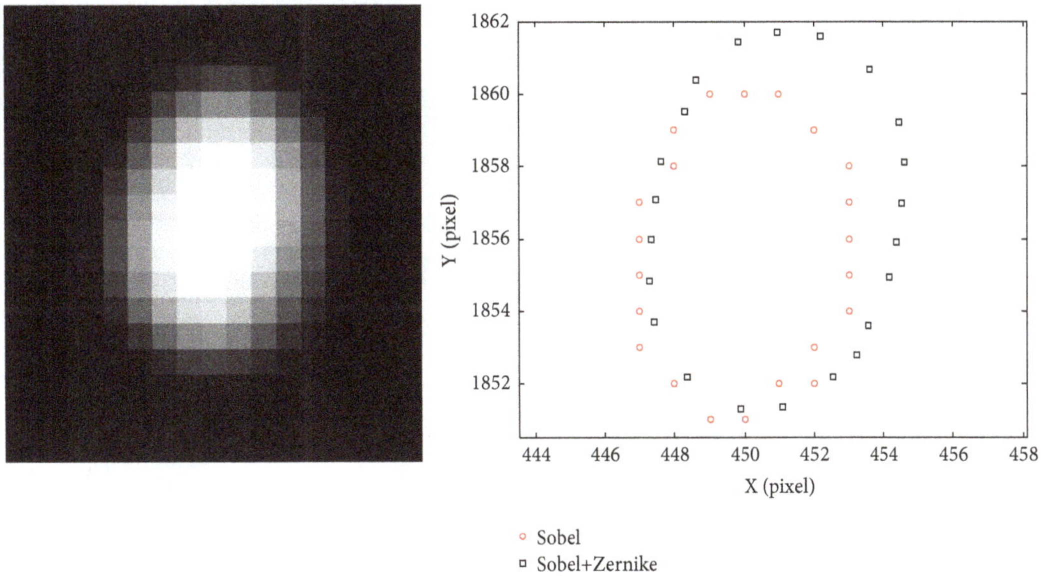

(b) An elliptical sun image and its edge-detection results

FIGURE 7: Two classic sun images and their edge-detection results.

(3) According to the subpixel edge points (x', y'), the center of circle is fitted and the value of the cost function J_c is calculated.

$$J_c(x_c, y_c, r) = \sum_{i=1}^{n} \left| (x_i' - x_c)^2 + (y_i' - y_c)^2 - r^2 \right|^2 \quad (22)$$

where v_i is residual. Suppose the initial values of the unknown parameters are $X_0 = [x_{c0} \ \ y_{c0} \ \ r_0]^T$; thus, (23) is linearized as

$$v_i = a_i \delta x_c + b_i \delta y_c + c_i \delta r - l_i \quad (23)$$

and

$$\begin{aligned} a_i &= 2x_{c0} - 2x_i' \\ b_i &= 2y_{c0} - 2y_i' \\ c_i &= r_0^2 - (x_i' - x_{c0})^2 - (y_i' - y_{c0})^2 \\ l_i &= r_0{}^2 - (x_i' - x_{c0})^2 - (y_i' - y_{c0})^2. \end{aligned} \quad (24)$$

The error equations of all the edge points can be written in the form of a matrix, as follows:

$$\boldsymbol{v} = \boldsymbol{A} \cdot \delta \boldsymbol{X} - \boldsymbol{l}, \quad (25)$$

where $\boldsymbol{v}$ is the residual vector, $\boldsymbol{A}$ is the coefficient matrix, $\delta\boldsymbol{X}$ is the correction vector corresponding to $\boldsymbol{X}_0$, and $\boldsymbol{l}$ is the free term vector. The expanded forms of the four matrices and vectors are

$$\boldsymbol{v} = \begin{bmatrix} v_1 & v_2 & \dots & v_n \end{bmatrix}^{\mathrm{T}}, \tag{26}$$

$$\boldsymbol{A} = \begin{bmatrix} a_1 & b_1 & c_1 \\ \vdots & \vdots & \vdots \\ a_n & b_n & c_n \end{bmatrix}, \tag{27}$$

$$\delta\boldsymbol{X} = \begin{bmatrix} \delta x_{\mathrm{c}} & \delta y_{\mathrm{c}} & \delta r \end{bmatrix}^{\mathrm{T}}, \tag{28}$$

$$\boldsymbol{l} = \begin{bmatrix} l_1 & l_2 & \dots & l_n \end{bmatrix}^{\mathrm{T}}. \tag{29}$$

We assume all the edge points have the same weight, and then $\delta\boldsymbol{X}$ is calculated by [23]

$$\delta\boldsymbol{X} = \left(\boldsymbol{A}^{\mathrm{T}}\boldsymbol{A}\right)^{-1}\boldsymbol{A}^{\mathrm{T}}\boldsymbol{l}. \tag{30}$$

The unknown parameters are updated by $\boldsymbol{X} = \boldsymbol{X}_0 + \delta\boldsymbol{X}$, and iterations are needed until the absolute values of $\delta\boldsymbol{X}$ are smaller than 0.001 pixels. Then, J_{c} can be calculated according to (22).

(4) According to the subpixel edge points, the ellipse's center is fitted and the value of the cost function J_{e} is calculated.

$$J_{\mathrm{e}}(A, B, C, D, E) = \sum_{i=1}^{n} \left| Ax_i'^2 + Bx_i'y_i' + Cy_i'^2 + Dx_i' + Ey_i' + 1 \right|^2 \tag{31}$$

A, B, C, D, and E are the basic parameters of the ellipse that need to be estimated, and the ellipse's center $(x_{\mathrm{e}}, y_{\mathrm{e}})$ can be calculated by

$$x_{\mathrm{e}} = \frac{BE - 2CD}{4AC - B^2}, \quad y_{\mathrm{e}} = \frac{BD - 2AE}{4AC - B^2}. \tag{32}$$

The error equation of each edge point is represented as

$$v_i = Ax_i'^2 + Bx_i'y_i' + Cy_i'^2 + Dx_i' + Ey_i' + 1. \tag{33}$$

Equation (33) is linear; therefore, the coefficient matrix $\boldsymbol{A}$ and free item vector $\boldsymbol{l}$ are written as

$$\boldsymbol{A} = \begin{pmatrix} x_1'^2 & x_1'y_1' & y_1'^2 & x_1' & y_1' \\ \vdots & \vdots & \vdots & \vdots & \vdots \\ x_n'^2 & x_n'y_n' & y_n'^2 & x_n' & y_n' \end{pmatrix}, \tag{34}$$

$$\boldsymbol{l} = \begin{bmatrix} -1 & -1 & \dots & -1 \end{bmatrix}^{\mathrm{T}}. \tag{35}$$

Suppose the vector of unknown parameters is $\boldsymbol{X} = \begin{bmatrix} A & B & C & D & E \end{bmatrix}^{\mathrm{T}}$, which can be calculated by $\boldsymbol{X} = (\boldsymbol{A}^{\mathrm{T}}\boldsymbol{A})^{-1}\boldsymbol{A}^{\mathrm{T}}\boldsymbol{l}$. Then, x_{e} and y_{e} are derived by (32), and we use (31) to calculate J_{e}.

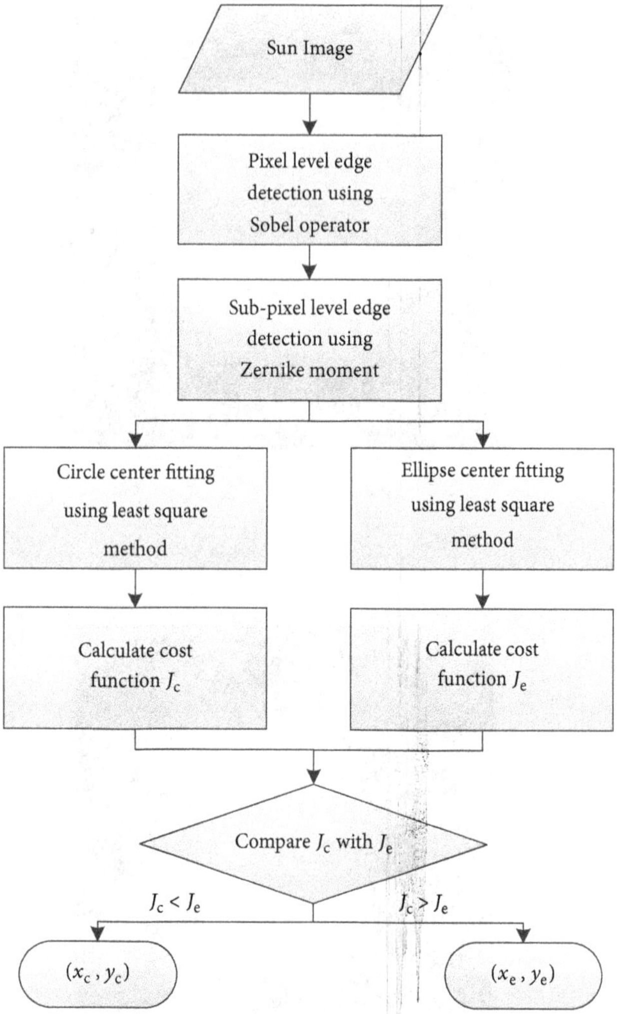

FIGURE 8: Flow chart of centroid extraction of the sun image.

(5) The shape of the sun image is judged by comparing J_{c} to J_{e}. If $J_{\mathrm{c}} < J_{\mathrm{e}}$, $(x_{\mathrm{c}}, y_{\mathrm{c}})$ should be adopted. Otherwise, $(x_{\mathrm{e}}, y_{\mathrm{e}})$ should be adopted.

(6) The precision is estimated. The root-mean-square error (RMSE) of the centroids can be calculated by

$$\mathrm{RMSE} = \sqrt{m_x^2 + m_y^2}. \tag{36}$$

m_x and m_y are the RMSE of centroid coordinates $(x_{\mathrm{c}}, y_{\mathrm{c}})$ or $(x_{\mathrm{e}}, y_{\mathrm{e}})$, which can be calculated according to the residual vector $\boldsymbol{v}$ and coefficient matrix $\boldsymbol{A}$.

Figure 8 is the flow chart of our centroid-extraction algorithm for the sun image.

3.4. Heading-Determination Algorithm. Without loss of generality, we consider the X axis of the prototype to be aligned with that of the charge-coupled device (CCD), and the dip angles of the electronic inclinometer are always adjusted to zero in the field test. Then, we take 4 steps to calculate the heading:

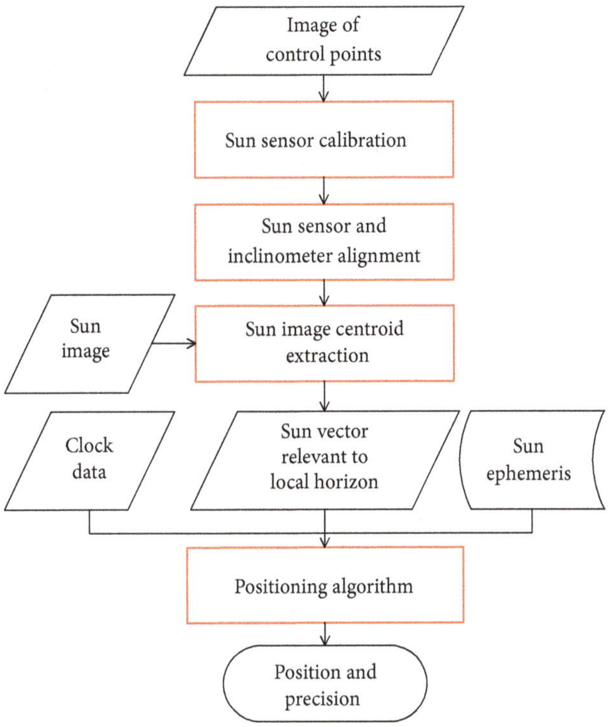

FIGURE 9: Flow chart of heading-determination algorithm.

(1) Extract the centroid of a sun image using the method mentioned in Section 3.3 and calculate the vector $\boldsymbol{S}_C$ of the sun with respect to the sun sensor frame according to the intrinsic parameters of the sun sensor.

(2) Rotate the camera coordinate system by $\boldsymbol{R}_X(\gamma)$ and $\boldsymbol{R}_Y(\psi)$; then calculate the vector of the sun in the transition frame $\boldsymbol{S}_T$ by

$$\boldsymbol{S}_T = \boldsymbol{R}_Y(\psi)\,\boldsymbol{R}_X(\gamma)\,\boldsymbol{S}_C = \begin{bmatrix} X_T \\ Y_T \\ Z_T \end{bmatrix}. \quad (37)$$

The azimuth A_T of the sun relative to the X axis of the new coordinate system can be calculated by $\boldsymbol{S}_T$ according to (7). We then use the Naval Observatory Vector Astronomy Software version 3.0 (NOVAS3.0) and DE405 ephemeris to calculate the sun's predicted azimuth A_H. The observation epoch in UTC is provided by local clock.

(3) Calculate the heading ω according to (3)and estimate the precision.

It is clear that only one sun image is needed to calculate the heading in our algorithm, which is beneficial for reducing the data-processing time. Figure 9 shows the flow chart of our heading-determination algorithm, where the contents of the red boxes are the key algorithms presented in Section 3.

4. Field Tests and Results

In order to test our algorithms and methods for high-precision heading determination, three field tests were conducted in central China's Henan province in 2017. The details of the tests are presented in this section, including the hardware configuration, test conditions, and test results.

4.1. Hardware Configuration. Our sun sensor is composed of a fish-eye lens, a CCD, and a filter. The fish-eye lens is AF DX made by Nikon, with 10.5-mm focal length and 180° field of view. The CCD is Alta U9000 made by Apogee, with a 3,056-by-3,056 array of 12-micron square pixels and a 7 square-inch and 3-inch thick body. The filter is mounted between the lens and CCD to weaken the light of the sun. The sun sensor can capture images of the sun without searching, which means that no rotating platform is needed.

The Leica Nivel230 electronic inclinometer is chosen to obtain the dip angles relative to local horizontal plane, and the precision is up to 1 arcsecond with a measurement range of -3.78 arcminutes to +3.78 arcminutes, a weight of 700 g, and a 3.5-square-inch and 2.7-inch thick body. We choose the SA.45s chip-scale atomic clock made by Microsemi with a month aging rate of 3^{-10} s, low power consumption of 120 mW, and volume of 17 cc. The ARK-1122F embedded computer with a dual Atom-core processor is used for processing data including sun images, dip angles, and time information.

We develop a prototype by integrating the sun sensor, inclinometer, and chip-scale atomic clock. Figure 10 shows a photograph of our prototype.

Each test was conducted by the following steps:

(1) Put the prototype on a pillar and adjust the legs of the prototype to make the outputs of the inclinometer close to zero.

(2) Keep the prototype static and continuously shoot images of the sun. During shooting, the inclinometer data and time data are collected together.

(3) Use the heading-determination algorithm presented in this paper to calculate the heading.

(4) Estimate the heading-determination precision.

4.2. Test Conditions. Detailed conditions of the three tests are presented in Table 4, including the date and time, observation times, sun elevation fluctuation, and weather.

The three tests were conducted in different months in order to analyze the impact of the sun elevation on the heading determination. On July 20, the maximum elevation of the sun was up to 75.7°, which means the sun imaging positions were close to the principal point on the image plane. On Oct 14 and Nov 11, the maximum elevation of the sun was 46.7° and 37.7°, respectively. This means that the sun imaging positions were away from the principal point on the image plane. Figure 11 shows the time series of the predicted elevation of the sun. Additionally, during the first and third tests, thin clouds or fog sheltered the sensor from the sun, which slightly impacts the observation of the sun.

TABLE 4: Test conditions.

Date	Beijing Time	Observations	Sun Elevation	Weather
July 20, 2017	12:05–13:01	252	74.3–75.7°	Thin Clouds
Oct 14, 2017	11:42–13:12	250	44.3–46.7°	Sunny
Nov 11, 2017	11:59–13:13	200	35.6–37.7°	Thin Fog

FIGURE 10: Actual picture of the prototype.

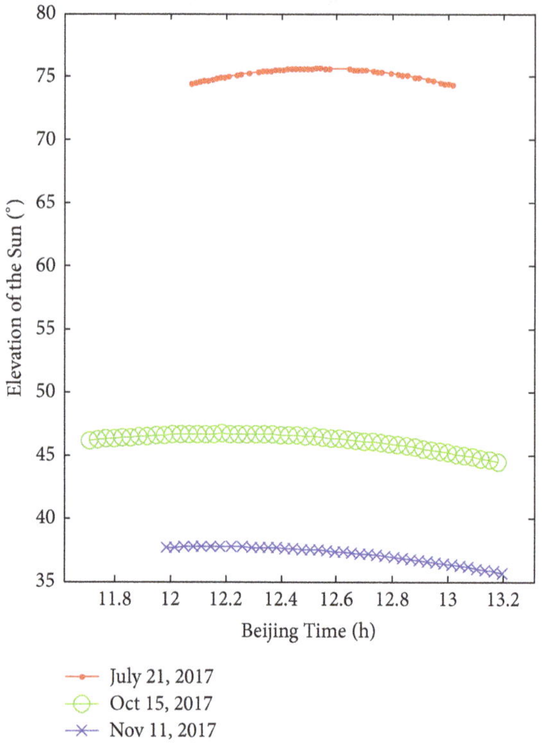

FIGURE 11: Times series of the predicted elevations of the sun.

4.3. Results. Because the real heading is difficult to obtain, we consider the mean value of the measured heading as a reference for precision estimation. Figure 12 depicts the time series of the heading errors of the three tests.

In the first test, the heading errors fluctuate greatly with an amplitude of 2.5 arcminutes. However, in the second and third tests, the heading errors become quite small, and the amplitude is only 1 arcminute. The standard deviations of the three tests are $\pm 0.97'$, $\pm 0.30'$, and $\pm 0.28'$, respectively. It is evident that the second and third tests achieve more precise heading results. It appears that the elevation of the sun has a great impact on the heading determination. Because the sun elevation has a strong correlation with the sun imaging position, Figure 13 directly shows the trajectories of the sun image centroids of the three tests on the image plane.

On July 20, the trajectory of the sun image centroids was close to the principal point with a mean distance of 220.6 pixels. However, the trajectories on Oct 14 and Nov 11 were 642.8 pixels and 785.8 pixels away from the principal point, respectively. Obviously, radial errors of the sun image centroids have no effect on the sun azimuth measurement, whereas it is linearly influenced by tangential errors. Because the tangential error of the sun image centroid is a small quantity, the error of the sun azimuth measurement can be expressed as follows:

$$e_{A_C} = \frac{e_c}{r}, \quad (38)$$

where e_c is the tangential error of sun image centroid and r is the polar distance of sun image centroid relative to the principal point. The RMSE series of the sun image centroids, which are calculated by fitting the residuals of the subpixel edge points of the sun image, are depicted in Figure 14.

Figure 14 shows that the RMSE of the sun image centroids are similar in the three tests, which indicates that the sun elevation is irrelevant to the sun image centroid extraction. The RMSE of several sun image centroids in the first and third tests are up to 0.1 pixels, probably due to the cloudy or foggy weather. However, the mean RMSE is about 0.065 pixels, which indicates the superiority of our sun image centroid-extraction algorithm. The tangential mean RMSE can be estimated by $0.065/\sqrt{2} = 0.046$pixels. Supposing that $e_c = 0.046$ pixels, $r_1 = 220.6$ pixels, $r_2 = 642.8$ pixels, and $r_3 = 785.8$ pixels, we can calculate the measurement errors of the sun azimuth. The results are $e_{A1_C} = 0.72'$, $e_{A2_C} = 0.25'$, and $e_{A3_C} = 0.20'$, which essentially coincide with the standard deviations of the heading error series in Figure 12.

Additionally, the heading error series of the three tests are not totally random, and some trend in the variation is

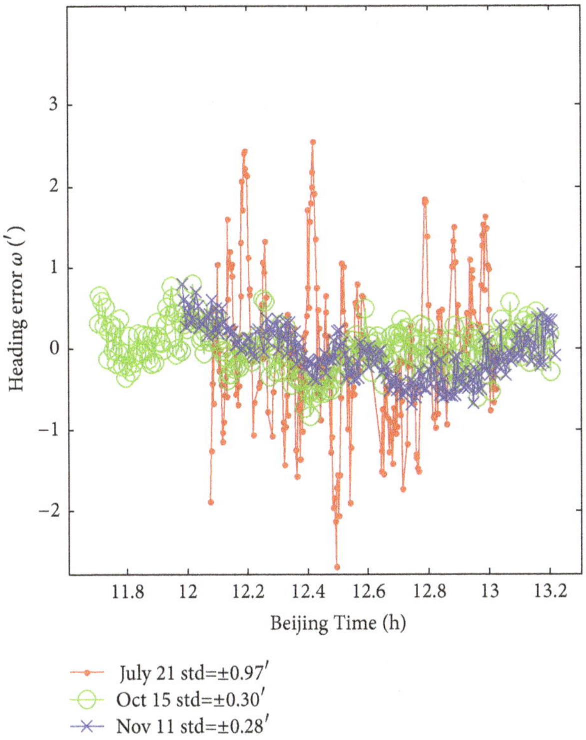

FIGURE 12: Time series of the heading errors.

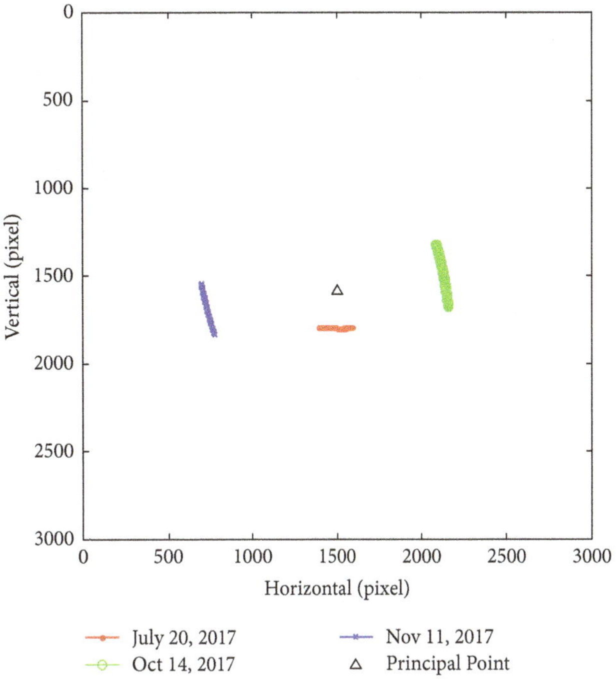

FIGURE 13: The trajectories of sun image centroids.

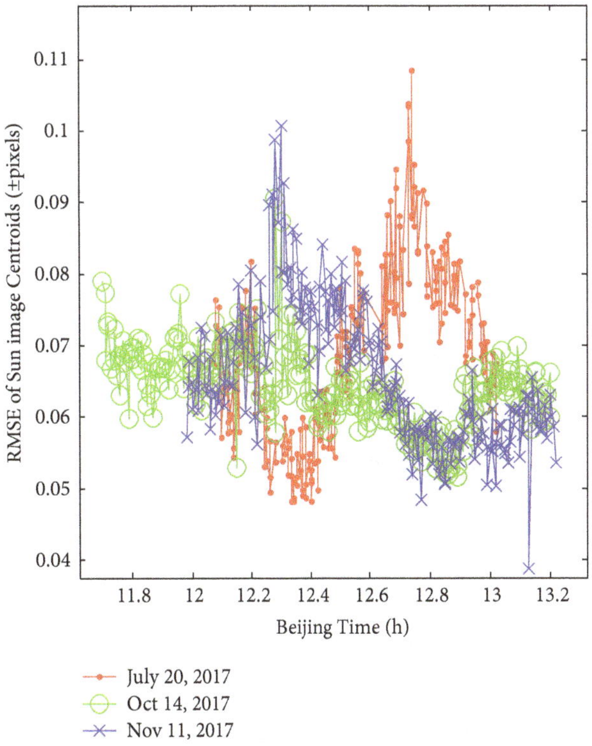

FIGURE 14: Mean square error series of sun image centroids.

obvious. This is mainly caused by the residual distortion after sun sensor calibration, alignment error of the sun sensor and inclinometer, and periodic change in the weather. As the minimum imaging period of the sun sensor is 10.8 s and the heading calculation time is less than 0.1 s, we conclude that just one sun image can produce a 1-arcminute heading using our prototype.

5. Conclusions

This paper attempts to improve the heading-determination precision to 1-arcminute level for Mars rovers using only one sun image. Algorithms for the sun sensor calibration, sun sensor and inclinometer alignment, sun image centroid extraction, and heading determination are presented in detail. A prototype is developed, and the results of three ground-based field tests indicate that the precision of heading reaches 0.28–0.97′ (1σ), which is the best reported precision for heading determination in Mars rovers so far. We not only improve the heading precision to the 1-arcminute level, but also reduce the observation time for the sun from 10 to 30 minutes to about 10 seconds.

However, some questions still need to be studied in the future. (1) When the outputs of the electronic inclinometer are not strictly adjusted to zero, we need a proper model to modify it. (2) Because the rover works on Mars for several or tens of years, we want to utilize the stars as control points to calibrate the sun sensor and inclinometer. (3) Dust on the sun sensor, resulting from dust storms on Mars, and clouds obscuring the view of the sun may produce irregular sun images, which have an impact on the precision of the heading. We are trying to use the robust estimation method to adjust the weight of the edge points to improve the centroid-extraction precision.

Appendix

A. Partial Derivatives of Vi with respect to the 12 Parameters

Partial derivatives of A_C and θ with respect to x_0 and y_0 are

$$\frac{\partial A_C}{\partial x_0} = \frac{-y + y_0}{r^2}, \quad \frac{\partial A_C}{\partial y_0} = \frac{x - x_0}{r^2} \tag{A.1}$$

$$\frac{\partial \theta}{\partial x_0} = \left(2 + 2k_1 S + 3k_2 S^2 + 4k_3 S^3\right) \frac{\partial S}{\partial r} \frac{\partial r}{\partial x_0}, \quad \frac{\partial \theta}{\partial y_0} = \left(2 + 2k_1 S + 3k_2 S^2 + 4k_3 S^3\right) \frac{\partial S}{\partial r} \frac{\partial r}{\partial y_0} \tag{A.2}$$

where $S = \arcsin(r/2f)$ and $\partial S/\partial r, \partial S/\partial f, \partial r/\partial x_0$, and $\partial r/\partial y_0$ are written as

$$\frac{\partial S}{\partial r} = \frac{1}{\sqrt{4f^2 - r^2}}, \quad \frac{\partial S}{\partial f} = \frac{-r}{f\sqrt{4f^2 - r^2}} \tag{A.3}$$

$$\frac{\partial r}{\partial x_0} = \frac{x_0 - x}{r}, \quad \frac{\partial r}{\partial y_0} = \frac{y_0 - y}{r} \tag{A.4}$$

Partial derivatives of A_C and θ with respect to f are

$$\frac{\partial A_C}{\partial f} = 0 \tag{A.5}$$

$$\frac{\partial \theta}{\partial f} = \left(2 + 2k_1 S + 3k_2 S^2 + 4k_3 S^3\right) \frac{\partial S}{\partial f} \tag{A.6}$$

Partial derivatives of A_C and θ with respect to k_i $(i = 1, 2, 3)$ are

$$\frac{\partial A_C}{\partial k_i} = 0, \quad \frac{\partial \theta}{\partial k_i} = \left(\arcsin\left(\frac{r}{2f}\right)\right)^{i+1} \tag{A.7}$$

Partial derivatives of $\mathbf{S}_C$ with respect to x_0 and y_0 are

$$\begin{aligned} \frac{\partial X_C}{\partial x_0} &= \cos\theta \cos A_C \cdot \frac{\partial \theta}{\partial x_0} - \sin\theta \sin A_C \cdot \frac{\partial A_C}{\partial x_0} \\ \frac{\partial Y_C}{\partial x_0} &= \cos\theta \sin A_C \cdot \frac{\partial \theta}{\partial x_0} + \sin\theta \cos A_C \cdot \frac{\partial A_C}{\partial x_0} \\ \frac{\partial Z_C}{\partial x_0} &= -\sin\theta \cdot \frac{\partial \theta}{\partial x_0} \end{aligned} \tag{A.8}$$

$$\begin{aligned} \frac{\partial X_C}{\partial y_0} &= \cos\theta \cos A_C \cdot \frac{\partial \theta}{\partial y_0} - \sin\theta \sin A_C \cdot \frac{\partial A_C}{\partial y_0} \\ \frac{\partial Y_C}{\partial y_0} &= \cos\theta \sin A_C \cdot \frac{\partial \theta}{\partial y_0} + \sin\theta \cos A_C \cdot \frac{\partial A_C}{\partial y_0} \\ \frac{\partial Z_C}{\partial y_0} &= -\sin\theta \cdot \frac{\partial \theta}{\partial y_0} \end{aligned} \tag{A.9}$$

Partial derivatives of $\mathbf{S}_C$ with respect to f and k_i $(i = 1, 2, 3)$ are

$$\begin{aligned} \frac{\partial X_C}{\partial f} &= \cos\theta \cos A_C \cdot \frac{\partial \theta}{\partial f} \\ \frac{\partial Y_C}{\partial f} &= \cos\theta \sin A_C \cdot \frac{\partial \theta}{\partial f} \\ \frac{\partial Z_C}{\partial f} &= -\sin\theta \cdot \frac{\partial \theta}{\partial f} \\ \frac{\partial X_C}{\partial k_i} &= \cos\theta \cos A_C \cdot \frac{\partial \theta}{\partial k_i} \\ \frac{\partial Y_C}{\partial k_i} &= \cos\theta \sin A_C \cdot \frac{\partial \theta}{\partial k_i} \\ \frac{\partial Z_C}{\partial k_i} &= -\sin\theta \cdot \frac{\partial \theta}{\partial k_i} \end{aligned} \tag{A.10}$$

Partial derivatives of $\mathbf{S}_0$ with respect to X_0, Y_0, and Z_0 are

$$\begin{aligned} \frac{\partial X}{\partial X_0} &= \frac{Y^2 + Z^2}{d} \\ \frac{\partial Y}{\partial X_0} &= -\frac{XY}{d} \\ \frac{\partial Z}{\partial X_0} &= -\frac{XZ}{d} \\ \frac{\partial X}{\partial Y_0} &= -\frac{XY}{d} \\ \frac{\partial Y}{\partial Y_0} &= \frac{X^2 + Z^2}{d} \\ \frac{\partial Z}{\partial Y_0} &= -\frac{YZ}{d} \\ \frac{\partial X}{\partial Z_0} &= -\frac{XZ}{d} \\ \frac{\partial Y}{\partial Z_0} &= -\frac{YZ}{d} \\ \frac{\partial Z}{\partial Z_0} &= \frac{X^2 + Y^2}{d} \end{aligned} \tag{A.11}$$

where $d = \sqrt{(X_H + X_0)^2 + (Y_H + Y_0)^2 + (Z_H + Z_0)^2}$. According to equations above, it is easy to obtain the partial derivatives of V_i with respect to 12 fish-eye sun sensor parameters.

B. Basic Rotation Matrix

Rotation matrix around X, Y, and Z axis can be represented by

$$\boldsymbol{R}_Z(\kappa) = \begin{pmatrix} \cos\kappa & \sin\kappa & 0 \\ -\sin\kappa & \cos\kappa & 0 \\ 0 & 0 & 1 \end{pmatrix} \quad \text{(B.1)}$$

$$\boldsymbol{R}_Y(\psi) = \begin{pmatrix} \cos\psi & 0 & -\sin\psi \\ 0 & 1 & 0 \\ \sin\psi & 0 & \cos\psi \end{pmatrix} \quad \text{(B.2)}$$

$$\boldsymbol{R}_Y(\psi) = \begin{pmatrix} \cos\psi & 0 & -\sin\psi \\ 0 & 1 & 0 \\ \sin\psi & 0 & \cos\psi \end{pmatrix} \quad \text{(B.3)}$$

C. Zernike 7×7 Models

We use Zernike 7×7 models deduced by Gao et al. as follows [24]:

$$M_{11}R = \begin{bmatrix} 0.0000 & -0.0150 & -0.0190 & 0.0000 & 0.0190 & 0.0150 & 0.0000 \\ -0.0224 & -0.0466 & -0.0233 & 0.0000 & 0.0233 & 0.0466 & 0.0224 \\ -0.0573 & -0.0466 & -0.0233 & 0.0000 & 0.0233 & 0.0466 & 0.0573 \\ -0.0690 & -0.0466 & -0.0233 & 0.0000 & 0.0233 & 0.0466 & 0.0690 \\ -0.0573 & -0.0466 & -0.0233 & 0.0000 & 0.0233 & 0.0466 & 0.0573 \\ -0.0224 & -0.0466 & -0.0233 & 0.0000 & 0.0233 & 0.0466 & 0.0224 \\ 0.0000 & -0.0150 & -0.0190 & 0.0000 & 0.0190 & 0.0150 & 0.0000 \end{bmatrix}$$

$$M_{11}I = \begin{bmatrix} 0.0000 & -0.0224 & -0.0573 & -0.0690 & -0.0573 & -0.0224 & 0.0000 \\ -0.0150 & -0.0466 & -0.0466 & -0.0466 & -0.0466 & -0.0466 & -0.0150 \\ -0.0190 & -0.0233 & -0.0233 & -0.0233 & -0.0233 & -0.0233 & -0.0190 \\ 0.0000 & 0.0000 & 0.0000 & 0.0000 & 0.0000 & 0.0000 & 0.0000 \\ 0.0190 & 0.0233 & 0.0233 & 0.0233 & 0.0233 & 0.0233 & 0.0190 \\ 0.0150 & 0.0466 & 0.0466 & 0.0466 & 0.0466 & 0.0466 & 0.0150 \\ 0.0000 & 0.0224 & 0.0573 & 0.0690 & 0.0573 & 0.0224 & 0.0000 \end{bmatrix}$$

$$M_{20} = \begin{bmatrix} 0.0000 & 0.0225 & 0.0394 & 0.0396 & 0.0394 & 0.0225 & 0.0000 \\ 0.0225 & 0.0271 & -0.0128 & -0.0261 & -0.0128 & 0.0271 & 0.0225 \\ 0.0394 & -0.0128 & -0.0528 & -0.0661 & -0.0528 & -0.0128 & 0.0394 \\ 0.0396 & -0.0261 & -0.0661 & -0.0794 & -0.0661 & -0.0261 & 0.0396 \\ 0.0394 & -0.0128 & -0.0528 & -0.0661 & -0.0528 & -0.0128 & 0.0394 \\ 0.0225 & 0.0271 & -0.0128 & -0.0261 & -0.0128 & 0.0271 & 0.0225 \\ 0.0000 & 0.0225 & 0.0394 & 0.0396 & 0.0394 & 0.0225 & 0.0000 \end{bmatrix} \quad \text{(C.1)}$$

For each pixel-level edge point (x, y) obtained by Sobel operator, we use $M_{11}R$, $M_{11}I$, and M_{20} for convolution operations. Then we get three important Zernike moments as $Z_{11}R$, $Z_{11}I$, and Z_{20}, and rotation angle ω is calculated by

$$\omega = \operatorname{atan}\frac{Z_{11}I}{Z_{11}R} \quad \text{(C.2)}$$

and the length l from center point to subpixel edge point is obtained by

$$l = \frac{Z_{20}}{Z_{11}R\cos\omega + Z_{11}I\sin\omega} \quad \text{(C.3)}$$

Hence we can get the coordinates of the subpixel edge point as follows:

$$\begin{bmatrix} x' \\ y' \end{bmatrix} = \begin{bmatrix} x \\ y \end{bmatrix} + \frac{7 \times l}{2}\begin{bmatrix} \cos\omega \\ \sin\omega \end{bmatrix} \quad \text{(C.4)}$$

Acknowledgments

This research was supported by the National Natural Science Foundation of China (NSFC) under Grant nos. 41704006 and 41504018 and funded by State Key Laboratory of Geo-Information Engineering under Grant no. SKLGIE2016-Z-2-1.

References

[1] W. Gong, "Discussions on localization capabilities of MSL and MER rovers," *Annals of GIS*, vol. 21, no. 1, pp. 69–79, 2015.

[2] K. S. Ali, C. A. Vanelli, J. J. Biesiadecki et al., "Attitude and position estimation on the Mars Exploration Rovers," in *Proceedings of the IEEE Systems, Man and Cybernetics Society, Proceedings - 2005 International Conference on Systems, Man and Cybernetics*, pp. 20–27, USA, October 2005.

[3] R. Volpe, "Mars rover navigation results using sun sensor heading determination," in *Proceedings of the IEEE/RSJ International Conference on Intelligent Robots and Systems (IROS'99)*, pp. 460–467, Kyongju, South Korea.

[4] A. Trebi-Ollennu, T. Huntsberger, Y. Cheng, E. T. Baumgartner, B. Kennedy, and P. Schenker, "Design and analysis of a sun sensor for planetary rover absolute heading detection," *IEEE Transactions on Robotics and Automation*, vol. 17, no. 6, pp. 939–947, 2001.

[5] M. C. Deans, D. Wettergreen, and D. Villa, "A Sun Tracker for Planetary Analog Rovers," in *Proceedings of the The 8th International Symposium on Artificial Intelligence, Robotics and Automation in Space*, p. 603, Munich, Germany, 2005.

[6] J. Enright, P. Furgale, and T. Barfoot, "Sun sensing for planetary rover navigation," in *Proceedings of the 2009 IEEE Aerospace Conference*, USA, March 2009.

[7] P. Furgale, J. Enright, and T. Barfoot, "Sun sensor navigation for planetary rovers: Theory and field testing," *IEEE Transactions on Aerospace and Electronic Systems*, vol. 47, no. 3, pp. 1631–1647, 2011.

[8] P. Yang, L. Xie, and J. Liu, "Simultaneous celestial positioning and orientation for the lunar rover," *Aerospace Science and Technology*, vol. 34, no. 1, pp. 45–54, 2014.

[9] M. Ilyas, B. Hong, K. Cho, S.-H. Baeg, and S. Park, "Integrated navigation system design for micro planetary rovers: Comparison of absolute heading estimation algorithms and nonlinear filtering," *Sensors*, vol. 16, no. 5, 2016.

[10] Y. Kim, W. Jung, and H. Bang, "Visual target tracking and relative navigation for unmanned aerial vehicles in a GPS-denied environment," *International Journal of Aeronautical and Space Sciences*, vol. 15, no. 3, pp. 258–266, 2014.

[11] A. Lambert, P. Furgale, T. D. Barfoot, and J. Enright, "Field testing of visual odometry aided by a sun sensor and inclinometer," *Journal of Field Robotics*, vol. 29, no. 3, pp. 426–444, 2012.

[12] X. Ning, L. Liu, J. Fang, and W. Wu, "Initial position and attitude determination of lunar rovers by INS/CNS integration," *Aerospace Science and Technology*, vol. 30, no. 1, pp. 323–332, 2013.

[13] Z. He, X. Wang, and J. Fang, "An innovative high-precision SINS/CNS deep integrated navigation scheme for the Mars rover," *Aerospace Science and Technology*, vol. 39, pp. 559–566, 2014.

[14] X. Ning, M. Gui, Y. Xu, X. Bai, and J. Fang, "INS/VNS/CNS integrated navigation method for planetary rovers," *Aerospace Science and Technology*, vol. 48, pp. 102–114, 2015.

[15] C.-H. Li, Y. Zheng, C. Zhang, Y.-L. Yuan, Y.-Y. Lian, and P.-Y. Zhou, "Astronomical vessel position determination utilizing the optical super wide angle lens camera," *Journal of Navigation*, vol. 67, no. 4, pp. 633–649, 2014.

[16] J. Yao, B. Han, and Y. Yang, "Applications of Lodrigues matrix in 3D coordinate transformation," *Geomatics and Information Science of Wuhan University*, vol. 31, no. 12, pp. 1094–1119, 2006.

[17] Y. Zhan, Y. Zheng, and C. Zhang, "Celestial Positioning with CCD Observing the Sun," in *China Satellite Navigation Conference (CSNC) 2013 Proceedings*, vol. 245 of *Lecture Notes in Electrical Engineering*, pp. 697–706, Springer Berlin Heidelberg, Berlin, Heidelberg, 2013.

[18] Z. Zhang, "A flexible new technique for camera calibration," *IEEE Transactions on Pattern Analysis and Machine Intelligence*, vol. 22, no. 11, pp. 1330–1334, 2000.

[19] M. Ilyas, K. Cho, S. H. Baeg, and S. Park, "Absolute Navigation Information Estimation for Planetary Rovers," in *Proceedings of Asian-Pacific Conference on Aerospace Technology and Science, , Jeju Island*, p. 1, 2015.

[20] Y. Zhan, Y. Zheng, C. Zhang, G. Ma, and Y. Luo, "Image centroid algorithms for sun sensors with super wide field of view," *Cehui Xuebao/Acta Geodaetica et Cartographica Sinica*, vol. 44, no. 10, pp. 1078–1084, 2015.

[21] P. Cui, F. Yue, and H. Cui, "Attitude and position determination scheme of lunar rovers basing on the celestial vectors observation," in *Proceedings of the 2007 IEEE International Conference on Integration Technology, ICIT 2007*, pp. 538–543, China, March 2007.

[22] P. Yang, L. Xie, and J.-L. Liu, "Zernike moment based high-accuracy sun image centroid algorithm," *Yuhang Xuebao/Journal of Astronautics*, vol. 32, no. 9, pp. 1963–1970, 2011.

[23] L. F. Sui, L. J. Song, and H. Z. Cai, *Error Theory and Foundation of Surveying Adjustment*, Surveying and Mapping Press, Beijing, China, 2010.

[24] S.-Y. Gao, M.-Y. Zhao, L. Zhang, and Y.-Y. Zou, "Improved algorithm about subpixel edge detection of image based on Zernike orthogonal moments," *Zidonghua Xuebao/ Acta Automatica Sinica*, vol. 34, no. 9, pp. 1163–1168, 2008.

2

Kilonova/Macronova Emission from Compact Binary Mergers

Masaomi Tanaka

National Astronomical Observatory of Japan, Mitaka, Tokyo 181-8588, Japan

Correspondence should be addressed to Masaomi Tanaka; masaomi.tanaka@nao.ac.jp

Academic Editor: WeiKang Zheng

We review current understanding of kilonova/macronova emission from compact binary mergers (mergers of two neutron stars or a neutron star and a black hole). Kilonova/macronova is emission powered by radioactive decays of r-process nuclei and it is one of the most promising electromagnetic counterparts of gravitational wave sources. Emission from the dynamical ejecta of $\sim 0.01 M_{\odot}$ is likely to have a luminosity of $\sim 10^{40}$–10^{41} erg s^{-1} with a characteristic timescale of about 1 week. The spectral peak is located in red optical or near-infrared wavelengths. A subsequent accretion disk wind may provide an additional luminosity or an earlier/bluer emission if it is not absorbed by the precedent dynamical ejecta. The detection of near-infrared excess in short GRB 130603B and possible optical excess in GRB 060614 supports the concept of the kilonova/macronova scenario. At 200 Mpc distance, a typical peak brightness of kilonova/macronova with $0.01 M_{\odot}$ ejecta is about 22 mag and the emission rapidly fades to >24 mag within ~10 days. Kilonova/macronova candidates can be distinguished from supernovae by (1) the faster time evolution, (2) fainter absolute magnitudes, and (3) redder colors. Since the high expansion velocity ($v \sim 0.1$–$0.2c$) is a robust outcome of compact binary mergers, the detection of smooth spectra will be the smoking gun to conclusively identify the gravitational wave source.

1. Introduction

Mergers of compact stars, that is, neutron star (NS) and black hole (BH), are promising candidates for direct detection of gravitational waves (GWs). On 2015 September 14, Advanced LIGO [1] has detected the first ever direct GW signals from a BH-BH merger (GW150914) [2]. This discovery marked the dawn of GW astronomy.

NS-NS mergers and BH-NS mergers are also important and leading candidates for the GW detection. They are also thought to be progenitors of short-hard gamma-ray bursts (GRBs [3–5]; see also [6, 7] for reviews). When the designed sensitivity is realized, Advanced LIGO [1], Advanced Virgo [8], and KAGRA [9] can detect the GWs from these events up to ~200 Mpc (for NS-NS mergers) and ~800 Mpc (for BH-NS mergers). Although the event rates are still uncertain, more than one GW event per year is expected [10].

Since localization only by the GW detectors is not accurate, for example, more than a few 10 deg^2 [11–14], identification of electromagnetic (EM) counterparts is essentially important to study the astrophysical nature of the GW sources. In the early observing runs of Advanced LIGO and Virgo, the localization accuracy can be >100 deg^2 [15–17]. In fact, the localization for GW150914 was about 600 deg^2 (90% probability) [18].

To identify the GW source from such a large localization area, intensive transient surveys should be performed (see, e.g., [19–24] for the case of GW150914). NS-NS mergers and BH-NS mergers are expected to emit EM emission in various forms. One of the most robust candidates is a short GRB. However, the GRB may elude our detection due to the strong relativistic beaming. Other possible EM signals include synchrotron radio emission by the interaction between the ejected material and interstellar gas [25–27] or X-ray emission from a central engine [28–31].

Among variety of emission mechanisms, optical and infrared (IR) emission powered by radioactive decay of r-process nuclei [32–37] is of great interest. This emission is called "kilonova" [34] or "macronova" [33] (we use the term of kilonova in this paper). Kilonova emission is thought to be promising: by advancement of numerical simulations, in particular numerical relativity [38–41], it has been proved that a part of the NS material is surely ejected from NS-NS and BH-NS mergers (e.g., [36, 42–49]). In the ejected material, r-process nucleosynthesis undoubtedly takes place (e.g., [35, 36, 49–56]). Therefore the emission powered by r-process nuclei is a natural outcome from these merger events.

Observations of kilonova will also have important implications for the origin of r-process elements in the Universe. The event rate of NS-NS mergers and BH-NS mergers will be measured by the detection of GWs. In addition, as described in this paper, the brightness of kilonova reflects the amount of the ejected r-process elements. Therefore, by combination of GW observations and EM observations, that is, "multi-messenger" observations, we can measure the production rate of r-process elements by NS-NS and BH-NS mergers, which is essential to understand the origin of r-process elements. In fact, importance of compact binary mergers in chemical evolution has been extensively studied in recent years [72–82].

This paper reviews kilonova emission from compact binary mergers. The primal aim of this paper is providing a guide for optical and infrared follow-up observations for GW sources. For the physical processes of compact binary mergers and various EM emission mechanisms, see recent reviews by Rosswog [83] and Fernández and Metzger [84]. First, we give overview of kilonova emission and describe the expected properties of the emission in Section 2. Then, we compare kilonova models with currently available observations in Section 3. Based on the current theoretical and observational understanding, we discuss prospects for EM follow-up observations of GW sources in Section 4. Finally, we give summary in Section 5. In this paper, the magnitudes are given in the AB magnitude unless otherwise specified.

2. Kilonova Emission

2.1. Overview. The idea of kilonova emission was first introduced by Li and Paczyński [32]. The emission mechanism is similar to that of Type Ia supernova (SN). The main differences are the following: (1) a typical ejecta mass from compact binary mergers is only an order of $0.01M_\odot$ ($1.4M_\odot$ for Type Ia SN), (2) a typical expansion velocity is as high as $v \sim 0.1$–$0.2c = 30,000$–$60,000\,\mathrm{km\,s^{-1}}$ ($\sim$10,000 $\mathrm{km\,s^{-1}}$ for Type Ia SN), and (3) the heating source is decay energy of radioactive r-process nuclei (^{56}Ni for Type Ia SN).

Suppose spherical, homogeneous, and homologously expanding ejecta with a radioactive energy deposition. A typical optical depth in the ejecta is $\tau = \kappa\rho R$, where κ is the mass absorption coefficient or "opacity" ($\mathrm{cm^2\,g^{-1}}$), ρ is the density, and R is the radius of the ejecta. Then, the diffusion timescale in the ejecta is

$$t_{\mathrm{diff}} = \frac{R}{c}\tau \simeq \frac{3\kappa M_{\mathrm{ej}}}{4\pi c v t}, \tag{1}$$

by adopting $M_{\mathrm{ej}} = (4\pi/3)\rho R^3$ (homogeneous ejecta) and $R = vt$ (homologous expansion).

When the dynamical timescale of the ejecta ($t_{\mathrm{dyn}} = R/v = t$) becomes comparable to the diffusion timescale, photons can escape from the ejecta effectively [85]. From the condition of $t_{\mathrm{diff}} = t_{\mathrm{dyn}}$, the characteristic timescale of the emission can be written as follows:

$$t_{\mathrm{peak}} = \left(\frac{3\kappa M_{\mathrm{ej}}}{4\pi c v}\right)^{1/2} \simeq 8.4\ \mathrm{days}\left(\frac{M_{\mathrm{ej}}}{0.01M_\odot}\right)^{1/2} \times\left(\frac{v}{0.1c}\right)^{-1/2}\left(\frac{\kappa}{10\,\mathrm{cm^2\,g^{-1}}}\right)^{1/2}. \tag{2}$$

The radioactive decay energy of mixture of r-process nuclei is known to have a power-law dependence $\dot{q}(t) \simeq 2\times10^{10}\,\mathrm{erg\,s^{-1}\,g^{-1}}(t/1\ \mathrm{day})^{-1.3}$ [34, 35, 54, 86–88]. By introducing a fraction of energy deposition (ϵ_{dep}), the total energy deposition rate (or the deposition luminosity) is $L_{\mathrm{dep}} = \epsilon_{\mathrm{dep}} M_{\mathrm{ej}}\dot{q}(t)$. A majority ($\sim$90%) of decay energy is released by β decay while the other 10% is released by fission [34]. For the β decay, about 25%, 25%, and 50% of the energy are carried by neutrinos, electrons, and γ-rays, respectively. Among these, almost all the energy carried by electrons is deposited, and a fraction of the γ-ray energy is also deposited to the ejecta. Thus, the fraction ϵ_{dep} is about 0.5 (see [89] for more details). The dashed line in Figure 1 shows the deposition luminosity L_{dep} for $\epsilon_{\mathrm{dep}} = 0.5$ and $M_{\mathrm{ej}} = 0.01M_\odot$.

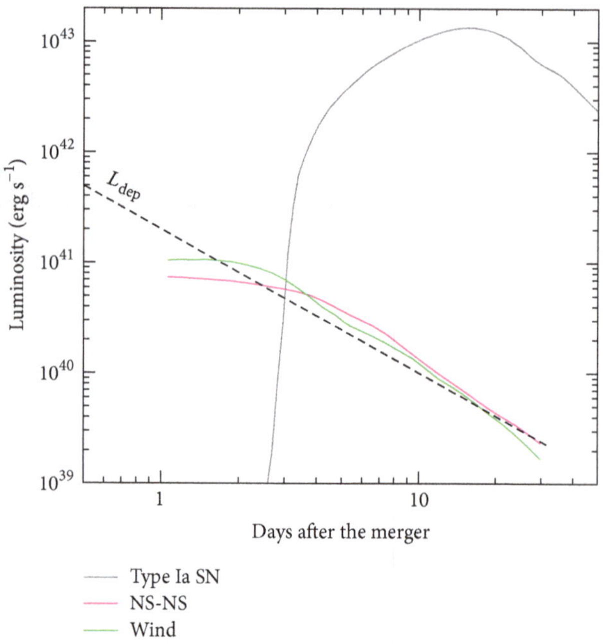

FIGURE 1: Bolometric light curves of a NS-NS merger model (red, $M_{\mathrm{ej}} = 0.01M_\odot$ [57, 58]) and a wind model (green, $M_{\mathrm{ej}} = 0.01M_\odot$) compared with a light curve of Type Ia SN model (gray, $M_{\mathrm{ej}} = 1.4M_\odot$). The black dashed line shows the deposition luminosity by radioactive decay of r-process nuclei ($\epsilon_{\mathrm{dep}} = 0.5$ and $M_{\mathrm{ej}} = 0.01M_\odot$).

Since the peak luminosity is approximated by the deposition luminosity at t_{peak} (so-called Arnett's law [85]), the peak luminosity of kilonova can be written as follows:

$$L_{peak} = L_{dep}(t_{peak}) = \epsilon_{dep} M_{ej} \dot{q}(t_{peak})$$
$$\simeq 1.3 \times 10^{40}\,\mathrm{erg\,s^{-1}} \times \left(\frac{\epsilon_{dep}}{0.5}\right)^{1/2} \left(\frac{M_{ej}}{0.01 M_{\odot}}\right)^{0.35} \times \left(\frac{v}{0.1c}\right)^{0.65} \left(\frac{\kappa}{10\,\mathrm{cm^2\,g^{-1}}}\right)^{-0.65}. \quad (3)$$

An important factor in this analysis is the opacity in the ejected material from compact binary mergers. Previously, the opacity had been assumed to be similar to that of Type Ia SN, that is, $\kappa \sim 0.1\,\mathrm{cm^2\,g^{-1}}$ (bound-bound opacity of iron-peak elements). However, recent studies [57, 90, 91] show that the opacity in the r-process element-rich ejecta is as high as $\kappa \sim 10\,\mathrm{cm^2\,g^{-1}}$ (bound-bound opacity of lanthanide elements). This finding largely revised our understanding of the emission properties of kilonova. As evident from (2) and (3), a higher opacity by a factor of 100 leads to a longer timescale by a factor of ~10 and a lower luminosity by a factor of ~20.

2.2. NS-NS Mergers. When two NSs merge with each other, a small part of the NSs is tidally disrupted and ejected to the interstellar medium (e.g., [36, 42]). This ejecta component is mainly distributed in the orbital plane of the NSs. In addition to this, the collision drives a strong shock, and shock-heated material is also ejected in a nearly spherical manner (e.g., [48, 92]). As a result, NS-NS mergers have quasi-spherical ejecta. The mass of the ejecta depends on the mass ratio and the eccentricity of the orbit of the binary, as well as the radius of the NS or equation of state (EOS, e.g., [48, 92–96]): a more uneven mass ratio and more eccentric orbit lead to a larger amount of tidally disrupted ejecta and a smaller NS radius leads to a larger amount of shock-driven ejecta.

The red line in Figure 1 shows the expected luminosity of a NS-NS merger model (APR4-1215 from Hotokezaka et al. [48]). This model adopts a "soft" EOS APR4 [97], which gives the radius of 11.1 km for a $1.35 M_{\odot}$ NS. The gravitational masses of two NSs are $1.2 M_{\odot} + 1.5 M_{\odot}$ and the ejecta mass is $0.01 M_{\odot}$. The light curve does not have a clear peak since the energy deposited in the outer layer can escape earlier. Since photons kept in the ejecta by the earlier stage effectively escape from the ejecta at the characteristic timescale (2), the luminosity exceeds the energy deposition rate at ~5–8 days after the merger.

Figure 2 shows multicolor light curves of the same NS-NS merger model (red line; see the right axis for the absolute magnitudes). As a result of the high opacity and the low temperature [90], the optical emission is greatly suppressed, resulting in an extremely "red" color of the emission. The red color is more clearly shown in Figure 3, where the spectral evolution of the NS-NS merger model is compared with the spectra of a Type Ia SN and a broad-line Type Ic SN. In fact, the peak of the spectrum is located at near-IR wavelengths [57, 90, 91].

Because of the extremely high expansion velocities, NS-NS mergers show feature-less spectra (Figure 3). This is a big contrast to the spectra of SNe (black and gray lines), where Doppler-shifted absorption lines of strong features can be identified. Even broad-line Type Ic SN 1998bw (associated with long-duration GRB 980425) showed some absorption features although many lines are blended. Since the high expansion velocity is a robust outcome of dynamical ejecta from compact binary mergers, the confirmation of the smooth spectrum will be a key to conclusively identify the GW sources.

The current wavelength-dependent radiative transfer simulations assume the uniform element abundances. However, recent numerical simulations with neutrino transport show that the element abundances in the ejecta becomes nonuniform [54, 92, 95, 96]. Because of the high temperature and neutrino absorption, the polar region can have higher electron fractions (Y_e or number of protons per nucleon), resulting in a wide distribution of Y_e in the ejecta. Interestingly the wide distribution of Y_e is preferable for reproducing the solar r-process abundance ratios [54, 56]. This effect can have a big impact on the kilonova emission: if the synthesis of lanthanide elements is suppressed in the polar direction, the opacity there can be smaller, and thus, the emission to the polar direction can be more luminous with an earlier peak.

2.3. BH-NS Mergers. Mergers of BH and NS are also important targets for GW detection (see [98] for a review). Although the event rate is rather uncertain [10], the number of events can be comparable to that of NS-NS mergers thanks to the stronger GW signals and thus larger horizon distances. BH-NS mergers in various conditions have been extensively studied by numerical simulations (e.g., [99–103]). In particular, for a low BH/NS mass ratio (or small BH mass) and a high BH spin, ejecta mass of BH-NS mergers can be larger than that of NS-NS mergers [59, 104–109]. Since the tidal disruption is the dominant mechanism of the mass ejection, a larger NS radius (or stiff EOS) gives a higher ejecta mass, which is opposite to the situation in NS-NS mergers, where shock-driven ejecta dominates.

Radiative transfer simulations in BH-NS merger ejecta show that kilonova emission from BH-NS mergers can be more luminous in optical wavelengths than that from NS-NS mergers [58]. The blue lines in Figure 2 show the light curve of a BH-NS merger model (APR4Q3a75 from Kyutoku et al. [59]), a merger of a $1.35 M_{\odot}$ NS and a $4.05 M_{\odot}$ BH with a spin parameter of $a = 0.75$. The mass of the ejecta is $M_{ej} = 0.01 M_{\odot}$. Since BH-NS merger ejecta are highly anisotropic and confined to a small solid angle, the temperature of the ejecta can be higher for a given mass of the ejecta, and thus, the emission tends to be bluer than in NS-NS mergers. Therefore, even if the bolometric luminosity is similar, the optical luminosity of BH-NS mergers can be higher than that of NS-NS mergers.

It is emphasized that the mass ejection from BH-NS mergers has a much larger diversity compared with NS-NS mergers, depending on the mass ratio, the BH spin, and its orientation. As a result, the expected brightness also has a

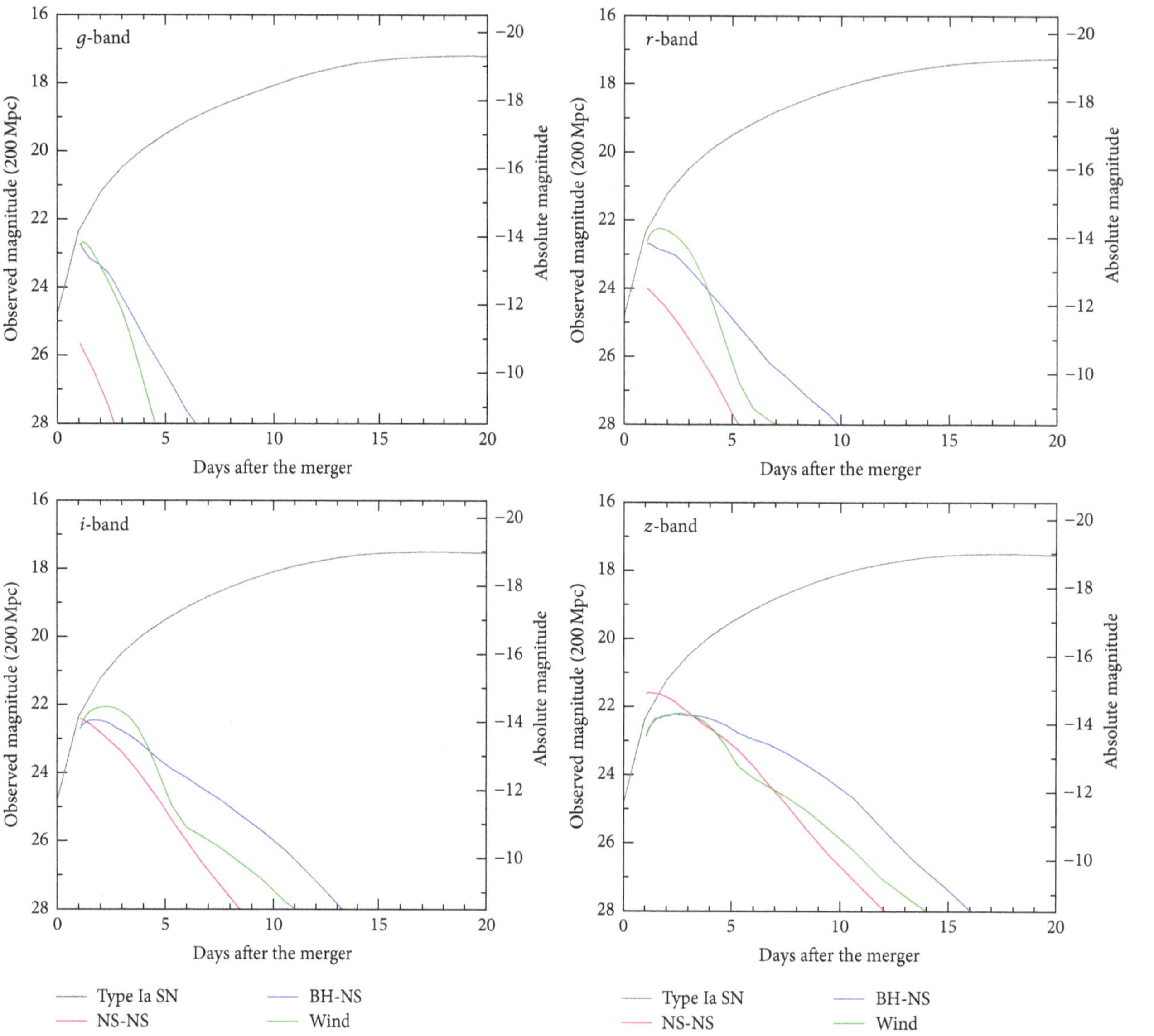

Figure 2: Expected observed magnitudes of kilonova models at 200 Mpc distance [57, 58]. The red, blue, and green lines show the models of NS-NS merger (APR4-1215, [48]), BH-NS merger (APR4Q3a75, [59]), and a wind model (this paper), respectively. The ejecta mass is $M_{ej} = 0.01M_{\odot}$ for these models. For comparison, light curve models of Type Ia SN are shown in gray. The corresponding absolute magnitudes are indicated in the right axis.

large diversity. See Kawaguchi et al. [110] for the expected kilonova brightness for a wide parameter space.

2.4. Wind Components. After the merger of two NSs, a hypermassive NS is formed at the center, and it subsequently collapses to a BH. During this process, accretion disk surrounding the central remnant is formed. A BH-accretion disk system is also formed in BH-NS mergers. From such accretion disk systems, an outflow or disk "wind" can be driven by neutrino heating, viscous heating, or nuclear recombination [56, 111–117]. A typical velocity of the wind is $v = 10{,}000$–$20{,}000\ \mathrm{km\,s^{-1}}$, slower than the precedent dynamical ejecta. Although the ejecta mass largely depends on the ejection mechanism, a typical mass is likely an order of $M_{ej} = 0.01M_{\odot}$ or even larger.

This wind component is another important source of kilonova emission [112, 113, 118–120]. The emission properties depend on the element composition in the ejecta. In particular, if a high electron fraction ($Y_e \gtrsim 0.25$) is realized by the neutrino emission from a long-lived hypermassive NS [118, 119] or shock heating in the outflow [115], synthesis of lanthanide elements can be suppressed in the wind. Then, the resulting emission can be bluer than the emission from the dynamical ejecta thanks to the lower opacity [57, 90]. This component can be called "blue kilonova" [84].

To demonstrate the effect of the low opacity, we show a simple wind model in Figures 1 and 2. In this model, we adopt a spherical ejecta of $M_{ej} = 0.01M_{\odot}$ with a density structure of $\rho \propto r^{-2}$ from $v = 0.01c$ to $0.1c$ (with the average velocity of $v \sim 20{,}000\ \mathrm{km\,s^{-1}}$). The elements in the ejecta are assumed

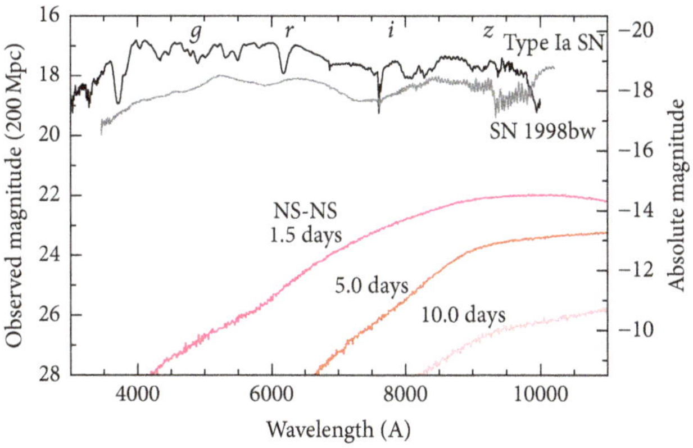

FIGURE 3: Expected observed spectra of the NS-NS merger model APR4-1215 ($M_{ej} = 0.01M_\odot$) compared with the spectra of normal Type Ia SN 2005cf [60–62] and broad-line Type Ic SN 1998bw [63, 64]. The spectra are shown in AB magnitudes (f_ν) at 200 Mpc distance. The corresponding absolute magnitudes are indicated in the right axis.

to be lanthanide-free: only the elements of $Z = 31–54$ are included with the solar abundance ratios. As shown by previous works [119], the emission from such a wind can peak earlier than that from the dynamical ejecta (Figure 1) and the emission is bluer (Figure 2).

Note that this simple model neglects the presence of the dynamical ejecta outside of the wind component. The effect of the dynamical ejecta is in fact important, because it works as a "lanthanide curtain" [119] absorbing the emission from the disk wind. Interestingly, as described in Section 2.2, the polar region of the dynamical ejecta can have a higher Y_e, and the "lanthanide curtain" may not be present in the direction. Also, in BH-NS mergers, the dynamical ejecta is distributed in the orbital plane, and disk wind can be directly observed from most of the lines of sight. If the wind component is dominant for kilonova emission and can be directly observed, the spectra are not as smooth as the spectra of dynamical ejecta because of the slower expansion [119]. More realistic simulations capturing all of these situations will be important to understand the emission from the disk wind.

3. Lessons from Observations

Since short GRBs are believed to be driven by NS-NS mergers or BH-NS mergers (see, e.g., [6, 7]), models of kilonova can be tested by the observations of short GRBs. As well known, SN component has been detected in the afterglow of long GRBs (see [121, 122] for reviews). If kilonova emission occurs, the emission can be in principle visible on top of the afterglow, but such an emission had eluded the detection for long time [123].

In 2013, a clear excess emission was detected in the near-IR afterglow of GRB 130603B [67, 68]. Interestingly, the excess was not visible in the optical data. Since this behavior nicely agrees with the expected properties of kilonova, the excess is interpreted to be the kilonova emission.

Figure 4(a) shows kilonova models compared with the observations of GRB 130603B. The observed brightness of the near-IR excess in GRB 130603B requires a relatively large ejecta mass of $M_{ej} \gtrsim 0.02M_\odot$ [67, 68, 73, 124]. As pointed out by Hotokezaka et al. [124], this favors a soft EOS for a NS-NS merger model (i.e., more shock-driven ejection) and a stiff EOS for a BH-NS merger model (i.e., more tidally driven ejection). Another possibility to explain the brightness may be an additional emission from the disk wind (green line in Figure 4; see [118, 119]).

Note that the excess was detected only at one epoch in one filter. Therefore, other interpretations are also possible, for example, emission by the external shock [125] or by a central magnetar [126, 127], or thermal emission from newly formed dust [128]. Importantly, a late-time excess is also visible in X-ray [129], and thus, the near-IR and X-ray excesses might be caused by the same mechanism, possibly the central engine [130, 131].

Another interesting case is GRB 060614. This GRB was formally classified as a long GRB because the duration is about 100 sec. However, since no bright SN was accompanied, the origin was not clear [132–135]. Recently the existence of a possible excess in the optical afterglow was reported [69, 70]. Figure 4(b) shows the comparison between GRB 060614 and the same sets of the models. If this excess is caused by kilonova, a large ejecta mass of $M_{ej} \sim 0.1M_\odot$ is required. This fact may favor a BH-NS merger scenario with a stiff EOS [69, 70]. It is however important to note that the emission from BH-NS merger has a large variation, and such an effective mass ejection requires a low BH/NS mass ratio and a high BH spin [110]. See also [136] for possible optical excess in GRB 050709, a genuine short GRB with a duration of 0.5 sec [137–140]. If the excess is attributed to kilonova, the required ejecta mass is $M_{ej} \sim 0.05M_\odot$.

Finally, an early brightening in optical data of GRB 080503 at $t \sim 1–5$ days can also be attributed to kilonova [141] although the redshift of this object is unfortunately unknown. Kasen et al. [119] give a possible interpretation with the disk wind model. Note that a long-lasting X-ray emission was also detected in GRB 080503 at $t \lesssim 2$ days, and it may favor a common mechanism for optical and X-ray emission [131, 142].

4. Prospects for EM Follow-Up Observations of GW Sources

Figure 2 shows the expected brightness of compact binary merger models at 200 Mpc (left axis). All the models assume a canonical ejecta mass of $M_{ej} = 0.01M_\odot$, and therefore, the emission can be brighter or fainter depending on the merger parameters and the EOS (see Section 2). Keeping this caveat in mind, typical models suggest that the expected kilonova brightness at 200 Mpc is about 22 mag in red optical wavelengths (i- or z-bands) at $t < 5$ days after the merger. The brightness quickly declines to >24 mag within $t \sim 10$ days after the merger. To detect this emission, we ultimately need 8 m class telescopes. Currently the wide-field capability for 8 m class telescopes is available only at the 8.2 m Subaru

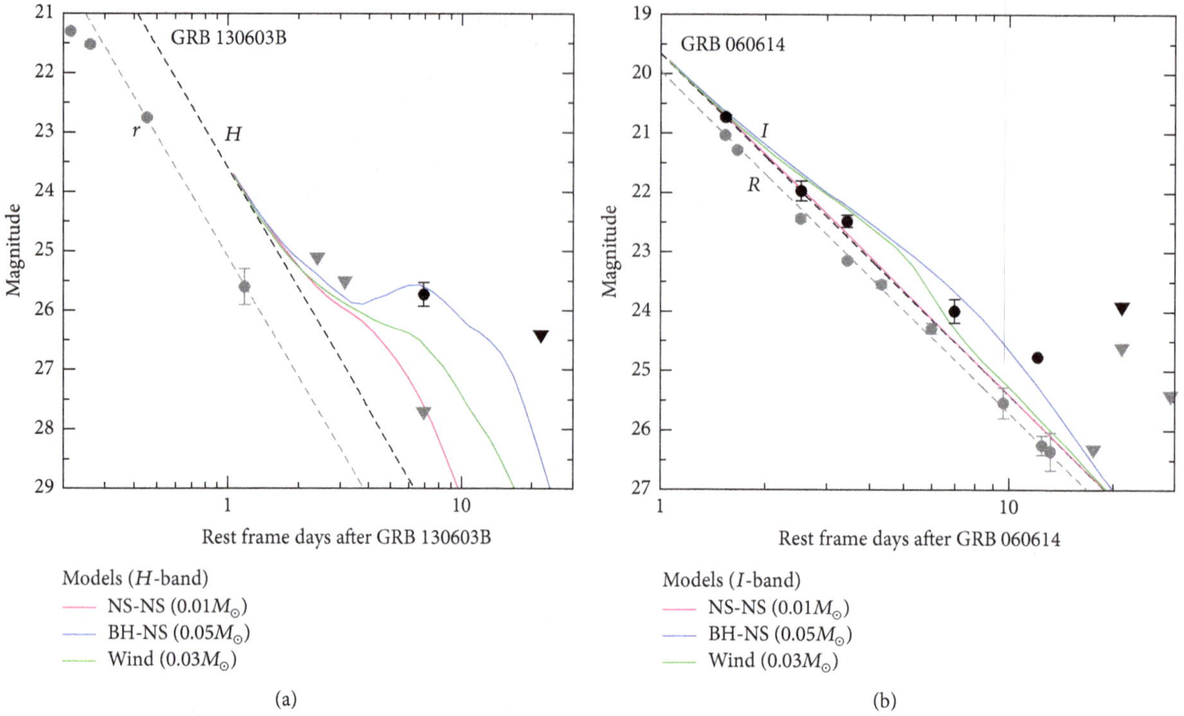

FIGURE 4: Comparison of kilonova models with GRB 130603B (a) and GRB 060614 (b). The models used in these plots are those with relatively high ejecta masses: APR4-1215 (NS-NS, $M_{ej} = 0.01M_{\odot}$ [48]), H4Q3a75 (BH-NS, $M_{ej} = 0.05M_{\odot}$ [59]), and a wind model with $M_{ej} = 0.03M_{\odot}$ (this paper). The H4Q3a75 model is a merger of a $1.35M_{\odot}$ NS and a $4.05M_{\odot}$ BH with a spin parameter of $a = 0.75$. This model adopts a "stiff" EOS H4 [65, 66] which gives a 13.6 km radius for $1.35M_{\odot}$ NS. For GRB 130603B, the afterglow component is assumed to be $f_{\nu} \propto t^{-2.7}$ [67, 68]. For GRB 060614, it is assumed to be $f_{\nu} \propto t^{-2.3}$ [69], which is a conservative choice (see [70] for a possibility of a steeper decline). The observed and model magnitudes for GRB 060614 are given in the Vega system as in the literature [70].

telescope: Subaru/Hyper Suprime-Cam (HSC) has the field of view (FOV) of 1.77 deg^2 [143, 144]. In future, the 8.4 m Large Synoptic Survey Telescope (LSST) with 9.6 deg^2 FOV will be online [145, 146]. Note that targeted galaxy surveys are also effective to search for the transients associated with galaxies [147, 148].

It is again emphasized that the expected brightness of kilonova can have a large variety. If the kilonova candidates seen in GRB 130603B ($M_{ej} \gtrsim 0.02M_{\odot}$) and GRB 060614 ($M_{ej} \sim 0.1M_{\odot}$) are typical cases (see Section 3), the emission can be brighter by ~1-2 mag. In addition, there are also possibilities of bright, precursor emission (e.g., [29, 130, 149]) which are not discussed in depth in this paper. And, of course, the emission is brighter for objects at closer distances. Therefore, surveys with small-aperture telescopes (typically with wider FOVs) are also important. See, for example, Nissanke et al. [13] and Kasliwal and Nissanke [16] for detailed survey simulations for various expected brightness of the EM counterpart.

A big challenge for identification of the GW source is contamination of SNe. NS-NS mergers and BH-NS mergers are rare events compared with SNe, and thus, much larger number of SNe are detected when optical surveys are performed over 10 deg^2 (see [21–23] for the case of GW150914). Therefore, it is extremely important to effectively select the candidates of kilonova from a larger number of SNe.

To help the classification, color-magnitude and color-color diagrams for the kilonova models and Type Ia SNe are shown in Figure 5. The numbers attached with the models are days after the merger while dots for SNe are given with 5-day interval. According to the current understanding, the light curves of kilonova can be characterized as follows.

(1) The timescale of variability should be shorter than that of SNe (Figure 2). This is robust since the ejecta mass from compact binary mergers is much smaller than SNe.

(2) The emission is fainter than SNe. This is also robust because of the smaller ejecta mass and thus the lower available radioactive energy (Figure 1).

(3) The emissions are expected to be redder than SNe. This is an outcome of a high opacity in the ejecta, but the exact color depends on the ejecta composition ([58, 90, 118, 119], Section 2).

Therefore, in order to effectively search for the EM counterpart of the GW source, multiple visits in a timescale of <10 days will be important so that the rapid time evolution can be captured. Surveys with multiple filters are also helpful to use color information. As shown in Figure 5, observed magnitudes of kilonovae at ~200 Mpc are similar to those of SNe at larger distances ($z \gtrsim 0.3$ for Type Ia SNe). Therefore,

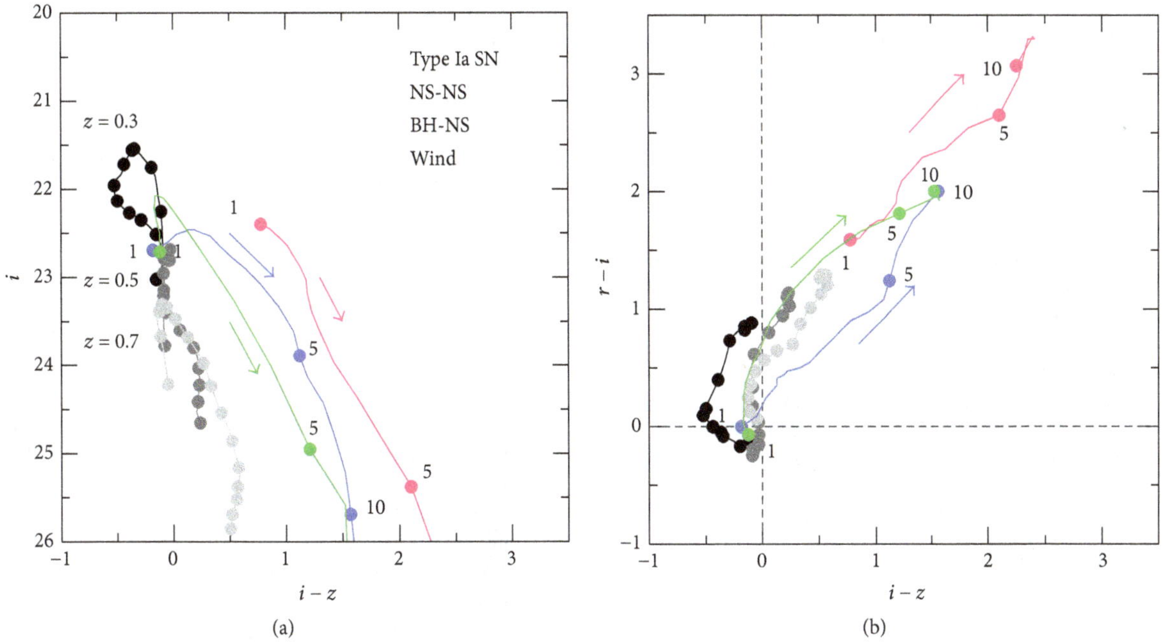

FIGURE 5: Color-magnitude diagram (a) and color-color diagram (b) for compact binary merger models ($M_{ej} = 0.01M_{\odot}$) at 200 Mpc compared with Type Ia SN with similar observed magnitudes (z = 0.3, 0.5, and 0.7). For Type Ia SN, we use spectral templates [71] with K-correction. The numbers for binary merger models show time from the merger in days while dots for Type Ia SN are given with 5-day interval.

if redshifts of the host galaxies are estimated, kilonova candidates can be further selected by the close distances and the intrinsic faintness.

5. Summary

The direct detection of GWs from GW150914 opened GW astronomy. To study the astrophysical nature of the GW sources, the identification of the EM counterparts is essentially important. In this paper, we reviewed the current understanding of kilonova emission from compact binary mergers.

Kilonova emission from the dynamical ejecta of $0.01M_{\odot}$ has a typical luminosity is an order of 10^{40}–10^{41} erg s^{-1} with the characteristic timescale of about 1 week. Because of the high opacity and the low temperature, the spectral peak is located at red optical or near-IR wavelengths. In addition to the emission from the dynamical ejecta, a subsequent disk wind can cause an additional emission which may peak earlier with a bluer color if the emission is not absorbed by the precedent ejecta.

The detection of excess in GRB 130603B (and possibly GRB 060614) supports the kilonova scenario. If the excesses found in these objects are attributed to the kilonova emission, the required ejecta masses are $M_{ej} \gtrsim 0.02M_{\odot}$ and $M_{ej} \sim 0.1M_{\odot}$, respectively. The comparison between such observations and numerical simulations gives important insight to study the progenitor of compact binary mergers and EOS of NS.

At 200 Mpc distance, a typical peak brightness of kilonova emission is about 22 mag in the red optical wavelengths (i- or z-bands). The emission quickly fades to >24 mag within ~10 days. To distinguish GW sources from SNe, observations with multiple visits in a timescale of <10 days are important to select the objects with rapid temporal evolution. The use of multiple filters is also helpful to select red objects. Since the extremely high expansion velocities ($v \sim 0.1$–$0.2c$) are unique features of dynamical mass ejection from compact binary mergers, detection of extremely smooth spectrum will be the smoking gun to conclusively identify the GW sources.

Competing Interests

The author declares that there is no conflict of interests regarding the publication of this paper.

Acknowledgments

The author thanks Kenta Hotokezaka, Yuichiro Sekiguchi, Masaru Shibata, Kenta Kiuchi, Shinya Wanajo, Koutarou Kyutoku, Kyohei Kawaguchi, Keiichi Maeda, Takaya Nozawa, and Yutaka Hirai for fruitful discussion on compact binary mergers, nucleosynthesis, and kilonova emission. The author also thanks Nozomu Tominaga, Tomoki Morokuma, Michitoshi Yoshida, Kouji Ohta, and the J-GEM collaboration for valuable discussion on EM follow-up observations. Numerical simulations presented in this paper were carried out with Cray XC30 at Center for Computational Astrophysics, National Astronomical Observatory of Japan. This research has been supported by the Grant-in-Aid for Scientific

Research of the Japan Society for the Promotion of Science (24740117, 15H02075) and Grant-in-Aid for Scientific Research on Innovative Areas of the Ministry of Education, Culture, Sports, Science and Technology (25103515, 15H00788).

References

[1] G. M. Harry and LIGO Scientific Collaboration, "Advanced LIGO: the next generation of gravitational wave detectors," *Classical and Quantum Gravity*, vol. 27, no. 8, Article ID 084006, 2010.

[2] B. P. Abbott, R. Abbott, T. D. Abbott et al. et al., "Observation of gravitational waves from a binary black hole merger," *Physical Review Letters*, vol. 116, no. 6, Article ID 061102, 2016.

[3] S. I. Blinnikov, I. D. Novikov, T. V. Perevodchikova, and A. G. Polnarev, "Exploding neutron stars in close binaries," *Soviet Astronomy Letters*, vol. 10, no. 3, pp. 177–179, 1984.

[4] D. Eichler, M. Livio, T. Piran, and D. N. Schramm, "Nucleosynthesis, neutrino bursts and γ-rays from coalescing neutron stars," *Nature*, vol. 340, no. 6229, pp. 126–128, 1989.

[5] B. Paczynski, "Gamma-ray bursters at cosmological distances," *The Astrophysical Journal*, vol. 308, pp. L43–L46, 1986.

[6] E. Berger, "Short-duration gamma-ray bursts," *Annual Review of Astronomy and Astrophysics*, vol. 52, pp. 43–105, 2014.

[7] E. Nakar, "Short-hard gamma-ray bursts," *Physics Reports*, vol. 442, no. 1–6, pp. 166–236, 2007.

[8] F. Acernese, M. Agathos, K. Agatsuma et al. et al., "Focus issue: advanced interferometric gravitational wave detectors," *Classical and Quantum Gravity*, vol. 32, no. 2, Article ID 024001, 2015.

[9] K. Somiya, "Detector configuration of KAGRA–the Japanese cryogenic gravitational-wave detector," *Classical and Quantum Gravity*, vol. 29, no. 12, Article ID 124007, 2012.

[10] J. Abadie, B. P. Abbott, R. Abbott et al., "Predictions for the rates of compact binary coalescences observable by ground-based gravitational-wave detectors," *Classical and Quantum Gravity*, vol. 27, Article ID 173001, 2010.

[11] B. P. Abbott, R. Abbott, T. D. Abbott et al., "Prospects for observing and localizing gravitational-wave transients with advanced LIGO and advanced virgo," *Living Reviews in Relativity*, vol. 19, article 1, 2016.

[12] L. Z. Kelley, I. Mandel, and E. Ramirez-Ruiz, "Electromagnetic transients as triggers in searches for gravitational waves from compact binary mergers," *Physical Review D*, vol. 87, no. 12, Article ID 123004, 16 pages, 2013.

[13] S. Nissanke, M. Kasliwal, and A. Georgieva, "Identifying elusive electromagnetic counterparts to gravitational wave mergers: an end-to-end simulation," *The Astrophysical Journal*, vol. 767, no. 2, article 124, 2013.

[14] S. Nissanke, J. Sievers, N. Dalal, and D. Holz, "Localizing compact binary inspirals on the sky using ground-based gravitational wave interferometers," *Astrophysical Journal*, vol. 739, no. 2, article 99, 2011.

[15] R. Essick, S. Vitale, E. Katsavounidis, G. Vedovato, and S. Klimenko, "Localization of short duration gravitational-wave transients with the early advanced ligo and virgo detectors," *Astrophysical Journal*, vol. 800, no. 2, article 81, 2015.

[16] M. M. Kasliwal and S. Nissanke, "On discovering electromagnetic emission from neutron star mergers: the early years of two gravitational wave detectors," *The Astrophysical Journal Letters*, vol. 789, no. 1, article L5, 2014.

[17] L. P. Singer, L. R. Price, B. Farr et al., "The first two years of electromagnetic follow-up with advanced ligo and virgo," *Astrophysical Journal*, vol. 795, no. 2, article 105, 2014.

[18] The LIGO Scientific Collaboration and the Virgo Collaboration, "Properties of the binary black hole merger GW150914," 2016, https://arxiv.org/abs/1602.03840.

[19] B. P. Abbott, R. Abbott, T. D. Abbott et al., "Localization and broadband follow-up of the gravitational-wave transient GW150914," 2016, https://arxiv.org/abs/1602.08492.

[20] P. A. Evans, J. A. Kennea, S. D. Barthelmy et al., "*Swift* follow-up of the gravitational wave source GW150914," *MNRAS Letters*, vol. 460, no. 1, pp. L40–L44, 2016.

[21] M. M. Kasliwal, S. B. Cenko, L. P. Singer et al., "iPTF search for anoptical counterpart to gravitational wave trigger GW150914," http://arxiv.org/abs/1602.08764.

[22] S. J. Smartt, K. C. Chambers, K. W. Smith et al., "Pan-STARRS and PESSTO search for the optical counterpart to the LIGO gravitational wave source GW150914," http://arxiv.org/abs/1602.04156.

[23] M. Soares-Santos, R. Kessler, E. Berger et al., "A dark energy camera search for anoptical counterpart to the first advanced LIGO gravitational wave event GW150914," http://arxiv.org/abs/1602.04198.

[24] T. Morokuma, M. Tanaka, Y. Asakura et al., "J-GEM follow-up observations to search for an optical counterpart of the first gravitational wave source GW150914," http://arxiv.org/abs/1605.03216.

[25] E. Nakar and T. Piran, "Detectable radio flares following gravitational waves from mergers of binary neutron stars," *Nature*, vol. 478, no. 7367, pp. 82–84, 2011.

[26] T. Piran, E. Nakar, and S. Rosswog, "The electromagnetic signals of compact binary mergers," *Monthly Notices of the Royal Astronomical Society*, vol. 430, no. 3, pp. 2121–2136, 2013.

[27] K. Hotokezaka and T. Piran, "Mass ejection from neutron star mergers: different components and expected radio signals," *Monthly Notices of the Royal Astronomical Society*, vol. 450, no. 2, pp. 1430–1440, 2015.

[28] T. Nakamura, K. Kashiyama, D. Nakauchi, Y. Suwa, T. Sakamoto, and N. Kawai, "Soft X-ray extended emissions of short gamma-ray bursts as electromagnetic counterparts of compact binary mergers: possible origin and detectability," *Astrophysical Journal*, vol. 796, no. 1, article 13, 2014.

[29] B. D. Metzger and A. L. Piro, "Optical and X-ray emission from stable millisecond magnetars formed from the merger of binary neutron stars," *Monthly Notices of the Royal Astronomical Society*, vol. 439, no. 4, pp. 3916–3930, 2014.

[30] S. Kisaka, K. Ioka, and T. Nakamura, "Isotropic detectable X-ray counterparts to gravitational waves from neutron star binary mergers," *The Astrophysical Journal Letters*, vol. 809, article L8, 2015.

[31] D. M. Siegel and R. Ciolfi, "Electromagnetic emission from long-lived binary neutron star merger remnants. II. light curves and spectra," *The Astrophysical Journal*, vol. 819, no. 1, p. 15, 2016.

[32] L.-X. Li and B. Paczyński, "Transient events from neutron star mergers," *The Astrophysical Journal*, vol. 507, no. 1, pp. L59–L62, 1998.

[33] S. R. Kulkarni, "Modeling supernova-like explosions associated with gamma-ray bursts with short durations," 2005, http://arxiv.org/abs/astro-ph/0510256.

[34] B. D. Metzger, G. Martínez-Pinedo, S. Darbha et al., "Electromagnetic counterparts of compact object mergers powered by the radioactive decay of r-process nuclei," *Monthly Notices of the Royal Astronomical Society*, vol. 406, no. 4, pp. 2650–2662, 2010.

[35] L. F. Roberts, D. Kasen, W. H. Lee, and E. Ramirez-Ruiz, "Electromagnetic transients powered by nuclear decay in the tidal tails of coalescing compact binaries," *The Astrophysical Journal Letters*, vol. 736, no. 1, article L21, 2011.

[36] S. Goriely, A. Bauswein, and H.-T. Janka, "R-process nucleosynthesis in dynamically ejected matter of neutron star mergers," *Astrophysical Journal Letters*, vol. 738, no. 2, article L32, 2011.

[37] B. D. Metzger and E. Berger, "What is the most promising electromagnetic counterpart of a neutron star binary merger?" *The Astrophysical Journal*, vol. 746, no. 1, p. 48, 2012.

[38] M. Shibata and K. Uryū, "Simulation of merging binary neutron stars in full general relativity: $\Gamma = 2$ case," *Physical Review D*, vol. 61, no. 6, Article ID 064001, 18 pages, 2000.

[39] M. Shibata, K. Taniguchi, and K. Uryū, "Merger of binary neutron stars with realistic equations of state in full general relativity," *Physical Review D*, vol. 71, no. 8, Article ID 084021, 2005.

[40] M. D. Duez, "Numerical relativity confronts compact neutron star binaries: a review and status report," *Classical and Quantum Gravity*, vol. 27, no. 11, Article ID 114002, 2010.

[41] J. A. Faber and F. A. Rasio, "Binary neutron star mergers," *Living Reviews in Relativity*, vol. 15, article 8, 2012.

[42] S. Rosswog, M. Liebendörfer, F.-K. Thielemann, M. B. Davies, W. Benz, and T. Piran, "Mass ejection in neutron star mergers," *Astronomy and Astrophysics*, vol. 341, no. 2, pp. 499–526, 1999.

[43] S. Rosswog, M. B. Davies, F.-K. Thielemann, and T. Piran, "Merging neutron stars: asymmetric systems," *Astronomy & Astrophysics*, vol. 360, pp. 171–184, 2000.

[44] M. Ruffert and H.-T. Janka, "Coalescing neutron stars—a step towards physical models III. Improved numerics and different neutron star masses and spins," *Astronomy and Astrophysics*, vol. 380, no. 2, pp. 544–577, 2001.

[45] S. Rosswog, "Mergers of neutron star-black hole binaries with small mass ratios: nucleosynthesis, gamma-ray bursts, and electromagnetic transients," *Astrophysical Journal*, vol. 634, no. 2, pp. 1202–1213, 2005.

[46] W. H. Lee and E. Ramirez-Ruiz, "The progenitors of short gamma-ray bursts," *New Journal of Physics*, vol. 9, article A17, 2007.

[47] S. Rosswog, "The dynamic ejecta of compact object mergers and eccentric collisions," *Royal Society of London Philosophical Transactions Series A*, vol. 371, no. 1992, Article ID 20272, 2013.

[48] K. Hotokezaka, K. Kiuchi, K. Kyutoku et al., "Mass ejection from the merger of binary neutron stars," *Physical Review D—Particles, Fields, Gravitation and Cosmology*, vol. 87, no. 2, Article ID 024001, 2013.

[49] A. Bauswein, S. Goriely, and H.-T. Janka, "Systematics of dynamical mass ejection, nucleosynthesis, and radioactively powered electromagnetic signals from neutron-star mergers," *Astrophysical Journal*, vol. 773, no. 1, article 78, 2013.

[50] J. M. Lattimer and D. N. Schramm, "Black-hole-neutron-star collisions," *Astrophysical Journal*, vol. 192, part 2, pp. L145–L147, 1974.

[51] J. M. Lattimer and D. N. Schramm, "The tidal disruption of neutron stars by black holes in close binaries," *The Astrophysical Journal*, vol. 210, pp. 549–567, 1976.

[52] C. Freiburghaus, S. Rosswog, and F.-K. Thielemann, "r-process in neutron star mergers," *The Astrophysical Journal*, vol. 525, no. 2, pp. L121–L124, 1999.

[53] O. Korobkin, S. Rosswog, A. Arcones, and C. Winteler, "On the astrophysical robustness of the neutron star merger r-process," *Monthly Notices of the Royal Astronomical Society*, vol. 426, no. 3, pp. 1940–1949, 2012.

[54] S. Wanajo, Y. Sekiguchi, N. Nishimura, K. Kiuchi, K. Kyutoku, and M. Shibata, "Production of all the r-process nuclides in the dynamical ejecta of neutron star mergers," *The Astrophysical Journal Letters*, vol. 789, no. 2, article L39, 2014.

[55] J. de Jesús Mendoza-Temis, M.-R. Wu, K. Langanke, G. Martínez-Pinedo, A. Bauswein, and H.-T. Janka, "Nuclear robustness of the r process in neutron-star mergers," *Physical Review C*, vol. 92, no. 5, Article ID 055805, 16 pages, 2015.

[56] O. Just, A. Bauswein, R. A. Pulpillo, S. Goriely, and H. T. Janka, "Comprehensive nucleosynthesis analysis for ejecta of compact binary mergers," *Monthly Notices of the Royal Astronomical Society*, vol. 448, no. 1, pp. 541–567, 2015.

[57] M. Tanaka and K. Hotokezaka, "Radiative transfer simulations of neutron star merger ejecta," *The Astrophysical Journal*, vol. 775, no. 2, article 113, 2013.

[58] M. Tanaka, K. Hotokezaka, K. Kyutoku et al., "Radioactively powered emission from black hole-neutron star mergers," *Astrophysical Journal*, vol. 780, no. 1, article 31, 2014.

[59] K. Kyutoku, K. Ioka, and M. Shibata, "Anisotropic mass ejection from black hole-neutron star binaries: diversity of electromagnetic counterparts," *Physical Review D*, vol. 88, no. 4, Article ID 041503, 2013.

[60] A. Pastorello, S. Taubenberger, N. Elias-Rosa et al., "ESC observations of SN 2005cf -I. Photometric evolution of a normal Type Ia supernova," *Monthly Notices of the Royal Astronomical Society*, vol. 376, no. 3, pp. 1301–1316, 2007.

[61] G. Garavini, S. Nobili, S. Taubenberger et al., "ESC observations of SN 2005cf. II. Optical spectroscopy and the high-velocity features," *Astronomy & Astrophysics*, vol. 471, no. 2, pp. 527–535, 2007.

[62] X. Wang, W. Li, A. V. Filippenko et al., "The golden standard type ia supernova 2005cf: observations from the ultraviolet to the near-infrared wavebands," *The Astrophysical Journal*, vol. 697, no. 1, pp. 380–408, 2009.

[63] K. Iwamoto, P. A. Mazzali, K. Nomoto et al., "A hypernova model for the supernova associated with the γ-ray burst of 25 April 1998," *Nature*, vol. 395, no. 6703, pp. 672–674, 1998.

[64] K. Iwamoto, P. A. Mazzali, K. Nomoto et al., "A hypernova model for the supernova associated with the γ-ray burst of 25 April 1998," *Nature*, vol. 395, no. 6703, pp. 672–674, 1998.

[65] N. K. Glendenning and S. A. Moszkowski, "Reconciliation of neutron-star masses and binding of the Λ in hypernuclei," *Physical Review Letters*, vol. 67, p. 2414, 1991.

[66] B. D. Lackey, M. Nayyar, and B. J. Owen, "Observational constraints on hyperons in neutron stars," *Physical Review D*, vol. 73, no. 2, Article ID 024021, 2006.

[67] N. R. Tanvir, A. J. Levan, A. S. Fruchter et al., "A 'kilonova' associated with the short-duration γ-ray burst GRB 130603B," *Nature*, vol. 500, no. 7464, pp. 547–549, 2013.

[68] E. Berger, W. Fong, and R. Chornock, "An r-process kilonova associated with the short-hard GRB 130603B," *The Astrophysical Journal Letters*, vol. 774, no. 2, article L23, 2013.

[69] B. Yang, Z.-P. Jin, X. Li et al., "A possible macronova in the late afterglow of the long-short burst GRB 060614," *Nature Communications*, vol. 6, article 7323, 2015.

[70] Z.-P. Jin, X. Li, Z. Cano, S. Covino, Y.-Z. Fan, and D.-M. Wei, "The light curve of the macronova associated with the long-short burst GRB 060614," *Astrophysical Journal Letters*, vol. 811, no. 2, article L22, 2015.

[71] P. Nugent, A. Kim, and S. Perlmutter, "K-corrections and extinction corrections for type Ia supernovae," *Publications of the Astronomical Society of the Pacific*, vol. 114, no. 798, pp. 803–819, 2002.

[72] D. Argast, M. Samland, F.-K. Thielemann, and Y.-Z. Qian, "Neutron star mergers versus core-collapse supernovae as dominant r-process sites in the early Galaxy," *Astronomy and Astrophysics*, vol. 416, no. 3, pp. 997–1011, 2004.

[73] T. Piran, O. Korobkin, and S. Rosswog, "Implications of GRB 130603B and itsmacronova for r-process nucleosynthesis," http://arxiv.org/abs/1401.2166.

[74] F. Matteucci, D. Romano, A. Arcones, O. Korobkin, and S. Rosswog, "Europium production: neutron star mergers versus core-collapse supernovae," *Monthly Notices of the Royal Astronomical Society*, vol. 438, no. 3, Article ID stt2350, pp. 2177–2185, 2014.

[75] T. Tsujimoto and T. Shigeyama, "Enrichment history of r-process elements shaped by a merger of neutron star pairs," *Astronomy and Astrophysics*, vol. 565, article L5, 2014.

[76] Y. Komiya, S. Yamada, T. Suda, and M. Y. Fujimoto, "The new model of chemical evolution of *r*-process elements based on the hierarchical galaxy formation. I. Ba and Eu," *The Astrophysical Journal*, vol. 783, no. 2, p. 132, 2014.

[77] G. Cescutti, D. Romano, F. Matteucci, C. Chiappini, and R. Hirschi, "The role of neutron star mergers in the chemical evolution of the Galactic halo," *Astronomy & Astrophysics*, vol. 577, article A139, 10 pages, 2015.

[78] B. Wehmeyer, M. Pignatari, and F.-K. Thielemann, "Galactic evolution of rapid neutron capture process abundances: the inhomogeneous approach," *Monthly Notices of the Royal Astronomical Society*, vol. 452, no. 2, pp. 1970–1981, 2015.

[79] Y. Ishimaru, S. Wanajo, and N. Prantzos, "Neutron star mergers as the origin of r-process elements in the galactic halo based on the sub-halo clustering scenario," *The Astrophysical Journal Letters*, vol. 804, no. 2, article L35, 2015.

[80] S. Shen, R. J. Cooke, E. Ramirez-Ruiz, P. Madau, L. Mayer, and J. Guedes, "The history of r-process enrichment in the milky way," *Astrophysical Journal*, vol. 807, no. 2, article 115, 2015.

[81] F. van de Voort, E. Quataert, P. F. Hopkins, D. Kereš, and C. Faucher-Giguere, "Galactic r-process enrichment by neutron star mergers in cosmological simulations of a Milky Way-mass galaxy," *Monthly Notices of the Royal Astronomical Society*, vol. 447, no. 1, pp. 140–148, 2015.

[82] Y. Hirai, Y. Ishimaru, T. R. Saitoh, M. S. Fujii, J. Hidaka, and T. Kajino, "Enrichment of r-process elements in dwarf spheroidal galaxies in chemo-dynamical evolution model," *The Astrophysical Journal*, vol. 814, no. 1, p. 41, 2015.

[83] S. Rosswog, "The multi-messenger picture of compact binary mergers," *International Journal of Modern Physics. D. Gravitation, Astrophysics, Cosmology*, vol. 24, no. 5, Article ID 1530012, 2015.

[84] R. Fernández and B. D. Metzger, "Electromagnetic signatures of neutron star mergers in the advanced LIGO era," 2015, http://arxiv.org/abs/1512.05435.

[85] W. D. Arnett, "Type I supernovae. I—analytic solutions for the early part of the light curve," *The Astrophysical Journal*, vol. 253, no. 2, pp. 785–797, 1982.

[86] S. Rosswog, O. Korobkin, A. Arcones, F.-K. Thielemann, and T. Piran, "The long-term evolution of neutron star merger remnants—I. The impact of r-process nucleosynthesis," *Monthly Notices of the Royal Astronomical Society*, vol. 439, no. 1, pp. 744–756, 2014.

[87] D. Grossman, O. Korobkin, S. Rosswog, and T. Piran, "The long-term evolution of neutron star merger remnants—II. Radioactively powered transients," *Monthly Notices of the Royal Astronomical Society*, vol. 439, no. 1, pp. 757–770, 2014.

[88] J. Lippuner and L. F. Roberts, "r-Process lanthanide production and heating rates in kilonovae," *The Astrophysical Journal*, vol. 815, no. 2, p. 82, 2015.

[89] K. Hotokezaka, S. Wanajo, M. Tanaka, A. Bamba, Y. Terada, and T. Piran, "Radioactive decay products in neutron star merger ejecta: heating efficiency and γ-ray emission," *Monthly Notices of the Royal Astronomical Society*, vol. 459, no. 1, pp. 35–43, 2016.

[90] D. Kasen, N. R. Badnell, and J. Barnes, "Opacities and spectra of the r-process ejecta from neutron star mergers," *Astrophysical Journal*, vol. 774, no. 1, article 25, 2013.

[91] J. Barnes and D. Kasen, "Effect of a high opacity on the light curves of radioactively powered transients from compact object mergers," *The Astrophysical Journal*, vol. 775, no. 1, p. 18, 2013.

[92] D. Radice, F. Galeazzi, J. Lippuner, L. F. Roberts, C. D. Ott, and L. Rezzolla, "Dynamical mass ejection from binary neutron star mergers," 2016, http://arxiv.org/abs/1601.02426.

[93] S. Rosswog, T. Piran, and E. Nakar, "The multimessenger picture of compact object encounters: binary mergers versus dynamical collisions," *Monthly Notices of the Royal Astronomical Society*, vol. 430, no. 4, pp. 2585–2604, 2013.

[94] C. Palenzuela, S. L. Liebling, D. Neilsen et al., "Effects of the microphysical equation of state in the mergers of magnetized neutron stars with neutrino cooling," *Physical Review D*, vol. 92, no. 4, Article ID 044045, 23 pages, 2015.

[95] Y. Sekiguchi, K. Kiuchi, K. Kyutoku, and M. Shibata, "Dynamical mass ejection from binary neutron star mergers: radiation-hydrodynamics study in general relativity," *Physical Review D*, vol. 91, no. 5, Article ID 064059, 2015.

[96] Y. Sekiguchi, K. Kiuchi, K. Kyutoku, M. Shibata, and K. Taniguchi, "Dynamical mass ejection from the merger of asymmetric binary neutron stars: radiation-hydrodynamics study in general relativity," http://arxiv.org/abs/1603.01918.

[97] A. Akmal, V. R. Pandharipande, and D. G. Ravenhall, "Equation of state of nucleon matter and neutron star structure," *Physical Review C—Nuclear Physics*, vol. 58, no. 3, pp. 1804–1828, 1998.

[98] M. Shibata and K. Taniguchi, "Coalescence of black hole-neutron star binaries," *Living Reviews in Relativity*, vol. 14, article 6, 2011.

[99] M. Shibata and K. Taniguchi, "Merger of binary neutron stars to a black hole: Disk mass, short gamma-ray bursts, and quasi-normal mode ringing," *Physical Review D*, vol. 73, Article ID 064027, 2006.

[100] Z. B. Etienne, J. A. Faber, Y. T. Liu, S. L. Shapiro, K. Taniguchi, and T. W. Baumgarte, "Fully general relativistic simulations of black hole-neutron star mergers," *Physical Review D*, vol. 77, no. 8, Article ID 084002, 2008.

[101] M. D. Duez, F. Foucart, L. E. Kidder, H. P. Pfeiffer, M. A. Scheel, and S. A. Teukolsky, "Evolving black hole-neutron star binaries in general relativity using pseudospectral and finite difference methods," *Physical Review D—Particles, Fields, Gravitation and Cosmology*, vol. 78, no. 10, Article ID 104015, 2008.

[102] K. Kyutoku, M. Shibata, and K. Taniguchi, "Gravitational waves from nonspinning black hole-neutron star binaries: dependence on equations of state," *Physical Review D*, vol. 82, Article ID 044049, 2010.

[103] K. Kyutoku, H. Okawa, M. Shibata, and K. Taniguchi, "Gravitational waves from spinning black hole-neutron star binaries: dependence on black hole spins and on neutron star equations of state," *Physical Review D*, vol. 84, no. 6, Article ID 064018, 2011.

[104] M. B. Deaton, M. D. Duez, F. Foucart et al., "Black hole-neutron star mergers with a hot nuclear equation of state: outflow and neutrino-cooled disk for a low-mass, high-spin case," *Astrophysical Journal*, vol. 776, no. 1, article 47, 2013.

[105] F. Foucart, M. B. Deaton, M. D. Duez et al., "Black-hole-neutron-star mergers at realistic mass ratios: equation of state and spin orientation effects," *Physical Review D—Particles, Fields, Gravitation and Cosmology*, vol. 87, no. 8, Article ID 084006, 2013.

[106] G. Lovelace, M. D. Duez, F. Foucart et al., "Massive disc formation in the tidal disruption of a neutron star by a nearly extremal black hole," *Classical and Quantum Gravity*, vol. 30, no. 13, Article ID 135004, 2013.

[107] F. Foucart, M. B. Deaton, M. D. Duez et al., "Neutron star-black hole mergers with a nuclear equation of state and neutrino cooling: dependence in the binary parameters," *Physical Review D*, vol. 90, no. 2, Article ID 024026, 2014.

[108] K. Kyutoku, K. Ioka, H. Okawa, M. Shibata, and K. Taniguchi, "Dynamical mass ejection from black hole-neutron star binaries," *Physical Review D—Particles, Fields, Gravitation and Cosmology*, vol. 92, no. 4, Article ID 044028, 2015.

[109] K. Kawaguchi, K. Kyutoku, H. Nakano, H. Okawa, M. Shibata, and K. Taniguchi, "Black hole-neutron star binary merger: dependence on black hole spin orientation and equation of state," *Physical Review D*, vol. 92, no. 2, Article ID 024014, 2015.

[110] K. Kawaguchi, K. Kyutoku, M. Shibata, and M. Tanaka, "Models of Kilonova/macronova emission from black hole-neutronstar mergers," http://arxiv.org/abs/1601.07711.

[111] L. Dessart, C. D. Ott, A. Burrows, S. Rosswog, and E. Livne, "Neutrino signatures and the neutrino-driven wind in binary neutron star mergers," *Astrophysical Journal*, vol. 690, no. 2, pp. 1681–1705, 2009.

[112] R. Fernández and B. D. Metzger, "Delayed outflows from black hole accretion tori following neutron star binary coalescence," *Monthly Notices of the Royal Astronomical Society*, vol. 435, no. 1, p. 502, 2013.

[113] A. Perego, S. Rosswog, R. M. Cabezón et al., "Neutrino-driven winds from neutron star merger remnants," *Monthly Notices of the Royal Astronomical Society*, vol. 443, no. 4, pp. 3134–3156, 2014.

[114] K. Kiuchi, K. Kyutoku, Y. Sekiguchi, M. Shibata, and T. Wada, "High resolution numerical relativity simulations for the merger of binary magnetized neutron stars," *Physical Review D—Particles, Fields, Gravitation and Cosmology*, vol. 90, no. 4, Article ID 041502, 2014.

[115] K. Kiuchi, Y. Sekiguchi, K. Kyutoku, M. Shibata, K. Taniguchi, and T. Wada, "High resolution magnetohydrodynamic simulation of black hole-neutron star merger: mass ejection and short gamma ray bursts," *Physical Review D*, vol. 92, no. 6, Article ID 064034, 8 pages, 2015.

[116] R. Fern, D. Kasen, B. D. Metzger, and E. Quataert, "Outflows from accretion discs formed in neutron star mergers: effect of black hole spin," *Monthly Notices of the Royal Astronomical Society*, vol. 446, no. 1, pp. 750–758, 2015.

[117] R. Fernández, E. Quataert, J. Schwab, D. Kasen, and S. Rosswog, "The interplay of disc wind and dynamical ejecta in the aftermath of neutron star-black hole mergers," *Monthly Notices of the Royal Astronomical Society*, vol. 449, no. 1, pp. 390–402, 2015.

[118] B. D. Metzger and R. Fernández, "Red or blue? A potential kilonova imprint of the delay until black hole formation following a neutron star merger," *Monthly Notices of the Royal Astronomical Society*, vol. 441, no. 4, pp. 3444–3453, 2014.

[119] D. Kasen, R. Fernández, and B. D. Metzger, "Kilonova light curves from the disc wind outflows of compact object mergers," *Monthly Notices of the Royal Astronomical Society*, vol. 450, no. 2, pp. 1777–1786, 2015.

[120] D. Martin, A. Perego, A. Arcones, F.-K. Thielemann, O. Korobkin, and S. Rosswog, "Neutrino-driven winds in the aftermath of a neutron star merger: nucleosynthesis and electromagnetic transients," *The Astrophysical Journal*, vol. 813, no. 1, p. 2, 2015.

[121] S. E. Woosley and J. S. Bloom, "The supernova-gamma-ray burst connection," *Annual Review of Astronomy and Astrophysics*, vol. 44, pp. 507–556, 2006.

[122] Z. Cano, S.-Q. Wang, Z.-G. Dai, and X.-F. Wu, "The observer's guide to the gamma-ray burst-supernovaconnection," https://arxiv.org/abs/1604.03549.

[123] D. A. Kann, S. Klose, B. Zhang et al., "The afterglows of *Swift*-era gamma-ray bursts. II. Type I GRB versus type II GRB optical afterglows," *The Astrophysical Journal*, vol. 734, no. 2, article 96, 2011.

[124] K. Hotokezaka, K. Kyutoku, M. Tanaka et al., "Progenitor models of the electromagnetic transient associated with the short gamma ray burst 130603B," *Astrophysical Journal Letters*, vol. 778, no. 1, article L16, 2013.

[125] Z.-P. Jin, D. Xu, Y.-Z. Fan, X.-F. Wu, and D.-M. Wei, "Is the late near-infrared bump in short-hard grb 130603B due to the Li-Paczynski kilonova?" *The Astrophysical Journal*, vol. 775, no. 1, p. L19, 2013.

[126] Y.-W. Yu, B. Zhang, and H. Gao, "Bright "Merger-nova" from the remnant of a neutron star binary merger: a signature of a newly born, massive, millisecond magnetar," *The Astrophysical Journal*, vol. 776, no. 2, p. L40, 2013.

[127] Y.-Z. Fan, Y.-W. Yu, D. Xu et al., "A supramassive magnetar central engine for GRB 130603B," *The Astrophysical Journal*, vol. 779, no. 2, p. L25, 2013.

[128] H. Takami, T. Nozawa, and K. Ioka, "Dust formation in macronovae," *The Astrophysical Journal Letters*, vol. 789, article L6, 2014.

[129] W. Fong, E. Berger, B. D. Metzger et al., "Short GRB 130603B: discovery of a jet break in the optical and radio afterglows, and a mysterious late-time X-ray excess," *Astrophysical Journal*, vol. 780, no. 2, article 118, 2014.

[130] S. Kisaka, K. Ioka, and H. Takami, "Energy sources and light curves of macronovae," *Astrophysical Journal*, vol. 802, no. 2, article 119, 2015.

[131] S. Kisaka, K. Ioka, and E. Nakar, "X-ray-powered macronovae," *The Astrophysical Journal*, vol. 818, no. 2, p. 104, 2016.

[132] N. Gehrels, J. P. Norris, S. D. Barthelmy et al., "A new γ-ray burst classification scheme from GRB 060614," *Nature*, vol. 444, no. 7122, pp. 1044–1046, 2006.

[133] J. P. U. Fynbo, D. Watson, C. C. Thöne et al., "No supernovae associated with two long-duration γ-ray bursts," *Nature*, vol. 444, no. 7122, pp. 1047–1049, 2006.

[134] M. D. Valle, G. Chincarini, N. Panagia et al., "An enigmatic long-lasting γ-ray burst not accompanied by a bright supernova," *Nature*, vol. 444, no. 7122, pp. 1050–1052, 2006.

[135] A. Gal-Yam, D. B. Fox, P. A. Price et al., "A novel explosive process is required for the γ-ray burst GRB 060614," *Nature*, vol. 444, no. 7122, pp. 1053–1055, 2006.

[136] Z.-P. Jin, K. Hotokezaka, X. Li et al., "The 050709 macronova and the GRB/macronovaconnection," https://arxiv.org/abs/1603.07869.

[137] J. S. Villasenor, D. Q. Lamb, G. R. Ricker et al., "Discovery of the short γ-ray burst GRB 050709," *Nature*, vol. 437, no. 7060, pp. 855–858, 2005.

[138] J. Hjorth, D. Watson, J. P. U. Fynbo et al., "The optical afterglow of the short γ-ray burst GRB 050709," *Nature*, vol. 437, no. 7060, pp. 859–861, 2005.

[139] D. B. Fox, D. A. Frail, P. A. Price et al., "The afterglow of GRB 050709 and the nature of the short-hard γ-ray bursts," *Nature*, vol. 437, no. 7060, pp. 845–850, 2005.

[140] S. Covino, D. Malesani, G. L. Israel et al., "Optical emission from GRB 050709: a short/hard GRB in a star-forming galaxy," *Astronomy & Astrophysics*, vol. 447, no. 2, pp. L5–L8, 2006.

[141] D. A. Perley, B. D. Metzger, J. Granot et al., "GRB 080503: implications of a naked short gamma-ray burst dominated by extended emission," *Astrophysical Journal*, vol. 696, no. 2, pp. 1871–1885, 2009.

[142] H. Gao, X. Ding, X.-F. Wu, Z.-G. Dai, and B. Zhang, "GRB 080503 late afterglow re-brightening: signature of a magnetar-powered merger-nova," *The Astrophysical Journal*, vol. 807, no. 2, p. 163, 2015.

[143] S. Miyazaki, Y. Komiyama, H. Nakaya et al., "HyperSuprime: project overview," in *Ground-based and Airborne Instrumentation for Astronomy*, vol. 6269 of *Proceedings of SPIE*, May 2006.

[144] S. Miyazaki, Y. Komiyama, H. Nakaya et al., "Hyper suprime-cam," in *Ground-based and Airborne Instrumentation for Astronomy IV*, vol. 8446 of *Society of Photo-Optical Instrumentation Engineers (SPIE) Conference Series*, 2012.

[145] Z. Ivezic, J. A. Tyson, E. Acosta et al., "LSST: from science drivers to reference design and anticipated data products," 2008, https://arxiv.org/abs/0805.2366.

[146] P. A. Abell, J. Allison, S. F. Anderson et al., "LSST science book,version 2.0," http://arxiv.org/abs/0912.0201.

[147] N. Gehrels, J. K. Cannizzo, J. Kanner, M. M. Kasliwal, S. Nissanke, and L. P. Singer, "Galaxy strategy for ligo-virgo gravitational wave counterpart searches," *The Astrophysical Journal*, vol. 820, no. 2, p. 136, 2016.

[148] L. P. Singer, H.-Y. Chen, D. E. Holz et al., "Going the distance: mapping host galaxies of LIGO and virgo sources in three dimensions using local cosmography and targeted follow-up," http://arxiv.org/abs/1603.07333.

[149] B. D. Metzger, A. Bauswein, S. Goriely, and D. Kasen, "Neutron-powered precursors of kilonovae," *Monthly Notices of the Royal Astronomical Society*, vol. 446, no. 1, pp. 1115–1120, 2014.

3

Modeling Kelvin–Helmholtz Instability in Soft X-Ray Solar Jets

Ivan Zhelyazkov,[1] Ramesh Chandra,[2] and Abhishek K. Srivastava[3]

[1] *Faculty of Physics, Sofia University, 1164 Sofia, Bulgaria*
[2] *Department of Physics, Kumaun University, Nainital 263001, India*
[3] *Department of Physics, Indian Institute of Technology, Banaras Hindu University, Varanasi 221005, India*

Correspondence should be addressed to Ivan Zhelyazkov; izh@phys.uni-sofia.bg

Academic Editor: Valery Nakariakov

Development of Kelvin–Helmholtz (KH) instability in solar coronal jets can trigger the wave turbulence considered as one of the main mechanisms of coronal heating. In this review, we have investigated the propagation of normal MHD modes running on three X-ray jets modeling them as untwisted and slightly twisted moving cylindrical flux tubes. The basic physical parameters of the jets are temperatures in the range of 5.2–8.2 MK, particle number densities of the order of 10^9 cm^{-3}, and speeds of 385, 437, and 532 km s^{-1}, respectively. For small density contrast between the environment and a given jet, as well as at ambient coronal temperature of 2.0 MK and magnetic field around 7 G, we have obtained that the kink ($m = 1$) mode propagating on moving untwisted flux tubes can become unstable in the first and second jets at flow speeds of ≅348 and 429 km s^{-1}, respectively. The KH instability onset in the third jet requires a speed of ≅826 km s^{-1}, higher than the observed one. The same mode, propagating in weakly twisted flux tubes, becomes unstable at flow speeds of ≅361 km s^{-1} for the first and of 443 km s^{-1} for the second jet. Except the kink mode, the twisted moving flux tube supports the propagation of higher ($m > 1$) MHD modes that can become unstable at accessible jets' speeds.

1. Introduction

Jets are considered to be ubiquitous confined plasma ejecta in the solar atmosphere. They have been extensively observed in the solar atmosphere in various wavebands, such as Hα [1, 2], Ca II H [3–5], EUV [6], and soft X-ray [7] in order to understand their multitemperature characteristics. X-ray jets were discovered by the Soft X-Ray Telescope (SXT) on board *Yohkoh* [8], as transient X-ray energy release and enhancement with apparent collimated ballistic motions of the plasma associated with the flares in X-ray bright points, emerging flux regions, or active regions (for details, see Shibata et al. [7]). As it has been pointed out by Shimojo et al. [9], jets from X-ray bright points in active regions most likely appear at the western edge of preceding sunspots and exhibit a recurrent plasma propulsion in the solar atmosphere. These X-ray jets are confined plasma dynamics with typical morphological properties, for example, (1–40) $\times 10^4$ km length, and the width of 5×10^3–10^5 km. Such jets possess apparent velocities of 10–1000 km s^{-1} and lifetime of 100–16,000 s [9]. The electron densities of the X-ray jets are of the orders of (0.7–4) $\times$ 10^9 cm^{-3}. Their temperatures lie in the range of 3–8 MK with an average temperature of 5.6 MK [10].

In terms of spatial location jets can be classified as polar jets [11] and active region jets [12]. A study of polar jet parameters based on *Hinode* XRT observations was carried out by Savcheva et al. [13] who showed that jets preferably occur inside the polar coronal holes. Culhane et al. [11] have found from *Hinode*'s Extreme-ultraviolet Imaging Spectrometer (EIS) 40$''$ slot observations of a polar coronal hole that jet temperature ranges from 0.4 to 5.0 MK. The jet velocities had typical values that are mostly less than the Sun's escape velocity (618 km s^{-1}); therefore, in consequence most of the jets fall back in the lower solar atmosphere after their triggering. Using the XTR on *Hinode*, Cirtain et al. [14] conclude that X-ray jets in polar coronal holes have two distinct velocities: one near the Alfvén speed (~800 km s^{-1}) and another near the sound speed (200 km s^{-1}). Moreover, they were the first to give an evidence for the propagation of Alfvén waves in solar X-ray jets. Kim et al. [15] presented the morphological and kinematic characteristics of three small-scale X-ray/EUV jets simultaneously observed by the *Hinode* XRT and the *Transition Region and Coronal Explorer (TRACE)*. While observing the coronal jets, for two different wavelength bands, they obtain matching characteristics for their projected speed (90–310 km s^{-1}), lifetime (100–2000 s), and size (1.1–5 × 10^5 km). Chifor at al. [12] have reported 2007 January 15/16 observations of a recurring jet situated on the west side of NOAA active region 10938. A strong blue-shifted component and an indication of a weak red-shifted component at the base of the jet were observed around T_e = 1.6 MK in these jets. The upflow velocities were observed exceeding up to 150 km s^{-1}. These jets were seen over a range of temperatures between 0.25 and 2.5 MK, while their estimated electron densities lie above 10^{11} cm^{-3} for the high-velocity upflow components. Yang et al. [16] presented simultaneous observations of three recurring jets in EUV and soft X-ray (SXR), which occurred in an active region on 2007 June 5. On comparing their morphological and kinematic properties, the authors have found that EUV and SXR jets had similar onset locations, directions, size, and terminal velocities. The three observed jets were having maximum Doppler velocities ranging from 25 to 121 km s^{-1} in the Fe XII λ195 line and from 115 to 232 km s^{-1} in the He II λ256 line. Extensive multi-instrument observations obtained simultaneously with the SUMER spectrometer on board the *Solar and Heliospheric Observatory (SoHO)*, with EIS and XRT on board *Hinode*, and with the Extreme-ultraviolet imagers (EUVI) of the Sun-Earth Connection Coronal and Heliospheric Investigation (SECCHI) instrument suite on board the Ahead and Behind *STEREO* spacecrafts were performed by Madjarska [17]. The dynamic process of X-ray jet formation and evolution has been derived in great detail. In particular, for the first time there was found spectroscopically a temperature of 12 MK (Fe XXIII 263.76 Å) and density of 4 × 10^{10} cm^{-3} in the quiet Sun. The author has clearly identified two types of upflows in which the first one was the collimated upflow along the open magnetic fields, and the second was the formation of a plasma cloud from the expelled bright point small-scale loops. Chandrashekhar et al. [18] studied the dynamics of two jets seen in a polar coronal hole with a combination of EIS and XRT/*Hinode* data. They found no evidence of helical motions in these events but detected a significant shift of the jet position in a direction normal to the jet axis, with a drift velocity of about 27 and 7 km s^{-1}, respectively.

The launch of the *Solar Dynamics Observatory (SDO)* [19] with the Atmospheric Imaging Assembly (AIA) [20, 21] opens a new page in observing the solar jets. Moschou et al. [22] have reported high cadence observations of solar coronal jets observed in the extreme-ultraviolet (EUV) 304 Å using Atmospheric Imaging Assembly (AIA) instrument on board *SDO*. They registered, in fact, coronal hole jets, with speeds of 94 to 760 km s^{-1} and lifetimes of the order of several tens of minutes. A detailed description of the dynamical behavior of a jet in an on-disk coronal hole observed with AIA/*SDO* was presented by Chandrashekhar et al. [23]. Their study reveals new evidence of plasma flows prior to the jet's initiation along the small-scale loops at the base of the jet. The authors have also found further evidence that flows along the jet consisting of multiple, quasi-periodic small-scale plasma ejections. In addition, spectroscopic analysis estimates temperature as Log 5.89 ± 0.08 K and electron densities as Log 8.75 ± 0.05 cm^{-3} in the observed jet. Measured properties of the registered transverse wave have provided evidence that strong damping of the wave occurred as it propagates along the jet with speeds of ~110 km s^{-1}. Using the magnetoseismological inversion, observed plasma, and wave parameters, the jet's magnetic field is estimated as B = 1.21 ± 0.2 G. Recently, Sterling et al. [24] have reported high-resolution X-ray and extreme-ultraviolet observations of 20 randomly selected X-ray jets that form in coronal holes at the solar polar caps. In each jet, converse to the widely accepted emerging magnetic flux model, a miniature version of the filament eruptions that initiated coronal mass ejections drove the jet producing reconnection process. Formation of a rotating jet during the filament eruption on 2013 April 10–11 on the base of multiwavelength and multiviewpoint observations with *STEREO*/SECCHI/EUVI and *SDO*/AIA was reported by Filippov et al. [25]. The confined eruption of the filament within a null-point topology, which is also known as an Eiffel tower magnetic field configuration, forms a twisted jet after magnetic reconnection near the null point. The sign of the helicity in the jet is observed the same as that of the sign of the helicity in the filament. It is noteworthy that the untwisting motion of the reconnected magnetic field lines gives rise to the accelerating plasma along the jet.

It is well established that the magnetic reconnection at different heights in the solar atmosphere plays a key role in triggering jet-like events. The magnetic reconnection between open and closed fields (standard reconnection scenario) is one of the well-known processes of the jet's occurrence [15, 26]. The jets emerging by this kind of mechanism are known as *standard jets* [3, 7, 27, 28]. Some other observational and simulation studies showed that the reconnection at the magnetic null in a fan-spine magnetic topology can also trigger jet-like events [29–32]. Eruptions of small arches, filaments, and flux ropes from within this type of magnetic field configurations can be responsible for reconnection and jets' occurring. These types of jets are known as *blowout jets* [28, 33, 34]. Moore et al. [35] used the full-disk He II 304 Å movies from the Atmospheric Imaging Assembly on *SDO* to study the cool (T ~ 10^5 K) component of X-ray jets observed in polar coronal holes by XRT. The AIA 304 Å movies revealed that most polar X-ray jets spin as they erupt. The authors examined 54 X-ray jets that were found in polar coronal holes in XRT movies sporadically taken during the first year of continuous operation of AIA (2010 May through 2011 April). These 54 jets were big and bright enough in the XRT images to be categorized as a standard jet or as a blowout jet. From the X-ray movies, 19 of the 54 jets appeared like standard jets, 32 appeared as blowout jets, and three were ambiguous and were not falling in any category. Moore et al. [36] have studied 14 large-scale solar coronal jets observed in Sun's pole. In EUV movies from the *SDO*/AIA, each jet was very similar to most X-ray and EUV jets erupting in coronal holes. However, each was exceptional in that it went higher than most of the standard coronal jets. They were detected in the outer corona beyond $2.2\,R_\odot$ in images as observed from the *Solar and Heliospheric Observatory/Large Angle Spectroscopic Coronagraph* (*LASCO*/C2 coronagraph [37]).

Schmieder et al. [38] proposed a new model between the standard and blowout models, where magnetic reconnection occurs in the bald patches around some twisted field lines, one of whose foot points is open. Pariat et al. [39] included the magnetic field inclination and photospheric field distribution and performed another 3D numerical MHD model for the two different types of jet events: standard and blowout jets. We note also that stereoscopic studies (multiple points of view) have been carried out by the EUVI/SECCHI imagers on board the twin *STEREO* spacecraft to estimate the expected speed, motion, and morphology of polar coronal jets [40]. To sum up, the flux emergence [41, 42] and flux cancellation [43, 44] are the two main triggering processes that are known to be responsible for jets' occurrence. In few observational and primarily numerical studies the wave-induced reconnection has also been suggested as a cause for the onset of jet-like events [42, 45–47].

We consider magnetically structured X-ray solar jets as moving cylindrical magnetic flux tubes that support the excitation/propagation of various kind of magnetohydrodynamic (MHD) oscillations and waves. While in static solar atmospheric plasma the propagating MHD modes are stable, the axial motion of the flux tubes engenders a velocity jump at the tube surface which can trigger a Kelvin–Helmholtz (KH) instability. The KH instability arises at the interface of two fluid layers that move with different speeds (see, e.g., Chandrasekhar [48])—then a strong velocity shear arises near the interface between these two fluids forming a vortex sheet. This vortex sheet becomes unstable to the spiral-like perturbations at small spatial scales [49]. In cylindrical geometry, when a magnetic flux tube is axially moving, such a vortex sheet is evolved near tube's boundary and it may become unstable against KH instability provided that the tube axial velocity exceeds a critical value [50]. Further on, in the nonlinear stage of KH instability, this vortex sheet causes the conversion of the directed flow energy into turbulent energy making an energy cascade at smaller spatial scales [51].

The KH instability studying in various solar jets over the past decade arose from the fact that KH vortices were observed in solar prominences [52–54], in Sweet–Parker current sheets [55], in a coronal streamer [56], and in coronal mass ejections [57–61]. All these observations stimulated the modeling of KH instability in moving twisted magnetic flux tubes in nonmagnetic environment [62], in magnetic tubes of partially ionized plasma [63], in spicules [64–66], in photospheric tubes [67], in high-temperature and cool surges [68, 69], in dark mottles [70], at the boundary of rising coronal mass ejections [60, 61, 71, 72], in rotating, tornado-like magnetized jets [49], and in a chromospheric jet (fast disappearance of rapid red-shifted and blue-shifted excursions alongside a larger scale Hα jet) [73]. A more extensive review on modeling the KH instability in solar atmosphere jets the reader can be seen in Zhelyazkov [74] and references therein.

The first modeling of KH instability in X-ray jets was carried out by Vasheghani Farahani et al. [75] who explored transverse wave propagation along the detected by Cirtain et al. [14] coronal hole soft X-ray jets. Vasheghani Farahani et al. analyzed analytically, in the limit of thin magnetic flux tube, the dispersion relation of the kink MHD mode and have obtained that this mode is unstable against the KH instability when the critical jet velocity is equal to $4.47v_\mathrm{A} = 3576\,\mathrm{km\,s^{-1}}$ ($v_\mathrm{A} = 800\,\mathrm{km\,s^{-1}}$ is the Alfvén speed inside the jet). Numerical solving of the same dispersion relation when considering the jet and its environment as cold magnetized plasmas, carried out by Zhelyazkov [76, 77], yielded a little bit lower critical flow speed for the instability onset; namely, $4.31v_\mathrm{A} = 3448\,\mathrm{km\,s^{-1}}$. The lowest critical jet speed of $4.025v_\mathrm{A} = 3220\,\mathrm{km\,s^{-1}}$ was derived by numerically solving the wave dispersion relation without any approximations, that is, treating both media as compressible plasmas. But even the latter critical jet speed is still too high for the KH instability to be detected/observed in coronal hole soft X-ray jets. The reason for obtaining such high critical speeds is the circumstance that Vasheghani Farahani et al. [75] assumed an electron number density of the order of $10^8\,\mathrm{cm^{-3}}$ and magnetic field strength of 10 G.

Here, we study the propagation of kink and higher MHD modes in standard active region soft X-ray jets, notably jets #8, #11, and #16 of Shimojo and Shibata's set of sixteen observed flares and jets [10], and have shown that with one order higher electron densities, $\sim 10^9\,\mathrm{cm^{-3}}$, and moderate magnetic field, ~ 7 G, MHD modes in high-speed jets, like these ones, can become unstable against the KH instability at accessible jets speeds, except for jet #16 which requires a higher flow velocity. In the next section we list the basic physical parameters of jet #11 and the topology of magnetic fields inside and outside the moving flux tube modeling jet and also derive the wave dispersion equation for both untwisted and twisted tubes. The physical parameters of other two jets (#8 and #16) will be provided *en route* in the next section. The numerical solving MHD wave dispersion relations and the discussion of the conditions under which the KH instability can develop in such moving structures are presented in Section 3. The last section summarizes the main results obtained in this article and outlooks our future studies of KH instability in more complex (rotating) solar atmosphere jets.

2. Geometry, Magnetic Field Topology, and MHD Wave Dispersion Relations

We consider the soft X-ray jet as a straight cylinder of radius a and density ρ_i embedded in a uniform field environment with density ρ_e. We study the propagation of MHD waves in two magnetic configurations, notably in untwisted and twisted flux tubes. For an untwisted tube magnetic fields in both media are homogeneous and directed along the z-axis of our cylindrical coordinate system (r, ϕ, z): $\mathbf{B}_\mathrm{i} = (0, 0, B_\mathrm{i})$ and $\mathbf{B}_\mathrm{e} = (0, 0, B_\mathrm{e})$, respectively (see Figure 1).

The magnetic field inside the twisted tube is helicoid, $\mathbf{B}_\mathrm{i} = (0, B_{\mathrm{i}\phi}(r), B_{\mathrm{i}z}(r))$, while outside the tube the magnetic field is

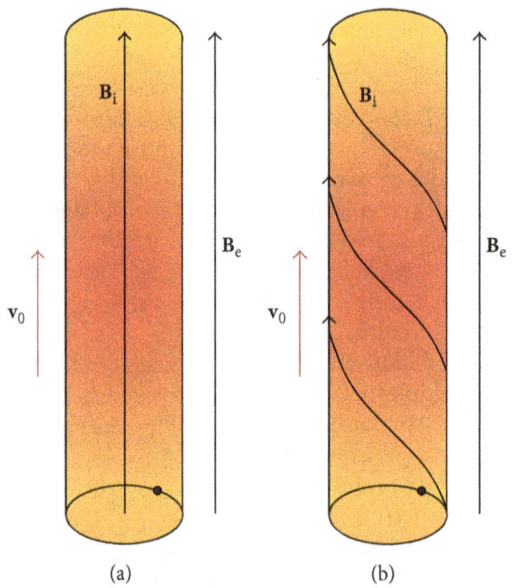

FIGURE 1: Equilibrium magnetic fields of a soft X-ray solar jet in an untwisted flux tube (a) and in a weakly twisted flux tube (b).

uniform and directed along the tube axis, $\mathbf{B}_e = (0, 0, B_e)$. Note that we assume a magnetic fields' equilibrium with uniform twist for which the magnetic field inside the tube is $\mathbf{B}_i = (0, Ar, B_{iz})$, where A and B_{iz} are constant. The parameter that characterizes the uniform magnetic field twist is the ratio $B_{i\phi}(a)/B_{iz} \equiv \varepsilon$, that is, $\varepsilon = Aa/B_{iz}$. Our frame of reference for studying the wave propagation in the jet is attached to the surrounding magnetoplasma—thus $\mathbf{v}_0 = (0, 0, v_0)$ represents the relative jet velocity, if there is any flow in the environment. The jump of the tangential velocity at the tube boundary then initiates the magnetic KH instability onset when the jump exceeds a critical value.

Before dealing with governing MHD equations, it is necessary to specify what kind of plasma each medium is (the moving tube and its environment). As seen from the Events List in [10], the electron density in jet #11 is $n_{jet} \equiv n_i = 2.9 \times 10^9\,\mathrm{cm}^{-3}$, the temperature is $T_{jet} \equiv T_i = 5.5$–$6.4\,\mathrm{MK}$, and jet speed (in their notation) is $V_{jet} = 437\,\mathrm{km\,s}^{-1}$. Our choice for environment magnetic field, electron density, and temperature is $B_e = 6.7\,\mathrm{G}$, $n_e = 2.6 \times 10^9\,\mathrm{cm}^{-3}$, and $T_e = 2.0\,\mathrm{MK}$, respectively. With $T_i = 5.5\,\mathrm{MK}$, the total pressure balance equation (equality of the sum of thermal and magnetic pressures in both media), that is,

$$p_i + \frac{B_i^2}{2\mu} = p_e + \frac{B_e^2}{2\mu}, \tag{1}$$

where μ is the magnetic permeability of vacuum, yields the following basic sound and Alfvén speeds in the jet and its environment: $c_{si} = 275\,\mathrm{km\,s}^{-1}$ and $v_{Ai} = 111\,\mathrm{km\,s}^{-1}$ and $c_{se} = 166\,\mathrm{km\,s}^{-1}$ and $v_{Ae} = 286\,\mathrm{km\,s}^{-1}$, respectively. Accordingly, the plasma betas in both media are $\beta_i = 7.369$ and $\beta_e = 0.403$—this implies that, in principle, one can treat the jet as incompressible plasma and its surrounding medium as cool magnetized plasma. When studying the MHD wave propagation in the untwisted magnetic flux tube we shall use two approaches, namely, of compressible plasmas in both media and the simplified limit of incompressible and cool plasmas—a similarity of dispersion curves patterns, obtained from corresponding dispersion relations, will eventually justify the usage of the second approach in particular in the case of twisted tube. There are two important input parameters in the modeling KH instability in moving magnetic flux tubes, which are the density contrast, $\rho_e/\rho_i \equiv \eta = 0.896$, and the ratio of the axial external and internal magnetic fields, $B_e/B_i \equiv b = 2.44$. For the twisted tube the second parameter has the form $b_{twist} = B_e/B_{iz}$.

In a system of cylindrical coordinates, the equilibrium physical variables (density, fluid velocity, and pressure) are functions of the radial coordinate r only. Then, their perturbations can be Fourier-analyzed putting them proportional to $\exp[\mathrm{i}(-\omega t + m\phi + k_z z)]$, where ω is the angular wave frequency (that, in general, can be a complex quantity), m is the mode number (a positive or negative integer), and k_z is the axial wave number. We can eliminate all except two of the perturbations (perturbation p_{tot} of the total (thermal + magnetic) pressure and the radial component ξ_r of the Lagrangian displacement $\boldsymbol{\xi}$) to get the following governing equations [78]:

$$D\frac{\mathrm{d}}{\mathrm{d}r}(r\xi_r) = C_1 r\xi_r - C_2 r p_{tot},$$
$$D\frac{\mathrm{d}p_{tot}}{\mathrm{d}r} = C_3\xi_r - C_1 p_{tot}. \tag{2}$$

The coefficients D, C_1, C_2, and C_3 are functions of the equilibrium variables ρ_0, $\mathbf{B}_0$, and $\mathbf{v}_0$ and of the Doppler-shifted frequency $\Omega = \omega - \mathbf{k}\cdot\mathbf{v}_0$ and have the following forms:

$$D = \rho_0\left(\Omega^2 - \omega_A^2\right)C_4,$$

$$C_1 = \frac{2B_{0\phi}}{\mu r}\left(\Omega^4 B_{0\phi} - \frac{m}{r} f_B C_4\right),$$

$$C_2 = \Omega^4 - \left(k_z^2 + \frac{m^2}{r^2}\right)C_4,$$

$$C_3 = \rho_0 D\left[\Omega^2 - \omega_A^2 + \frac{2B_{0\phi}}{\mu\rho_0}\frac{d}{dr}\left(\frac{B_{0\phi}}{r}\right)\right] + 4\Omega^4\left(\frac{B_{0\phi}^2}{\mu r}\right)^2 - \rho_0 C_4 \frac{4B_{0\phi}^2}{\mu r^2}\omega_A^2, \quad (3)$$

where

$$C_4 = \left(c_s^2 + c_A^2\right)\left(\Omega^2 - \omega_c^2\right),$$

$$f_B = \frac{m}{r}B_{0\phi} + \mathbf{k}\cdot\mathbf{B}_0,$$

$$\omega_A^2 = \frac{f_B^2}{\mu\rho_0}, \quad (4)$$

$$\omega_c^2 = \frac{c_s^2}{c_s^2 + c_A^2}\omega_A^2.$$

Here ω_A is the Alfvén frequency and ω_c is the cusp frequency; the other notation is standard.

Eliminating ξ_r from (2), one obtains the well-known second-order ordinary differential equation [79–81]

$$\frac{d^2 p_{tot}}{dr^2} + \left[\frac{C_3}{rD}\frac{d}{dr}\left(\frac{rD}{C_3}\right)\right]\frac{dp_{tot}}{dr} + \left[\frac{C_3}{rD}\frac{d}{dr}\left(\frac{rC_1}{C_3}\right) + \frac{1}{D^2}\left(C_2C_3 - C_1^2\right)\right]p_{tot} = 0. \quad (5)$$

By means of the solutions to (5) in both media, one can find the corresponding expressions for ξ_r, and after merging these solutions, together with those for p_{tot}, through appropriate boundary conditions at the interface $r = a$, one can derive the dispersion relation of the normal modes propagating in the moving magnetic flux tube.

2.1. Dispersion Relation of MHD Modes in an Untwisted Flux Tube. In an untwisted magnetic flux tube, the coefficient $C_1 = 0$, while $C_3 = \rho_0 D(\Omega^2 - \omega_A^2)$—then (5) takes the form

$$\frac{d^2 p_{tot}}{dr^2} + \frac{1}{r}\frac{dp_{tot}}{dr} - \left(m_0^2 + \frac{m^2}{r^2}\right)p_{tot} = 0, \quad (6)$$

where

$$m_0^2 = -\frac{\left(\Omega^2 - k_z^2c_s^2\right)\left(\Omega^2 - k_z^2v_A^2\right)}{\left(c_s^2 + v_A^2\right)\left(\Omega^2 - \omega_c^2\right)}. \quad (7)$$

The cusp frequency, ω_c, is usually expressed via the so-called tube speed, c_T, notably $\omega_c = k_z c_T$, where [82]

$$c_T = \frac{c_s v_A}{\sqrt{c_s^2 + v_A^2}}. \quad (8)$$

The solutions for p_{tot} can be written in terms of modified Bessel functions: $I_m(m_{0i}r)$ inside the jet and $K_m(m_{0e}r)$ in its surrounding plasma. We note that wave attenuation coefficients, m_{0i} and m_{0e}, in both media are calculated from (7) with replacing the sound and Alfvén speeds with the corresponding values for each medium. Recall that, in evaluating m_{0e}, the wave frequency is not Doppler-shifted—it is simply ω. By expressing the Lagrangian displacements ξ_{ir} and ξ_{er} in both media via the derivatives of corresponding Bessel functions and by applying the boundary conditions for continuity of the pressure perturbation p_{tot} and ξ_r across the interface, $r = a$, one obtains the dispersion relation of normal MHD modes propagating in a flowing compressible jet surrounded by a static compressible plasma [65, 83, 84]

$$\frac{\rho_e}{\rho_i}\left(\omega^2 - k_z^2v_{Ae}^2\right)m_{0i}\frac{I_m'(m_{0i}a)}{I_m(m_{0i}a)} - \left[(\omega - \mathbf{k}\cdot\mathbf{v}_0)^2 - k_z^2v_{Ai}^2\right]m_{0e}\frac{K_m'(m_{0e}a)}{K_m(m_{0e}a)} = 0. \quad (9)$$

Due to the flowing plasma, the wave frequency is Doppler-shifted inside the jet. We recall that for the kink mode ($m = 1$) one defines the so-called kink speed [82]

$$c_k = \left(\frac{\rho_i v_{Ai}^2 + \rho_e v_{Ae}^2}{\rho_i + \rho_e}\right)^{1/2} = \left(\frac{1 + B_e^2/B_i^2}{1 + \rho_e/\rho_i}\right)^{1/2} v_{Ai}, \quad (10)$$

which, as seen, is independent of sound speeds and characterizes the propagation of transverse perturbations. We will show that notably the kink mode can become unstable against KH instability.

When the jet is considered as incompressible plasma and its environment as a cool one, the MHD wave dispersion relation (9) keeps its form, but the two attenuation coefficients $m_{0i,e}$ become much simpler, namely,

$$m_{0i} = k_z,$$

$$m_{0e} = \frac{\left(k_z^2v_{Ae}^2 - \omega^2\right)^{1/2}}{v_{Ae}}, \quad (11)$$

respectively.

2.2. Dispersion Relation of MHD Normal Modes in a Twisted Flux Tube. Inside the tube ($r \leqslant a$), where $B_{i\phi} = Ar$, the quantities f_B and ω_{Ai} take the forms

$$f_B = mA + k_z B_{iz},$$

$$\omega_{Ai} = \frac{mA + k_z B_{iz}}{\sqrt{\mu\rho_i}}, \quad (12)$$

respectively. For incompressible plasma, we redefine (without the loss of generality) the coefficients D, C_1, C_2, and C_3 by dividing them by C_4 to obtain

$$D = \rho\left(\Omega^2 - \omega_A^2\right),$$

$$C_1 = -\frac{2mB_\phi}{\mu r^2}\left(\frac{m}{r}B_\phi + k_z B_z\right),$$

$$C_2 = -\left(\frac{m^2}{r^2} + k_z^2\right), \quad (13)$$

$$C_3 = D^2 + D\frac{2B_\phi}{\mu}\frac{d}{dr}\left(\frac{B_\phi}{r}\right) - \frac{4B_\phi^2}{\mu r^2}\rho\omega_A^2.$$

Radial displacement ξ_r is expressed through the total pressure perturbation as

$$\xi_r = \frac{D}{C_3}\frac{dp_{\rm tot}}{dr} + \frac{C_1}{C_3}p_{\rm tot}. \qquad (14)$$

The solution to this equation obviously depends upon the magnetic field and density profile.

With aforementioned coefficients D, C_1, C_2, and C_3, evaluated for the jet's medium, (5) reduces to the modified Bessel equation

$$\left[\frac{d^2}{dr^2} + \frac{1}{r}\frac{d}{dr} - \left(m_{0i}^2 + \frac{m^2}{r^2}\right)\right]p_{\rm tot} = 0, \qquad (15)$$

where

$$m_{0i}^2 = k_z^2\left[1 - \frac{4A^2\omega_{\rm Ai}^2}{\mu\rho_{\rm i}\left(\Omega^2 - \omega_{\rm Ai}^2\right)^2}\right]. \qquad (16)$$

The solution to (15) bounded at the tube axis is

$$p_{\rm tot}(r \leqslant a) = \alpha_{\rm i} I_m(m_{0i}r), \qquad (17)$$

where I_m is the modified Bessel function of order m and $\alpha_{\rm i}$ is a constant. Lagrangian displacement $\xi_{\rm ir}$, by using (14) can be written as

$$\xi_{\rm ir} = \frac{\alpha_{\rm i}}{r}\left\{\frac{\left(\Omega^2 - \omega_{\rm Ai}^2\right)m_{0i}rI_m'(m_{0i}r)}{\rho_{\rm i}\left(\Omega^2 - \omega_{\rm Ai}^2\right)^2 - 4A^2\omega_{\rm Ai}^2/\mu} - \frac{2mA\omega_{\rm Ai}I_m(m_{0i}r)/\sqrt{\mu\rho_{\rm i}}}{\rho_{\rm i}\left(\Omega^2 - \omega_{\rm Ai}^2\right)^2 - 4A^2\omega_{\rm Ai}^2/\mu}\right\}, \qquad (18)$$

where the prime sign means a differentiation with respect the Bessel function argument.

For the cool environment with a straight-line magnetic field $B_{ez} = B_{\rm e}$ and homogeneous density $\rho_{\rm e}$, the C_{1-3} and D coefficients take the form

$$D = \rho\left(\omega^2 - \omega_{\rm A}^2\right),$$
$$C_1 = 0,$$
$$C_2 = -\left[\frac{m^2}{r^2} + k_z^2\left(1 - \frac{\omega^2}{\omega_{\rm A}^2}\right)\right], \qquad (19)$$
$$C_3 = D^2.$$

The total pressure perturbation outside the tube obeys the same Bessel equation as (15), but m_{0i}^2 is replaced by

$$m_{0e}^2 = k_z^2\left(1 - \frac{\omega^2}{\omega_{\rm Ae}^2}\right), \qquad (20)$$

which coincides with the attenuation coefficient in the cool environment of an untwisted flux tube. The solution bounded at infinity now is

$$p_{\rm tot}(r > a) = \alpha_{\rm e} K_m(m_{0e}r), \qquad (21)$$

where K_m is the modified Bessel function of order m and $\alpha_{\rm e}$ is a constant.

In this case, the Lagrangian displacement can be written as

$$\xi_{er} = \frac{\alpha_{\rm e}}{r}\frac{m_{0e}rK_m'(m_{0e}r)}{\rho_{\rm e}\left(\omega^2 - \omega_{\rm Ae}^2\right)}, \qquad (22)$$

and the Alfvén frequency is simplified to

$$\omega_{\rm Ae} = \frac{k_z B_{ez}}{\sqrt{\mu\rho_{\rm e}}} = k_z v_{\rm Ae}. \qquad (23)$$

Here, $v_{\rm Ae} = B_{\rm e}/\sqrt{\mu\rho_{\rm e}}$ is the Alfvén speed in the surrounding magnetized plasma.

The boundary conditions which merge the solutions of the Lagrangian displacement and total pressure perturbation inside and outside the twisted magnetic flux tube have the forms [85]

$$\xi_{\rm ir}\big|_{r=a} = \xi_{er}\big|_{r=a},$$
$$p_{\rm tot\,i} - \frac{B_{\rm i\phi}^2}{\mu a}\xi_{\rm ir}\bigg|_{r=a} = p_{\rm tot\,e}\big|_{r=a}, \qquad (24)$$

where total pressure perturbations $p_{\rm tot\,i}$ and $p_{\rm tot\,e}$ are given by (17) and (21), respectively. With the help of these boundary conditions we derive the dispersion relation of the normal MHD modes propagating along a twisted magnetic flux tube with axial mass flow $\mathbf{v}_0$

$$\frac{\left(\Omega^2 - \omega_{\rm Ai}^2\right)F_m(m_{0i}a) - 2mA\omega_{\rm Ai}/\sqrt{\mu\rho_{\rm i}}}{\left(\Omega^2 - \omega_{\rm Ai}^2\right)^2 - 4A^2\omega_{\rm Ai}^2/\mu\rho_{\rm i}} = \frac{P_m(m_{0e}a)}{(\rho_{\rm e}/\rho_{\rm i})\left(\omega^2 - \omega_{\rm Ae}^2\right) + A^2P_m(m_{0e}a)/\mu\rho_{\rm i}}, \qquad (25)$$

where, as we already mentioned, $\Omega = \omega - \mathbf{k}\cdot\mathbf{v}_0$ is the Doppler-shifted wave frequency in the moving medium,

$$F_m(m_{0i}a) = \frac{m_{0i}aI_m'(m_{0i}a)}{I_m(m_{0i}a)},$$
$$P_m(m_{0e}a) = \frac{m_{0e}aK_m'(m_{0e}a)}{K_m(m_{0e}a)}. \qquad (26)$$

This dispersion equation is similar to the dispersion equation of normal MHD modes in a twisted flux tube surrounded by incompressible plasma [65]—the only difference is that, there in (10), $\kappa_{\rm e} \equiv m_{0e} = k_z$.

3. Numerical Solutions and Results

Firstly we shall study the dispersion characteristics of the kink ($m = 1$) mode in untwisted moving magnetic flux tube (in two approaches, notably (i) considering the jet and its surrounding plasma as compressible media and (ii) treating the jet as an incompressible medium while its environment is assumed to be cool plasma) and, later on, explore the same thing for the kink ($m = 1$) and higher ($m > 1$) MHD modes propagating in a twisted moving flux tube. Two dispersion relations (9) and (25) are transcendent equations in which the wave frequency, ω, is a complex quantity: $\mathrm{Re}(\omega) + \mathrm{Im}(\omega)$, while the axial wave number, k_z, is a real variable. The appearance of KH instability (i.e., $\mathrm{Im}(\omega) > 0$) is determined primarily by the jet velocity and in searching for a critical/threshold value of it, we will gradually change velocity magnitude, v_0, from zero to that critical value (and beyond). Our numerical task is to solve the dispersion relation in complex variables, obtaining the real and imaginary parts of the wave frequency, or as is usually done, of the wave phase velocity $v_{\mathrm{ph}} = \omega/k_z$, as functions of the axial wave number, k_z, at various magnitudes of the velocity shear between the soft X-ray jet and its environment, $\mathbf{v}_0$.

For numerical solving the wave dispersion relations, it is practically to normalize all variables and we do that normalizing the wave length, $\lambda = 2\pi/k_z$, to the tube radius, a, that implies a dimensionless wave number $k_z a$. All speeds are normalized with respect to the Alfvén speed inside the jet, v_{Ai}. For normalizing the Alfvén speed in the ambient coronal plasma, v_{Ae}, we need the density contrast, η, and the ratio of the magnetic fields $b = B_{\mathrm{e}}/B_{\mathrm{i}}$, to find $v_{\mathrm{Ae}}/v_{\mathrm{Ai}} = b/\sqrt{\eta}$. The normalization of sound speeds in both media requires the specification of the reduced plasma betas, $\widetilde{\beta}_{\mathrm{i,e}} = c_{\mathrm{si,e}}^2/v_{\mathrm{Ai,e}}^2$. In the dimensionless computations, the flow speed, v_0, will be presented by the Alfvén Mach number $M_{\mathrm{A}} = v_0/v_{\mathrm{Ai}}$.

3.1. Kelvin–Helmholtz Instability in Untwisted Flux Tubes. Among the various MHD wave spectra that exist in a static magnetic field flux tube of compressible plasma, surrounded by compressible medium, the most interesting for us is the kink-speed wave whose speed for jet #11, according to (10), is equal to

$$c_{\mathrm{k}} = \sqrt{\frac{1+b^2}{1+\eta}}\, v_{\mathrm{Ai}} = 212.7\,\mathrm{km\,s^{-1}} \tag{27}$$

or in dimensionless form $\dfrac{c_{\mathrm{k}}}{v_{\mathrm{Ai}}} = 1.9166.$

The normal modes propagating in a static homogeneously magnetized flux tube can be pure surface waves, pseudo-surface (body) waves, or leaky waves (see Cally [86]). The type of the wave crucially depends on the ordering of the basic speeds in both media (the flux tube and its surrounding plasma), more specifically of sound and Alfvén speeds as well as corresponding tube speeds. In our case the ordering is

$$v_{\mathrm{Ai}} < c_{\mathrm{se}} < c_{\mathrm{si}} < v_{\mathrm{Ae}}, \tag{28}$$

which with $c_{\mathrm{Ti}} \cong 103\,\mathrm{km\,s^{-1}}$ and $c_{\mathrm{Te}} \cong 143.6\,\mathrm{km\,s^{-1}}$ after normalizing them with respect to the external sound speed, c_{se}, yields (in Cally's notation)

$$\begin{aligned} A &= 0.6687, \\ C &= 1.6566, \\ A_{\mathrm{e}} &= 1.7229, \\ C_{\mathrm{T}} &= 0.62, \\ C_{\mathrm{Te}} &= 0.8649. \end{aligned} \tag{29}$$

Since $A < C$ and $A < C_{\mathrm{Te}}$ along with $A < C_{\mathrm{Te}} < C$, according to Cally's classification, the kink mode in a rest flux tube must be pure surface mode type S_+^- (for detail, see Cally [86]). For an S_+^- type wave V should lie between A and C: $A \leqslant V \leqslant C$, and also $V < C_{\mathrm{Te}}$. In our normalization, the normalized wave velocity should be bracketed between 1 and $\sqrt{\widetilde{\beta}_{\mathrm{i}}} = 2.478$, that is, $1 \leqslant \omega/(k_z v_{\mathrm{Ai}}) \leqslant 2.478$. The typical wave dimensionless wave velocity $\omega/(k_z v_{\mathrm{Ai}})$ is the normalized kink speed $c_{\mathrm{k}}/v_{\mathrm{Ai}} = 1.9166$, which, as expected, lies between 1 and 2.478. Concerning the relation between normalized wave phase velocity and external tube speed, in our case we have the opposite inequality; that is, $V > C_{\mathrm{Te}}$, that in Cally's normalization reads as $1.2805 > 0.8649$ and in ours as $1.9166 > 1.2937$. The reason for that unexpected change in the mutual relations between two velocities is the relatively bigger value of parameter b (=2.44) that yields a larger magnitude of c_{k}, than, for example, in the case when b is close to 1—then the inequality $V < C_{\mathrm{Te}}$ is satisfied (see, e.g., Zhelyazkov et al. [87]).

A reasonable question is how the flow will change the dispersion characteristics of the kink ($m = 1$) mode. Calculations show that the flow shifts upwards the kink-speed dispersion curve and splits it into two separate curves [65]. (A similar duplication is observed for the tube-speed v_{Ti} dispersion curves, too.) The evolution of the pair of kink-speed dispersion curves can be seen in Figure 2(a)—the input parameters for solving the dispersion equation (9) of the kink mode ($m = 1$) traveling in a moving flux tube of compressible plasma surrounded by compressible coronal medium are $\eta = 0.896$, $b = 2.44$, $\widetilde{\beta}_{\mathrm{i}} = 6.141$, and $\widetilde{\beta}_{\mathrm{e}} = 0.336$. During calculations the Alfvén Mach number, M_{A}, was varied from zero (static plasma) to values at which we obtained unstable solutions. Let us first note that at $M_{\mathrm{A}} = 0$ we get the kink-speed dispersion curve which at a very small dimensionless wave number $k_z a$ (=0.005) yields the value of the normalized wave phase velocity equal to 1.9168—very close to the previously estimated magnitude of 1.9166. This observation implies that our code for solving the wave dispersion relation is correct. For small Alfvén Mach numbers the pair of kink-speed modes travel with velocities $M_{\mathrm{A}} \mp c_{\mathrm{k}}/v_{\mathrm{Ai}}$ [65] (in that case, their dispersion curves go practically parallel). At higher M_{A}, however, the behavior of each curve of the pair $M_{\mathrm{A}} \mp c_{\mathrm{k}}/v_{\mathrm{Ai}}$ turns out to be completely different. As seen from Figure 2(a), for $M_{\mathrm{A}} \geqslant 3.785$ both kink curves break and merge forming a family of semiclosed dispersion curves. A further increasing in M_{A} leads to a separation of those curves in opposite directions. This behavior of the kink ($m = 1$) mode dispersion curves signals us that we are in a range of the $\mathrm{Re}(v_{\mathrm{ph}}/v_{\mathrm{Ai}})$–$k_z a$-plane where one can expect the occurrence of KH instability. A prediction of the Alfvén Mach number M_{A} at which the instability will occur can be found from the inequality [88]

$$|m|\, M_{\mathrm{A}}^2 > \left(1 + \frac{1}{\eta}\right)\left(|m|\, b^2 + 1\right) \tag{30}$$

that gives for the kink ($m = 1$) mode $M_{\mathrm{A}} > 3.839$. In our case the KH instability starts at that M_{A}, for which the left-hand side semiclosed curve disappears—this happens at

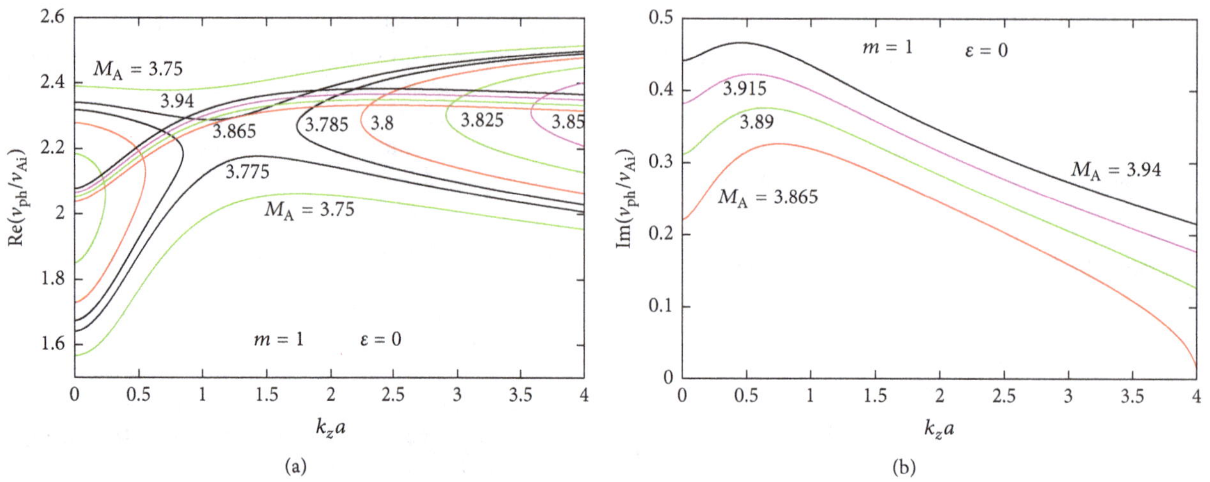

FIGURE 2: (a) Dispersion curves of stable and unstable kink ($m = 1$) MHD mode propagating in a moving untwisted magnetic flux tube of compressible plasma (modeling jet #11) at $\eta = 0.896$ and $b = 2.44$. Unstable dispersion curves located *above the middle of the plot* have been calculated for four values of the Alfvén Mach number $M_A = 3.865$, 3.89, 3.915, and 3.94. (b) The normalized growth rates of the unstable mode for the same values of M_A. Red curves in both plots correspond to the onset of KH instability.

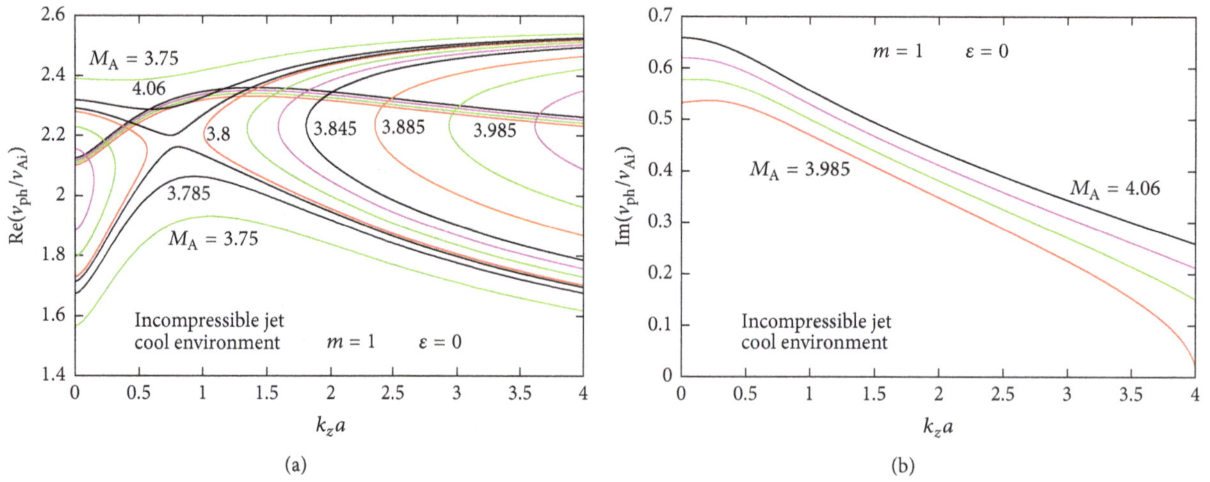

FIGURE 3: (a) Dispersion curves of stable and unstable kink ($m = 1$) MHD mode propagating in a moving untwisted magnetic flux tube of incompressible plasma surrounded by cool medium at the same input parameters as in Figure 2. The dispersion curves of unstable modes located *above the middle of the plot* have been calculated for $M_A = 3.985$, 4.01, 4.035, and 4.06. Alfvén Mach numbers associated with the nonlabeled black, green, and purple curves are equal to 3.796, 3.815, and 3.83, respectively, while those of far right green and purple semiclosed curves have magnitudes of 3.925 and 3.965. (b) Similar plots as in Figure 2.

$M_A^{cr} = 3.865$. The growth rates of unstable kink modes are plotted in Figure 2(b). The red curves in all diagrams denote the marginal dispersion/growth rate curves: for values of the Alfvén Mach number smaller than M_A^{cr} the kink mode is stable; otherwise it becomes unstable and the instability is of the KH type. With $M_A^{cr} = 3.865$ the kink mode will be unstable when the velocity of the moving flux tube is higher than 429 km s^{-1}—a speed, which is below than the observationally measured jet speed of 437 km s^{-1}. We would like to underline that all the stable kink modes in the untwisted ($\varepsilon = 0$) moving tube modeling jet #11 are pure surface modes while the unstable ones are not—the latter becomes partly surface and partly leaky modes (their external attenuation coefficients, m_{0e}s, are complex quantities with positive imaginary parts). This circumstance means that wave energy is radiated outward in the surrounding medium, which allows us to claim the KH instability plays a dual role: once in its nonlinear stage the instability can trigger wave turbulence and simultaneously the propagating KH-mode is radiating its energy outside.

When numerically solving (9) for the kink ($m = 1$) mode treating the jet's plasma as incompressible medium and its environment as cool plasma, we obtain dispersion curves' and growth rates' patterns very similar to those shown in Figure 2 (see Figure 3). Computations show, as expected, that the stable kink mode is a nonleaky surface mode, while the

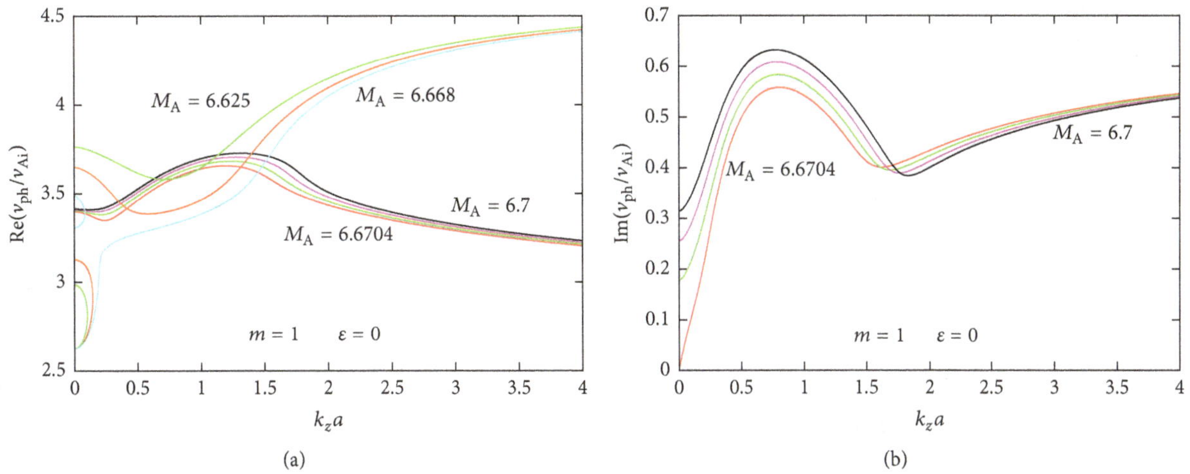

FIGURE 4: (a) Dispersion curves of stable and unstable kink ($m = 1$) MHD mode propagating in a moving untwisted magnetic flux tube of compressible plasma (modeling jet #8) at $\eta = 0.963$ and $b = 4.56$. Unstable mode's dispersion curves are pictured *in the middle of the plot* for $M_A = 6.6704$, 6.68, 6.69, and 6.7. (b) The corresponding normalized growth rates of unstable modes for the same values of M_A.

unstable one possesses a real internal attenuation coefficient, $m_{0i} = k_z$, not changed by the instability, but the external one, m_{0e}, becomes a complex quantity with positive real and negative imaginary parts. Now the threshold Alfvén Mach number is a little bit bigger and yields a critical jet speed for the instability onset of $\cong 442\,\mathrm{km\,s^{-1}}$, being with $5\,\mathrm{km\,s^{-1}}$ higher than the observationally measured $437\,\mathrm{km\,s^{-1}}$. We note that, as a rule, jet's incompressible plasma approximation yields slightly higher threshold Alfvén Mach numbers than the model of compressible media. This ascertainment shows how sensitive to jet's and its environment's media treatment the occurrence of KH instability of the kink mode in our jet is —one can expect instability onset at an accessible jet speed only if both media are considered as compressible magnetized plasmas.

The basic physical parameters of jet #8, namely $n_i = 4.1 \times 10^9\,\mathrm{cm^{-3}}$ and $T_i = 5.2$ MK, for a low density contrast of $\eta = 0.963$ assuming as before that the temperature of surrounding plasma is $T_e = 2.0$ MK and the background magnetic field, B_e, is equal to 7 G, yield the following sound and Alfvén speeds: $c_{se} = 166\,\mathrm{km\,s^{-1}}$, $c_{si} = 267.5\,\mathrm{km\,s^{-1}}$, $v_{Ae} = 242.8\,\mathrm{km\,s^{-1}}$, and $v_{Ai} = 52.22\,\mathrm{km\,s^{-1}}$. Accordingly, plasma betas in both media are $\beta_i = 31.5$ and $\beta_e = 0.56$, while the magnetic fields ratio $b = 4.563$ defines $B_i = 1.53$ G. The speed ordering, $v_{Ai} < c_{se} < v_{Ae} < c_{si}$, according to Cally's [86] classification, tells us that in a rest magnetic flux tube the propagating wave must be a pure surface mode of type S_+^-. The numerical computations confirm this, as well as reproduce the normalized kink speed of 3.3342 within three places after the decimal point. With the flow inclusion, the behavior of the pair kink-speed dispersion curves in the region of the instability onset, as seen from Figure 4(a), is completely different; more specifically, the lower semiclosed kink-speed dispersion curves at $M_A = 6.625$ and 6.65 (the orange curve) correspond to pseudosurface (body) waves, while the higher kink-speed dispersion curves are associated with surface waves. Note that at $M_A = 6.668$ the surface wave dispersion curves brakes in two parts: the right-hand one merges with the lower kink-speed curve at $k_z a = 0.2337$ (see the blue line) while its left-hand side forms a narrow semiclosed dispersion curve of pure surface mode. The threshold Alfvén Mach number is equal to 6.6704 (the prediction one according to (30) is 6.67) which implies that the critical flow speed for KH instability onset is equal to $348.3\,\mathrm{km\,s^{-1}}$—a value far below the jet speed of $385\,\mathrm{km\,s^{-1}}$. This circumstance guarantees that even if one considers the jet medium as incompressible plasma and its environment as cool plasma we will get an accessible jet's speed for instability onset.

In a similar way, calculating the sound and Alfvén speeds for jet #16 and its environment (with $n_i = 1.0 \times 10^9\,\mathrm{cm^{-3}}$, $T_i = 7.4$ MK, $\eta = 0.95$, $T_e = 2.0$ MK, and $B_e = 6.7$ G) we get $c_{se} = 166\,\mathrm{km\,s^{-1}}$, $c_{si} \cong 319\,\mathrm{km\,s^{-1}}$, $v_{Ae} \cong 474\,\mathrm{km\,s^{-1}}$, $v_{Ai} = 350\,\mathrm{km\,s^{-1}}$, and plasma betas $\beta_i \cong 1.0$ and $\beta_e = 0.14$. The magnetic fields ratio, b, is equal to 1.32 giving $B_i \cong 5.1$ G. In this case, the speed ordering is $c_{se} < c_{si} < v_{Ai} < v_{Ae}$ and the kink ($m = 1$) mode propagating in a rest magnetic flux tube is a pseudosurface (body) wave of type B_+^- (see Table I in [86]). The numerical calculations confirm this mode type and yield a normalized value of the kink speed very close to the predicted one of 1.1858. Now the pattern of stable kink-speed dispersion curves is more complicated (see Figure 5(a))—while the regular lower kink-speed curves correspond to pseudosurface (body) waves, the higher kink-speed dispersion curves have the form of narrow and long semiclosed up to $M_A = 2.3$ loops and are in fact bulk waves: both attenuation coefficients are purely imaginary numbers. As expected, the dispersion curves of unstable kink mode possess complex attenuation coefficients (with positive real and imaginary parts for m_{0i} and positive real and negative imaginary part for m_{0e}). The threshold Alfvén Mach number being equal to 2.359305 determines a critical flow velocity of $825.8\,\mathrm{km\,s^{-1}}$ for KH instability onset which is much higher than the jet #16 speed of $532\,\mathrm{km\,s^{-1}}$. This consideration shows that even a relatively high-speed observed soft X-ray jet

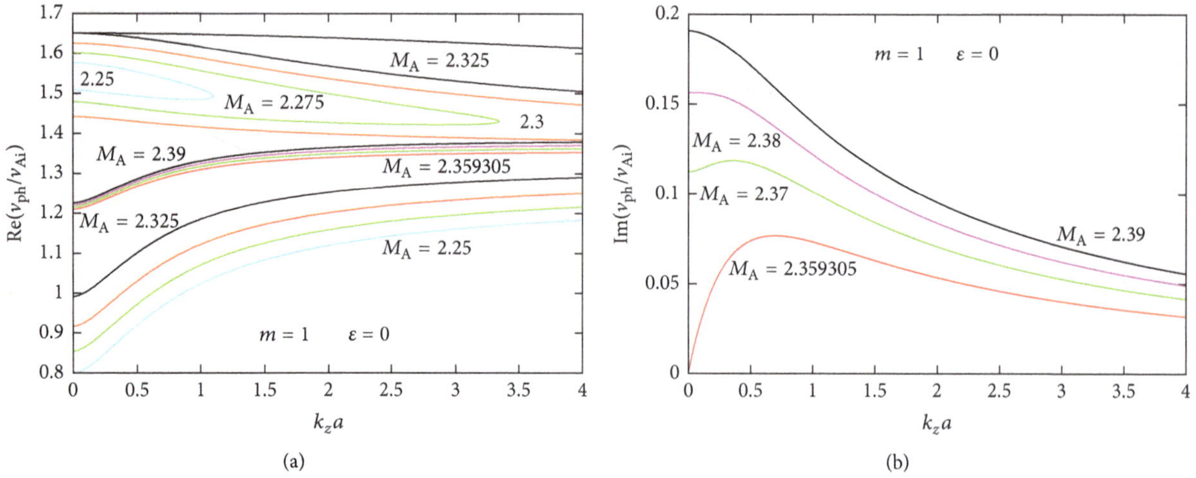

FIGURE 5: (a) Dispersion curves of stable and unstable kink ($m = 1$) MHD mode propagating in a moving untwisted magnetic flux tube of compressible plasma (modeling jet #16) at $\eta = 0.95$ and $b = 1.32$. Dispersion curves of unstable modes located *in the middle of the plot* have been calculated for $M_A = 2.359305, 2.37, 2.38$, and 2.39. (b) Plots of corresponding normalized growth rates.

cannot become unstable—the reason for that in this case is the low density of the jet's plasma.

3.2. Kelvin-Helmholtz Instability in Twisted Flux Tubes. A natural question that immediately raises is how the twist of the internal magnetic field, $\mathbf{B}_i$, will change the conditions for instability occurrence and the shape of the dispersion curves. When we model the soft X-ray jet as a moving twisted magnetic flux tubes, there are two types of instabilities that can develop in the jet: namely, kink instability due to the twist of the magnetic field and KH instability owing to the tangential discontinuity of plasma velocity at the tube boundary. According to Dungey and Loughhead [89], the kink instability will occur provided $B_{i\phi}(a) > 2B_{iz}$, or in our notation, as $\varepsilon > 2$. Further on, in order to prevent the developing of kink instability, we will consider weakly twisted flux tubes, $\varepsilon < 1$, and, thus, only KH instability can take place in such a jet configuration. When numerically solving dispersion equation (25) we need the normalization of the local Alfvén frequencies $\omega_{Ai,e}$. The normalization of these frequencies is performed by multiplying each of them by the tube radius, a, and dividing by the Alfvén speed $v_{Ai} = B_{iz}/\sqrt{\mu\rho_i}$ to get

$$\frac{a\omega_{Ai}}{v_{Ai}} = m\frac{Aa}{B_{iz}} + k_z a = m\varepsilon + k_z a,$$
$$\frac{a\omega_{Ae}}{v_{Ai}} = k_z a\frac{B_e/B_{iz}}{\sqrt{\eta}} = k_z a\frac{b_{twist}}{\sqrt{\eta}}, \quad (31)$$

where $b_{twist} = B_e/B_{iz} = b\sqrt{1+\varepsilon^2}$. For small values of ε, we will take $b_{twist} \cong b$.

We begin our calculations for the kink ($m = 1$) mode in the moving twisted tube (modeling jet #11) with $\eta = 0.896$, $b_{twist} \cong b = 2.44$, and $\varepsilon = 0.025$. The results of numerical task are presented in Figure 6, from which one sees that the kink mode dispersion and growth rate curves are very similar to those shown in Figure 3. The critical flow velocity for emerging KH instability now is $v_0^{cr} = 442.6\,\mathrm{km\,s^{-1}}$, calculated by using the "reduced" Alfvén speed $v_{Ai}/\sqrt{1+\varepsilon^2} = 110.96\,\mathrm{km\,s^{-1}}$. KH instability of the ($m = -1$) mode will occur if the jet speed exceeds $442\,\mathrm{km\,s^{-1}}$, of which value is beyond the observationally derived jet speed of $437\,\mathrm{km\,s^{-1}}$. We note that while the wave attenuation coefficient of the $m = 1$ mode propagating in untwisted moving magnetic flux tube of incompressible plasma surrounded by a cool medium is not changed by the flow, in the twisted flux tube (in the same approximations) that attenuation coefficient becomes a complex number with positive imaginary part, like the attenuation coefficient in the cool environment.

According to the instability criterion (30), the occurrence of KH instability of higher MHD modes would require lower threshold Alfvén Mach numbers (and accordingly lower critical jet speeds) than those of the kink ($m = \pm 1$) mode. Our computations show, however, that, with $\varepsilon = 0.025$, the threshold Alfvén Mach numbers are generally still high to initiate the KH instability. Their magnitudes for the $m = 2$, 3, and 4 modes are 3.965, 3.943, and 3.926, respectively. The corresponding critical velocities for instability onset are accordingly equal to 440.0, 437.5, and $435.6\,\mathrm{km\,s^{-1}}$. As seen, only the $m = 4$ MHD mode can become unstable against KH instability. A clutch of unstable dispersion curves and normalized growth rates are shown in Figure 7. Here we observe a new phenomenon: the KH instability starts at some critical $k_z a$-number along with the corresponding threshold Alfvén Mach number. That critical dimensionless wave number for the $m = 4$ MHD mode (look at Figure 7) is equal to 0.528. If we assume that jet #11 has a width $\Delta\ell = 5 \times 10^3$ km, then the critical wavelength for a KH instability emergence is $\lambda_{cr}^{m=4} \cong 29.8$ Mm.

A substantial decrease in the threshold Alfvén Mach number in moving twisted magnetic flux tubes can be obtained by increasing the magnetic field twist parameter, say taking, for instance, $\varepsilon = 0.4$. The dispersion curves and normalized wave growth rates of the kink ($m = \pm 1$)

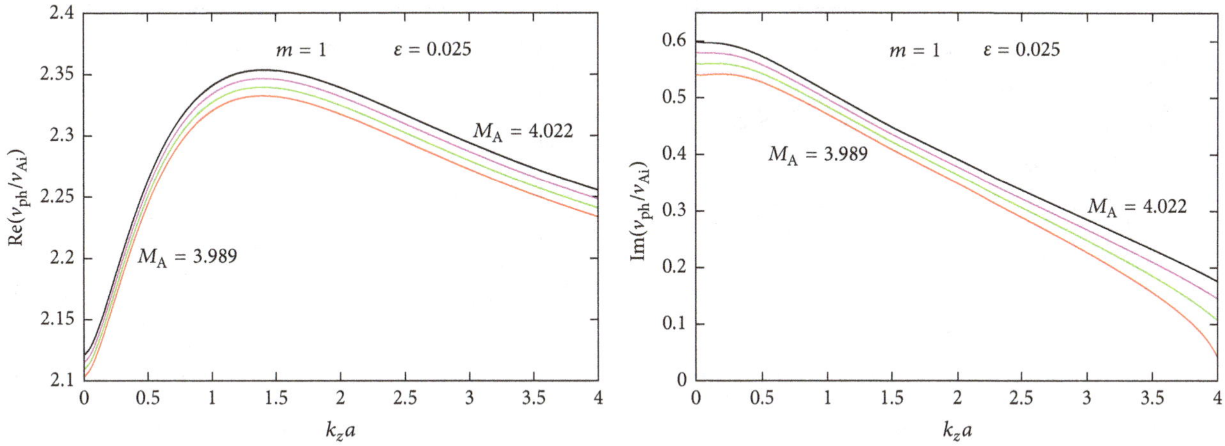

(a) (b)

FIGURE 6: (a) Dispersion curves of unstable kink ($m = 1$) MHD mode propagating in a moving twisted flux tube of incompressible plasma (modeling jet #11) at $\eta = 0.896$ and $\varepsilon = 0.025$ and for $M_A = 3.989$, 4.0 (green curve), 4.011 (purple curve), and 4.022. (b) The normalized growth rates for the same values of M_A.

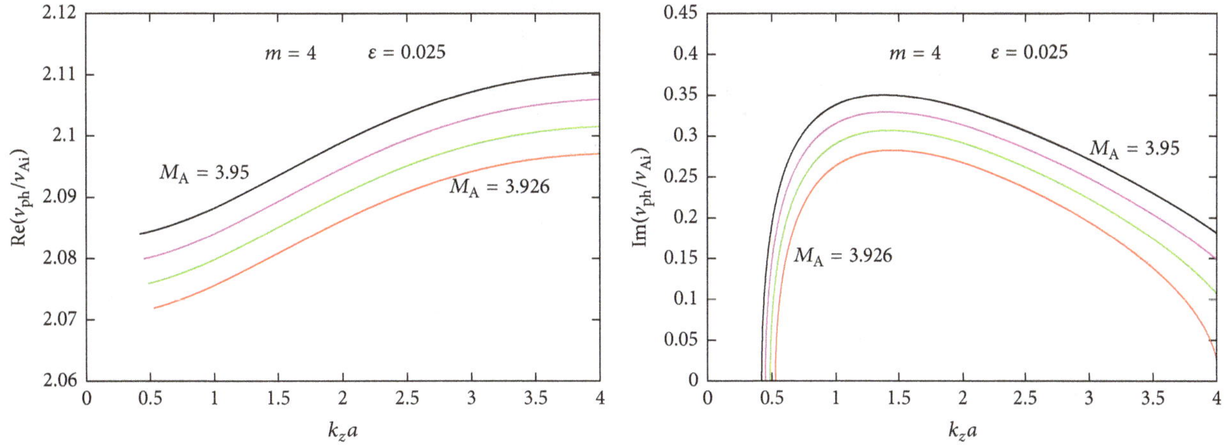

(a) (b)

FIGURE 7: (a) Dispersion curves of unstable ($m = 4$) MHD mode propagating in a moving twisted flux tube of incompressible plasma at the same parameters as in Figure 6 for $M_A = 3.926$, 3.934 (green curve), 3.942 (purple curve), and 3.95. (b) The normalized growth rates for the same values of M_A.

mode are plotted in Figures 8 and 9. One is immediately seeing the rather complicated forms of both the dispersion curves and dimensionless growth rates of the $m = -1$ kink mode—such a complication was absent at $\varepsilon = 0.025$. We note that computations were performed at $b_{\text{twist}} = b/\sqrt{1+\varepsilon^2}$, and the reference "reduced" Alfvén speed inside the jet now is $v_{Ai}/\sqrt{1+\varepsilon^2} = 103\,\text{km s}^{-1}$. With this Alfvén speed the critical velocities for the instability occurrence are equal to $416\,\text{km s}^{-1}$ (for the $m = 1$ mode) and $\cong 408\,\text{km s}^{-1}$ for the $m = -1$ mode, respectively.

In solving dispersion equation (25) for the $m = 2$ MHD mode, we start with a little bit bigger Alfvén Mach number than that predicted by instability criterion (30) and have obtained three different kinds of dispersion curves and normalized growth rates (see Figure 10), notably curves with distinctive change of the normalized wave phase velocity in a relatively narrow $k_z a$-interval (the orange and blue curves in Figure 10(a)), two piece-wise dispersion curves (the purple and green curves in the same plot), and one almost linear dispersion curve (the red one which is in fact the marginal dispersion curve obtained for the threshold Alfvén Mach number equal to 4.062). Not less interesting are the dimensionless growth rate curves corresponding to these three kinds of dispersion curves. For the threshold Alfvén Mach number of 4.062 we obtained one instability window and instability starts at the critical dimensionless wave number of 1.944, or equivalently at $\lambda_{\text{cr}}^{m=2} \cong 8.0$ Mm. For the slightly superthreshold Alfvén Mach numbers of 4.075 and 4.1 we got two different instability windows (see the green and purple curves in Figure 10(b)) and a further increase in M_A leads to merging of those two instability windows—see the blue and orange curves in the same plot. The critical flow velocity at which KH instability arises is equal to $\cong 418\,\text{km s}^{-1}$.

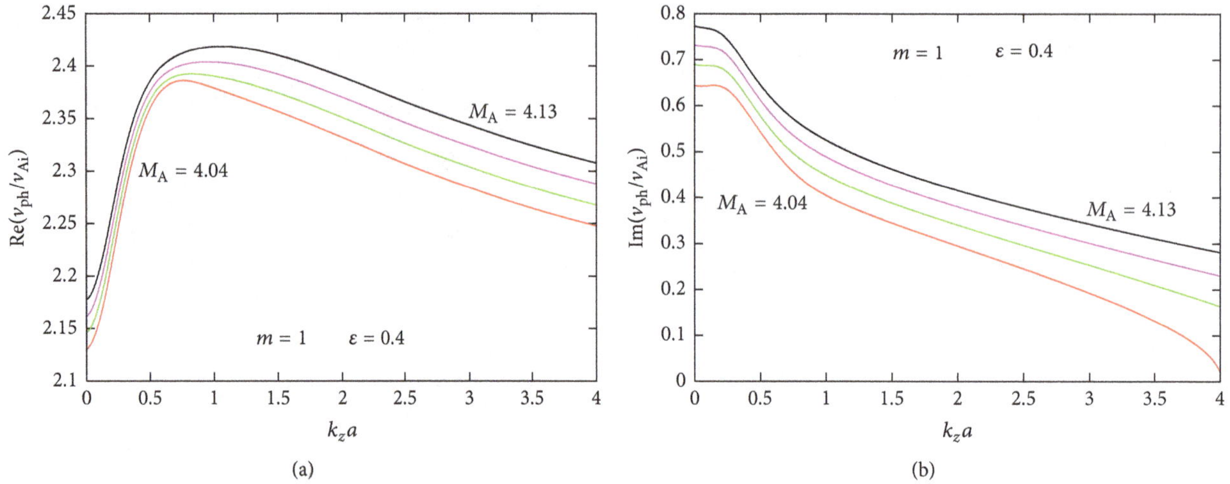

FIGURE 8: (a) Dispersion curves of unstable ($m = 1$) MHD mode propagating in a moving twisted flux tube of incompressible plasma (modeling jet #11) at $\eta = 0.896$ and $\varepsilon = 0.4$ and for $M_A = 4.04$, 4.07 (green curve), 4.1 (purple curve), and 4.13. (b) The normalized growth rates for the same values of M_A.

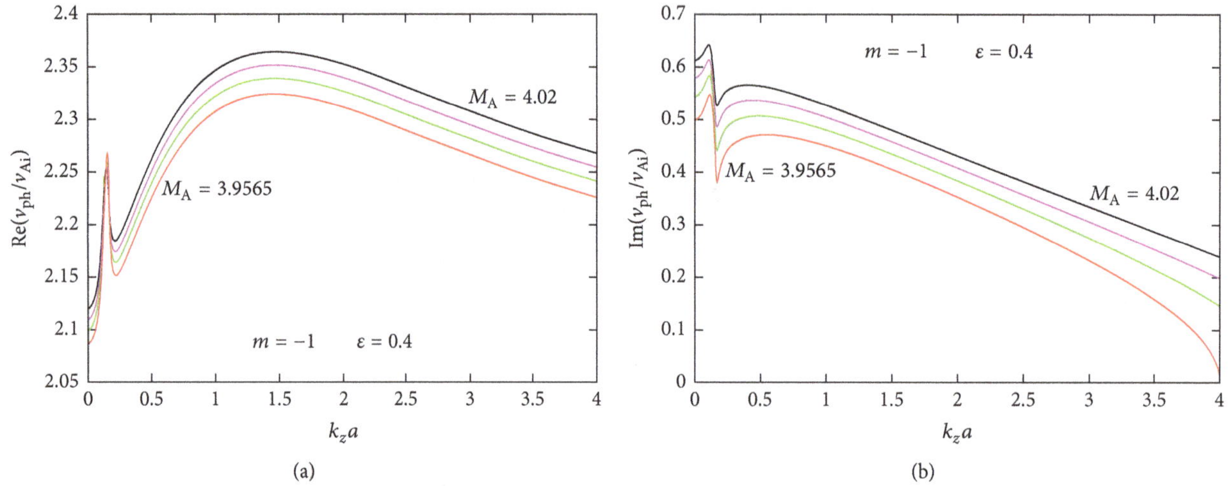

FIGURE 9: (a) Dispersion curves of unstable ($m = -1$) MHD mode propagating in a moving twisted flux tube of incompressible plasma (modeling jet #11) at the same parameters as in Figure 8 and for $M_A = 3.9565$, 3.98 (green curve), 4.0 (purple curve), and 4.02. (b) The normalized growth rates for the same values of M_A.

We have skipped the complicated dispersion and growth rate curves pictured in Figure 10 for the $m = 3$ and $m = 4$ MHD modes—we have calculated only a few curves for Alfvén Mach numbers close to the corresponding threshold ones (see Figures 11 and 12). One can observe that for the magnetic field twist parameter $\varepsilon = 0.4$ the critical dimensionless wave numbers are shifted far on the right—their values for both modes are equal to 3.619 and 4.512, respectively, that yield $\lambda_{cr}^{m=3} \cong 4.3$ Mm and $\lambda_{cr}^{m=4} \cong 3.5$ Mm. The critical jet speeds at which KH instability emerges are equal correspondingly to ≅422 and 424 km s^{-1}. We note that all the threshold Alfvén Mach numbers of higher MHD modes ($m \geqslant 2$) traveling on the moving twisted magnetic flux tube are larger than the predicted ones, but while for $\varepsilon = 0.025$ they (Alfvén Mach numbers) are decreasing with increasing the mode number, at $\varepsilon = 0.4$ we have just the opposite ordering. We would like to notice that finding the marginal dispersion and growth rate curves of the higher modes at $\varepsilon = 0.4$ turns out to be a laborious computational task.

To finish our survey on KH instability of MHD modes in moving twisted magnetic flux tubes, we will briefly consider how a weak twist of the internal magnetic field, $\varepsilon = 0.025$, will change the critical jet speeds for the occurrence of KH instability in jet #8. We are not going to graphically present the dispersion curves and growth rates of unstable modes because simply there is nothing special in their shape—in the incompressible plasma jet approximation and cool environment at small magnetic field twist the curves look, more or less, similar. Recall that at density contrast of 0.963 and magnetic fields ratio $b = 4.56$, the threshold Alfvén Mach number for the kink ($m = 1$) mode is equal to 6.9047,

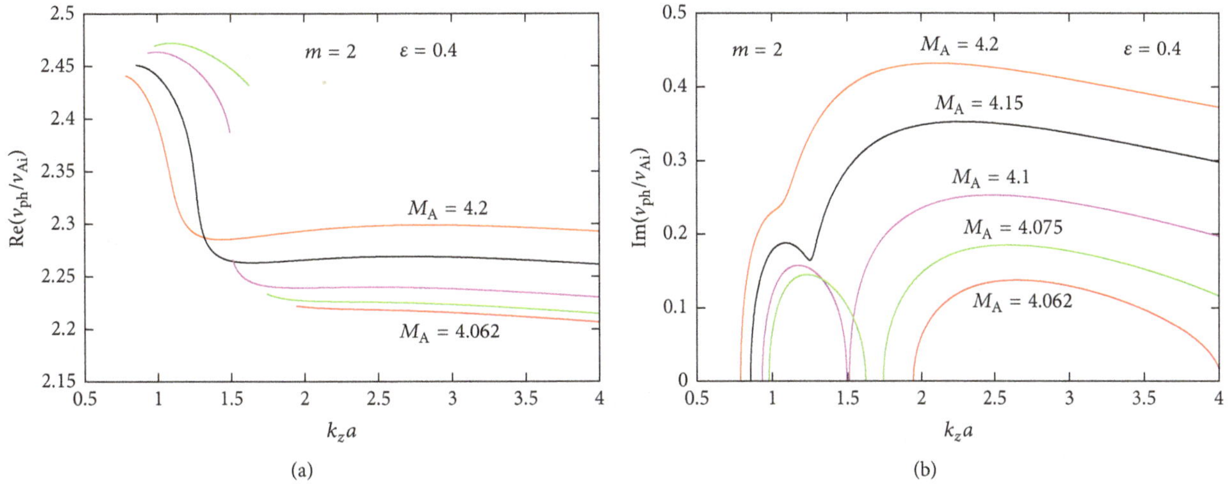

FIGURE 10: (a) Dispersion curves of unstable ($m = 2$) MHD mode propagating in a moving twisted flux tube of incompressible plasma (modeling jet #11) at the same parameters as in Figure 8 and for $M_A = 4.2$, 4.15, 4.1, 4.075, and 4.062. (b) The normalized growth rates for the same values of M_A.

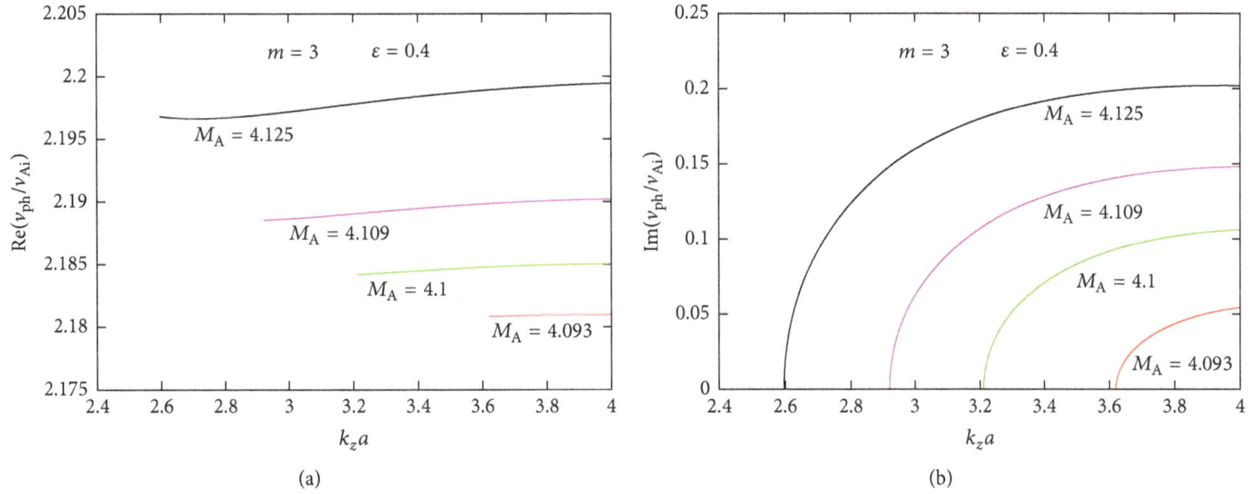

FIGURE 11: (a) Dispersion curves of unstable ($m = 3$) MHD mode propagating in a moving twisted flux tube of incompressible plasma (modeling jet #11) at the same parameters as in Figure 8 and for $M_A = 4.093$, 4.1, 4.109, and 4.125. (b) The normalized growth rates for the same values of M_A.

which for the reference Alfvén speed of 52.22 km s^{-1} yields a critical flow velocity of 360.6 km s^{-1} that is lower than the jet speed of 385 km s^{-1}. The KH instability of $m = 2$, 3, and 4 MHD modes should occur at flow velocities of ≅259, 357, and 355.5 km s^{-1} and start at critical dimensionless wave numbers $(k_z a)_{cr}$ equal, respectively, to 0.054, 0.119, and 0.205.

4. Summary and Conclusion

In this paper, we have explored the conditions under which MHD waves with various mode numbers ($m = \pm 1$, 2, 3, and 4) propagating along untwisted and twisted magnetic flux tubes (representing three observed X-ray solar jets) can become unstable against the KH instability. We have used two models of a soft X-ray jet, namely, considering the jet and its surrounding coronal medium as compressible magnetized plasmas and another simplified model representing the jet as an incompressible medium and its environment as cool coronal plasma. A comparison of the dispersion curves and normalized wave growth rates of the kink ($m = 1$) mode calculated on the base of these two models shows that one obtains similar dispersion curves' and growth rates' patterns (see Figures 2 and 3). This circumstance justifies the usage of the second, simplified, model in studying the propagation characteristics of MHD modes in moving twisted magnetic flux tubes. We must, however, immediately underline that the onset of KH instability in incompressible magnetic flux tube always requires slightly greater threshold Alfvén Mach numbers and correspondingly higher critical jet speeds.

Our numerical computations show the following:

(i) The flow does not change the nature of propagating stable MHD modes being pure surface waves in a

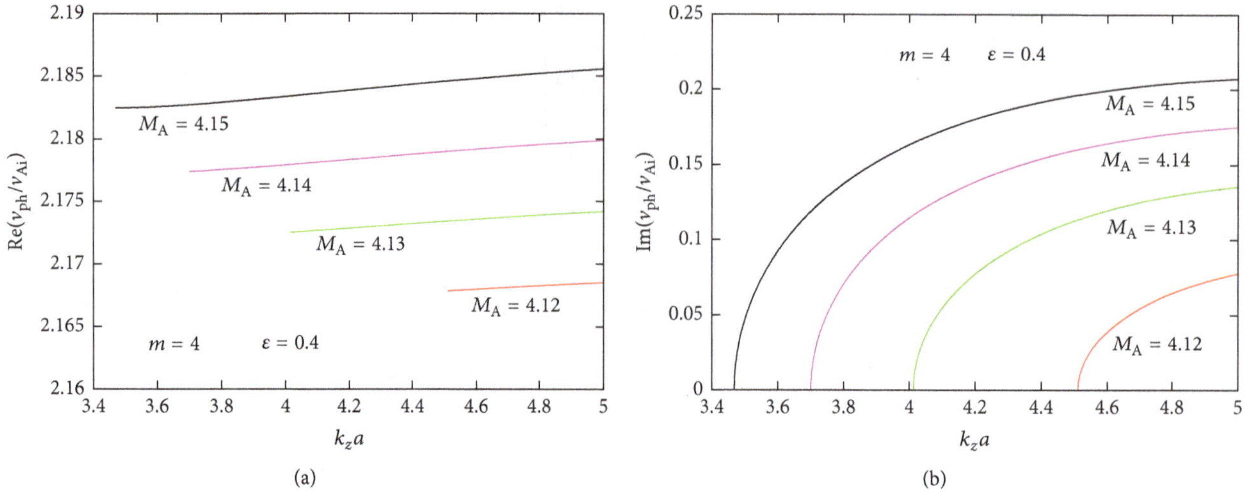

FIGURE 12: (a) Dispersion curves of unstable ($m = 4$) MHD mode propagating in a moving twisted flux tube of incompressible plasma (modeling jet #11) at the same parameters as in Figure 8 and for $M_A = 4.12, 4.13, 4.14$, and 4.15. (b) The normalized growth rates for the same values of M_A.

rest magnetic flux tube, but unstable ones due to high enough jet speed become partly surface and partly leaky waves. That is why it is more appropriate to term them *generalized surface waves.*

(ii) The KH instability of the kink mode ($m = \pm 1$) starts at Alfvén Mach numbers very close to the predicted ones. Numerically found threshold M_As for the higher MHD modes are, in fact, greater than their initial evaluations. Accessible flow velocities for instability onset in jet #11 have been derived for moving untwisted magnetic flux tube of compressible plasma (429 km s^{-1}) and for the $m = 4$ MHD mode propagating in weakly twisted magnetic flux tube ($\varepsilon = 0.025$) of incompressible plasma surrounded by cool coronal medium (435.6 km s^{-1}). A nonnegligible decrease of the critical jet speed for all considered modes running in twisted moving flux tube is obtained when the magnetic field twist parameter is $\varepsilon = 0.4$—then the critical velocities of the $m = \pm 1$, 2, 3, and 4 modes are equal to 416, ≅408, 418, ≅422, and 424 km s^{-1}, respectively. The instability onset of all considered MHD modes in jet #8 occurs at flow velocities, in general, lower than the jet speed of 385 km s^{-1}—their magnitudes for the kink ($m = 1$) mode running on untwisted or slightly twisted magnetic flux tube (with magnetic field twist parameter $\varepsilon = 0.025$) are equal to 348.3 and 360.6 km s^{-1}, respectively. The corresponding critical flow velocities for the higher ($m = 2$, 3, and 4) MHD modes turn out to lie between aforementioned speeds, namely, being equal to ≅359, 35,7 and 355.5 km s^{-1}. Jet #16 possessing the highest flow speed of 532 km s^{-1} in Shimojo and Shibata's Events List [10] is, however, stable against the KH instability—the critical flow velocity at which the kink ($m = 1$) mode would become unstable propagating in untwisted magnetic flux tube is equal to ≅826 km s^{-1}—much higher than the jet speed. The critical velocities in twisted tubes are even larger, equal to 852.4, 848.3, 844.8, and 842.6 km s^{-1} for the $m = 1$–4 MHD modes, respectively.

(iii) The instability onset for a given mode m is very sensitive to the main input parameters: density contrast, η, magnetic field ratio, b, the magnetic field twist ε, and especially the background magnetic field B_e. If we assume that, for jet #11 $B_e = 10$ G, Alfvén speeds in the two media have new values, notably $v_{Ai} = 320.38$ km s^{-1} and $v_{Ae} = 427.5$ km s^{-1}. With these new Alfvén speeds the input parameters for solving wave dispersion relations (9) and (25) have changed and more specifically $\widetilde{\beta}_i = 0.7376$, $\widetilde{\beta}_e = 0.1506$, and $b = 1.2636$. Now the critical jet velocity for the instability onset dramatically increases—it becomes 751 km s^{-1} for the kink mode in untwisted tube and 774 km s^{-1} in weakly twisted ($\varepsilon = 0.025$) flux tube.

(iv) If we accept the Paraschiv's et al. [90] suggestion that plasma density in the region crossed by coronal jets has to be the sum $n_{cor} + n_{jet}$ or, on average, two times higher than the plasma in the surrounding corona n_{cor}; then the total pressure balance equation, again for jet #11, can be satisfied at $B_e = 10$ G and we have the following set of input parameters: $\eta = 0.437$, $v_{Ai} = 105.7$ km s^{-1}, $v_{Ae} = 427.5$ km s^{-1}, $\widetilde{\beta}_i = 6.7782$, $\widetilde{\beta}_e = 0.1506$, and $b = 3.781$. With these parameters the critical flow velocities for KH instability occurrence reduce to 560 km s^{-1} for the kink mode in untwisted flux tube and to 575 km s^{-1} for the same mode in weakly twisted ($\varepsilon = 0.025$) flux tube. Note that within this approach one obtains critical jet speeds that are approximately with 200 km s^{-1} less than those calculated for the same background magnetic field of 10 G. Here one raises the big question: "which approach, the standard one or that of Paraschiv's et al. [90], will be appropriate for studying/modeling the KH instability in solar X-ray (and other) jets?" The perfect answer to this question will be obtained when we will try to model a really observationally detected KH instability in soft X-ray jets like that of Foullon et al. [57].

It is intriguing to see what critical jet speeds we will obtain for a numerically modeled solar X-jet—a case in point is the study of Miyagoshi and Yokoyama [91]. These authors have presented MHD numerical simulations of solar X-ray jets based on magnetic reconnection model that includes chromospheric evaporation. Peculiar to their study is that total pressure balance equation excludes the magnetic pressure inside the jet saying that it is very weak; that is (in their notation),

$$p_{\text{jet}} = p_{\text{cor}} + \frac{B^2}{2\mu}, \quad (32)$$

where $B \equiv B_e$ is the background (coronal) magnetic field. Supposing that $n_{\text{cor}} = 10^9\ \text{cm}^{-3}$, $T_{\text{cor}} = 10^6\ \text{K}$, $B = 10\ \text{G}$, and $n_{\text{jet}} = 4.5 \times 10^9\ \text{cm}^{-3}$, they obtain from their total balance equation the temperature of the jet, $T_{\text{jet}} = 6.7 \times 10^6\ \text{K}$—indeed a reasonable value. However, to satisfy the standard total pressure balance (1) (assuming a very small, but finite value for the magnetic field in the jet, B_i), with the aforementioned values for both the temperature and jet density, we were forced to double the coronal density, that is, to take $n_{\text{cor}} \equiv n_e = 2.0 \times 10^9\ \text{cm}^{-3}$. In such a case the sound and Alfvén speeds in the jet and its environment are $c_{\text{si}} \cong 304\ \text{km s}^{-1}$, $c_{\text{se}} = 117\ \text{km s}^{-1}$, $v_{\text{Ai}} = 47.67\ \text{km s}^{-1}$, and $v_{\text{Ae}} = 487.5\ \text{km s}^{-1}$. The ordering of sound and Alfvén speeds is the same as for the #11 X-ray jet in the Shimojo and Shibata paper [10] that implies the propagation of pure surface stable modes. The magnetic field inside the jet is $B_i \cong 1.47\ \text{G}$—hence the magnetic fields ratio is $b = 6.817$. Thus the input parameters for solving (9) and (25) are $\eta = 0.444$, $\widetilde{\beta}_i = 40.5872$, and $\widetilde{\beta}_e = 0.0695$ and, as we already found, $b = 6.817$. Note that this relatively high value of b would suspect a rather big value of the threshold Alfvén Mach number—the instability criterion yields $M_A > 12.42$. Numerically found threshold M_A for initiating KH instability of the kink ($m = 1$) mode in moving untwisted magnetic flux tube of compressible plasma is 12.435 that yields a critical velocity of $607\ \text{km s}^{-1}$. For the case of weakly twisted flux tube ($\varepsilon = 0.025$) in the approximation incompressible jet and cool plasma environment we obtained for the same kink mode a critical velocity of $620\ \text{km s}^{-1}$. The critical speeds for the higher ($m = 2$–4) modes are accordingly of 615, 609, and $606\ \text{km s}^{-1}$.

The three examples of coronal X-jets embedded in a background magnetic field of 10 G show that KH instability of basically the kink ($m = 1$) mode can occur if jets' speeds lie in the range of 500–$700\ \text{km s}^{-1}$. Such velocities are generally accessible for high-speed jets; lower critical speeds between 400 and $500\ \text{km s}^{-1}$ can be observed/detected at moderate external magnetic fields in the range of 6–7 G. This requirement is in agreement with Pucci et al. [34] evaluation that the magnetic field inside a standard X-ray jet would be equal to 2.8 G, while for a blowout jet that value should be around 4.5 G. To be honest, we can say that the most of soft X-ray solar jets with speeds below 400–$450\ \text{km s}^{-1}$ are stable against the KH instability—this instability can develop only in relatively dense high-speed jets. Note, however, that each soft X-ray jet is a unique event—it requires a separate careful exploration of the instability conditions. Arising KH instability in turn might trigger wave turbulence considered as an effective mechanism for coronal heating.

Our simplified model of studying the possibilities for emerging of KH instability in fast soft X-ray jets has to be improved by assuming a radial density gradient or some magnetic field and flow velocity shears. A challenge also is the modeling of the same instability in rotating and oscillatory swaying coronal hole X-ray jets.

Competing Interests

The authors declare that they have no competing interests.

Acknowledgments

This work was supported by the Bulgarian Science Fund and the Department of Science & Technology, Government of India Fund, under Indo-Bulgarian Bilateral Project DNTS/INDIA 01/7, /Int/Bulgaria/P-2/12. The authors are indebted to Snezhana Yordanova for drawing one figure.

References

[1] J. M. Beckers, "Solar spicules," *Annual Review of Astronomy and Astrophysics*, vol. 10, no. 1, pp. 73–100, 1972.

[2] J. R. Roy, "The magnetic properties of solar surges," *Solar Physics*, vol. 28, no. 1, pp. 95–114, 1973.

[3] K. Shibata, T. Nakamura, T. Matsumoto et al., "Chromospheric anemone jets as evidence of ubiquitous reconnection," *Science*, vol. 318, no. 5856, pp. 1591–1594, 2007.

[4] C. Chifor, H. Isobe, E. Mason et al., "Magnetic flux cancellation associated with a recurring solar jet observed with Hinode, RHESSI, and STEREO/EUVI," *Astronomy & Astrophysics*, vol. 491, no. 1, pp. 279–288, 2008.

[5] N. Nishizuka, M. Shimizu, T. Nakamura et al., "Giant chromospheric anemone jet observed with *Hinode* and comparison with magnetohydrodynamic simulations: Evidence of propagating Alfvén waves and magnetic reconnection," *The Astrophysical Journal Letters*, vol. 683, no. 1, pp. L83–L86, 2008.

[6] D. Alexander and L. Fletcher, "High-resolution observations of plasma jets in the solar corona," *Solar Physics*, vol. 190, no. 1-2, pp. 167–184, 1999.

[7] K. Shibata, Y. Ishido, L. Acton et al., "Observations of X-ray jets with the Yohkoh soft X-ray telescope," *Publications of the Astronomical Society of Japan*, vol. 44, no. 5, pp. L173–L179, 1992.

[8] S. Tsuneta, L. Acton, M. Bruner et al., "The soft X-ray telescope for the SOLAR-A mission," *Solar Physics*, vol. 136, no. 1, pp. 37–67, 1991.

[9] M. Shimojo, S. Hashimoto, K. Shibata, T. Hirayama, H. S. Hudson, and L. W. Acton, "Statistical study of solar X-ray jets observed with the Yohkoh soft X-ray telescope," *Publications of*

the Astronomical Society of Japan, vol. 48, no. 1, pp. 123–136, 1996.

[10] M. Shimojo and K. Shibata, "Physical parameters of solar X-ray jets," *The Astrophysical Journal*, vol. 542, no. 2, pp. 1100–1108, 2000.

[11] L. Culhane, L. K. Harra, D. Baker et al., "Hinode EUV study of jets in the Sun's south polar corona," *Publications of the Astronomical Society of Japan*, vol. 59, supplement 3, pp. S751–S756, 2007.

[12] C. Chifor, P. R. Young, H. Isobe et al., "An active region jet observed with Hinode," *Astronomy & Astrophysics*, vol. 481, no. 1, pp. L57–L60, 2008.

[13] A. Savcheva, J. Cirtain, E. E. DeLuca et al., "A study of polar jet parameters based on Hinode XRT observations," *Publications of the Astronomical Society of Japan*, vol. 59, supplement 3, pp. S771–S778, 2007.

[14] J. W. Cirtain, L. Golub, L. Lundquist et al., "Evidence for Alfvén waves in solar X-ray jets," *Science*, vol. 318, no. 5856, pp. 1580–1582, 2007.

[15] Y.-H. Kim, Y.-J. Moon, Y.-D. Park et al., "Small-scale X-ray/EUV jets seen in Hinode XRT and TRACE," *Publications of the Astronomical Society of Japan*, vol. 59, supplement 3, pp. S763–S769, 2007.

[16] L.-H. Yang, Y.-C. Jiang, J.-Y. Yang, Y. Bi, R.-S. Zheng, and J.-C. Hong, "Observations of EUV and soft X-ray recurring jets in an active region," *Research in Astronomy and Astrophysics*, vol. 11, no. 10, pp. 1229–1242, 2011.

[17] M. S. Madjarska, "Dynamics and plasma properties of an X-ray jet from SUMER, EIS, XRT, and EUVI A & B simultaneous observations," *Astronomy & Astrophysics*, vol. 526, no. 2, article A19, 2011.

[18] K. Chandrashekhar, A. Bemporad, D. Banerjee, G. R. Gupta, and L. Teriaca, "Characteristics of polar coronal hole jets," *Astronomy & Astrophysics*, vol. 561, article A104, 2014.

[19] W. D. Pesnell, B. J. Thompson, and P. C. Chamberlin, "The *Solar Dynamics Observatory* (SDO)," *Solar Physics*, vol. 275, no. 1, pp. 3–15, 2012.

[20] J. R. Lemen, A. M. Title, D. J. Akin et al., "The *Atmospheric Imaging Assembly* (AIA) on the *Solar Dynamics Observatory* (SDO)," *Solar Physics*, vol. 275, no. 1, pp. 17–40, 2012.

[21] P. Boerner, C. Edwards, J. Lemen et al., "Initial calibration of the *Atmospheric Imaging Assembly* (AIA) on the *Solar Dynamics Observatory* (SDO)," *Solar Physics*, vol. 275, no. 1, pp. 41–66, 2012.

[22] S. P. Moschou, K. Tsinganos, A. Vourlidas, and V. Archontis, "SDO observations of solar jets," *Solar Physics*, vol. 284, no. 2, pp. 427–438, 2013.

[23] K. Chandrashekhar, R. J. Morton, D. Banerjee, and G. R. Gupta, "The dynamical behaviour of a jet in an on-disk coronal hole observed with AIA/SDO," *Astronomy & Astrophysics*, vol. 562, article A98, 2014.

[24] A. C. Sterling, R. L. Moore, D. A. Falconer, and M. Adams, "Small-scale filament eruptions as the driver of X-ray jets in solar coronal holes," *Nature*, vol. 523, no. 7561, pp. 437–440, 2015.

[25] B. Filippov, A. K. Srivastava, B. N. Dwivedi et al., "Formation of a rotating jet during the filament eruption on 2013 April 10-11," *Monthly Notices of the Royal Astronomical Society*, vol. 451, no. 1, pp. 1117–1129, 2015.

[26] M. Shimojo, K. Shibata, T. Yokoyama, and K. Hori, "One-dimensional and pseudo-two-dimensional hydrodynamic simulations of solar X-ray jets," *Astrophysical Journal*, vol. 550, no. 2, pp. 1051–1063, 2001.

[27] S. Kamio, H. Hara, T. Watanabe et al., "Velocity structure of jets in a coronal hole," *Publications of the Astronomical Society of Japan*, vol. 59, supplement 3, pp. S757–S762, 2007.

[28] R. L. Moore, J. W. Cirtain, A. C. Sterling, and D. A. Falconer, "Dichotomy of solar coronal jets: Standard jets and blowout jets," *The Astrophysical Journal*, vol. 720, no. 1, pp. 757–770, 2010.

[29] E. Pariat, S. K. Antiochos, and C. R. DeVore, "A model for solar polar jets," *The Astrophysical Journal*, vol. 691, no. 1, pp. 61–74, 2009.

[30] B. Filippov, L. Golub, and S. Koutchmy, "X-ray jet dynamics in a polar coronal hole region," *Solar Physics*, vol. 254, no. 2, pp. 259–269, 2009.

[31] E. Pariat, S. K. Antiochos, and C. R. Devore, "Three-dimensional modeling of quasi-homologous solar jets," *Astrophysical Journal*, vol. 714, no. 2, pp. 1762–1778, 2010.

[32] F. Jiang, J. Zhang, and S. Yang, "Interaction between an emerging flux region and a pre-existing fan-spine dome observed by IRIS and SDO," *Publications of the Astronomical Society of Japan*, vol. 67, no. 4, article 78, 2015.

[33] A. C. Sterling, L. K. Harra, and R. L. Moore, "Fibrillar chromospheric spicule-like counterparts to an extreme-ultraviolet and soft X-ray blowout coronal jet," *The Astrophysical Journal*, vol. 722, no. 2, pp. 1644–1653, 2010.

[34] S. Pucci, G. Poletto, A. C. Sterling, and M. Romoli, "Physical parameters of standard and blowout jets," *Astrophysical Journal*, vol. 776, article 16, 2013.

[35] R. L. Moore, A. C. Sterling, D. A. Falconer, and D. Robe, "The cool component and the dichotomy, lateral expansion, and axial rotation of solar X-ray jets," *The Astrophysical Journal*, vol. 769, no. 2, article 134, 2013.

[36] R. L. Moore, A. C. Sterling, and D. A. Falconer, "Magnetic untwisting in solar jets that go into the outer corona in polar coronal holes," *Astrophysical Journal*, vol. 806, no. 1, article 11, 2015.

[37] G. E. Brueckner, R. A. Howard, M. J. Koomen et al., "The Large Angle Spectroscopic Coronagraph (LASCO)," *Solar Physics*, vol. 162, no. 1, pp. 357–402, 1995.

[38] B. Schmieder, Y. Guo, F. Moreno-Insertis et al., "Twisting solar coronal jet launched at the boundary of an active region," *Astronomy and Astrophysics*, vol. 559, article A1, 2013.

[39] E. Pariat, K. Dalmasse, C. R. DeVore, S. K. Antiochos, and J. T. Karpen, "Model for straight and helical solar jets: I. Parametric studies of the magnetic field geometry," *Astronomy and Astrophysics*, vol. 573, article 130, 2015.

[40] S. Patsourakos, E. Pariat, A. Vourlidas, S. K. Antiochos, and J. P. Wuelser, "*STEREO* SECCHI stereoscopic observations constraining the initiation of polar coronal jets," *The Astrophysical Journal*, vol. 680, no. 1, pp. L73–L76, 2008.

[41] C. Gontikakis, V. Archontis, and K. Tsinganos, "Observations and 3D MHD simulations of a solar active region jet," *Astronomy & Astrophysics*, vol. 506, no. 3, pp. L45–L48, 2009.

[42] R. Chandra, G. R. Gupta, S. Mulay, and D. Tripathi, "Sunspot waves and triggering of homologous active region jets," *Monthly Notices of the Royal Astronomical Society*, vol. 446, no. 4, pp. 3741–3748, 2015.

[43] J. Chae, H. Wang, C.-Y. Lee, P. R. Goode, and U. Schühle, "Photospheric magnetic field changes associated with transition

region explosive events," *Astrophysical Journal*, vol. 497, no. 2, pp. L109–L112, 1998.

[44] L. R. B. Rubio and C. Beck, "Magnetic flux cancellation in the moat of sunspots: results from simultaneous vector spectropolarimetry in the visible and the infrared," *Astrophysical Journal*, vol. 626, no. 2, pp. L125–L128, 2005.

[45] L. Heggland, B. De Pontieu, and V. H. Hansteen, "Observational signatures of simulated reconnection events in the solar chromosphere and transition region," *The Astrophysical Journal*, vol. 702, no. 1, pp. 1–18, 2009.

[46] D. E. Innes, R. H. Cameron, and S. K. Solanki, "EUV jets, type III radio bursts and sunspot waves investigated using SDO/AIA observations," *Astronomy and Astrophysics*, vol. 531, article L13, 2011.

[47] A. K. Srivastava and K. Murawski, "Observations of a pulse-driven cool polar jet by SDO/AIA," *Astronomy & Astrophysics*, vol. 534, article A62, 2011.

[48] S. Chandrasekhar, *Hydrodynamic and Hydromagnetic Stability*, The International Series of Monographs on Physics, Clarendon Press, Oxford, UK, 1961.

[49] T. V. Zaqarashvili, I. Zhelyazkov, and L. Ofman, "Stability of rotating magnetized jets in the solar atmosphere. I. Kelvin–Helmholtz instability," *Astrophysical Journal*, vol. 813, no. 2, article 123, 2015.

[50] D. Ryu, T. W. Jones, and A. Frank, "The magnetohydrodynamic Kelvin-Helmholtz instability: A three-dimensional study of nonlinear evolution," *The Astrophysical Journal*, vol. 545, no. 1, pp. 475–493, 2000.

[51] S. A. Maslowe, "Shear flow instabilities and transition," in *Hydromagnetic Instabilities and the Transition to Turbulence*, H. L. Swinney and J. P. Gollub, Eds., pp. 181–228, Springer, Berlin, Germany, 1985.

[52] T. E. Berger, G. Slater, N. Hurlburt et al., "Quiescent prominence dynamics observed with the Hinode solar optical telescope. I. Turbulent upflow plumes," *The Astrophysical Journal*, vol. 716, no. 2, pp. 1288–1307, 2010.

[53] M. Ryutova, T. Berger, Z. Frank, T. Tarbell, and A. Title, "Observation of plasma instabilities in quiescent prominences," *Solar Physics*, vol. 267, no. 1, pp. 75–94, 2010.

[54] D. Martínez-Gómez, R. Soler, and J. Terradas, "Onset of the Kelvin-Helmholtz instability in partially ionized magnetic flux tubes," *Astronomy and Astrophysics*, vol. 578, article A104, 2015.

[55] N. F. Loureiro, A. A. Schekochihin, and D. A. Uzdensky, "Plasmoid and Kelvin-Helmholtz instabilities in Sweet-Parker current sheets," *Physical Review E: Statistical, Nonlinear, and Soft Matter Physics*, vol. 87, no. 1, article 013102, 2013.

[56] L. Feng, B. Inhester, and W. Q. Gan, "Kelvin–Helmholtz instability of a coronal streamer," *Astrophysical Journal*, vol. 774, no. 2, article 141, 2013.

[57] C. Foullon, E. Verwichte, V. M. Nakariakov, K. Nykyri, and C. J. Farrugia, "Magnetic Kelvin–Helmholtz instability at the Sun," *Astrophysical Journal Letters*, vol. 729, no. 1, article L8, 2011.

[58] L. Ofman and B. J. Thompson, "*SDO*/AIA observation of Kelvin–Helmholtz instability in the solar corona," *The Astrophysical Journal Letters*, vol. 734, no. 1, article L11, 2011.

[59] C. Foullon, E. Verwichte, K. Nykyri, M. J. Aschwanden, and I. G. Hannah, "Kelvin–Helmholtz instability of the CME reconnection outflow layer in the low corona," *The Astrophysical Journal*, vol. 767, no. 2, article 170, 2013.

[60] U. V. Möstl, M. Temmer, and A. M. Veronig, "The Kelvin–Helmholtz instability at coronal mass ejection boundaries in the solar corona: Observations and 2.5D MHD simulations," *Astrophysical Journal Letters*, vol. 766, no. 1, article 12, 2013.

[61] K. Nykyri and C. Foullon, "First magnetic seismology of the CME reconnection outflow layer in the low corona with 2.5-D MHD simulations of the Kelvin-Helmholtz instability," *Geophysical Research Letters*, vol. 40, no. 16, pp. 4154–4159, 2013.

[62] T. V. Zaqarashvili, A. J. Díaz, R. Oliver, and J. L. Ballester, "Instability of twisted magnetic tubes with axial mass flows," *Astronomy & Astrophysics*, vol. 516, article A84, 2010.

[63] R. Soler, A. J. Díaz, J. L. Ballester, and M. Goossens, "Kelvin–Helmholtz instability in partially ionized compressible plasmas," *The Astrophysical Journal*, vol. 749, no. 2, article 163, 2012.

[64] T. V. Zaqarashvili, "Solar spicules: recent challenges in observations and theory," in *Proceedings of the 3rd School and Workshop on Space Plasma Physics*, I. Zhelyazkov and T. Mishonov, Eds., vol. 1356 of *AIP Conference Proceedings*, pp. 106–116, AIP Publishing LLC, Melville, NY, USA, 2011.

[65] I. Zhelyazkov, "Magnetohydrodynamic waves and their stability status in solar spicules," *Astronomy & Astrophysics*, vol. 537, article A124, 2012.

[66] A. Ajabshirizadeh, H. Ebadi, R. E. Vekalati, and K. Molaverdikhani, "The possibility of Kelvin-Helmholtz instability in solar spicules," *Astrophysics and Space Science*, vol. 357, no. 1, article 33, 2015.

[67] I. Zhelyazkov and T. V. Zaqarashvili, "Kelvin-Helmholtz instability of kink waves in photospheric twisted flux tubes," *Astronomy & Astrophysics*, vol. 547, article A14, 2012.

[68] I. Zhelyazkov, R. Chandra, A. K. Srivastava, and T. Mishonov, "Kelvin–Helmholtz instability of magnetohydrodynamic waves propagating on solar surges," *Astrophysics and Space Science*, vol. 356, no. 2, pp. 231–240, 2015.

[69] I. Zhelyazkov, T. V. Zaqarashvili, R. Chandra, A. K. Srivastava, and T. Mishonov, "Kelvin–Helmholtz instability in solar cool surges," *Advances in Space Research*, vol. 56, no. 12, pp. 2727–2737, 2015.

[70] I. Zhelyazkov, R. Chandra, and A. K. Srivastava, "Can magnetohydrodynamic waves traveling on solar dark mottles become unstable?" *Bulgarian Journal of Physics*, vol. 42, no. 1, pp. 68–87, 2015.

[71] I. Zhelyazkov, T. V. Zaqarashvili, and R. Chandra, "Kelvin–Helmholtz instability in coronal mass ejecta in the lower corona," *Astronomy and Astrophysics*, vol. 574, 2015.

[72] I. Zhelyazkov, R. Chandra, and A. K. Srivastava, "Kelvin–Helmholtz instability in coronal mass ejections and solar surges," in *Proceedings of the 5th School and Workshop on Space Plasma Physics*, I. Zhelyazkov and T. Mishonov, Eds., vol. 1714 of *AIP Conference Proceedings*, AIP Publishing LLC, Melville, NY, USA, article 030005, 2016.

[73] D. Kuridze, V. Henriques, M. Mathioudakis et al., "The dynamics of rapid redshifted and blueshifted excursions in the solar Hα line," *The Astrophysical Journal*, vol. 802, no. 1, article 26, 2015.

[74] I. Zhelyazkov, "On modeling the Kelvin–Helmholtz instability in solar atmosphere," *Journal of Astrophysics and Astronomy*, vol. 36, no. 1, pp. 233–254, 2015.

[75] S. Vasheghani Farahani, T. Van Doorsselaere, E. Verwichte, and V. M. Nakariakov, "Propagating transverse waves in soft X-ray coronal jets," *Astronomy & Astrophysics*, vol. 498, no. 2, pp. L29–L32, 2009.

[76] I. Zhelyazkov, "Review of the magnetohydrodynamic waves and their stability in solar spicules and X-ray jets," in *Topics in Magnethydrodynamics*, L. Zheng, Ed., chapter 6, InTech, Rijeka, Croatia, 2012.

[77] I. Zhelyazkov, "Kelvin–Helmholtz instability of kink waves in photospheric, chromospheric, and X-ray solar jets," in *Proceedings of the 4th School and Workshop on Space Plasma Physics*, I. Zhelyazkov and T. Mishonov, Eds., vol. 1551, pp. 150–164, AIP Publishing LLC, Melville, NY, USA, 2013.

[78] M. Goossens, J. V. Hollweg, and T. Sakurai, "Resonant behaviour of MHD waves on magnetic flux tubes. III. Effect of equilibrium flow," *Solar Physics*, vol. 138, no. 2, pp. 233–255, 1992.

[79] K. Hain and R. Lüst, "Zur Stabilität zylindersymmetrischer Plasmakonfigurationen mit Volumenströmen," *Zeitschrift für Naturforschung*, vol. 13, pp. 936–940, 1958.

[80] J. P. Goedbloed, "Stabilization of magnetohydrodynamic instabilities by force-free magnetic fields: I. Plane plasma layer," *Physica*, vol. 53, no. 3, pp. 412–444, 1971.

[81] T. Sakurai, M. Goossens, and J. V. Hollweg, "Resonant behaviour of MHD waves on magnetic flux tubes. I. Connection formulae at the resonant surfaces," *Solar Physics*, vol. 133, no. 2, pp. 227–245, 1991.

[82] P. M. Edwin and B. Roberts, "Wave propagation in a magnetic cylinder," *Solar Physics*, vol. 88, no. 1-2, pp. 179–191, 1983.

[83] M. Terra-Homem, R. Erdélyi, and I. Ballai, "Linear and nonlinear MHD wave propagation in steady-state magnetic cylinders," *Solar Physics*, vol. 217, no. 2, pp. 199–223, 2003.

[84] V. M. Nakariakov, "MHD oscillations in solar and stellar coronae: current results and perspectives," *Advances in Space Research*, vol. 39, no. 12, pp. 1804–1813, 2007.

[85] K. Bennett, B. Roberts, and U. Narain, "Waves in twisted magnetic flux tubes," *Solar Physics*, vol. 185, no. 1, pp. 41–59, 1999.

[86] P. S. Cally, "Leaky and non-leaky oscillations in magnetic flux tubes," *Solar Physics*, vol. 103, no. 2, pp. 277–298, 1986.

[87] I. Zhelyazkov, R. Chandra, and A. K. Srivastava, "Kelvin-Helmholtz instability in an active region jet observed with Hinode," *Astrophysics and Space Science*, vol. 361, no. 2, article 51, 2016.

[88] T. V. Zaqarashvili, Z. Vörös, and I. Zhelyazkov, "Kelvin-Helmholtz instability of twisted magnetic flux tubes in the solar wind," *Astronomy & Astrophysics*, vol. 561, article A62, 2014.

[89] J. W. Dungey and R. E. Loughhead, "Twisted magnetic fields in conducting fluids," *Australian Journal of Physics*, vol. 7, no. 1, pp. 5–13, 1954.

[90] A. R. Paraschiv, A. Bemporad, and A. C. Sterling, "Physical properties of solar polar jets: A statistical study with Hinode XRT data," *Astronomy & Astrophysics*, vol. 579, article A96, 2015.

[91] T. Miyagoshi and T. Yokoyama, "Magnetohydrodynamic numerical simulations of solar X-ray jets based on the magnetic reconnection model that includes chromospheric evaporation," *The Astrophysical Journal*, vol. 593, no. 2, pp. L133–L136, 2003.

4

Intelligent Cognitive Radio Models for Enhancing Future Radio Astronomy Observations

Ayodele Abiola Periola and Olabisi Emmanuel Falowo

Communication Research Group, Department of Electrical Engineering, University of Cape Town, Rondebosch, Cape Town, South Africa

Correspondence should be addressed to Ayodele Abiola Periola; periola@crg.ee.uct.ac.za

Academic Editor: Alberto J. Castro-Tirado

Radio astronomy organisations desire to optimise the terrestrial radio astronomy observations by mitigating against interference and enhancing angular resolution. Ground telescopes (GTs) experience interference from intersatellite links (ISLs). Astronomy source radio signals received by GTs are analysed at the high performance computing (HPC) infrastructure. Furthermore, observation limitation conditions prevent GTs from conducting radio astronomy observations all the time, thereby causing low HPC utilisation. This paper proposes mechanisms that protect GTs from ISL interference without permanent prevention of ISL data transmission and enhance angular resolution. The ISL transmits data by taking advantage of similarities in the sequence of observed astronomy sources to increase ISL connection duration. In addition, the paper proposes a mechanism that enhances angular resolution by using reconfigurable earth stations. Furthermore, the paper presents the opportunistic computing scheme (OCS) to enhance HPC utilisation. OCS enables the underutilised HPC to be used to train learning algorithms of a cognitive base station. The performances of the three mechanisms are evaluated. Simulations show that the proposed mechanisms protect GTs from ISL interference, enhance angular resolution, and improve HPC utilisation.

1. Introduction

Astronomy is the scientific study of the universe by analysing astronomy source signals. Astronomy source signals can be received by either ground or space telescopes. Astronomy observations can also be categorised based on the signal source. Astronomy source signals that are observed by ground telescopes (GTs) can arise from optical, radio, and gravitational waves. In addition, astronomical source signals from X-ray, infra-red, and ultraviolet radiation are observed from space telescopes.

Radio and gravitational astronomy observations are complementary [1–3] and help in understanding the universe. The spectrum access of terrestrial radio astronomy observations is influenced by wavelength variation due to red shift and blue shift. Wavelength variation necessitates that GTs should have access to significant bandwidth resources. Terrestrial radio astronomy observations experience interference from the radio waves radiated by intersatellite links (ISLs) of low earth orbiting satellites. The use of ISLs is projected to increase due to small satellite proliferation [4–7].

Hence, a solution that protects terrestrial radio astronomy observations from interfering ISLs is needed. Such a solution can be designed using the cognitive radio (CR). The CR differentiates users based on their priority to access the radio spectrum. The two types of users recognised by a CR are primary users and secondary users with higher and lower spectrum access priority, respectively. The CR can be used to design interference protection schemes by using the interweaving or underlay spectrum sharing model. The interweaving spectrum sharing model enables secondary users to share access to the radio spectrum with primary users. The sharing is realised by informing secondary users of the spectrum access epochs of primary users. Being aware of primary user spectrum access epochs, the secondary users can use the spectrum when primary users are absent.

A CR based interweaving spectrum sharing framework that protects GTs from ISL interference is proposed in [8].

In [8], the primary user is the GT, while the secondary user is the ISL. The information on the epochs of astronomy source observation is accessible to the satellite and is used to determine the ISL activation epoch and duration. The CR is used to deactivate the ISL for a given duration, while terrestrial radio astronomy observations are ongoing. The results in [8] require further investigation to examine relations between GT interference protection and the similarities in the patterns of the observed astronomy sources.

Furthermore, organisations desiring to conduct radio astronomy observations can be classified based on GT availability. Some organisations can afford to construct their own GTs, while others convert unused earth stations to GTs. The conversion of unused earth stations is feasible due to the increasing use of fibre optic cables instead of satellites for broadband Internet access [9–14]. The discussion in [9–14] focuses on utilising converted unused satellite earth stations as GTs but does not consider the presence of terrestrial wireless networks in the destination electromagnetic environment of converted earth stations. In addition, the use of converted telescopes should also enhance the angular resolution of the terrestrial radio astronomy observations. Hoare and Rawlings [10] propose the use of a multimode telescope for satellite communications and terrestrial radio astronomy observations. The dynamic use of a multimode GT should enhance the angular resolution. The reconfigurable CR can be used to enhance the angular resolution when multimode GTs are used in terrestrial radio astronomy. The GT realised via conversion is also susceptible to interference when its destination environment comprises terrestrial wireless networks. Therefore, GTs require interference protection mechanisms.

In addition, terrestrial radio astronomy organisations also seek to maximise high performance computing (HPC) infrastructure utilisation. According to Barbosa et al. [15], the Atacama Large Millimetre/Submillimetre array (ALMA)'s HPC is underutilised due to observation limitation conditions. HPC utilisation can be improved by using techniques such as time multiplexing. However, other multiplex techniques such as duty cycle division multiplex which outperform time multiplexing have been proposed [16–18]. Therefore, a duty cycle multiplex scheme that can enhance HPC utilisation is required.

This paper addresses two goals for terrestrial radio astronomy organisations using converted GTs. It proposes mechanisms that optimise the conduct of terrestrial radio astronomy observations by avoiding interference and improving angular resolution. The paper also proposes a mechanism that enhances HPC utilisation. This paper makes the following contributions:

(1) It proposes an optimisation framework for terrestrial radio astronomy observations. The optimisation framework protects terrestrial radio astronomy observations from ISL interference and enhances the angular resolution of terrestrial radio astronomy organisation. The paper analyses additional data sets from the Karoo Array Telescope [19] to investigate the range of ISL transmit duration permissible without causing interference to GTs. The angular resolution is also enhanced by using CR enabled multimode GTs for terrestrial radio astronomy observations.

(2) It proposes an intelligent framework that uses similarities in astronomy source observation data for proactive interference avoidance between ISLs and GTs.

(3) The paper proposes the opportunistic computing scheme (OCS) that uses a duty cycle multiplex to enhance HPC utilisation. This paper investigates OCS's success probability and the terrestrial wireless network throughput as a function of the number of GTs.

The remainder of this paper is organised as follows. Section 2 discusses the related literature. Section 3 focuses on problem definition. Section 4 presents the proposed mechanisms. Section 5 discusses the simulation results. Section 6 concludes the paper.

2. Related Work

This section is divided into two parts. The first part addresses issues related to optimising terrestrial radio astronomy observations. It discusses literature focusing on the interference protection of GTs and improving angular resolution. The second part discusses the improvement of HPC utilisation.

2.1. Optimising Terrestrial Radio Astronomy Observations. Interference-free spectrum access and the improvement of angular resolution are important goals in the conduct of terrestrial radio astronomy observations. The goal of interference-free spectrum access can be achieved via spectrum reservation [20–24]. Spectrum reservation aims to ensure that new services do not encroach into bands dedicated for terrestrial radio astronomy observations. However, spectrum reservation faces interference challenges from new services which are ignorant of terrestrial radio astronomy observations. Interference mitigation measures such as restricting satellites from GT sky region are proposed in [24]. The solution in [24] does not consider the forwarding of data via ISLs through the satellite network. A restriction of satellites increases latency when the shortest path through the satellite network lies in the GT sky region. The proliferation of small satellite constellations [4–7] that use ISLs poses interference risks to GTs.

Another area of innovation in terrestrial radio astronomy observations is the conversion of unused earth stations to GTs. The increasing use of optical cables has been recognised to make some satellite earth stations redundant [9–14]. GTs realised from converted earth stations have been used in the UK [9, 10], Mozambique [11], Ghana [12, 13], and New Zealand [14]. The conversion of unused satellite earth stations enables the reuse of satellite installations and reduces astronomy infrastructure cost. However, the converted earth stations have been those without a tracking system.

It can be inferred from [9–14] that the satellite earth stations to be converted have been left unused for a long period of time. During the idle time of unused earth stations, the roll-out of terrestrial wireless networks in concerned areas is not unlikely, thereby exposing GTs to terrestrial wireless network interference as inferred from [25–28]. Therefore, converted earth stations require interference protection mechanisms when they are in the vicinity of terrestrial wireless networks.

The use of additional earth stations alongside converted earth stations increases the number of GTs and the baseline. The increase in the number of GTs and baseline gives an opportunity to improve the angular resolution of terrestrial radio astronomy observation. In the absence of additional GTs, the terrestrial radio astronomy organisation has a fixed baseline. The angular resolution can be improved in a GT array with a dynamic baseline.

Furthermore, Woodburn et al. [14] recognise that the Goonhilly-3 GT can be used for satellite communications and terrestrial radio astronomy. Such a GT reduces costs due to dish and transponder reuse. A dual purpose GT can be opportunistically used to increase the baseline of terrestrial radio astronomy observations for a given period of time. However, the dual purpose GT intended for satellite communications and terrestrial radio astronomy observations requires mode switching mechanisms. The mode switching mechanism determines the epochs where the dual purpose GT can be used for receiving packets or radio astronomy data. The dual GT cannot be concurrently used for communications and astronomy observations. This is because the high transmit power in communication signals interferes with the astronomy source radio signal. The dual GT should also host mechanisms that protect it from terrestrial wireless network interference.

In addressing the challenges of interference mitigation and enhancing angular resolution, this paper considers the CR as suitable. CR spectrum sharing models are suitable for the design of interference protection schemes for terrestrial radio astronomy observations. In [29], the CR application considers the underlay spectrum sharing model and prevents interference by limiting terrestrial wireless network transmit power. A reduction of the terrestrial wireless network transmit power is proposed in [29] because astronomy source transmit power cannot be controlled. Investigations in [29] show that the desired coexistence between terrestrial wireless networks and GTs is infeasible. This is because of the large transmit power of terrestrial wireless networks compared to the very low received signal strength of astronomy source signals. However, the CR supports other spectrum sharing models such as the interweaving spectrum sharing model.

In addition, the CR benefits from the reconfigurable software defined radio. Being reconfigurable, the CR can be used to design a dual GT with a dynamic baseline. The inclusion of a dual GT in a manner that increases the baseline enhances observatory angular resolution. The inclusion of the CR in the dual GT enables the design of a GT that responds to the requirement of improving the angular resolution. Being suitable for designing solutions that enhance interference mitigation and angular resolution, the CR can be used to design technical solutions that enhance terrestrial radio astronomy observations [30]. However, the use of the CR in this aspect requires further consideration.

2.2. High Performance Computing Infrastructure Utilisation. The HPC is used to process the astronomy source radio waves that are received by GTs. It is connected to the GTs via optic fibre links. The terrestrial radio astronomy organisation should maximally utilise the HPC. Barbosa et al. [15] point out that the ALMA HPC has a 38% utilisation as indicated in the ALMA Cycle 0 report. This results in low power efficiency, since the HPC is powered all the time [15].

The resulting HPC underutilisation can be addressed by using multiplexing techniques. Multiplexing techniques have been used in wireless communications for sharing bandwidth resources and suitable for enhancing HPC utilisation. The use of multiplexing proposed in [15] can also be extended to accommodate data processing from cognitive terrestrial wireless networks. Such an application can enhance the CR autonomous capability as seen in [31, 32]. The CR in [31, 32] has a limited autonomous capacity because it does not autogenerate and train new learning mechanisms. Generative artificial intelligence [32] can enable the CR to autogenerate and train new intelligent mechanisms. The autogenerated learning mechanisms can be used to determine CR transmission parameters after training [33, 34]. The discussion in [31, 32] has not considered using HPC's unused computational resources to train autogenerated learning mechanisms. Andreani [35] presents results that can be used to estimate the GT nonobservation time fraction in ALMA Cycle 3 spanning the period from October 2015 to August 2016. The discussion in [35] also identifies factors causing GT observation limitations such as opacity and phase stability. The occurrence of observation limitations affects GTs observation and is independent of interference from ISLs or terrestrial wireless networks. From the estimated observation fraction presented in [35], the ALMA HPC is left unutilised for 51.4% of the time that it is powered. Though, the ALMA underutilisation is noted to have reduced by 10.6%, an underutilised capacity of 48.6% still exists.

3. Problem Definition

This section describes the challenges being addressed for a terrestrial radio astronomy organisation that uses GTs realised from converted unused earth stations. It is divided into two parts. The first part focuses on the optimisation goals. In the first part, the challenges discussed are those of interference avoidance and enhancing angular resolution. The second part describes the problem of enhancing HPC utilisation.

3.1. Optimisation: Defined Challenges for Terrestrial Radio Astronomy Observations. The considered scenario comprises low earth orbit satellites, GTs, cognitive base stations, and the HPC. The satellites are connected using ISLs. These entities have the following capabilities:

(1) Satellites: they are located in the low earth orbit and have a shortest path routing and station keeping algorithms

(2) High performance computing (HPC) infrastructure: the HPC is peta-scale and general purpose and is

shielded from radio frequency interference. It allocates computational units to each GT and has access to high speed Internet links. The HPC can also determine when observation limitation conditions cause HPC underutilisation

(3) Ground telescopes (GTs): GTs are installed after launching the satellite constellation. They are connected to the HPC by optic fibre links. They present the observed astronomy source radio signals to the HPC. Each GT allocated HPC computational units. The allocated HPC computational units are used to process signals received from the GT

(4) Cognitive base station (CBS): the CBS is a massive multiantenna system and is the central entity in terrestrial wireless network. It incorporates generative artificial intelligence, autogenerates learning mechanisms, and is connected to the HPC via high speed Internet links. The CBS uses the orthogonal frequency division multiplex-space division multiple access technology and receives signal streams from multiple terrestrial wireless network subscribers. Individual subscriber signals are extracted from the multiplexed signal by an artificial neural network multiuser detector. The multiuser detection aims to reduce user bit error rate. The CBS receives subscriber bit error rate via the control channel. It compares subscriber bit error rate with a predefined bit error rate threshold. The CBS autogenerates new artificial neural network multiuser detectors when user bit error rate exceeds the bit error rate threshold. It keeps existing artificial neural network multiuser detectors and examines their suitability in different future contexts

Let α, G, ϕ, and θ denote the set of satellites, GTs, HPC computational units allocated to GTs, and the utilisation of HPC computational units allocated to GTs, respectively.

$$\begin{aligned} \alpha &= \{\alpha_1, \dots, \alpha_i\}, \\ G &= \{G_1, \dots, G_m\}, \\ \phi &= \{\phi_1, \dots, \phi_m\}, \\ \theta &= \{\theta_1, \dots, \theta_m\}, \end{aligned} \tag{1}$$

where i and m are the maximum numbers of ISLs and GTs, respectively.

In addition, let η, υ, f_η, and f_G be the sets of satellite sky region, GT sky region, ISL frequency, and GT frequency, respectively.

$$\begin{aligned} \eta &= (\eta_1, \dots, \eta_i), \\ \upsilon &= (\upsilon_1, \dots, \upsilon_m), \\ f_\eta &= \left(f_{\eta_1}, \dots, f_{\eta_i}\right), \\ f_G &= \left(f_{G_1}, \dots, f_{G_m}\right). \end{aligned} \tag{2}$$

Satellites that are interconnected via ISLs and in the sky region of GTs cause intermodulation interference to terrestrial radio astronomy observations. The spectral main and side band signals are transmitted in the ISL antenna main lobe and side lobes, respectively. The presence of multiple satellites in the low earth orbit leads to the existence of multiple side lobes alongside each ISL main lobe due to nonideal satellite side lobe suppression. The ISL signal interferes with the astronomy source radio signal because the ISL's main and side lobe signals have a higher power than the astronomy source radio signal power.

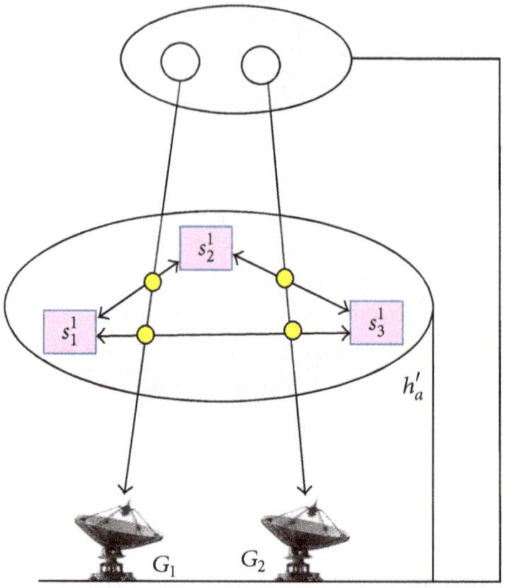

Figure 1: Interference between ISLs and GBTs.

Interference arises between ISLs and GTs when $\upsilon_c = \eta_d$, $f_{G_c} = f_{\eta_d}$, $c \in \{1, \dots, m\}$, and $d \in \{1, \dots, i\}$. A scenario showing the occurrence of an interference point can be seen in Figure 1. In Figure 1, astronomy sources and satellites have altitudes, h'_a and h'_s, respectively, where $h'_a > h'_s$. The operating frequencies of GTs, G_1, G_2, are f_{G_1} and f_{G_2}, respectively. The satellite sky region has three satellites, s_1^1, s_2^1, and s_3^1, which are connected with ISLs. ISLs s_1^1-s_2^1, s_2^1-s_3^1, and s_1^1-s_3^1 transmit on f_{η_1}, f_{η_2}, and f_{η_3}, respectively. The interference point in Figure 1 arises via either additive or multiplicative or other combinations of radio astronomy and the ISL signals.

The interference arises because of the nonlinear combination of ISL and astronomy source signals. ISL signals have intermodulation products that are radiated through the satellite antenna's main and side lobes. The radiated signals comprise intermodulation products and have a higher power than the astronomy source signals. The stronger ISL radiated signals cause interference with the weaker astronomy source signal. In the problem formulation here, the paper aims to demonstrate that ISL signals cause interference to astronomy source signals. The ISL signal is recognised to comprise multiple intermodulation products; the aim here is not to list these products but to demonstrate that their presence interferes with terrestrial radio astronomy organisations.

Let $\psi^{\eta}_{\alpha_a, G_c}$ denote the ISL signal from satellite α_a, $a \in \{1, \dots, i\}$, traversing η over G_c. In addition, let γ^{υ}_{p} denote the astronomy source radio signal p received by G_c in υ, respectively. In the absence of interference protection, the GT receives the signal, $s_{p,\upsilon_c}^{\alpha_a,G_c,\eta_d}$, given as

$$s_{p,\upsilon_c}^{\alpha_a,G_c,\eta_d} = A + B,$$

$$A = \sum_{a=1}^{i}\sum_{c=1}^{m}\psi_{\alpha_a,G_c}^{\eta_a} + \sum_{p=1}^{P}\sum_{c=1}^{m}\gamma_p^{\upsilon_c} + \prod_{a=1}^{i}\prod_{c=1}^{m}\psi_{\alpha_a,G_c}^{\eta_a} \times \prod_{p=1}^{P}\prod_{c=1}^{m}\gamma_p^{\upsilon_c}, \quad (3)$$

$$B = \sum_{a=1}^{i}\sum_{c=1}^{m}\psi_{\alpha_a,G_c}^{\upsilon_c} + \prod_{p=1}^{P}\prod_{c=1}^{m}\gamma_p^{\upsilon_c} + \prod_{a=1}^{i}\prod_{c=1}^{m}\psi_{\alpha_a,G_c}^{\eta_a}.$$

A and B are intermodulation products arising from the additive and multiplicative components of $\psi_{\alpha_a,G_c}^{\eta}$ and γ_p^{υ}. The additive and multiplicative components arise due to the combinations of different ISL signal components alongside components of the astronomy source radio signal.

Furthermore, let $N_G(t)$, $D(G,t)$, and $\lambda_G(t)$ denote the sets of (1) terrestrial wireless networks in GT vicinity at time t, (2) GTs baseline at time t, and (3) GT observation wavelength at time t, respectively.

$$\begin{aligned} t &= \{t_1,\dots,t_r\}, \\ N_G(t) &= \{N_{G_1}(t),\dots,N_{G_j}(t)\}, \\ D(G,t) &= \{(D(G_1,t)),\dots,(D(G_m,t))\}, \\ \lambda_G(t) &= \{\lambda_{G_1}(t),\dots,\lambda_{G_m}(t)\}. \end{aligned} \quad (4)$$

The angular resolution, $R(m-4,t_1)$, at time t_1 given $(m-4)$ GTs is

$$R(m-4,t_1) = \sum_{c=1}^{m-4}\frac{(\lambda_{G_c},t_1)}{\max(D(G_c,t_1))}. \quad (5)$$

Given that the baseline can be increased by using an additional GT such that $\max(D(G_{m-3},t_1)) > \max(D(G_{m-4},t_1))$, the angular resolution, $R(m-3,t_1)$, is

$$R(m-3,t_1) = \sum_{c=1}^{m-3}\frac{(\lambda_{G_c},t_1)}{\max(D(G_c,t_1))}. \quad (6)$$

However, the terrestrial astronomy organisation requires an algorithm that enables it to increase its baseline by using an additional GT. The access to the additional GT ensures that $\max(D(G_{m-3},t_1)) > \max(D(G_{m-4},t_1))$ without constructing a new GT. In addition, let $f_G(t)$ be the set of terrestrial wireless network frequencies in GT vicinity:

$$f_G(t) = \{f_{G_1}(t),\dots,f_{G_j}(t)\}, \quad (7)$$

where j is the maximum number of terrestrial wireless network channels. Interference arises between the terrestrial wireless network and GTs when $f_{G_{j'}}(t) = 1/\lambda_{G_c}(t)$, $j' \in \{1,\dots,j\}$. Hence, GTs require an interference protection mechanism.

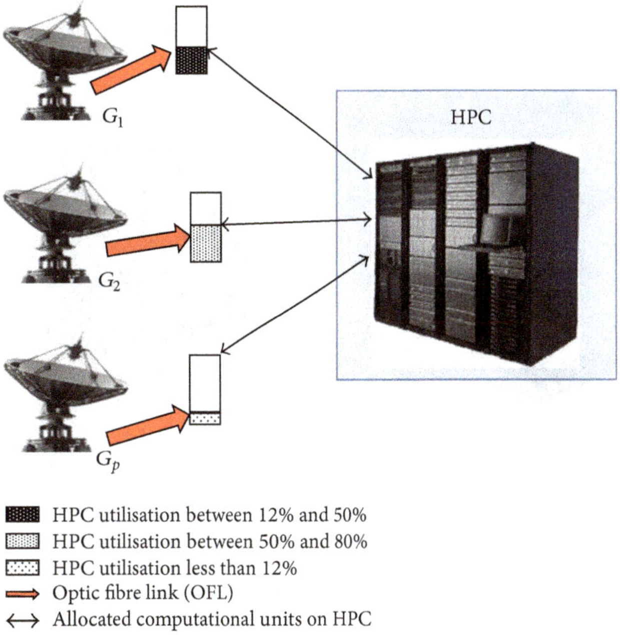

FIGURE 2: HPC underutilisation.

The discussion above has not considered the processing of the signals received by each GT. Each GT allocated ϕ_c HPC computational units, each having utilisation θ_c. The relations between GTs and the HPC for a case of HPC underutilisation and near optimal utilisation are shown in Figures 2 and 3, respectively.

Figure 2 shows the case where p, $p \in \{1,\dots,m\}$, GTs utilise the HPC. The first, second, and pth GTs have utilisation lying within 12% to 50% to 80% and less than 12%, respectively. In this case, $\phi_{1,\mathrm{res}}$ HPC computational units are unutilised.

Figure 3 shows the case where p GTs have near optimal HPC utilisation because all GTs utilise their computational units by up to 80%. In this case, HPC's underutilised computational unit is $\phi_{2,\mathrm{res}}$ and $\phi_{2,\mathrm{res}} < \phi_{1,\mathrm{res}}$.

The scenario given in Figure 2 can be described as

$$\begin{aligned} \theta_2 &> \theta_1 > \theta_p, \\ \phi_1 &= \phi_2 = \phi_p, \\ \sum_{p=1}^{p}\phi_p(1-\theta_p) &= \phi_{1,\mathrm{res}}. \end{aligned} \quad (8)$$

The scenario given in Figure 3 can be described as

$$\begin{aligned} \theta_2 &= \theta_1 = \theta_p, \\ \phi_1 &= \phi_2 = \phi_p, \\ \sum_{p=1}^{p}\phi_p(1-\theta_p) &= \phi_{2,\mathrm{res}}. \end{aligned} \quad (9)$$

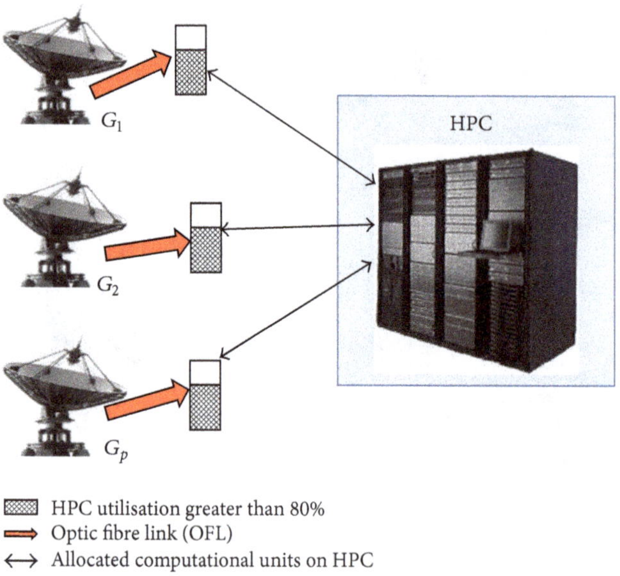

FIGURE 3: Near optimal HPC utilisation.

Given that ϕ_{th} is the threshold computational unit required for HPC sharing, the underutilised HPC can be shared with external applications if $\phi_{1,res} < \phi_{th}$ or $\phi_{2,res} < \phi_{th}$. However, the HPC requires a mechanism to verify when $\phi_{1,res} < \phi_{th}$ or $\phi_{2,res} < \phi_{th}$.

4. Proposed Schemes for Enhancing Terrestrial Radio Astronomy Observation

This section presents the proposed schemes and consists of two parts. The first part discusses the optimisation schemes. The optimisation scheme protects terrestrial radio astronomy observations from ISL interference and enhances angular resolution. The second presents the opportunistic computing scheme (OCS) proposed to enhance HPC utilisation.

The proposed schemes incorporate the CR that is reconfigurable and can make decisions for different contexts of a given application. Though most CR applications focus on terrestrial wireless networks, CR capabilities can enhance terrestrial radio astronomy observations goals. The CR acquires environmental awareness via sensing, makes inferences using sensed results, determines reconfiguration options, and executes the reconfiguration decisions.

4.1. Proposed Optimisation Mechanisms. The optimisation of terrestrial radio astronomy observations aims to achieve interference mitigation and enhance angular resolution. In this paper, we propose a CR interference mitigation framework that extends [8] by considering the similarities in the observation order of astronomy sources.

The interference mitigation framework is located on the satellite, assumes that the astronomy organisation has a database of the epochs of previously observed astronomy sources, and comprises three entities. The entities are as follows:

(1) Cognitive reasoner (CRE): the cognitive reasoner receives two sets of information from the terrestrial radio astronomy organisation. The first set comprises the right ascension RA, declination D, observation frequency OF, duration ODu, and dates ODt. These are held in the tuple $(\mathrm{RA}, D, \mathrm{OF}, \mathrm{ODu}, \mathrm{ODt})$. The second set comprises similarly observed sources, G with right ascension RA, and total observation duration D. The information on RA and D is held in S. The information on G, S, and I is held in the tuple (G, S, I). The CRE uses the information in $(\mathrm{RA}, D, \mathrm{OF}, \mathrm{ODu}, \mathrm{ODt})$ and (G, S, I) to determine the ISL activation epochs and duration. The tuples $(\mathrm{RA}, D, \mathrm{OF}, \mathrm{ODu}, \mathrm{ODt})$ and (G, S, I) are the first and second tuples, respectively

(2) Cognitive ISL deactivator (CSLA): the CSLA receives CRE outputs and uses these to determine the ISL transmission status and duration

(3) Plan acquisition channel (PAC): the PAC is a control channel that enables communications between the satellite's CRE and the terrestrial radio astronomy organisation. The first and second tuples are transmitted to the CRE from the terrestrial radio astronomy organisation via the PAC

The consideration of similarities implies that satellite does not have to analyse similar patterns all the time, thereby increasing ISL duration without interfering with GTs. The satellite does not have to analyse similar patterns all the time. Therefore, using similarity information can prevent interference to ongoing terrestrial radio astronomy observations while increasing ISL transmit duration.

The framework's flowchart is shown in Figure 4. As shown in Figure 4, the astronomy organisation transmits the first and second tuples via the PAC to the satellite via the PAC prior to commencing an observation. The CRE receives the first and second tuples and analyses them to determine the interference free ISL transmit epochs and duration. The CRE also determines whether a new similar pattern of observed sources is in the data received via the PAC. The new similar pattern is then used to update the second tuple on the satellite. The CSLA receives the CRE outputs and uses them to configure the ISL transmission status and duration.

Besides interference protection, optimising terrestrial radio astronomy observations also requires enhancing angular resolution. CR's reconfigurability is a useful feature in this regard. This paper applies a CR user classification to terrestrial radio astronomy observations. Terrestrial radio astronomy observations can be conducted using either primary GTs or secondary GTs. A primary GT is a GT that is designed for terrestrial radio astronomy observations only. The secondary GT is a GT that is capable of multiple applications. It can be used for other applications besides terrestrial radio astronomy observations. The Goonhilly-3 GT [14] intended for satellite communications and radio astronomy observations is an example of a secondary GT. This paper extends [14] by considering the CR as being suitable for designing a secondary GT.

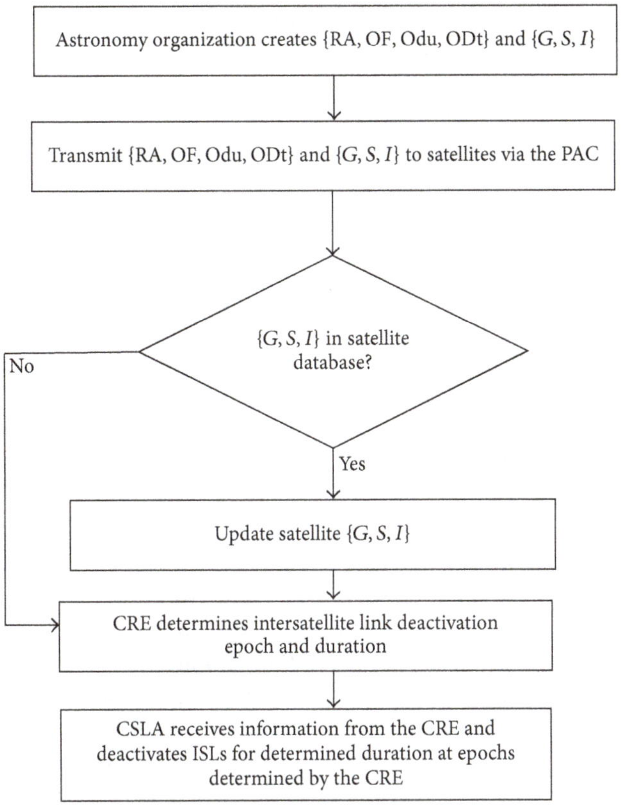

FIGURE 4: Flowchart of the learning framework.

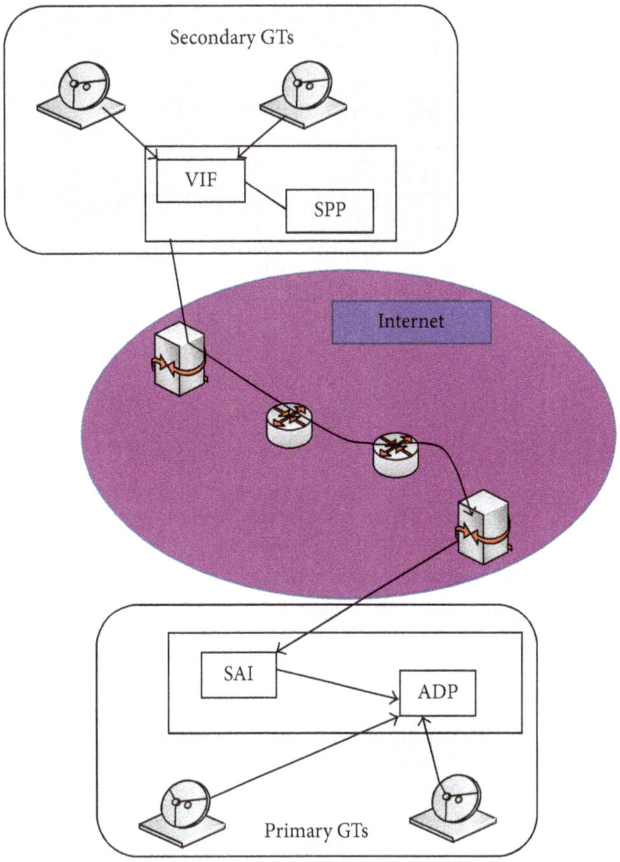

FIGURE 5: Interaction between primary and secondary GT.

The secondary GT incorporates a CR with mode switching and spectrum sensing mechanisms. The mode switching mechanisms enable the secondary GT to process communication packets and radio astronomy signals. This paper proposes a framework that enables interactions between terrestrial radio astronomy organisations (with primary GTs) and secondary GTs. Primary and secondary GTs interact via the Internet. Secondary GTs belong to other organisations. Radio astronomy data observed by the secondary GT are transmitted to the terrestrial radio astronomy organisation via the Internet.

The framework proposed to enhance the angular resolution has two ends belonging to the terrestrial astronomy organisation and the organisation that owns the secondary GTs. Each end has two entities. The entities at the terrestrial radio astronomy organisations are as follows:

(1) Astronomy data processor (ADP): the ADP processes radio astronomy data from primary and secondary GTs

(2) Satellite astronomy interface (SAI): the SAI holds information on the terrestrial radio astronomy organisation's observation objectives. It accesses accessible secondary GTs information capability via the Internet

The entities at the organisation that owns the secondary GTs are as follows:

(1) Visibility interface (VIF): the VIF collates information on the satellite visibility epoch and duration. The VIF accesses radio astronomy observation requirements from the SAI and determines suitable secondary GTs. In addition, the VIF obtains radio astronomy data from secondary GTs that are used for terrestrial radio astronomy observations. The obtained data is sent to the SAI via the Internet. The SAI sends radio astronomy data received from the VIF to the ADP

(2) Satellite packet processor (SPP): the SPP processes satellite communication packets when secondary GTs are used for satellite communications.

In this paper, the secondary GTs are owned by satellite communication network operators.

Relations between the ADP, SAI, VIF, and SPP are shown in Figure 5. In Figure 5, the secondary GT's packets and satellite visibility epochs are transmitted to the VIF. The VIF forwards packets to the SPP. It obtains and transmits satellite visibility epochs to the SAI via the Internet. CES data are processed at the ADP. Secondary GT's astronomy data are also sent to the ADP via the Internet. The SAI is aware of terrestrial radio astronomy observation goals and checks the SVB for available secondary GTs using SAI information. A bidirectional link exists between the SAI and the VIF.

Data obtained from the SAI are used by the VIF to select secondary GTs that satisfy the terrestrial radio astronomy organisation's observation objectives. The selection is done at the VIF. The selection algorithm considers the desired observation longitude, L_s^{υ}, latitude, l_s^{υ}, and frequency, f_{υ}, of υ. The information on L_s^{υ}, l_s^{υ}, and f_{υ} is held in T_1^{υ}. The selection procedure considers the zth secondary GT's longitude, L_e^z, latitude, l_e^z, and CR frequencies, f_e^z. Information on L_e^z, l_e^z, and f_e^z is held in T_2^z. The SAI selects the secondary GTs for which the Euclidean distance $d(T_1^{\upsilon}, T_2^z)$ is minimum.

$$d\left(T_1^{\upsilon}, T_2^z\right) = \min \sum_{c=1}^{m} \sum_{z=1}^{V} \sqrt{\left(L_s^{\upsilon_c} - L_e^z\right)^2 + \left(l_s^{\upsilon_c} - l_e^z\right)^2 + \left(f_{\upsilon_c} - f_e^z\right)^2}. \quad (10)$$

The secondary GT hosts a cyclostationary detector that can perfectly differentiate between radio astronomy and terrestrial wireless network signals. The perfect differentiation is achieved because terrestrial wireless networks use modulation schemes with known cyclostationary signatures. The secondary GT stops radio astronomy observations when terrestrial wireless signals are detected. The probability $P_c(\upsilon, z, d(T_1^{\upsilon}, T_2^z))$ that using z secondary GTs helps the terrestrial radio astronomy organisation to realise its observation objectives is

$$P_c\left(\upsilon, z, d\left(T_1^{\upsilon}, T_2^z\right)\right) = 1 - e^{-(\upsilon+z)d(T_1^{\upsilon}, T_2^z)}. \quad (11)$$

The use of secondary GTs also enhances the angular resolution. A smaller angular resolution is more beneficial. The angular resolution, θ_1, when the terrestrial radio astronomy organisation does not use secondary GTs is

$$\theta_1 = \sum_{c=1}^{m} \frac{\lambda_c}{b_{\max}}, \quad (12)$$

where λ_c is the observation wavelength of the cth GT and $b_{\max}$ is the observatory baseline.

In the case where there are terrestrial wireless networks in the vicinity of primary and secondary GTs, primary and secondary GTs experience interference. The resulting interference reduces the baseline's contribution in enhancing the angular resolution. The angular resolution is degraded because the electromagnetic radiation pattern of the terrestrial wireless network infiltrates that of the astronomical source being received by GTs. Given that the baseline is reduced by ϖ, the angular resolution θ_2 when secondary GTs are not incorporated and primary GTs experience terrestrial wireless network interference is

$$\theta_2 = \sum_{c=1}^{m} \frac{\lambda_c}{b_{\max}(1 - \varpi)}. \quad (13)$$

The use of secondary GTs alongside primary GTs improves the angular resolution, while using additional secondary GTs increases the baseline. In this case, the baseline is increased by α'. Assuming that both primary and secondary GTs are unaffected by terrestrial wireless network interference, the angular resolution θ_3 is

$$\theta_3 = \sum_{c=1}^{m} \frac{\lambda_c}{b_{\max}(1 + \alpha')} + \sum_{z=1}^{B'} \frac{\lambda_z}{b_{\max}(1 + \alpha')}, \quad (14)$$

where λ_z is the z_{th} secondary GT's observation wavelength and B' is the maximum number of secondary GTs.

In the event that primary and secondary GTs have terrestrial wireless networks in their vicinity. it is considered that primary and secondary GTs do not incorporate and incorporate an ideal cyclostationarity detector, respectively. The angular resolution, θ_4, is

$$\theta_4 = \sum_{c=1}^{m} \frac{\lambda_c}{b_{\max}(1 + \alpha')(1 - \varpi)} + \sum_{z=1}^{B'} \frac{\lambda_z}{b_{\max}(1 + \alpha')}. \quad (15)$$

4.2. Opportunistic Computing Scheme (OCS). The proposed OCS is a synergy between the CBS and HPC. The CBS uses artificial neural network multiuser detectors to ensure low bit error rate signal reception. The multiuser detectors are developed by training the neural networks with different bits of known modulated signals and user bit patterns for different channel states and multiantenna configuration. The cyberphysical system has two Internet entities that interact with the CBS and the HPC. These entities are as follows:

(1) Neural resource monitor (NRM): the NRM monitors CBS's usage of computational resources. It determines when the CBS resources are insufficient for developing newly autogenerated artificial neural network multiuser detectors. The NRM receives training data and instructions from the CBS and sends them to the TRM

(2) Training resource monitor (TRM): the TRM receives CBS training data and instructions from the NRM. It sends CBS training data and instructions of artificial neural network multiuser to underutilised HPC. The HPC executes CBS training instructions when observation limitation conditions results in HPC underutilisation by the GTs.

The OCS system showing relations between the CBS and the HPC at epochs t_1 and t_2 is shown in Figure 6. The CBS autogenerates new artificial neural network multiuser detectors when the obtained bit error rate exceeds the predefined threshold.

As shown in Figure 6, the CBS initially has four artificial neural network multiuser detectors, that is, brain like learning mechanisms at epoch t_1. The achieved bit error rate exceeds the predefined threshold bit error rate at epoch t_1. Hence, the development of new artificial neural network multiuser detectors is required. The CBS autogenerates new artificial neural network multiuser detectors at t_2. However, at t_2, the CBS does not have sufficient computational units to train the two autogenerated artificial neural network multiuser detectors. The OCS success probability is formulated using the probability that there are sufficient HPC computational

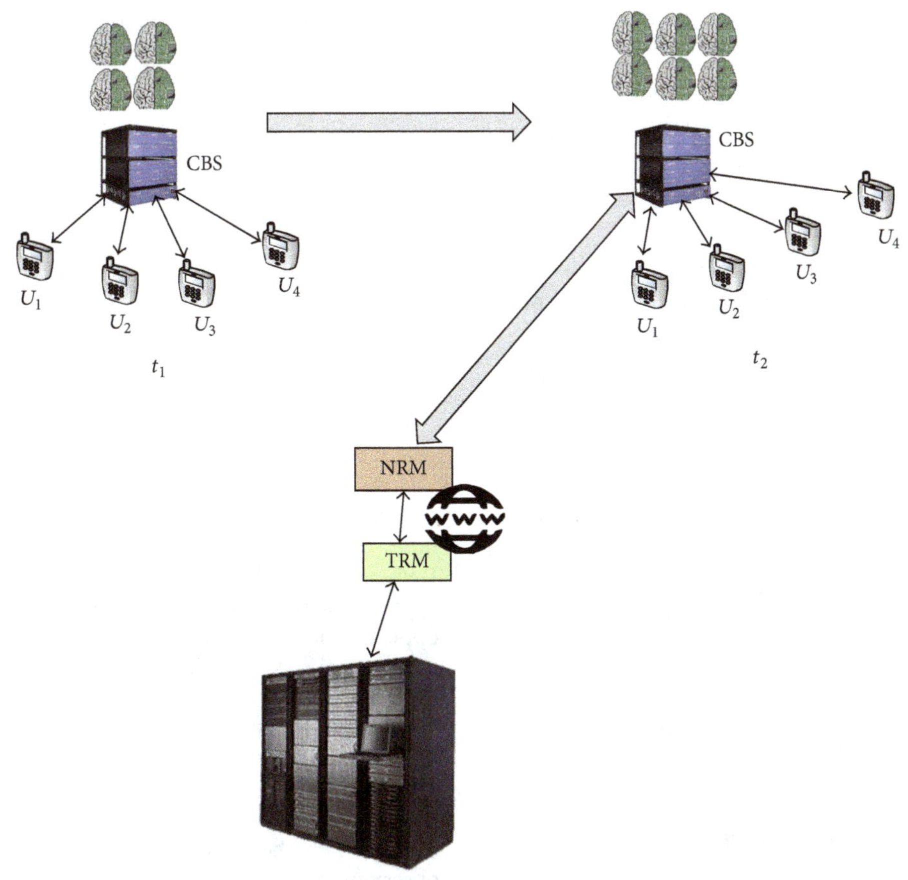

FIGURE 6: Evolution of states of artificial neural network multiuser detectors at epochs t_1 and t_2.

units for c GTs at time t, $P_c(c,t)$, and the probability of CCM failure for c telescopes at time t, $P_a(c,t)$. OCS fails in the following cases:

(1) There is TRM failure incident, though there are sufficient HPC computational resources.

(2) The TRM does not fail but there are insufficient HPC computational resources.

The OCS success probability, $P_{\text{OCS}}(c,t)$, can be obtained as follows:

$$P_{\text{OCS}}(c,t) = 1 - \left(P_a(c,t) \times P_c(c,t) + \left(1 - P_a(c,t)\right) \times \left(1 - P_c(c,t)\right)\right). \quad (16)$$

$P_{\text{OCS}}(c,t)$ is evaluated using the newly modified Weibull function [36] to model $P_c(c,t)$ and $P_a(c,t)$. OCS influences terrestrial wireless network throughput. The throughput is formulated to investigate OCS's ability to develop an artificial neural network multiuser detector that enhances signal reception by reducing the bit error rate when executing multiuser detection. A high OCS success execution probability results in a multiuser detector that reduces the number of corrupted bits received per second by the terrestrial wireless network subscriber. The reduction in the number of received corrupted bits increases the number of noncorrupted bits received per second by each subscriber, thereby enhancing terrestrial wireless network throughput.

In formulating the terrestrial wireless network throughput when OCS is used, we consider a scenario where CRs transmit to the CBS with transmit power, p_{11}^{tr}, over a channel with gain, h_{11}^{tr}. The transmitting CR experiences interference from neighbouring users. The interfering channel's gains and powers are given as $(h_{12}^{\text{int}}, h_{13}^{\text{int}}, \ldots, h_{1n}^{\text{int}})$ and $(p_{12}^{\text{int}}, p_{13}^{\text{int}}, \ldots, p_{1n}^{\text{int}})$, respectively. The development of an ideal artificial neural network multiuser detector occurs when $P_{\text{OCS}} = 1$; when $P_{\text{OCS}} < 1$, the artificial neural network multiuser detector is nonideal due to interference effects. Shannon throughput C when the terrestrial wireless network subscriber transmits on one channel is

$$C = \log_2\left(1 + \frac{\left(h_{11}^{\text{tr}}\right)^2 \times p_{11} \times P_{\text{OCS}}}{\sum_{n'=1}^{n}\left(h_{1n'}^{\text{int}}\right)^2 \times p_{1n'}^{\text{int}}}\right). \quad (17)$$

The subscriber can also use multiple channels with each channel having own interfering subscribers. The use of artificial neural network multiuser detectors developed via OCS reduces the number of received corrupted bits and enhances the signal to interference ratio, thereby improving throughput. Let $P_{\text{OCS}}^{n'}$ be the probability of successfully executing OCS for channel n'. In addition, let $h_{11}^{n'}$ and $p_{11}^{n'}$ be

TABLE 1: Analysis results of the spectrum utilisation and transmit opportunity.

S/N	Observation day	Observation duration	Spectrum utilisation (%)	Transmit opportunity (%)
1 (considered in [8])	16/02/13	28740	33.3	66.7
2 (considered in [8])	21/01/13	14700	17	83
3	14/10/12-15/10/12	31080	36	64
4	16/10/12	15180	17.57	82.43
5	28/10/12	36931	42.74	52.76
6	06/11/12-07/11/12	40800	57.22	42.78
7	14/11/12-15/11/12	36119	41.80	58.20
8	05/02/13	9846	11.36	88.64
9	11/02/13	18740	21.74	78.26
10	23/02/13	21730	25.15	74.85

the CR transmit channel gain and power over n', respectively. Similarly, let $h_{1i'}^{n'}$ and $p_{1i'}^{n'}$ be the interfering channel gain and power of user i' over n'. The throughput, C_1, when the CR transmits over N' channels, each of capacity B Hz, is

$$C = N' B \log_2 \left(1 + \frac{\sum_{n'=1}^{N'} \left(h_{11}^{n'}\right)^2 \times p_{11}^{n'} \times P_{\text{OCS}}^{n'}}{\sum_{n'=1}^{N'} \sum_{i'=1}^{m'} \left(h_{1i'}^{n'}\right)^2 \times p_{1i'}^{n'}} \right). \quad (18)$$

5. Performance Investigation

This section discusses the simulation results for the proposed mechanisms. It is divided into two parts. The first part presents results of terrestrial radio astronomy observation optimisation. It presents results on the spectrum usage analysis, similar observation strings of astronomy sources, and angular resolution. The second part investigates OCS's success execution probability and how OCS enhances terrestrial wireless network throughput.

5.1. Spectrum Usage Analysis. This section presents results on the spectrum usage of radio astronomy observations and examines ISL back-to-back duration and similar astronomy source observation strings. The spectrum utilisation and transmit opportunities are computed for the following days: 21/01/13, 16/02/13, 14/10/12, 15/10/12, 16/10/12, 28/10/12, 07/11/12, 14/11/12, 15/11/12, 05/02/13, 11/02/13, 23/02/12, and 06/11/12. The data used for 06/11/12 in [8] describes observations conducted between 17:15:54.8 and 23:37:42.6, while here it concerns observation made between 16:12:24.8 and 23:59:45.2. Data analysis results are shown in Table 1.

The average spectrum utilisation is computed using data in [8] and is also recalculated using new data. It is estimated to be 25.6% for data solely used in [8] and 29.1% when new data is incorporated, respectively. The average spectrum utilisation of radio astronomy observation increases by 3.5% when additional data is incorporated. The transmit opportunity Y decreases by 3.5%. The observation day, observation duration, spectrum utilisation, and transmit opportunity for previously considered data sets and the additional eight samples are presented in Table 1.

The standard deviation before and after the inclusion of more data for generalisation is evaluated to be 8.2% and 11.9%, respectively. Hence, the daily transmit duration is approximately one-ninth of the maximum obtained transmit opportunities. These transmit opportunities are exclusive of opportunities existing during the conduct of radio astronomy observations.

We also analyse the back-to-back duration D to determine the ISL transmit duration, while terrestrial radio astronomy observations are ongoing. The ISL transmit opportunities arise due to observation switching events. Switching occurs when the astronomy organisation has just finished observing an astronomy source and is about to commence the observation of another source. The switching results in a period during which terrestrial radio astronomy observations are not conducted. These periods are potential interference-free ISL transmit opportunities and are repeated for astronomy source observation patterns. The intelligent framework is used to determine the transmit duration.

In analysing D, we use extra data for observations conducted on 06/11/12 (period 1) and 07/11/12 (period 2) in addition to that of 16/02/13 (period 3) used in [8]. The observation in periods 1, 2, and 3 has data for 30, 85, and 95 observation epochs, respectively. In presenting data analysis results, the periods are classified as (1) early morning, 00:00:00.0–06:00:00.0, (2) morning, 08:42:05.3–12:02:55.5, (3) mid afternoon, 12:02:55.6–15:07:5.7, and (4) late afternoon, 15:07:15.8–18:34:55.7. The observation durations in these epochs are shown in Table 2.

The plots for the back-to-back connection for observations conducted in periods 1, 2, and 3 are shown in Figures 7, 8, and 9, respectively. From the results in Figures 7, 8, and 9, it can be observed that the analysis of additional data shows that ISL transmission can benefit from transmit opportunities due to the back-to-back connection duration. Further analysis shows that the average ISL back-to-back connection duration for periods 1, 2, and 3 is 49.5 seconds, 58.8 seconds, and 43.7 seconds, respectively.

The similar astronomy source observations strings for different observation dates are as follows:

(1) Source string 1: a_1 = PKS 1934-638, b_1 = PKS J0010-4153, c_1 = PKS J0022+0014, d_1 = PKS J0024-4202,

TABLE 2: Observation epochs used to investigate the back-to-back connection duration.

Observation epochs/periods	Early morning	Morning	Mid afternoon	Late afternoon
Period 1, 85 epochs				00:01:20.2–02:52:35.3
Period 2, 95 epochs, 00:01:20.2–02:52:35.3				
Period 3, 95 epochs	08:42:05.3–12:02:55.5	12:02:55.6–15:07:15.7		

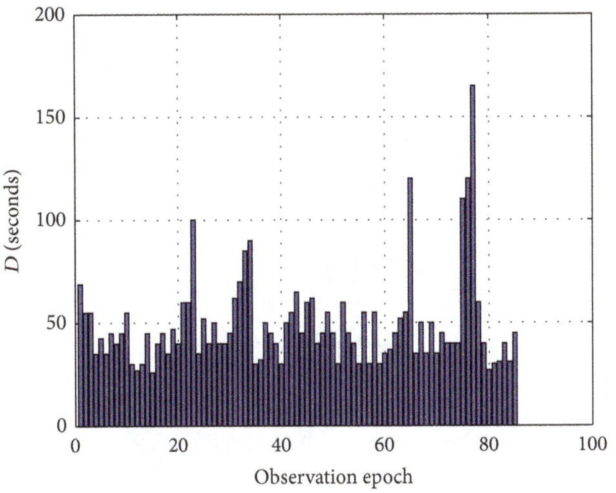

FIGURE 7: Back-to-back duration using period 1 observation data.

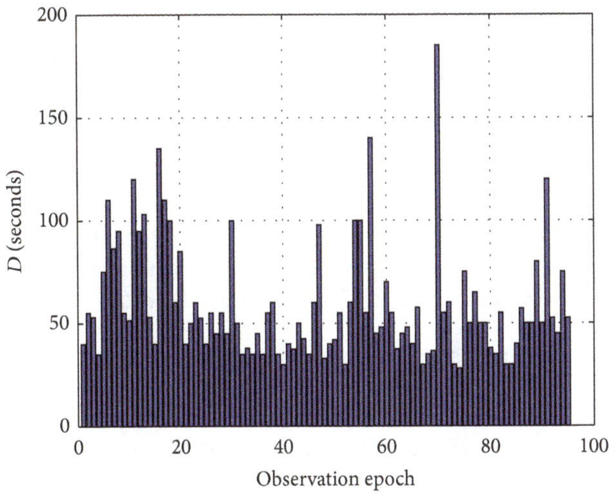

FIGURE 8: Back-to-back duration using period 2 observation data.

e_1 = PKS J0042-4414, f_1 = PKS J0059+0006, g_1 = PKS J0044-3530, and h_1 = 3C348

(2) Source string 2: a_2 = PKS J0240-2309, b_2 = PKS J0252-7104, c_2 = PKS J0303-6211, d_2 = PKS J0309-6058, e_2 = PKS J0318+1628, f_2 = PKS J0323+0534, g_2 = PKS J0351-2744, h_2 = PKS J0405-1308, i_2 = PKS J0409-1757, and j_2 = 3C123

(3) Source string 3: a_3 = PKS J0408-6544, b_3 = PKS J0420-6544, c_3 = PKS J0440-4333, d_3 = PKS J0442-0017,

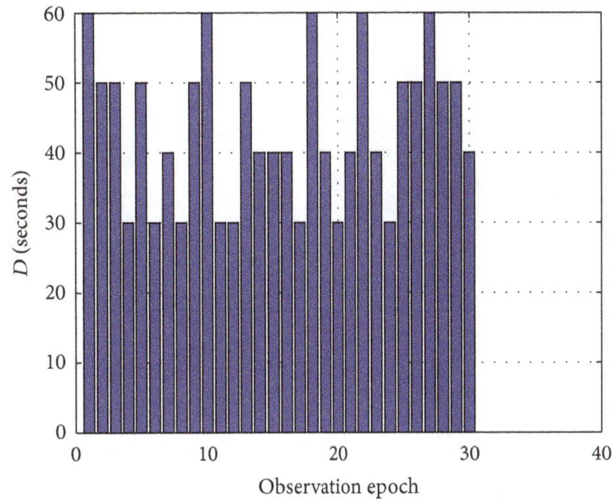

FIGURE 9: Back-to-back duration using period 3 observation data.

e_3 = PKS J0444-2809, f_3 = PKS J0453-2807, g_3 = PKS J0519-4546, h_3 = PKS J0534+1927, i_3 = PKS J0635-7516, j_3 = PKS J0744-0629, and k_3 = PKS J0831-1951

Source strings 1, 2, and 3 are observed on 14/11/2012, 28/10/2012, 06/11/2012, and 14/11/2012, 06/11/2012 and 14/11/2012, and 14/11/2012 and 06/11/2012, respectively. Assuming that similarity analysis in the intelligent framework takes up to 600 seconds, the achievable increment in the ISL transmit duration is 726 seconds. This increment is applicable to observations conducted on 14/11/2012, 28/10/2012, 06/11/2012, and 14/11/2012, 06/11/2012 and 14/11/2012, and 14/11/2012 and 06/11/2012, respectively.

Source substrings (a_1-g_1), (a_1-e_1), (a_2-i_2), and (a_2-d_2) repeatedly occur on 16/02/2012, 05/02/2013, 28/10/2012, and 06/11/2012 and 06/11/2012 and 05/02/2013. The transmit durations associated with substrings (a_1-g_1), (a_1-e_1), (a_2-i_2), and (a_2-d_2) are 360 seconds, 320 seconds, 340 seconds, and 190 seconds, respectively.

Source substrings (a_3-h_3), (i_3-k_3), (a_3-g_3), and (i_3-j_3) repeatedly occur on 28/10/2012 (2 epochs) and 06/11/2012 (2 epochs), respectively. The transmit durations associated with substrings (a_3-h_3), (i_3-k_3), (a_3-g_3), and (i_3-j_3) are 360 seconds, 160 seconds, 300 seconds, and 120 seconds, respectively. These strings and substrings show that the observations of some astronomy sources are repeated. Therefore, the interference protection framework is feasible. ISLs can exploit these transmit opportunities when they are aware of the astronomy database. In the event that the intelligent framework is

Table 3: Cost figures used to simulate ownership costs.

S/N	Component	Cost (USD)	Reference
1	Conversion of unused earth station	100,000	Hoare and Rawlings [10]
2	Internet link from CCE to SVB	1,000	
3	Control software per telescope	20,000 (20% of conversion cost)	Kemball and Cornwell [37]
4	Cyclostationary sensing module	11,413	LeMay [38]

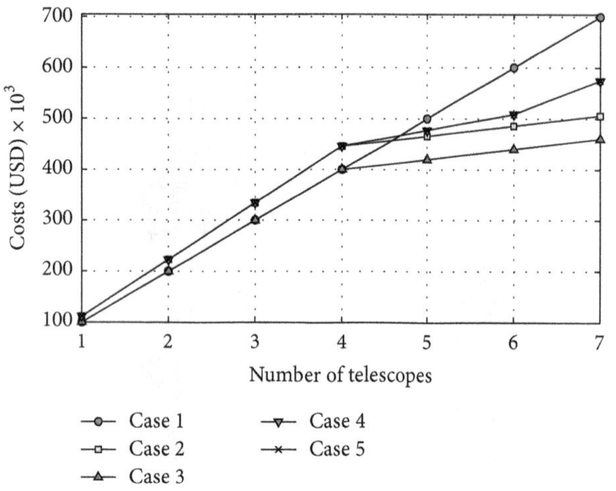

Figure 10: Terrestrial radio astronomy organisation ownership costs.

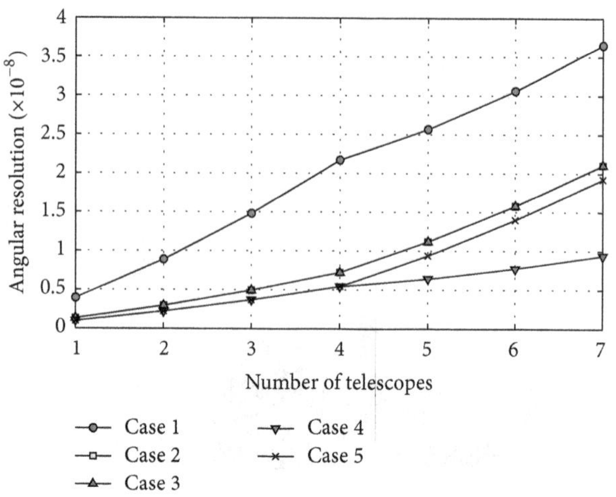

Figure 11: Terrestrial radio astronomy organisation angular resolution.

not incorporated, the maximum and minimum transmit durations are 49.5 seconds and 43.7 seconds, respectively. These maximum and minimum values describe the range of the back-to-back connection duration in the absence of the intelligent framework. The range of the average back-to-back connection duration is 43.7–49.5 seconds. When the proposed intelligent framework is used, the range of the average back-to-back connection duration is 769.7–775.5 seconds.

In addition, the costs of ownership and angular resolution of the terrestrial radio astronomy organisation that makes use of primary and secondary GTs are simulated and shown in Figures 10 and 11, respectively. The simulation is conducted for five cases, that is, Cases 1, 2, 3, 4, and 5. The cost of ownership is computed using the parameters given in Table 3.

The cases can be described as follows.

Case 1. The terrestrial radio astronomy organisation has seven primary GTs that are observed in the IEEE UHF band. The primary GTs do not have cyclostationary detectors and are not susceptible to terrestrial wireless network interference. This case describes the kind of scenario found in [11] because it does not use secondary GTs. The baseline is 2100 km.

Case 2. The terrestrial radio astronomy organisation uses four primary GTs and three secondary GTs. The secondary GTs are capable of processing satellite communications packets and astronomy source radio signals. Both primary and secondary GTs operated in the IEEE UHF band. The inclusion of secondary GTs doubles the baseline. Primary GTs incorporate a cyclostationary detection mechanism. Secondary GTs are not in terrestrial wireless network vicinity.

Case 3. The terrestrial radio astronomy organisation uses four primary GTs and three secondary GTs operating in the IEEE UHF and IEEE C bands, respectively. Primary GTs do not incorporate a cyclostationary detector for interference protection. Secondary GTs are unaffected by terrestrial wireless network interference. The baseline is 4200 km.

Case 4. The terrestrial radio astronomy organisation uses four primary GTs and three secondary GTs that operate in the IEEE UHF band. The primary and secondary GTs incorporate cyclostationary detectors. The baseline is 4200 km.

Case 5. The terrestrial radio astronomy organisation uses four primary GTs and three secondary GTs that operate in the IEEE UHF band and IEEE C band, respectively. The maximum separating distance between primary and secondary GTs is 4200 km. Both primary and secondary GTs incorporate the cyclostationary detection mechanism.

As shown in Figure 10, the use of secondary GTs alongside primary GTs reduces the terrestrial radio astronomy organisation's ownership costs. In the case where the number of GTs is between five and seven, the cost of ownership is maximum in Case 1 compared to Cases 2, 3, 4, and 5, respectively. When compared to the cost in Case 1, the use of secondary GTs in Cases 2, 3, 4, and 5 reduces ownership costs by 17.9%, 25.6%,

12.6%, and 12.6%, respectively. The costs in Cases 4 and 5 are equal because these cases are differentiated only by the GT observation frequency. The simulation parameters shown in Table 3 are not observation frequency dependent.

It can also be seen that the cost in Case 1 is lowest when there are up to four primary GTs compared to Cases 2, 3, 4, and 5, respectively. This is because the primary GTs in Case 1 do not have any cyclostationary module, multimode control software or Internet link costs. The cyclostationary module increases the cost in Cases 4 and 5. The inclusion of the control software and Internet link increases the cost in Cases 2, 3, 4, and 5. Therefore, the increased cost is due to the incorporation of the features proposed in this paper. The cost of adding the incorporated features increases the cost for the first four GTs for Cases 2, 4, and 5 when compared to Case 1. It does result in an increase for the cost comparison between Cases 1 and 3. This is because, in Cases 1 and 3, the first four GTs are primary GTs with similar functionalities. The average increase in costs of Cases 2, 4, and 5 compared to Case 1 is observed to be the same and equals 11.4%. The increase in cost is equal because only the cyclostationary module is added to the primary GT in Cases 2, 4, and 5 when compared to Case 1.

Further analysis of the results presented in Figure 11 shows that the angular resolution in Cases 2, 3, 4, and 5 outperforms that of Case 1 by 67.5%, 59.2%, 75%, and 66.7% on average, respectively. It can be seen that the opportunistic use of secondary GTs enhances the angular resolution. The incorporation of the cyclostationary module in Cases 4 and 5 also enhances the angular resolution because of the interference protection capability.

The improvement in angular resolution is larger when primary and secondary GTs use the IEEE UHF band as seen in Cases 2 and 4 because of the shorter wavelength. Nevertheless, secondary GT incorporation enhances angular resolution. Therefore, using a secondary GT that incorporates the cyclostationary detector improves the angular resolution when its inclusion increases the baseline. Hence, terrestrial radio astronomy organisations should combine primary and secondary GTs to reduce ownership costs and improve angular resolution.

5.2. Opportunistic Computing Scheme (OCS). The probability of success of the opportunistic computing scheme (OCS-SEP) is also investigated. OCS-SEP is dependent on the number of GTs and HPC computational units (CUs). The OCS-SEP is investigated for different number of GTs and HPC CUs. The simulation result is shown in Figure 12. An increase in the CU from 10 Kbits to 100 Kbits improves the OCS-SEP for the same number of GTs. An increase in CUs from 10 Kbits to 100 Kbits improves the OCS-SEP from 0.0387 to 0.9087. The OCS-SEP also improves when the CU increases from 100 Kbits to 1000 Kbits. An increase in the CU from 100 Kbits to 1000 Kbits increases the OCS-SEP from 0.0726 to 0.9085. Hence, the availability of more HPC CUs improves the OCS-SEP.

The influence of OCS on terrestrial wireless network throughput is also investigated. The terrestrial wireless network throughput is that obtained when OCS execution is

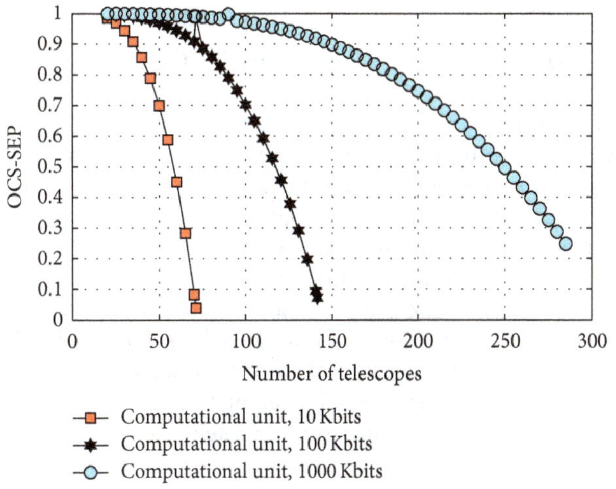

FIGURE 12: OCS-SEP for varying computational units and number of telescopes.

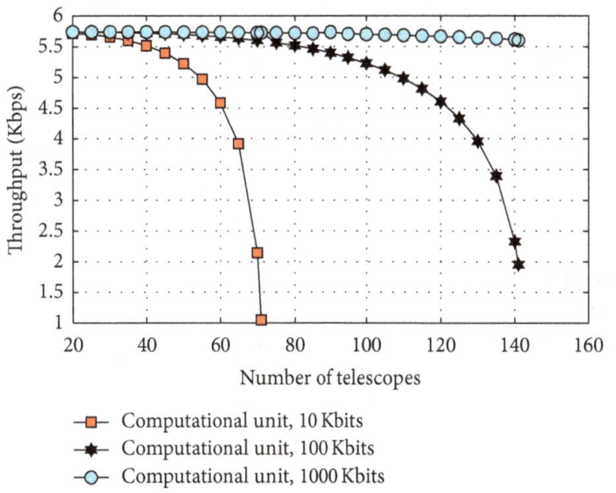

FIGURE 13: Relationship between terrestrial wireless network throughput and number of telescopes.

used to develop artificial neural network multiuser detectors. The simulation environment considers a scenario where three CRs share a channel to improve spectrum utilisation. The simulated OCS throughput is shown in Figure 13.

As seen in Figure 13, the throughput reduces with a decreasing HPC CU. The lowest throughput is obtained at epochs where available CUs cannot support the number of functional GTs. The achievable throughput is minimum when the HPC has 10 Kbits and there are 71 GTs and when the HPC has 100 Kbits and there are 141 GTs.

Further analysis shows that the CBS throughput is enhanced by 60.4% on average when the HPC CU is 1000 Kbits compared to when the HPC CU is 10 Kbits for up to 71 telescopes. The CBS throughput is enhanced by 59.3% on average when the HPC has 100 Kbits compared to when the HPC has 10 Kbits with up to 71 GTs. When there are up to 141 GTs, the increase in HPC CU from 100 Kbits to 1000 Kbits improves throughput by an average of 37.7%.

6. Conclusion

This paper addresses challenges affecting the future conduct of terrestrial radio astronomy observations. The terrestrial radio astronomy organisations being considered use ground telescopes realised by converting unused earth stations. The challenges are those of optimising terrestrial radio astronomy observations and enhancing high performance computing infrastructure utilisation. The optimisation goals aim to protect terrestrial radio astronomy observations from inter-satellite interference and enhance angular resolution. The interference mitigation framework utilises the similarities in the order of observed astronomy sources. The use of similarities protects ground telescopes from intersatellite link interference and increases intersatellite link connection duration. The paper also proposes the use of secondary telescopes to enhance angular resolution. In addition, the paper proposes the opportunistic computing scheme to enhance high performance computing infrastructure utilisation. The opportunistic computing scheme is a synergy between radio astronomy observations and cognitive base stations. It enhances cognitive base station autonomy. Investigations show that the intersatellite links that use cognitive radios with the proposed intelligent framework have an interference-free connection duration lying between 43.7 seconds and 49.5 seconds. The interference-free intersatellite link transmission duration is increased when the similarity in radio astronomy observation patterns is considered. Analysis also shows that the opportunistic computing scheme enables the realisation of cognitive base stations. In addition, the opportunistic computing scheme enhances terrestrial wireless network throughput. It is also shown that the use of secondary telescopes enhances angular resolution by up to 59% and reduces costs by up to 12.6%.

Competing Interests

The authors declare that they have no competing interests.

Acknowledgments

The authors would like to thank the researchers at the Communications Research Group, Department of Electrical Engineering, University of Cape Town, for their useful comments. The authors acknowledge the financial support of the National Research Foundation, South Africa (NRF SA), Telkom South Africa, Jasco/TeleSciences, and the Department of Trade and Industry/Technology and Human Resources Programme (DTI/THRIP).

References

[1] F. Combes, "The square kilometer array: cosmology, pulsars and other physics with the SKA," in *Proceedings of the 2nd International Summer School on Intelligent Signal Processing for Frontier Research and Industry*, pp. 1–16, Paris, France, July 2014.

[2] R. Deane, Z. Paragi, M. Jarvis et al., "Multiple supermassive black hole systems: SKA's future leading role," Proceedings of Science, Advancing Astrophysics with Square Kilometre Array, pp. 1–11, https://arxiv.org/abs/1501.01238.

[3] T. E. Bell, "Waiting for gravity: gravitational waves would give astronomers an unprecedented view into acts of astronomical violence," *IEEE Spectrum*, vol. 43, no. 7, pp. 40–46, 2006.

[4] T. Pultarova, "Telecommunications-Space Tycoons go head to head over mega satellite network," *Engineering and Technology*, vol. 10, no. 2, p. 20, 2015.

[5] M. DePasquade and J. Bradford, *Space Works: Nano/Micro Satellite Market Assessment, Revision C*, 2013.

[6] M. Poblet, "Affordable telecommunications: a new digital economy is calling," *Australian Journal of Telecommunications and the Digital Economy*, vol. 1, no. 1, article 12, 2013.

[7] http://www.nanosats.eu/.

[8] A. A. Periola and O. E. Falowo, "Interference protection of radio astronomy services using cognitive radio spectrum sharing models," in *Proceedings of the European Conference on Networks and Communications*, pp. 301–305, Paris, France, July 2015.

[9] H. R. Klockner, S. Rawlings, I. Heywood et al., "Goonhilly: a new site for e-Merlin and the EVN," in *Proceedings of the European VLBI Network Symposium and EVN Users Meeting: VLBI and the New Generation of Radio Arrays*, pp. 1–10, Manchester, UK, September 2010, http://www.jive.nl.

[10] M. Hoare and S. Rawlings, "Recycling for radio astronomy," *Astronomy and Geophysics*, vol. 53, no. 1, pp. 1.19–1.21, 2012.

[11] D. Barbosa, M. Bergano, V. A. R. M. Ribeiro et al., "Design, environmental and sustainability constraints of new African observatories: the example of the Mozambique Radio Astronomy Observatory," in *Proceedings of the AFRICON*, pp. 1–5, Pointe-Aux-Piments, Mauritius, September 2013.

[12] M. J. Gaylard, M. F. Bietenholz, L. Combrinck et al., "An African VLBI network of radio telescopes," http://arxiv.org/abs/1405.7214.

[13] B. D. Asabere, M. J. Gaylard, C. Horellou, H. Winkler, and T. Jarrett, "Radio Astronomy in Africa: the case of Ghana," pp. 1–6, 2015, http://arxiv.org/pdf/1503.08850.pdf.

[14] L. Woodburn, T. Natusch, S. Weston et al., "Conversion of a New Zealand 30-metre telecommunications antenna into a radio telescope," *Publications of the Astronomical Society of Australia*, vol. 32, article e017, 14 pages, 2015.

[15] D. Barbosa, A. J. Boonstra, R. Aguiar, A. V. Ardene, De. S. Vela, and L. V. Montnegro, "A Sustainable approach to large ICT Science based infrastructures; the case for Radio Astronomy," in *Proceedings of the IEEE International Energy Conference (ENERGYCON '14)*, pp. 668–674, Cavtat, Croatia, May 2014.

[16] M. N. Derahman, K. Dimyati, A. M. Mohammadi, and M. K. Abdullah, "Improvement of decision making protocol for Duty Cycle Division Multiplexing (DCDM) system," in *Proceedings of the 2nd International Conference on Future Networks (ICFN '10)*, pp. 155–158, Sanya, China, January 2010.

[17] A. Malekmohammadi, M. K. Abdullah, A. F. Abas, G. A. Mahdiraji, and M. Mokhtar, "Absolute Polar Duty Cycle Division Multiplexing (APDCDM); technique for wireless communications," in *Proceedings of the International Conference on Computer and Communication Engineering (ICCCE '08)*, pp. 617–620, Kuala Lumpur, Malaysia, May 2008.

[18] K. Rele, D. Roberson, B. Zhang et al., "A two-tiered cognitive radio system for interference identification in 2.4 GHz ISM band," in *Proceedings of the 7th IEEE Consumer Communications and Networking Conference (CCNC '10)*, pp. 1–5, IEEE, Las Vegas, Nev, USA, January 2010.

[19] P. Woudt, R. P. Fender, R. Armstrong, and C. Carignan, "Early science with the Karoo Array Telescope test array KAT-7," *South African Journal of Science*, vol. 109, no. 7-8, 2 pages, 2013.

[20] V. Pankonin and R. M. Price, "Radio astronomy and spectrum management: the impact of WARC-78," *IEEE Transactions on Communications*, vol. 29, no. 8, pp. 1228–1237, 1941.
[21] P. J. Waterman, "Conducting radio astronomy in the EMC environment," *IEEE Transactions on Electromagnetic Compatibility*, vol. 26, no. 1, pp. 29–33, 1984.
[22] D. R. DeBoer, S. L. C. Pol, M. M. Davis et al., "Radio frequencies: policy and management," *IEEE Transactions on Geoscience and Remote Sensing*, vol. 51, no. 10, pp. 4897–4907, 2013.
[23] T. E. Gergely, "Spectrum access for the passive services: the past and the future," *Proceedings of the IEEE*, vol. 102, no. 3, pp. 393–398, 2014.
[24] T. E. Gergely, "The SKA, RFI and ITU regulations," in *Proceedings of the Workshop on Mitigation of Radio Frequency Interference in Radio Astronomy*, pp. 1–10, Penticton, Canada, July 2004.
[25] I. Kucuk, "Radio astronomy in turkey: site selection studies for radio quiet zones," in *Proceedings of the RFI Mitigation Workshop*, Groningen, The Netherlands, March 2010.
[26] R. Umar, Z. Z. Abidin, Z. A. Ibrahim, Z. Rosli, and N. Noorazlan, "Selection of radio astronomical observation sites and its dependence on human generated RFI," *Research in Astronomy and Astrophysics*, vol. 14, no. 2, pp. 241–248, 2014.
[27] R. Umar, Z. Z. Abidin, and Z. A. Ibrahim, "The importance of site selection for radio astronomy," *Journal of Physics: Conference Series*, vol. 539, no. 1, Article ID 012009, 2014.
[28] N. H. Sabri, R. Umar, W. Z. A. Wan Mokhtar et al., "Preliminary study of vehicular traffic effect on radio signal for radio," *Jurnal Teknologi*, vol. 75, no. 1, pp. 313–318, 2015.
[29] M. J. Bentum, A. J. Boonstra, and W. A. Baan, "Impact of cognitive radio on radio astronomy," in *Proceedings of the RFI Mitigation Workshop*, pp. 1–7, Groningen, The Netherlands, March 2010.
[30] J. M. Ford and K. D. Buch, "Mitigation techniques in radio astronomy," in *Proceedings of the IEEE International Geoscience and Remote Sensing Symposium*, pp. 231–234, Québec, Canada, July 2014.
[31] H. Asadi, H. Volos, M. M. Marefat, and T. Bose, "Metacognition and the next generation of cognitive radio engines," *IEEE Communications Magazine*, vol. 54, no. 1, pp. 76–82, 2016.
[32] T. V. D. Zant, M. Kouw, and L. Schomaker, "Generative artificial intelligence," in *Philosophy and Theory of Artificial Intelligence*, V. C. Muller, Ed., vol. 5, pp. 103–120, 2012.
[33] K. Tsakgaris, A. Bantouna, and P. Demestichas, "Self-organizing maps for advanced learning in cognitive radio systems," *Computers and Electrical Engineering*, vol. 38, no. 4, pp. 852–870, 2012.
[34] K. P. Bagadi and S. Das, "Multiuser detection in SDMA-OFDM wireless communication system using complex multilayer perceptron neural network," *Wireless Personal Communications*, vol. 77, no. 1, pp. 21–39, 2014.
[35] P. Andreani, *ALMA Cycle 3 Proposer's Guide and Capabilities*, Doc 3.2, Ver 1.9, 2015.
[36] S. J. Almalki and J. Yuan, "A new modified Weibull distribution," *Reliability Engineering and System Safety*, vol. 111, pp. 164–170, 2013.
[37] A. J. Kemball and T. J. Cornwell, "A simple model of software costs for the square kilometre array," *Experimental Astronomy*, vol. 17, no. 1–3, pp. 317–327, 2004.
[38] B. LeMay, Agilent Technologies Introduces Basic Spectrum Analyzer for Budget-Driven Applications, http://www.agilent.com/about/newsroom/presrel/2012/30nov-em12148.html.

5

Analysis of the Conformally Flat Approximation for Binary Neutron Star Initial Conditions

In-Saeng Suh,[1] Grant J. Mathews,[2] J. Reese Haywood,[2] and N. Q. Lan[2,3,4]

[1] *Center for Astrophysics, Department of Physics and Center for Research Computing, University of Notre Dame, Notre Dame, IN 46556, USA*
[2] *Center for Astrophysics, Department of Physics, University of Notre Dame, Notre Dame, IN 46556, USA*
[3] *Hanoi National University of Education, 136 Xuan Thuy, Hanoi, Vietnam*
[4] *Joint Institute for Nuclear Astrophysics (JINA), University of Notre Dame, Notre Dame, IN 46556, USA*

Correspondence should be addressed to Grant J. Mathews; gmathews@nd.edu

Academic Editor: Ignazio Licata

The spatially conformally flat approximation (CFA) is a viable method to deduce initial conditions for the subsequent evolution of binary neutron stars employing the full Einstein equations. Here we analyze the viability of the CFA for the general relativistic hydrodynamic initial conditions of binary neutron stars. We illustrate the stability of the conformally flat condition on the hydrodynamics by numerically evolving ~100 quasicircular orbits. We illustrate the use of this approximation for orbiting neutron stars in the quasicircular orbit approximation to demonstrate the equation of state dependence of these initial conditions and how they might affect the emergent gravitational wave frequency as the stars approach the innermost stable circular orbit.

1. Introduction

The epoch of gravitational wave astronomy has now begun with the first detection [1, 2] of the merger of binary black holes by Advanced LIGO [3]. Now that the first ground based gravitational wave detection has been achieved, observations of binary neutron star mergers should soon be forthcoming. This is particularly true as other second generation observatories such as Advanced VIRGO [4] and KAGRA [5] will soon be online. In addition to binary black holes, neutron star binaries are thought to be among the best candidate sources gravitational radiation [6, 7]. The number of such systems detectable by Advanced LIGO is estimated [7–14] to be of order several events per year based upon observed close binary-pulsar systems [15, 16]. There is a difference between neutron star mergers and black hole mergers; however, in that neutron star mergers involve the complex evolution of the matter hydrodynamic equations in addition to the strong gravitational field equations. Hence, one must carefully consider both the hydrodynamic and field evolution of these systems.

To date there have been numerous attempts to calculate theoretical templates for gravitational waves from compact binaries based upon numerical and/or analytic approaches (see, e.g., [17–26]). However, most approaches utilize a combination of post-Newtonian (PN) techniques supplemented with quasicircular orbit calculations and then applying full GR for only the last few orbits before disruption. In this paper we analyze the hydrodynamic evolution in the spatially conformally flat metric approximation (CFA) as a means to compute stable initial conditions beyond the range of validity of the PN regime, that is, near the last stable orbits. We establish the numerical stability of this approach based upon many orbit simulations of quasicircular orbits. We show that one must follow the stars for several orbits before a stable quasicircular orbit can be achieved. We also illustrate the equation of state (EoS) dependence of the initial conditions and associated gravitational wave emission.

When binary neutron stars are well separated, the post-Newtonian (PN) approximation is sufficiently accurate [27]. In the PN scheme, the stars are often treated as point masses, either with or without spin. At third order, for example, it has been estimated [28–30] that the error due to assuming the stars are point masses is less than one orbital rotation [28] over the ~16,000 cycles that pass through the LIGO detector frequency band [7]. Nevertheless, it has been noted in many works [25, 31–42] that relativistic hydrodynamic effects might be evident even at the separation (~10–100 km) relevant to the LIGO window.

Indeed, the templates generated by PN approximations, unless carried out to fifth and sixth order [28, 29], may not be accurate unless the finite size and proper fluid motion of the stars are taken into account. In essence, the signal emitted during the last phases of inspiral depends upon the finite size and the equation of state (EoS) through the tidal deformation of the neutron stars and the cut-off frequency when tidal disruption occurs.

Numeric and analytic simulations [43–51] of binary neutron stars have explored the approach to the innermost stable circular orbit (ISCO). While these simulations represent some of the most accurate to date, simulations generally follow the evolution for a handful of orbits and are based upon initial conditions of quasicircular orbits obtained in the conformally flat approximation. Accurate templates of gravitational radiation require the ability to stably and reliably calculate the orbit initial conditions. The CFA provides a means to obtain accurate initial conditions near the ISCO.

The spatially conformally flat approximation to GR was first developed in detail in [32]. That original formulation, however, contained a mathematical error first pointed out by Flanagan [52] and subsequently corrected in [34]. This error in the solution to the shift vector led to a spurious NS crushing prior to merger. The formalism discussed below is for the corrected equations. Here, we discuss the hydrodynamic solutions as developed in [31–34, 53, 54]. This CFA formalism includes much of the nonlinearity inherent in GR and leads set of coupled, nonlinear, elliptic field equations that can be evolved stably. We also note that an alternative spectral method solution to the CFA configurations was developed by [55, 56], and approaches beyond the CFA have also been proposed [48]. However, our purpose here is to clarify the viability of the hydrodynamic solution without the imposition of a Killing vector or special symmetry. This approach is the most adaptable, for example, to general initial conditions such as that of arbitrarily elliptical orbits and/or arbitrarily spinning neutron stars.

Here, we summarize the original CFA approach and associated general relativistic hydrodynamics formalism developed in [32, 34, 53, 54] and illustrate that it can produce stable initial conditions anywhere between the post-Newtonian to ISCO regimes. We quantify how long this method takes to converge to quasiequilibrium and demonstrate the stability by subsequently integrating up to ~100 orbits for a binary neutron star system. We also analyze the EoS dependence of these quasicircular initial orbits and show how these orbits can be used to make preliminary estimates [57] of the gravitational wave signal for the initial conditions.

This paper is organized as follows. In Section 2 the basic method is summarized and in Section 3 a number of code tests are performed in the quasiequilibrium circular orbit limit to demonstrate the stability of the technique. The EoS dependence of the initial conditions and associated gravitational wave frequency are analyzed in Section 4. Conclusions are presented in Section 5.

2. Method

2.1. Field Equations. The solution of the field equations and hydrodynamic equations of motion were first solved in three spatial dimensions and explained in detail in the 1990s in [31, 32] and subsequently further reviewed in [53, 58]. Here, we present a brief summary to introduce the variables relevant to the present discussion.

One starts with the slicing of space-time into the usual one-parameter family of hypersurfaces separated by differential displacements in a time-like coordinate as defined in the (3 + 1) ADM formalism [59, 60].

In Cartesian x, y, z isotropic coordinates, proper distance is expressed as

$$ds^2 = -\left(\alpha^2 - \beta_i\beta^i\right)dt^2 + 2\beta_i dx^i dt + \phi^4\delta_{ij}dx^i dx^j, \quad (1)$$

where the lapse function α describes the differential lapse of proper time between two hypersurfaces. The quantity β_i is the shift vector denoting the shift in space-like coordinates between hypersurfaces. The curvature of the metric of the 3-geometry is described by a position-dependent conformal factor ϕ^4 times a flat-space Kronecker delta ($\gamma_{ij} = \phi^4\delta_{ij}$). This conformally flat condition (together with the maximal slicing gauge, $\mathrm{tr}\{K_{ij}\} = 0$) requires [60]

$$2\alpha K_{ij} = D_i\beta_j + D_j\beta_i - \frac{2}{3}\delta_{ij}D_k\beta^k, \quad (2)$$

where K_{ij} is the extrinsic curvature tensor and D_i are 3-space covariant derivatives. This conformally flat condition on the metric provides a numerically valid initial solution to the Einstein equations. The vanishing of the Weyl tensor for a stationary system in three spatial dimensions guarantees that a conformally flat solution to the Einstein equations exists.

One consequence of this conformally flat approximation to the three-metric is that the emission of gravitational radiation is not explicitly evolved. Nevertheless, one can extract the gravitational radiation signal and the back reaction via a multipole expansion [32, 61]. An application to the determination of the gravitational wave emission from the quasicircular orbits computed here is given in [57]. The advantage of this approximation is that conformal flatness stabilizes and simplifies the solution to the field equations.

As a third gauge condition, one can choose separate coordinate transformations for the shift vector and the hydrodynamic grid velocity to separately minimize the field and matter motion with respect to the coordinates. This set of gauge conditions is key to the present application. It allows one to stably evolve up to hundreds and even thousands of binary orbits without the numerical error associated with the frequent advocating of fluid through the grid.

Given a distribution of mass and momentum on some manifold, then one first solves the constraint equations of general relativity at each time for a fixed distribution of matter. One then evolves the hydrodynamic equations to the next time step. Thus, at each time slice a solution to the relativistic field equations and information on the hydrodynamic evolution is obtained.

The solutions for the field variables ϕ, α, and β^i reduce to simple Poisson-like equations in flat space. The Hamiltonian constraint [60] is used to solve for the conformal factor ϕ [32, 62]

$$\nabla^2\phi = -2\pi\phi^5\left[W^2\left(\rho\left(1+\epsilon\right)+P\right)-P+\frac{1}{16\pi}K_{ij}K^{ij}\right]. \quad (3)$$

In the Newtonian limit, the RHS is dominated [32] by the proper matter density ρ, but in strong fields and compact neutron stars there are also contributions from the internal energy density ϵ, pressure P, and extrinsic curvature. The source is also significantly enhanced by the generalized curved-space Lorentz factor W

$$W = \alpha U^t = \left[1+\frac{\sum U_i^2}{\phi^4}\right]^{1/2}, \quad (4)$$

where U^t is the time component of the relativistic four velocity and U_i are the covariant spatial components. This factor, W, becomes important near the last stable orbit as the specific kinetic energy of the stars rapidly increases.

In a similar manner [32, 62], the Hamiltonian constraint, together with the maximal slicing condition, provides an equation for the lapse function,

$$\nabla^2\left(\alpha\phi\right) = 2\pi\alpha\phi^5\left[3W^2\left[\rho\left(1+\epsilon\right)+P\right]-2\rho\left(1+\epsilon\right) + 3P+\frac{7}{16\pi}K_{ij}K^{ij}\right]. \quad (5)$$

Finally, the momentum constraints yields [60] an elliptic equation for the shift vector [34, 52],

$$\nabla^2\beta^i = \frac{\partial}{\partial x^i}\left(\frac{1}{3}\nabla\cdot\beta\right)+4\pi\rho_3^i, \quad (6)$$

where

$$\rho_3^i = 4\alpha\phi^4 S_i + \frac{1}{4\pi}\frac{\partial\ln\left(\alpha/\phi^6\right)}{\partial x^j}\left(\frac{\partial\beta^i}{\partial x^j}+\frac{\partial\beta^j}{\partial x^i}-\frac{2}{3}\delta_{ij}\frac{\partial\beta^k}{\partial x^k}\right). \quad (7)$$

Here S_i are the spatial components of the momentum density one-form as defined below.

We note that, in early applications of this approach, the source for the shift vector contained a spurious term due to an incorrect transformation between contravariant and covariant forms of the momentum density as was pointed out in [34, 52]. As illustrated in those papers, this was the main reason why early hydrodynamic calculations induced a controversial additional compression on stars causing them to collapse to black holes prior to inspiral [31]. This problem no longer exists in the formulation summarized here.

2.2. Relativistic Hydrodynamics. To solve for the fluid motion of the system in curved-space time it is convenient to use an Eulerian fluid description [63]. One begins with the perfect fluid stress-energy tensor in the Eulerian observer rest frame,

$$T_{\mu\nu} = Pg_{\mu\nu}+\left(\rho\left(1+\epsilon\right)+P\right)U_\mu U_\nu, \quad (8)$$

where U_μ is the relativistic four velocity one-form.

By introducing the usual set of Lorentz contracted state variables it is possible to write the relativistic hydrodynamic equations in a form which is reminiscent of their Newtonian counterparts [63]. The hydrodynamic state variables are the coordinate baryon mass density

$$D = W\rho; \quad (9)$$

the coordinate internal energy density

$$E = W\rho\epsilon; \quad (10)$$

the spatial three velocity

$$V^i = \alpha\frac{U_i}{\phi^4 W}-\beta^i; \quad (11)$$

and the covariant momentum density

$$S_i = \left(D+E+PW\right)U_i. \quad (12)$$

In terms of these state variables, the hydrodynamic equations in the CFA are as follows: the equation for the conservation of baryon number takes the form

$$\frac{\partial D}{\partial t} = -6D\frac{\partial\log\phi}{\partial t}-\frac{1}{\phi^6}\frac{\partial}{\partial x^j}\left(\phi^6 DV^j\right). \quad (13)$$

The equation for internal energy evolution becomes

$$\frac{\partial E}{\partial t} = -6\left(E+PW\right)\frac{\partial\log\phi}{\partial t}-\frac{1}{\phi^6}\frac{\partial}{\partial x^j}\left(\phi^6 EV^j\right) - P\left[\frac{\partial W}{\partial t}+\frac{1}{\phi^6}\frac{\partial}{\partial x^j}\left(\phi^6 WV^j\right)\right]. \quad (14)$$

Momentum conservation takes the form

$$\frac{\partial S_i}{\partial t} = -6S_i\frac{\partial\log\phi}{\partial t}-\frac{1}{\phi^6}\frac{\partial}{\partial x^j}\left(\phi^6 S_i V^j\right)-\alpha\frac{\partial P}{\partial x^i} + 2\alpha\left(D+E+PW\right)\left(W-\frac{1}{W}\right)\frac{\partial\log\phi}{\partial x^i} + S_j\frac{\partial\beta^j}{\partial x^i}-W\left(D+E+PW\right)\frac{\partial\alpha}{\partial x^i} - \alpha W\left(D+\Gamma E\right)\frac{\partial\chi}{\partial x^i}, \quad (15)$$

where the last term in (15) is the contribution from the radiation reaction potential χ as defined in [32, 57]. In the construction of quasistable orbit initial conditions, this term is set to zero. Including this term would allow for a calculation of the orbital evolution via gravitational wave emission in the CFA. However, there is no guarantee that this is a sufficiently accurate solution to the exact Einstein equations. Hence, the CFA is useful for the construction of initial conditions.

2.3. Angular Momentum and Orbital Frequency. In the quasi-circular orbit approximation (neglecting angular momentum in the radiation field), this system has a Killing vector corresponding to rotation in the orbital plane. Hence, for these calculations the angular momentum is well defined and given by an integral over the space-time components of the stress-energy tensor [64]; that is,

$$J^{ij} = \int \left(T^{i0}x^j - T^{j0}x^i\right) dV. \tag{16}$$

Aligning the z-axis with the angular momentum vector then gives

$$J = \int \left(xS^y - yS^x\right) dV. \tag{17}$$

To find the orbital frequency detected by a distant observer corresponding to a fixed angular momentum we employ a slightly modified derivation of the orbital frequency compared to that of [53]. In asymptotically flat coordinates the angular frequency detected by a distant observer is simply the coordinate angular velocity; that is, one evaluates

$$\omega \equiv \frac{d\phi}{dt} = \frac{U^\phi}{U^0}. \tag{18}$$

In the ADM conformally flat (3+1) curved space, our only task is then to deduce U^ϕ from code coordinates. For this we make a simple polar coordinate transformation keeping our conformally flat coordinates, so

$$U^\phi = \Lambda^\phi_\nu U^\nu = \frac{xU^y - yU^x}{x^2 + y^2}. \tag{19}$$

Now, the code uses covariant four velocities, $U_i = g_{i\nu}U^\nu = \beta_i U^0 + \phi^4 U^i$. This gives $U^i = U_i\beta_i(W/\alpha)/\phi^4$. Finally, one must density weight and volume average ω over the fluid differential volume elements. This gives

$$\omega = \frac{\int d^3x\phi^2 (D + \Gamma E)\left[(\alpha/W)\left(xU_y - yU_x\right) - \left(x\beta_y - y\beta_x\right)\right]/\left(x^2 + y^2\right)}{\int d^3x\phi^6 (D + \Gamma E)}. \tag{20}$$

This form differs slightly from that of [53] but leads to the similar results.

A key additional ingredient, however, is the implementation of a grid three velocity V^i_G that minimizes the matter motion with respect to U_i and β_i. Hence, the total angular frequency to a distant observer $\omega_{\rm tot} = \omega + \omega_G$, and in the limit of rigid corotation, $\omega_{\rm tot} \to \omega_G$, where $\omega_G = xV^y + yV^x$.

For the orbit calculations illustrated here we model corotating stars, that is, no spin in the corotating frame. This minimizes matter motion on the grid. However, we note that there is need at the present time of initial conditions for arbitrarily spinning neutron stars and the method described here is equally capable of supplying those initial conditions.

As a practical approach the simulation [32] of initial conditions is best run first with viscous damping in the hydrodynamics for sufficiently long time (a few thousand cycles) to relax the stars to a steady state. Subsequently, one can run with no damping. In the present illustration we examine stars at large separation which are in quasiequilibrium circular orbits and stable hydrodynamic configurations. In the initial relaxation phase the orbits are circularized by damping any radial velocity components. During the evolution, the rigorous conservation of angular momentum is imposed by adjusting the orbital angular velocity in (20) such that (17) remains constant. The simulated orbits described in this work span the time from the last several minutes up to orbit inspiral. Here, we illustrate the stability of the multiple orbit hydrodynamic simulation and examine where the initial conditions for the strong field orbit dynamics computed here deviates from the post-Newtonian regime.

3. Code Validation

3.1. Code Tests. To evolve stars at large separation distance it is best [53] to decompose the grid into a high resolution domain with a fine matter grid around the stars and a coarser domain with an extended grid for the fields. Figure 1 shows a schematic of this decomposition from [54].

As noted in [53] it is best to keep the number of zones across each star between 25 and 40 [54]. This keeps the error in the numerics below 0.5%. It has also been pointed out [53] that an artificial viscosity (AV) shock capturing scheme has an advantage over Riemann solvers because only about half as many zones are required to accurately resolve the stars when an AV scheme is employed compared to a Riemann solver.

The time steps dt_n are taken as the minimum of the time step as determined by several conditions. Each condition is also multiplied by a number less than one to accommodate the nonlinear nature of the equations.

The first condition is known as the Courant condition, that is, a search over all zones i for the zone with the minimum sound crossing time:

$$dt_1 = \min\left(\frac{dx^i_b}{C^i_s}\right), \tag{21}$$

where C^i_s is the sound speed in the ith zone.

The Newtonian sound speed is given by the variation of pressure with density. In relativity the wave speed is given instead by the adiabatic derivative of the pressure with respect

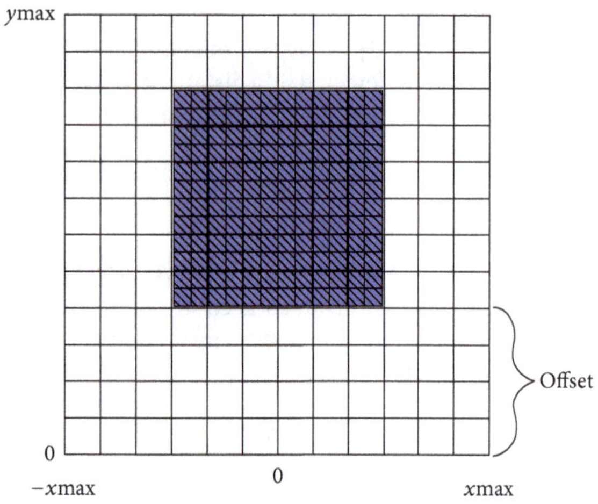

FIGURE 1: Schematic representation of the field and hydrodynamics grid used in the simulations. The inner blue grid represents the higher resolution matter grid and the outer white grid represents the field grid.

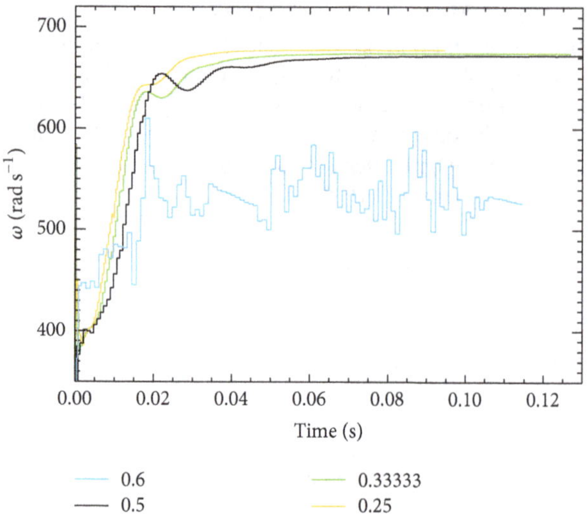

FIGURE 2: Comparison of the orbital angular velocity ω versus time for different values of the Courant parameter k. As can be seen, the simulations with $k = 0.25$–0.5 result in stable runs that converge to the same value, implying that a smaller k or equivalently a smaller δt is not necessary and would only use extra CPU time. For comparison, we plot a simulation with $k = 0.6$ to show that the stability is lost for $k > 0.5$.

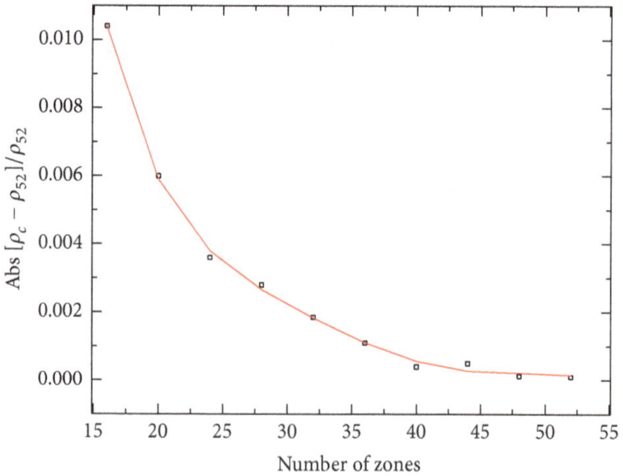

FIGURE 3: Plot of the error in the central density versus the number of zones across the star. It is clear that there is only a 1% error with ≈15 zones across the star. Increasing the number of zones across the star so that there are >35 zones across the star produces less than a 0.1% error.

to the relativistic inertial density. In terms of relativistic variables the local sound speed in zone i then becomes [53]

$$C_s = \sqrt{\frac{\Gamma_i (\Gamma_i - 1) E_i}{D_i + \Gamma_i E_i}}. \tag{22}$$

The second condition is a search for the zone with minimum time for material to flow across a zone

$$dt_2 = \min\left(\frac{dx_a^i}{|V^i|}\right). \tag{23}$$

This constraint is introduced to ensure stability and accuracy in the numerical advection calculation.

The third condition is introduced to maintain stability of the artificial viscosity algorithm. The viscous equations are analogous to a diffusion equation in four velocity with a diffusion coefficient $D \approx k_1 dx^i |\delta U^i|$, where $\delta U^i \equiv U^{i+1} - U^i$. We then can define a minimum viscous diffusion time across a zone derived from the stability condition for explicit diffusion equations

$$dt_3 = \frac{1}{4}\min\left(\frac{W^i dx_b^i}{|\delta U^i|}\right). \tag{24}$$

The time step dt is then assigned to be some fraction (referred to as the Courant parameter) of the minimum of these three conditions

$$dt = k \min(dt_1, dt_2, dt_3), \tag{25}$$

Obviously, smaller values for k increase the accuracy of the calculation but also increase the computation time.

Figure 2 shows a plot of orbital velocity versus time for various values of the Courant parameter. This figure establishes that the routines for the hydrodynamics are stable (e.g., changing the Courant condition has little to effect) as long as $k \leq 0.5$.

Figure 3 illustrates the difference between the central density ρ_c and the central density (ρ_{52}) at the highest resolution of 52 zones for single isolated stars. This is expressed as the fractional error as a function of the number of zones across a star. This plot was calculated using the relatively soft MW EoS, that is, the zero temperature and zero neutrino chemical potential limit of the EoS that has previously been used to model core-collapse supernovae [32, 53, 65].

This figure illustrates that here is only a 1% error in central density with ≈15 zones across the star, while increasing the

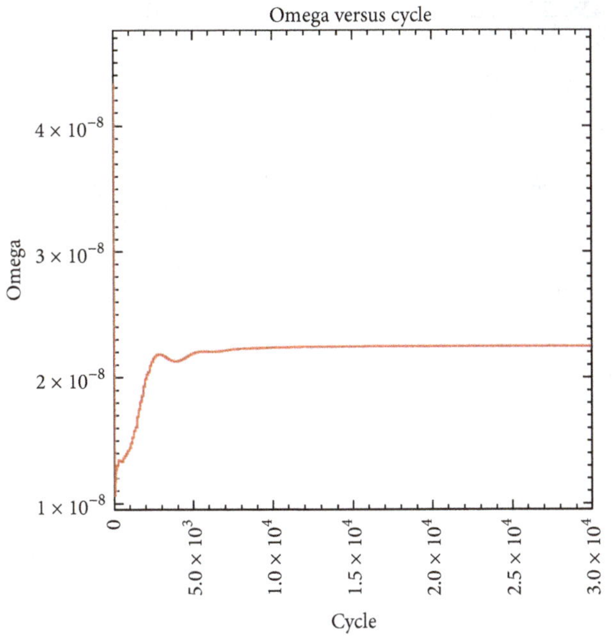

FIGURE 4: Plot of the orbital angular velocity, ω in geometrized units (cm^{-1}) versus cycle number. When ω stops changing the simulation has reached a circular binary orbit solution. This run was extended to over 30,000 cycles, corresponding to ≈20 orbits.

number of zones across the star to >35 produces less than a 0.1%. In the illustrations below we maintain $k = 0.5$ and ≈25 zones across each star as the best choice for both speed and accuracy needed to compute stable orbital initial conditions.

3.2. Orbit Stability. As an illustration of the orbit stability Figure 4 shows results from a simulation [54] in which the angular momentum was fixed at $J = 2.7 \times 10^{11}\,\mathrm{cm}^2$ and the Courant parameter set to $k = 0.5$. For this orbital calculation the MW EoS was employed and each star was fixed at a baryon mass of $M_B = 1.54\,\mathrm{M}_\odot$ and a gravitational mass in isolation of $M_G = 1.40\,\mathrm{M}_\odot$.

Figure 4 shows the evolution of the orbital angular velocity ω versus computational cycle for the first 30,000 code cycles corresponding to ≈20 orbits. The stars were initially placed on the grid using a solution of the TOV equation in isotropic coordinates for an isolated star. The stars were initially set to be corotating but were allowed to settle into their binary equilibrium. Notice that it takes ~5,000 cycles, corresponding to ~3 orbits, just to approach the quasiequilibrium binary solution. Indeed, the stars continued to gradually compact and slightly increase in orbital frequency until ~10 orbits; afterward, the stars were completely stable.

Figure 5 shows the contours of the lapse function α (roughly corresponding to the gravitational potential) and corresponding density profiles at cycle numbers, 0, 5200, and 25800 (≈0, 5, and 19 orbits). Figure 6 shows the contours of central density and the orientation of the binary orbit corresponding to these cycle numbers. One can visibly see from these figures the relaxation of the stars after the first few orbits and the stability of the density profiles after multiple orbits.

We note, however, that this orbit is on the edge of the ISCO. As such it could be unstable to inspiral even after many orbits. Figures 7 and 8 further illustrate this point. In these simulations various angular momenta were computed with a slightly higher neutron star mass ($M_b = 1.61\,\mathrm{M}_\odot$ and $M_g = 1.44\,\mathrm{M}_\odot$), but the same MW EoS. In this case the binary system was followed for nearly 100 orbits.

Figure 7 illustrates orbital angular frequency versus cycle number for three representative angular momenta bracketing the ISCO. The orbital separation for the lowest angular momentum ($J = 2.7 \times 10^{11}\,\mathrm{cm}^{-2}$) shown on Figure 7 is just inside the ISCO. Hence, even though it requires about 10 orbits before inspiral, the orbit is eventually unstable. Similarly, Figure 8 shows the central density versus number of orbits for 6 different angular momenta. Here one can see that orbits with $J \geq 3.0 \times 10^{11}\,\mathrm{cm}^{-2}$ are stable. Indeed, for these cases, after about the first 3 orbits the orbits continue with almost no discernible change in orbit frequency or central density.

As mentioned previously, the numerical relativistic neutron binary simulations of [43] all start with initial data that are subsequently evolved in a different manner compared to those with which they were created. One conclusion that may be drawn from the above set of simulations, however, is that the initial data must be evolved for ample time (>3 orbit) for the stars to reach a true quasiequilibrium binary configuration. That has not always been done in the literature.

4. Sensitivity of Initial Condition Orbital Parameters to the Equation of State

4.1. Equations of State. One hope in the forthcoming detection of gravitational waves is that a sensitivity exists to the neutron star equation of state. For illustration we utilize several representative equations of state often employed in the literature. These span a range from relatively soft to stiff nuclear matter. These are used to illustrate the EoS dependence of the initial conditions. One EoS often employed is that of a polytrope, that is, $p = K\rho^\Gamma$, with $\Gamma = 2$, where in cgs units, $K = 0.0445(c^2/\rho_n)$, and $\rho_n = 2.3 \times 10^{14}\,\mathrm{g\,cm}^{-3}$. These parameters, with $\rho_c = 4.74 \times 10^{14}\,\mathrm{g\,cm}^{-3}$, produce an isolated star having a radius = 17.12 km and baryon mass = $1.5\,\mathrm{M}_\odot$. As noted in previous sections we utilize the zero temperature and zero neutrino chemical potential MW EoS [32, 53, 65]. The third is the equation of state developed by Lattimer and Swesty [66] with two different choices of compressibility, one having compressibility $K = 220\,\mathrm{MeV}$ and the other having $K = 375\,\mathrm{MeV}$. We denote these as LS 220 and LS 375. The fourth EoS has been developed by Glendenning [67]. This EoS has $K = 240\,\mathrm{MeV}$, which is close to the experimental value [68]. We denote this EoS as GLN. Table 1 illustrates [54] the properties of isolated neutron stars generated with each EoS. For each case the baryon mass was chosen to obtain a gravitational mass for each star of $1.4\,\mathrm{M}_\odot$.

In Figure 9 we plot the equation of state index Γ versus density, ρ, for the various EoSs considered here. These are

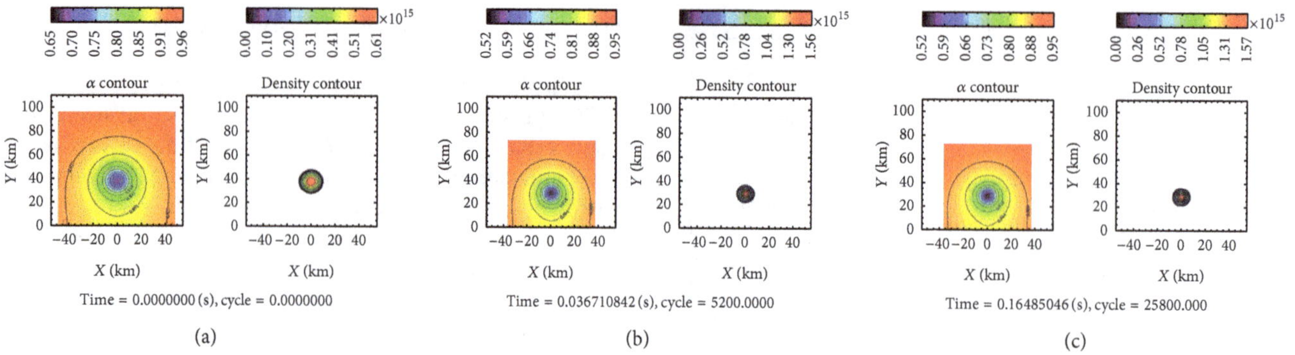

FIGURE 5: Contours of the lapse function (left) and central density (right) at cycle numbers 0 (a), 5,200 (b), and 25,800 (c) corresponding to roughly 0, 5, and 19 orbits.

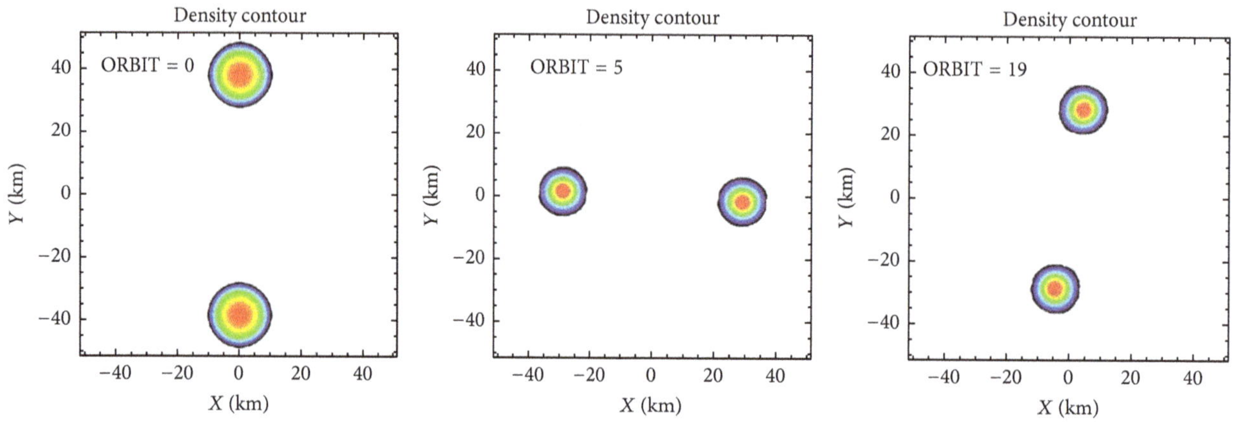

FIGURE 6: Contours of the central density for the binary system at the approximate number of orbits as labelled.

TABLE 1: Table presenting central density, baryon mass, and gravitational mass for the five adopted equations of state.

EoS	ρ_c ($\times 10^{15}$ g cm^{-3})	M_B ($M_\odot$)	M_G ($M_\odot$)
GLN	1.56	1.54	1.40
MW	1.39	1.54	1.40
LS 220	0.698	1.54	~1.40
LS 375	0.492	1.54	~1.40
$\Gamma = 2$ polytrope	0.474	1.50	1.40

compared to the simple polytropic $\Gamma = 2$ EoS often employed in the literature.

4.2. EoS Dependence of the Initial Condition Orbit Parameters. Table 2 summarizes the initial condition orbit parameters [54] at various fixed angular momenta for the various equations of state. In the case of orbits unstable to merger, this table lists the orbit parameters just before inspiral. These orbits span a range in specific angular momenta J/M_0^2 of ~5 to 10. The equations of state listed in Tables 1 and 2 are in approximate order of increasing stiffness from the top to the bottom.

As expected, the central densities are much higher for the relatively soft MW and GLN equations of state. Also, the orbit angular frequencies are considerably lower for the extended mass distributions of the stiff equations of state than for the more compact soft equations of state. These extended mass distributions induce a sensitivity of the emergent gravitational wave frequencies and amplitude due to the strong dependence of the gravitational wave frequency to the quadrupole moment of the mass distribution.

4.3. Gravitational Wave Frequency. The physical processes occurring during the last orbits of a neutron star binary are currently a subject of intense interest. As the stars approach their final orbits it is expected that the coupling of the orbital motion to the hydrodynamic evolution of the stars in the strong relativistic fields could provide insight into various physical properties of the coalescing system [58, 69]. In this regard, careful modeling of the initial conditions is needed which includes both the nonlinear general relativistic and hydrodynamic effects as well as a realistic neutron star equation of state.

Figure 10 shows the EoS sensitivity of the gravitational wave frequency $f = \omega/\pi$ as a function of proper separation d_p between the stars for the various orbits and equations of state summarized in Table 2. These are compared with the circular orbit condition in the (post)$^{5/2}$-Newtonian approximation, hereafter PN, analysis of reference [70]. In that paper

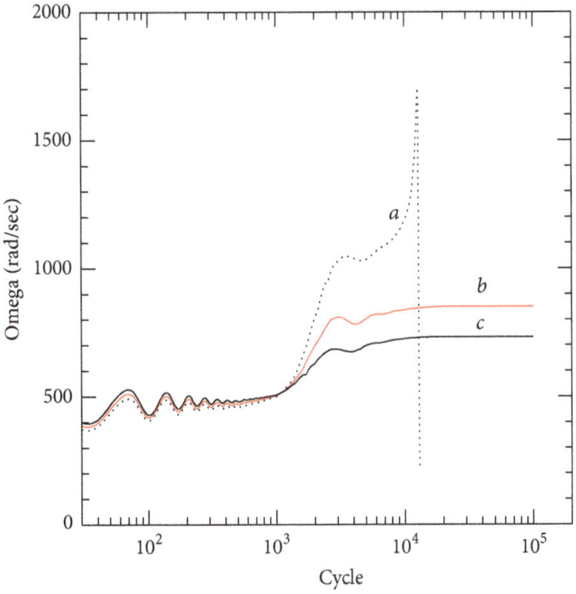

FIGURE 7: Plot of the orbital angular velocity, ω, versus cycle. When ω stops changing with time the simulation has reached a circular binary orbit solution. The run (a for $J = 2.7 \times 10^{11}$ cm^2) goes over ~10 obits and then becomes unstable to inspiral and merger after ~10^4 cycles. The stable two runs (b for $J = 2.8 \times 10^{11}$ cm^{-2} and c for $J = 2.9 \times 10^{11}$ cm^{-2}) were run for 100,000 cycles and ≈100 orbits.

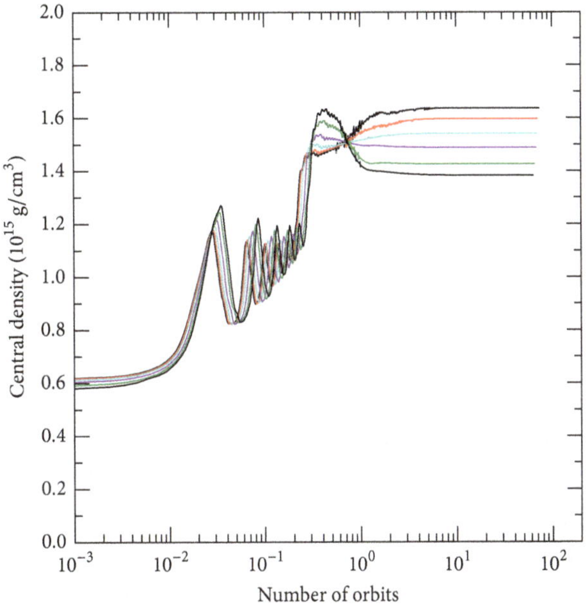

FIGURE 8: Plot of the central density, ρ_c, versus the number of orbit. The solid lines from top to bottom are for $J = 3.0, 3.2, 3.4, 3.6, 3.8, 4.0 \times 10^{11}$ cm^2.

a search was made for the inner most stable circular orbit in the absence of radiation reaction terms in the equations of motion. This is analogous to the calculations performed here which also analyzes orbit stability in the absence of radiation reaction.

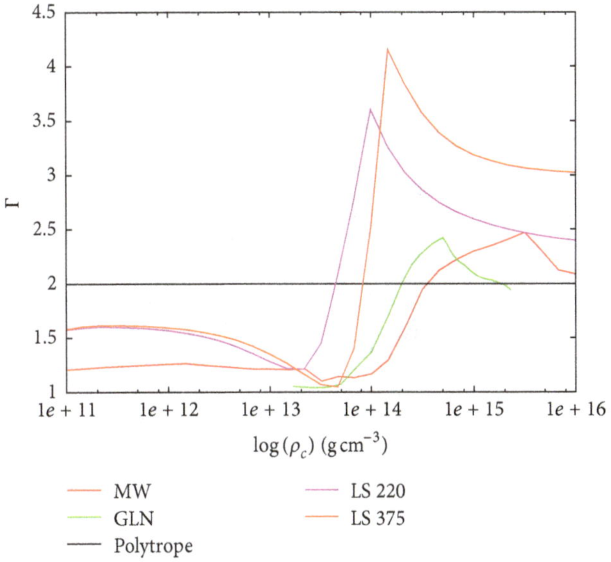

FIGURE 9: EoS index Γ versus central density for various equations of state. Large Γ implies a stiff EoS.

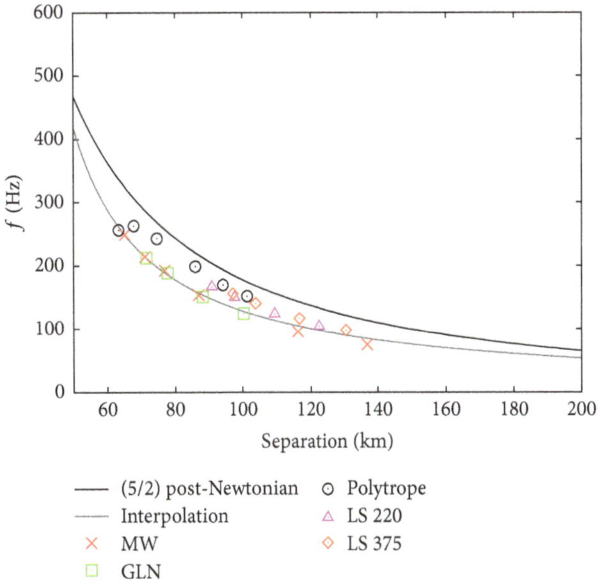

FIGURE 10: Computed gravitational wave frequency, f, versus proper separation for each EoS as labelled. The black line corresponds to the (post)$^{5/2}$-Newtonian estimate. Frequencies obtained from the stiff and polytropic equations of state do not deviate by more than ~10% from the PN prediction until a frequency greater than ~300 Hz. The grey line is an extrapolation of the frequencies obtained using the soft MW and GLN EoSs. These begin to deviate by more than 10% from the PN prediction at a frequency of ~100 Hz.

In the (post)$^{5/2}$-Newtonian equations of motion, a circular orbit is derived by setting time derivatives of the separation, angular frequency, and the radial acceleration to zero. This leads to the circular orbit condition [70]

$$\omega_0^2 = \frac{mA_0}{d_h^3}, \tag{26}$$

TABLE 2: Orbital parameters for each EoS.

EoS	J (cm^2)	ω (rad s^{-1})	d_p (km)	d_c (km)	M_{ADM} ($M_\odot$)	ρ_c (g cm^{-3})
GLN	2.7×10^{11}	666.5	71.62	57.67	1.390	1.73×10^{15}
	2.8×10^{11}	592.34	77.82	62.81	1.391	1.69×10^{15}
	3.0×10^{11}	475.05	88.06	73.53	1.394	1.61×10^{15}
	3.2×10^{11}	391.75	100.34	84.31	1.396	1.56×10^{15}
MW	2.6×10^{11}	780.92	65.22	51.52	1.391	1.67×10^{15}
	2.7×10^{11}	671.85	71.18	57.24	1.393	1.62×10^{15}
	2.8×10^{11}	602.80	76.94	61.86	1.394	1.60×10^{15}
	3.0×10^{11}	482.30	86.91	72.36	1.396	1.55×10^{15}
	3.5×10^{11}	300.46	116.13	100.8	1.399	1.44×10^{15}
	3.8×10^{11}	235.72	136.93	119.74	1.401	1.39×10^{15}
LS 220	2.7×10^{11}	523.59	90.77	77.34	1.403	7.18×10^{14}
	2.8×10^{11}	472.08	97.53	83.08	1.404	7.14×10^{14}
	3.0×10^{11}	389.96	109.78	94.84	1.405	7.06×10^{14}
	3.2×10^{11}	327.04	122.51	107.10	1.407	6.98×10^{14}
LS 375	2.7×10^{11}	490.09	97.09	83.92	1.404	5.00×10^{14}
	2.8×10^{11}	442.40	103.95	90.04	1.405	4.98×10^{14}
	3.0×10^{11}	366.67	116.65	102.50	1.406	4.95×10^{14}
	3.2×10^{11}	307.80	130.72	115.60	1.407	4.92×10^{14}
Polytrope	1.8×10^{11}	804.70	63.30	51.20	1.395	6.78×10^{14}
	2.1×10^{11}	826.03	67.85	55.18	1.396	7.00×10^{14}
	2.3×10^{11}	762.37	74.64	61.72	1.397	6.55×10^{14}
	2.5×10^{11}	624.33	85.87	72.71	1.399	6.24×10^{14}
	2.6×10^{11}	532.83	94.04	80.45	1.400	6.17×10^{14}
	2.7×10^{11}	477.19	101.34	86.95	1.400	6.05×10^{14}

where ω_0 is the circular orbit frequency and $m = 2M_G^0$, d_h is the separation in harmonic coordinates, and A_0 is a relative acceleration parameter which for equal mass stars becomes

$$A_0 = 1 - \frac{3}{2}\frac{m}{d_h}\left[3 - \frac{77}{8}\frac{m}{d_h} + (\omega_0 d_h)^2\right] + \frac{7}{4}(\omega_0 d_h)^2. \quad (27)$$

Equations (26) and (27) can be solved to find the orbit angular frequency as a function of harmonic separation d_h. The gravitational wave frequency is then twice the orbit frequency, $f = \omega_0/\pi$.

Although this is a gauge-dependent comparison, for illustration we show in Figure 10 the calculated gravitational wave frequency compared to the PN expectation as a function of proper binary separation distance up to 200 km. One should keep in mind, however, that there is some uncertainty in associating proper distance with the PN parameter (m/r). Hence, a comparison with the PN results is not particularly meaningful. It is nevertheless instructive to consider the difference in the numerical results as the stiffness of the EoS is varied. The polytropic and stiff EoSs begin to deviate (by >10%) from the softer equations of state (MW and GLN) for a gravitational wave frequency as low as ~100 Hz and more or less continue to deviate as the stars approach the ISCO at higher frequencies.

Indeed, a striking feature of Figure 10 is that as the stars approach the ISCO, the frequency varies more slowly with diminishing separation distance for the softer equations of state. A gradual change in frequency can mean more orbits in the LIGO window and hence a stronger signal to noise (cf. [57]).

Also, for a soft EoS the orbit becomes unstable to inspiral at a larger separation. At least part of the difference between the soft and stiff EoSs can be attributed to the effects of the finite size of the stars which are more compact for the soft equations of state [37].

We note that, for comparable angular momenta, our results are consistent with the EoS sensitivity study of [37] based upon a set of equations of state parameterized by a segmented polytropic indices and an overall pressure scale. Their calculations, however, were based upon two independent numerical relativity codes. The similarity of their simulations to the results in Table 2 further confirms the broad validity of the CFA approach when applied to initial conditions. For example, their orbit parameters are summarized in Table II of [37]. Their softest EoS is the Bss221 which corresponds to an adiabatic index of $\Gamma = 2.4$ for the core and a baryon mass of 1.501 $M_\odot$ and an ADM mass of 1.338 $M_\odot$ per star for a specific angular momentum of 1.61×10^{11} cm^2 (in our units) with a corresponding gravitational wave frequency of 530 Hz at a proper separation of 46 km. This EoS is comparable to the polytropic, MW, and GLN EoSs shown on Figure 10. For example, our closest orbit with the $\Gamma = 2$ polytropic EoS corresponds to a specific angular momentum of 1.8×10^{11} cm^2 and an ADM mass of 1.39 $M_\odot$ compared to their ADM mass of 1.34 $M_\odot$ at $J = 1.6 \times 10^{11}$ cm^2 for the same baryon mass of 1.5 $M_\odot$. Although, for the softer EoSs, their results are for a closer orbit than the numerical points given on Figure 10, an extension of the grey line fit to the numerical simulations of the soft EoSs would predict a frequency of 540 Hz at the same proper separation of 46 km compared to 530 Hz in the Bss221 simulation of [37].

The main parameter characterizing the last stable orbit in the post-Newtonian calculation is the ratio of coordinate separation to total mass (in isolation) d_h/m. The analogous quantity in our nonperturbative simulation is proper separation to gravitational mass, d_P/m. The separation corresponding to the last stable orbit in the post-Newtonian analysis does not occur until the stars have approached $6.03m$. For $M_G^0 = 1.4\,M_\odot$ stars, this would correspond to a separation distance of about 25 km. In the results reported here the last stable orbit occurs somewhere just below $7.7m_G^0$ at a proper separation distance of $d_P \approx 30$ km for both the polytropic and the MW stars.

5. Conclusions

The relativistic hydrodynamic equilibrium in the CFA remains as a viable approach to calculate the initial conditions for calculations of binary neutron stars. In this paper we have illustrated that one must construct initial conditions that have run for at least several orbits before equilibrium is guaranteed. We have demonstrated that beyond the first several orbits the equations are stable over many orbits (~100). We also have shown that such multiple orbit simulations are valuable as a means to estimate the location of the ISCO prior to a full dynamical calculation. Moreover, we have examined the sensitivity of the initial condition orbit parameters and initial gravitational wave frequency to the equation of state. We have illustrated how the initial condition orbital properties (e.g., central densities, orbital velocities, and binding energies) and location of the ISCO are significantly affected by the stiffness of the EoS.

Competing Interests

The authors declare that there is no conflict of interests regarding the publication of this paper.

Acknowledgments

The work at the University of Notre Dame (Grant J. Mathews) was supported by the US Department of Energy under Nuclear Theory Grant DE-FG02-95-ER40934 and by the University of Notre Dame Center for Research Computing. One of the authors (N. Q. Lan) was also supported in part by the National Science Foundation through the Joint Institute for Nuclear Astrophysics (JINA) at UND and in part by the Vietnam Ministry of Education (MOE). N. Q. Lan would also like to thank the Yukawa Institute for Theoretical Physics for their hospitality during a visit where part of this work was done.

References

[1] P. B. Abbot, R. Abbott, T. D. Abbott et al., "Observation of gravitational waves from a binary black hole merger," *Physical Review Letters*, vol. 116, no. 6, Article ID 061102, 2016.

[2] P. B. Abbot, R. Abbott, T. D. Abbott et al., "GW151226: observation of gravitational waves from a 22-solar-mass binary black hole coalescence," *Physical Review Letters*, vol. 116, Article ID 241103, 2016.

[3] J. Asai, B. P. Abbottl, R. Abbott et al., "Advanced LIGO," *Classical and Quantum Gravity*, vol. 32, no. 7, Article ID 074001, 2015.

[4] F. Acernese, M. Agathos, K. Agatsuma et al., "Advanced Virgo: a 2nd generation interferometric gravitational wave detector," *Classical and Quantum Gravity*, vol. 32, no. 2, Article ID 024001, 2015.

[5] K. Somiya, "Detector configuration of KAGRA–the Japanese cryogenic gravitational-wave detector," *Classical and Quantum Gravity*, vol. 29, no. 12, 2012.

[6] LIGO Scientific Collaboration, http://www.ligo.org.

[7] K. S. Thorne, "Compact stars in binaries," in *Proceedings of the IAU Symposium*, J. van Paradijs, E. P. J. van den Heuvel, and E. Kuulkers, Eds., vol. 165, p. 153, Kluwer Academic Publishers, 1996.

[8] G. M. Harry, "Advanced LIGO: the next generation of gravitational wave detectors," *Classical and Quantum Gravity*, vol. 27, no. 8, Article ID 084006, 2010.

[9] F. A. Rasio and S. L. Shapiro, "Black holes of $D = 5$ supergravity," *Classical and Quantum Gravity*, vol. 16, no. 1, p. 1, 1999.

[10] M. Bailes, "Pulsar velocities," in *Proceedings of the IAU Symposium, Compact Stars in Binaries*, J. van Paradijs, E. P. J. van den Heuvel, and E. Kuulkers, Eds., vol. 165, p. 213, Hague, the Netherlands, August 1994.

[11] A. V. Tutukov and L. R. YungelSon, "The merger rate of neutron star and black hole binaries," *Monthly Notices of the Royal Astronomical Society*, vol. 260, no. 3, pp. 675–678, 1993.

[12] E. S. Phinney, "The rate of neutron star binary mergers in the universe—minimal predictions for gravity wave detectors," *The Astrophysical Journal*, vol. 380, pp. L17–L21, 1991.

[13] V. Kalogera, C. Kim, D. R. Lorimer et al., "The cosmic coalescence rates for double neutron star binaries," *The Astrophysical Journal*, vol. 601, pp. L179–L182, 2004, Erratum to: *The Astrophysical Journal*, vol. 614, pp. L137-L138, 2004.

[14] J. Abadie, B. P. Abbott, R. Abbott et al., "Predictions for the rates of compact binary coalescences observable by ground-based gravitational-wave detectors," *Classical and Quantum Gravity*, vol. 27, no. 17, Article ID 173001, 2010.

[15] M. Burgay, N. D'Amico, A. Possenti et al., "An increased estimate of the merger rate of double neutron stars from observations of a highly relativistic system," *Nature*, vol. 426, no. 6966, pp. 531–533, 2003.

[16] J. M. Lattimer, "The nuclear equation of state and neutron star masses," *Annual Review of Nuclear and Particle Science*, vol. 62, no. 1, pp. 485–515, 2012.

[17] T. A. Apostolatos, "Construction of a template family for the detection of gravitational waves from coalescing binaries," *Physical Review D*, vol. 54, no. 4, pp. 2421–2437, 1996.

[18] S. Droz and E. Poisson, "Gravitational waves from inspiraling compact binaries: second post-Newtonian waveforms as search templates," *Physical Review D*, vol. 56, no. 8, pp. 4449–4454, 1997.

[19] B. S. Sathyaprakash, "Mother templates for gravitational wave chirps," *Classical and Quantum Gravity*, vol. 17, no. 23, pp. L157–L162, 2000.

[20] A. Buonanno, Y. Chen, and M. Vallisneri, "Detection template families for gravitational waves from the final stages of binary-black-hole inspirals: nonspinning case," *Physical Review D*, vol. 67, no. 2, Article ID 024016, 2003.

[21] S. Bose, "Search templates for stochastic gravitational-wave backgrounds," *Physical Review D*, vol. 71, no. 8, Article ID 082001, 7 pages, 2005.

[22] C. D. Ott, A. Burrows, L. Dessart, and E. Livne, "A New Mechanism for Gravitational-Wave Emission in Core-Collapse Supernovae," *Physical Review Letters*, vol. 96, no. 20, Article ID 201102, 2006.

[23] P. Ajith, N. Fotopoulos, S. Privitera, A. Neunzert, N. Mazumder, and A. J. Weinstein, "Effectual template bank for the detection of gravitational waves from inspiralling compact binaries with generic spins," *Physical Review D*, vol. 89, no. 8, Article ID 084041, 2014.

[24] F. Pannarale, E. Berti, K. Kyutoku, B. D. Lackey, and M. Shibata, "Gravitational-wave cutoff frequencies of tidally disruptive neutron star-black hole binary mergers," *Physical Review D*, vol. 92, no. 8, Article ID 081504, 2015.

[25] M. Agathos, J. Meidam, W. Del Pozzo et al., "Constraining the neutron star equation of state with gravitational wave signals from coalescing binary neutron stars," *Physical Review D—Particles, Fields, Gravitation and Cosmology*, vol. 92, no. 2, Article ID 023012, 2015.

[26] J. A. Clark, A. Bauswein, N. Stergioulas, and D. Shoemaker, "Observing gravitational waves from the post-merger phase of binary neutron star coalescence," *Classical and Quantum Gravity*, vol. 33, no. 8, 2016.

[27] B. Allen, W. G. Anderson, P. R. Brady, D. A. Brown, and J. D. Creighton, "FINDCHIRP: an algorithm for detection of gravitational waves from inspiraling compact binaries," *Physical Review D*, vol. 85, no. 12, 2012.

[28] L. Blanchet, "Gravitational radiation from post-Newtonian sources and inspiralling compact binaries," *Living Reviews in Relativity*, vol. 5, article 3, 2002.

[29] L. Blanchet, "Gravitational radiation from post-newtonian sources and inspiralling compact binaries," *Living Reviews in Relativity*, vol. 17, article no. 2, 2014.

[30] C. K. Mishra, K. Arun, and B. R. Iyer, "Third post-Newtonian gravitational waveforms for compact binary systems in general orbits: instantaneous terms," *Physical Review D*, vol. 91, no. 8, Article ID 084040, 2015.

[31] J. R. Wilson and G. J. Mathews, "Instabilities in close neutron star binaries," *Physical Review Letters*, vol. 75, no. 23, article 4161, 1995.

[32] J. R. Wilson, G. J. Mathews, and P. Marronetti, "Relativistic numerical model for close neutron-star binaries," *Physical Review D - Particles, Fields, Gravitation and Cosmology*, vol. 54, no. 2, article 1317, 1996.

[33] G. J. Mathews, P. Marronetti, and J. R. Wilson, "Relativistic hydrodynamics in close binary systems: analysis of neutron-star collapse," *Physical Review D*, vol. 58, no. 4, Article ID 043003, 1998.

[34] G. J. Mathews and J. R. Wilson, "Revised relativistic hydrodynamical model for neutron-star binaries," *Physical Review D*, vol. 61, no. 12, Article ID 127304, 2000.

[35] L. Baiotti, T. Damour, B. Giacomazzo, A. Nagar, and L. Rezzolla, "Analytic modeling of tidal effects in the relativistic inspiral of binary neutron stars," *Physical Review Letters*, vol. 105, no. 26, Article ID 261101, 2010.

[36] S. Bose, S. Ghosh, and P. Ajith, "Systematic errors in measuring parameters of non-spinning compact binary coalescences with post-Newtonian templates," *Classical and Quantum Gravity*, vol. 27, no. 11, Article ID 114001, 2010.

[37] J. S. Read, L. Baiotti, J. D. E. Creighton et al., "Matter effects on binary neutron star waveforms," *Physical Review D*, vol. 88, no. 4, Article ID 044042, 2013.

[38] A. Maselli, L. Gualtieri, and V. Ferrari, "Constraining the equation of state of nuclear matter with gravitational wave observations: tidal deformability and tidal disruption," *Physical Review D*, vol. 88, no. 10, Article ID 104040, 2013.

[39] A. Bauswein and N. Stergioulas, "Unified picture of the post-merger dynamics and gravitational wave emission in neutron star mergers," *Physical Review D*, vol. 91, no. 12, Article ID 124056, 2015.

[40] A. Bauswein, N. Stergioulas, and H.-T. Janka, "Neutron-star properties from the postmerger gravitational wave signal of binary neutron stars," *Physics of Particles and Nuclei*, vol. 46, no. 5, pp. 835–838, 2015.

[41] C. L. Fryer, K. Belczynski, E. Ramirez-Ruiz, S. Rosswog, G. Shen, and A. W. Steiner, "The fate of the compact remnant in neutron star mergers," *Astrophysical Journal*, vol. 812, no. 1, article no. 24, 2015.

[42] T. Dietrich, N. Moldenhauer, N. K. Johnson-McDaniel et al., "Binary neutron stars with generic spin, eccentricity, mass ratio, and compactness: quasi-equilibrium sequences and first evolutions," *Physical Review D—Particles, Fields, Gravitation and Cosmology*, vol. 92, no. 12, Article ID 124007, 2015.

[43] M. D. Duez, P. Marronetti, S. L. Shapiro, and T. W. Baumgarte, "Hydrodynamic simulations in $\mathbf{3+1}$ general relativity," *Physical Review. D. Third Series*, vol. 67, no. 2, 024004, 22 pages, 2003.

[44] P. Marronetti, M. D. Duez, S. L. Shapiro, and T. W. Baumgarte, "Dynamical determination of the innermost stable circular orbit of binary neutron stars," *Physical Review Letters*, vol. 92, no. 14, Article ID 141101, 2004.

[45] M. Miller, P. Gressman, and W. Suen, "Towards a realistic neutron star binary inspiral: initial data and multiple orbit evolution in full general relativity," *Physical Review D*, vol. 69, no. 6, Article ID 064026, 2004.

[46] M. Miller, "Accuracy requirements for the calculation of gravitational waveforms from coalescing compact binaries in numerical relativity," *Physical Review D*, vol. 71, no. 10, Article ID 104016, 2005.

[47] M. Miller, "General-relativistic decompression of binary neutron stars during dynamic inspiral," *Physical Review D*, vol. 75, no. 2, Article ID 024001, 2007.

[48] K. Uryu, F. Limousin, J. L. Friedman, E. Gourgoulhon, and M. Shibata, "Binary neutron stars: equilibrium models beyond spatial conformal flatness," *Physical Review Letters*, vol. 97, no. 17, Article ID 171101, 2006.

[49] K. Kiuchi, Y. Sekiguchi, M. Shibata, and K. Taniguchi, "Long-term general relativistic simulation of binary neutron stars collapsing to a black hole," *Physical Review D*, vol. 80, no. 6, Article ID 064037, 2009.

[50] S. Bernuzzi, A. Nagar, T. Dietrich, and T. Damour, "Modeling the dynamics of tidally interacting binary neutron stars up to the merger," *Physical Review Letters*, vol. 114, no. 16, Article ID 161103, 2015.

[51] R. De Pietri, A. Feo, F. Maione, and F. Löffler, "Modeling equal and unequal mass binary neutron star mergers using public codes," *Physical Review D*, vol. 93, no. 6, Article ID 064047, 2015.

[52] É. É. Flanagan, "Possible explanation for star-crushing effect in binary neutron star simulations," *Physical Review Letters*, vol. 82, no. 7, pp. 1354–1357, 1999.

[53] J. R. Wilson and G. J. Mathews, *Relativistic numerical hydrodynamics*, Cambridge Monographs on Mathematical Physics, Cambridge University Press, Cambridge, UK, 2003.

[54] J. R. Haywood, *Numerical relativistic hydrodynamic simulations of neutron stars [Ph.D. thesis]*, University of Notre Dame, 2006.

[55] E. Gourgoulhon, P. Grandclément, K. Taniguchi, J. Marck, and S. Bonazzola, "Quasiequilibrium sequences of synchronized and irrotational binary neutron stars in general relativity: method and tests," *Physical Review D*, vol. 63, no. 6, Article ID 064029, 2001.

[56] K. Taniguchi, E. Gourgoulhon, and S. Bonazzola, "Quasiequilibrium sequences of synchronized and irrotational binary neutron stars in general relativity. II. Newtonian limits," *Physical Review D*, vol. 64, Article ID 064012, 2001.

[57] N. Q. Lan, I.-S. Suh, G. J. Mathews, and J. R. Haywood, "Gravitational waveforms from multiple-orbit simulations of binary neutron stars," *Communications in Physics*, vol. 25, no. 4, pp. 299–308, 2015.

[58] G. J. Mathews and J. R. Wilson, "Binary–induced neutron star compression, heating, and collapse," *The Astrophysical Journal*, vol. 482, no. 2, pp. 929–941, 1997.

[59] R. Arnowitt, S. Deser, and C. W. Misner, "Republication of: the dynamics of general relativity," *General Relativity and Gravitation*, vol. 40, no. 9, pp. 1997–2027, 2008.

[60] J. W. York Jr. and L. Smarr, *Sources of Gravitational Radiation*, Cambridge University Press, Cambridge, UK, 1979.

[61] K. S. Thorne, "Multipole expansions of gravitational radiation," *Reviews of Modern Physics*, vol. 52, no. 2, pp. 299–339, 1980.

[62] C. R. Evans, *A method for numerical relativity: simulation of axisymmetric gravitational collapse and gravitational radiation generation [Ph.D. thesis]*, University of Texas at Austin, 1984.

[63] J. R. Wilson, *Sources of Gravitational Radiation ed Smarr L L*, Cambridge University Press, Cambridge, UK, 1979.

[64] C. W. Misner, K. S. Thorne, and J. A. Wheeler, *Gravitation*, W. H. Freeman and Co., San Francisco, Calif, USA, 1973.

[65] R. W. Mayle, M. Tavani, and J. R. Wilson, "Pions, supernovae, and the supranuclear matter density equation of state," *Astrophysical Journal*, vol. 418, no. 1, pp. 398–404, 1993.

[66] J. M. Lattimer and F. D. Swesty, "A generalized equation of state for hot, dense matter," *Nuclear Physics A*, vol. 535, no. 2, pp. 331–376, 1991.

[67] N. K. Glendenning, *Compact Stars: Nuclear Physics, Particle Physics, and General Relativity*, Springer, New York, NY, USA, 1996.

[68] U. Garg, "The isoscalar giant dipole resonance: a status report," *Nuclear Physics A*, vol. 731, pp. 3–14, 2004.

[69] C. Cutler, T. A. Apostolatos, L. Bildsten et al., "The last three minutes: issues in gravitational-wave measurements of coalescing compact binaries," *Physical Review Letters*, vol. 70, no. 20, article 2984, 1993.

[70] L. E. Kidder, C. M. Will, and A. G. Wiseman, "Spin effects in the inspiral of coalescing compact binaries," *Physical Review D*, vol. 47, no. 10, pp. R4183–R4187, 1993.

6

A Decade of GRB Follow-Up by BOOTES in Spain (2003–2013)

Martin Jelínek,[1,2] Alberto J. Castro-Tirado,[2,3] Ronan Cunniffe,[2] Javier Gorosabel,[2,4,5] Stanislav Vítek,[6] Petr Kubánek,[7,8] Antonio de Ugarte Postigo,[2] Sergey Guziy,[2] Juan C. Tello,[2] Petr Páta,[6] Rubén Sánchez-Ramírez,[2] Samantha Oates,[2] Soomin Jeong,[2,9] Jan Štrobl,[1] Sebastián Castillo-Carrión,[10] Tomás Mateo Sanguino,[11] Ovidio Rabaza,[12] Dolores Pérez-Ramírez,[13] Rafael Fernández-Muñoz,[14] Benito A. de la Morena Carretero,[15] René Hudec,[1,6] Víctor Reglero,[8] and Lola Sabau-Graziati[16]

[1] *Astronomický Ústav AV ČR, Ondřejov (ASÚ AV ČR), Ondřejov, Czech Republic*
[2] *Instituto de Astrofísica de Andalucía- (IAA-) CSIC, 18008 Granada, Spain*
[3] *Departamento de Ingeniería de Sistemas y Automática (Unidad Asociada al CSIC), Universidad de Málaga, 29010 Málaga, Spain*
[4] *Unidad Asociada Grupo Ciencia Planetarias UPV/EHU-IAA/CSIC, Departamento de Física Aplicada I, E.T.S. de Ingeniería, Universidad del País Vasco (UPV)/EHU, Alameda de Urquijo s/n, 48013 Bilbao, Spain*
[5] *Ikerbasque, Basque Foundation for Science, Alameda de Urquijo 36-5, 48008 Bilbao, Spain*
[6] *České Vysoké Učení Technické, Fakulta Elektrotechnická (ČVUT-FEL), Praha, Czech Republic*
[7] *Fyzikální ústav AV ČR, Na Slovance 2, 182 21 Praha 8, Czech Republic*
[8] *Image Processing Laboratory, Universidad de Valencia, Burjassot, Valencia, Spain*
[9] *Institute for Science and Technology in Space, Natural Science Campus, Sungkyunkwan University, Suwon 440-746, Republic of Korea*
[10] *Universidad de Málaga, Campus de Teatinos, Málaga, Spain*
[11] *Departamento de Ingeniería de Sistemas y Automática, Universidad de Huelva, E.P.S. de La Rábida, Huelva, Spain*
[12] *Department of Civil Engineering, University of Granada, 18071 Granada, Spain*
[13] *Universidad de Jaén, Campus las Lagunillas, 23071 Jaén, Spain*
[14] *Instituto de Hortofruticultura Subtropical y Mediterránea "La Mayora" (IHSM-CSIC), Algarrobo, 29750 Málaga, Spain*
[15] *Estación de Sondeos Atmosféricos (ESAt) de El Arenosillo (CEDEA-INTA), Mazagón, Huelva, Spain*
[16] *División de Ciencias del Espacio, INTA, Torrejón de Ardoz, Madrid, Spain*

Correspondence should be addressed to Martin Jelínek; mates@iaa.es

Academic Editor: Dean Hines

This article covers ten years of GRB follow-ups by the Spanish BOOTES stations: 71 follow-ups providing 23 detections. Follow-ups by BOOTES-1B from 2005 to 2008 were given in a previous article and are here reviewed and updated, and additional detection data points are included as the former article merely stated their existence. The all-sky cameras CASSANDRA have not yet detected any GRB optical afterglows, but limits are reported where available.

Dedicated to the memory of Dolores Pérez-Ramírez and Javier Gorosabel, who passed away while this paper was in preparation

1. Introduction

Ever since the discovery of Gamma-ray bursts (GRB) in 1967 [1], it was hoped to discover their counterparts at other wavelengths. The early GRB-related transient searching methods varied (wide-field optical systems as well as deep searches were being employed) but, given the coarse gamma-ray-based GRB localizations provided, generally lacked either sensitivity or good reaction time. The eventual discovery of GRB optical counterparts was done only when an X-ray follow-up telescope was available on the BeppoSAX satellite [2]. The optical afterglow could then be searched for with a large telescope in a small errorbox provided by the discovery of the X-ray afterglow. The first optical afterglow of a gamma-ray burst was discovered this way in 1997 [3].

Since then, astronomers have been trying to minimize the time delay between receiving the position and the start of observations—by both personal dedication and by automating the telescope reaction. The ultimate step in automation, to minimize the time delay, is a full robotization of the observatory to eliminate any human intervention in the follow-up process. This way, the reaction time can be minimized from ~10-minute limit that can be achieved with a human operated telescope to below 10 seconds. With improvements in computational methods and in image processing speed, blind (non-follow-up) wide-field methods are starting to be practical in the search for optical transients. Although limited in magnitude range, they have already provided important observations of the optical emission simultaneous to the gamma-ray production of a GRB [4].

Since 1997, the robotic telescope network BOOTES has been part of the effort to follow up gamma-ray burst events [5]. As of now, the network of robotic telescopes BOOTES consists of six telescopes around the globe, dedicated primarily to GRB afterglow follow-up. We present the results of our GRB follow-up programme by two telescopes of the network—BOOTES-1B and BOOTES-2—and by the respective stationary very-wide-field cameras (CASSANDRA). This text covers eleven years of GRB follow-ups: 71 follow-ups providing 21 detections.

Different instruments have been part of BOOTES during the years in question: a 30 cm telescope which was used for most of the time at BOOTES-1 station but at periods also at BOOTES-2, the fast-moving 60 cm telescope at BOOTES-2 (Telma), and also two all-sky cameras, CASSANDRA1 at BOOTES-1 and CASSANDRA2 at BOOTES-2. Results from CASSANDRAs are included where available, without paying attention to the complete sample.

This article is a follow-up of a previous article, that is, Jelínek et al. [6], which provided detailed description of evolution of BOOTES-1B, and analysis of efficiency of a system dedicated to GRB follow-up based on real data obtained during four years between 2005 and 2008. This work is a catalogue of BOOTES-1B and BOOTES-2 GRB observations between 2003 and 2013; it is complete in providing information about successfully followed up events but does not provide analysis of missed triggers as did the previous article.

1.1. BOOTES-1B. BOOTES-1 observatory is located at the atmospheric sounding station at El Arenosillo, Huelva, Spain (at lat.: 37°06′14″N, long.: 06°44′02″W). Over time, distinct system configurations were used, including also two 8-inch S-C telescopes, as described in Jelínek et al. [6]; the primary instrument of BOOTES-1B is a $D = 30$ cm Schmidt-Cassegrain optical tube assembly with a CCD camera. Prior to June 15, 2007, Bessel *VRI* filters were being used as noted with the observations, any observations obtained after this date have been obtained without filter (*C* or clear). We calibrate these observations against *R*-band, which, in the case of no color evolution of the optical counterpart, is expected to result in a small (~0.1 mag) constant offset in magnitude.

1.2. BOOTES-2. BOOTES-2 is located at CSIC's experimental station La Mayora (Instituto de Hortofruticultura Subtropical y Mediterránea- (IHSM-) CSIC) (at lat.: 36°45′33″N, long.: 04°02′27″W), 240 km from BOOTES-1. It was originally equipped with an identical 30 cm Schmidt-Cassegrain telescope to that at BOOTES-1B. In 2007 the telescope was upgraded to a lightweight 60 cm Ritchey-Chrètien telescope on a fast-slewing NTM-500 mount, both provided by Astelco. The camera was upgraded at the same time to an Andor iXon 1024 × 1024 EMCCD, and in 2012 the capabilities were extended yet again to low resolution spectroscopy, by the installation of the imaging spectrograph COLORES of our own design and construction [7]. Bessel magnitudes are calibrated to Vega system, SDSS to AB.

2. Optical Follow-Up of GRB Events

Here we will detail the individual results for each of the 23 events followed up and detected in 2003–2013. Each GRB is given a short introductory paragraph as a reminder of the basic observational properties of the event. Although we do not discuss the properties at other wavelengths, we try to include a comprehensive reference of literature relevant to each burst. As GCN reports usually summarise the relevant GCN circular traffic, we have omitted the raw GCN circulars except for events for which a GCN report or other more exhaustive paper is unavailable.

Further 48 follow-ups which resulted in detection limits are included in Tables 1 and 2 but are not given any further attention.

One by one, we show all the successful follow-ups that these telescopes have performed during the first ten years of the *Swift* era and since the transition of the BOOTES network to the RTS-2 [14] observatory control system, which was for the first time installed at BOOTES-2 in 2003 and during the summer of 2004 at BOOTES-1.

GRB 050525A (A Bright Low-Redshift ($z = 0.606$) Localized by Swift [15]). Plenty of optical observations were obtained, including the signature of the associated supernova sn2005nc [8, 16].

GRB 050525A was the first BOOTES-1B burst for which a detection was obtained. The telescope started the first exposure 28 s after receiving the notice, 383 s after the GRB trigger. An optical afterglow with $V \simeq 16$ was detected. A weak detection of a bright GRB implied a reexamination of observing strategies employed by BOOTES. The largest, 30 cm telescope was changed to make *R*-band imaging instead of using the field spectrograph to greatly improve sensitivity in terms of limiting magnitude. The 20 cm telescopes were still observing with $V + I$ filters (for details see [6]); see Table 3.

Table 1: BOOTES-1B GRBs in a table.

GRB	ΔT	Number of points	Result	Ref.
030913	2 h		$V > 17.5, C > 12$	
050215B	22 m		$V > 16.5, I > 15.0$	
050505	47 m		$V > 19$	
050509A	64 m		$V > 14.9$	
050509B	62 s		$V > 11.5$	
050525A	12 m†	1	16.5 ± 0.4	[8]
050528	71 s		$V > 13.8, I > 13.0$	
050824	10 m	4	$R = 18.2 \pm 0.3$	[9]
050904	2 m		$R > 18.2$	[10]
050922C	4 m	3	$R = 14.6 \pm 0.4$	
051109A	55 s	6	$R = 15.7 \pm 0.4$	
051211B	42 s		$R > 14$	
051221B	4 m		$R > 16$	
060421	61 s		$R > 14$	
061110B	11 m		$R > 18$	
071101	55 s		$C > 17.0$	
071109	59 s		$C > 13.0$	
080330	6 m	6	$C = 16.5 \pm 0.2$	
080413A	61 s	61	$C \simeq 13.3$	
080430	34 s	1	$C \simeq 15.5$	
080603B	1 h	11	$C \simeq 17.4$	[11]
080605	44 s	28	$C \simeq 14.7$	[12]
081003B	41 s		$C > 17.6$	
090313	12 h	1	$C \simeq 18.3$	
090519	99 s		$C > 17.6$	
090813	53 s	1	$C \simeq 17.9$	
090814A	3 m†		$C > 15.8$	
090814B	53 s†		$C > 17.5$	
090817	24 m		$C > 16.7$	
100906A	106 s		$C > 16.5$	
110205A	102 s	16	$C \sim 14$	[13]
110212A	50 s		$C > 13.0$	
110213A	15 h	1	$C = 18.3 \pm 0.2$	
110411A	24 s		$C > 17.8$	
111016A	1.25 h		$C > 17.8$	
120326A	40 m	1	$C \sim 19.5$	
120327A	41 m†	6	$C = 17.5$	
120328A	7.5 m		$C > 16$	
120521C	11.7 m		$C > 20.5$	
120711B	107 s		$C > 18.2$	
120729A	10 h		$C > 19.0$	
121017A	79 s		$C > 19.0$	
121024A	40 m	1	$C = 18.2 \pm 0.5$	
121209A	42 s†		$C > 16.5$	
130122A	28 m		$C > 18.4$	

Note. † marks alerts covered in real time by wide-field camera CASSANDRA-1.

This burst was covered in real time by both all-sky cameras of BOOTES (CASSANDRA1 and CASSANDRA2), providing an unfiltered limit of >9.0 [17].

Table 2: BOOTES-2 GRBs in a table.

GRB	ΔT	Number of points	Result	Ref.
080603B		20	$R \simeq 17.4$	[11]
080605		5	$R \simeq 14.7$	[12]
090817	145 s		$R > 18.3$	
090904A	86 s		$R > 16.1$	
091202	5.5 h		$R > 18.3$	
100219A	6.3 h		$C > 18.3$	
100418A	1.8 h	11	$C = 19.3$	
100522A	625 s		$C > 15.5$	
100526A	4 h		$r' > 14$	
100614A	6.9 m		$C > 18$	
100901A	10 h	10	$C = 17.52 \pm 0.08$	
100915A	106 s		$C > 16.5$	
101020A	5.1 h		$r' > 18.0$	
101112A	595 s	15	$C = 15.5$	
110106B	10.3 m		$C > 16.5$	
110205A	15 m	13	$R \sim 14$	[13]
110212A	32 m		$R > 16.5$	
110223A	228 s		$R > 17.6$	
120729A	13.25 h		$R > 19.4$	
120805A	25 m		$R > 18.5$	
120816A	66 m		$R > 18$	
121001A	32 m		$I > 19.7$	
121017A	3 m		$C > 18.5, i' > 19.5$	
130418A	1.5 h	21	$C = 16.8 \pm 0.06$	
130505A	11.94 h	1	$R_C = 19.26 \pm 0.06$	
130606A	13 m	21	$i' = 16.7 \pm 0.3$	
130608A	2.3 h		$C > 18.8$	
130612A	4.8 m		$C > 18.6$	
130806A	40 s		$C > 18.3$	
131202A	4.25 h		$i' > 19.7$	

Table 3: GRB 050525A: observing log of BOOTES-1B.

ΔT [h]	exp [s]	mag	dmag	Filter
0.195	39×10 s	16.51	0.39	R

Note. Published by Resmi et al. [8].

BOOTES observation of this GRB is included in Resmi et al. [8].

GRB 050824 (A Dim Burst Detected by Swift). The optical afterglow of this GRB is discovered with the 1.5 m telescope at OSN; redshift $z = 0.83$ as determined by VLT [9].

BOOTES-1B was the first telescope to observe this optical transient, starting 636 s after the trigger with $R \simeq 17.5$. The weather was not stable and the focus not perfect, but BOOTES-1B worked as expected. In the end, several hours of data were obtained. BOOTES observation of this GRB is included in Sollerman et al. [9]; see Table 4 and Figure 1.

GRB 050922C. A *Swift* short and intense long burst [18, 19] was observed also by *HETE2* [20]. Optical afterglow is mag ~ 15; $z = 2.198$ [21].

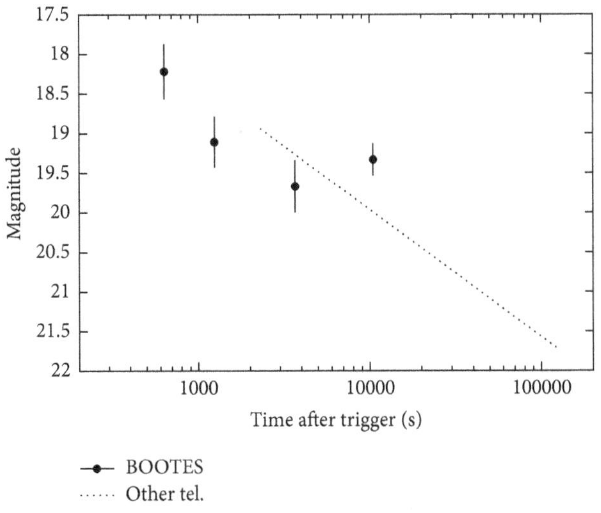

FIGURE 1: The optical light curve of GRB 050824; the optical light curve represents the behaviour seen by Sollerman et al. [9].

TABLE 4: GRB 050824: observing log of BOOTES-1B.

ΔT [h]	exp [s]	mag	dmag	Filter
0.1763	2×300 s	18.22	0.35	R
0.3462	8×300 s	19.11	0.32	R
1.0249	22×300 s	19.67	0.33	R
2.9091	31×300 s	19.33	0.20	R

Note. Published by Sollerman et al. [9].

TABLE 5: GRB 050922C: observing log of BOOTES-1B.

ΔT [h]	exp [s]	mag	dmag	Filter
0.0694	40	14.58	0.35	R
0.3752	900	17.01	0.39	R
0.6193	900	18.53	0.59	R

Due to clouds, the limiting magnitude of BOOTES-1B dropped from ~17.0 for a 30 s exposure to merely 12.9. The afterglow was eventually detected with the R-band camera (at the 30 cm telescope) during gaps between passing clouds. The first weak detection was obtained 228 s after the GRB trigger and gave $R \simeq 14.6$; see Table 5.

GRB 051109A (A Burst Detected by Swift [23]). The optical afterglow was mag ~ 15, and the redshift was determined to be $z = 2.346$ [24]. The optical lightcurve was published by Mirabal et al. [22].

At BOOTES-1B the image acquisition started 54.8 s after the burst with the 30 cm telescope in R-band and one of the 20 cm telescopes in I-band [25]. There were still a number of performance problems—most importantly synchronization between cameras such that when the telescope position was to be changed, both cameras had to be idle. As the 30 cm telescope was taking shorter exposures, extra exposures could have been made while waiting for the longer exposures being taken at the 20 cm to finish. The 20 cm detection is, after critical revision, only at the level of 2-σ. The R-band

TABLE 6: GRB 051109A: observing log of BOOTES-1B.

ΔT [s]	exp [s]	mag	dmag	Filter
59.7	10	15.67	0.35	R
122.2	74	16.02	0.19	R
257.9	41	16.65	0.41	R
756.6	205	17.18	0.22	R
1021.5	313	17.68	0.26	R
508.4	908	16.98	0.54	I

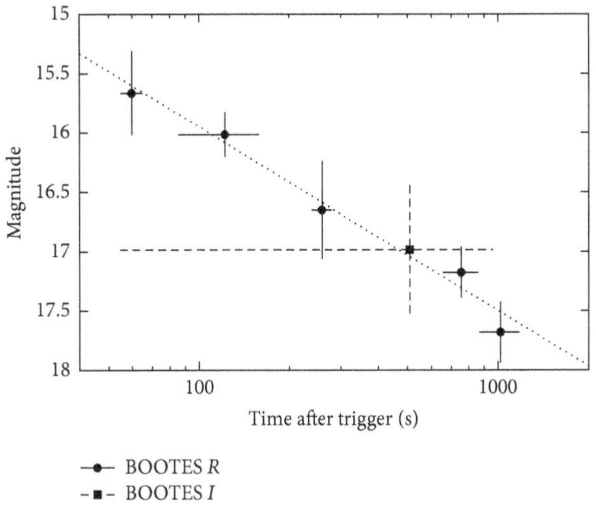

FIGURE 2: The optical light curve of GRB 051109A. The dotted line represents the optical decay observed by Mirabal et al. [22].

observation shows the object until about 20 minutes after the GRB, when it becomes too dim to measure in the vicinity of a 17.5 m nearby star. Mean decay rate observed by BOOTES is $\alpha = 0.63 \pm 0.06$ ($F_{\mathrm{opt}} \sim t^{-\alpha}$).

The relatively shallow decay observed by BOOTES is in close agreement with what was observed several minutes later by the 2.4 m MDM ($\alpha = 0.62 \pm 0.03$) and according to an unofficial report [26] there was a decay change later, by about 3 h after the burst to $\alpha = 0.89 \pm 0.05$; see Table 6 and Figure 2.

GRB 080330 (A Rather Bright Long Burst Detected by Swift). Afterglow was reported to be detected by UVOT, TAROT, ROTSE-III, Liverpool Telescope, and GROND. Spectroscopic redshift was measured as $z = 1.51$ by the NOT [28].

This GRB happened during the first day recommissioning of BOOTES-1B after its move from the BOOTES-2 site at La Mayora. The GCN client was not yet operational and at the time of the GRB we were focusing the telescope. The first image was obtained 379 s after the GRB trigger and the optical afterglow was detected with magnitude ~16.3 on the first image. A bug in the centering algorithm caused a loss of part subsequent data. Further detections were obtained starting 21 min after the GRB when the problem was fixed.

The light curve (as seen by [27]) seems to show an optical flare and then a possible hydrodynamic peak. The data of BOOTES, however, trace only the final part of this

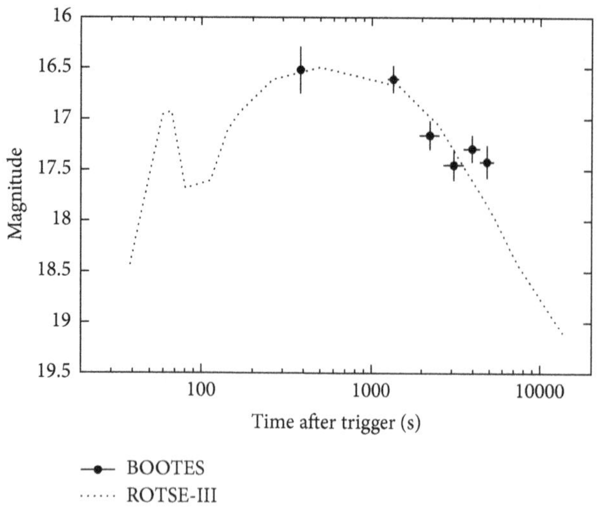

FIGURE 3: The optical light curve of GRB 080330. The dotted line shows the light curve as seen by ROTSE-III [27].

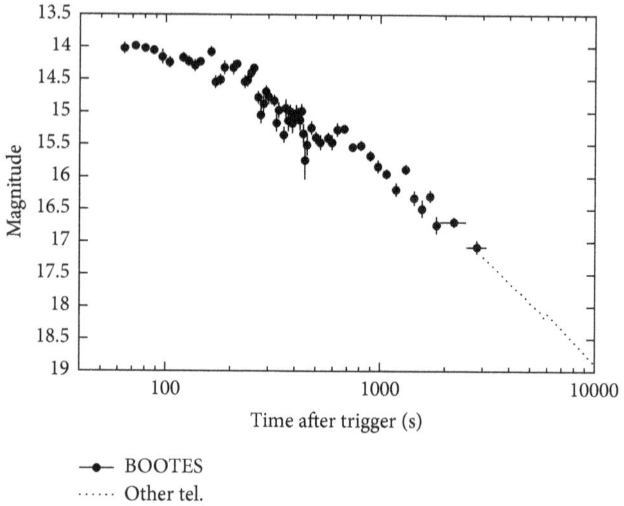

FIGURE 4: The optical light curve of GRB 080413A.

TABLE 7: GRB 080330: observing log of BOOTES-1B.

ΔT [h]	exp [s]	mag	dmag	Filter
0.1061	7	16.52	0.23	Clear
0.3752	210	16.61	0.13	Clear
0.6193	588	17.16	0.14	Clear
0.8547	825	17.45	0.15	Clear
1.0915	862	17.29	0.13	Clear
1.3384	905	17.42	0.16	Clear

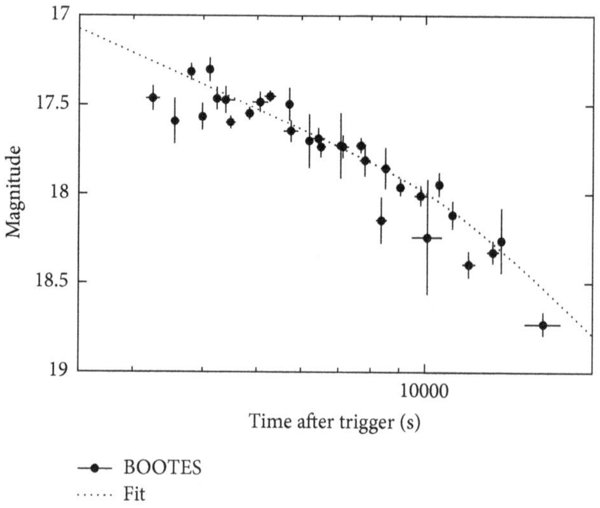

FIGURE 5: The optical light curve of GRB 080603B [11].

behaviour, where the decay accelerates after passing through the hydrodynamic peak; see Table 7 and Figure 3.

GRB 080413A. A rather bright GRB was detected by *Swift* and also by *Suzaku*-WAM; optical afterglow was detected by ROTSE-III [27]; and redshift $z = 2.433$ was detected by VLT+UVES [29].

BOOTES-1B started obtaining images of the GRB 080413A just 60.7 s after the trigger (46.3 s after reception of the alert). An $R \simeq 13.3$ magnitude decaying optical afterglow was found ([30], *Jelínek et al., in prep.*); see Figure 4.

GRB 080430 (A Burst Detected by Swift). It was a widely observed, low-redshift $z \simeq 0.75$ optical afterglow with a slowly decaying optical afterglow [31]. It was observed also at very high energies by *MAGIC* without detection [32].

BOOTES-1B obtained the first image of this GRB 34.4 s after the trigger. An optical transient was detected on combined unfiltered images with a magnitude $\simeq 15.5$ [33].

GRB 080603B. A long GRB localized by *Swift* is detected also by *Konus*-Wind and by *INTEGRAL* [34]. Bright optical afterglow was observed. Extensive follow-up was carried out. Redshift is $z = 2.69$ [35].

This GRB happened in Spain during sunset. We obtained first useful images starting one hour after the trigger. An $R \simeq 17.4$ optical transient was detected with both BOOTES-1B and BOOTES-2; see Figure 5. BOOTES observation of this GRB is included in Jelínek et al. [11]; see Figure 5.

GRB 080605 (A Long Burst Detected by Swift [36]). The host was found to be a metal enriched star forming galaxy at redshift $z = 1.64$ [37] and exhibited the 2175 Å extinction feature [38].

GRB 080605 was observed by both BOOTES-1B (28 photometric points) and BOOTES-2 (5 photometric points) starting 44 s after the trigger. A rapidly decaying optical afterglow ($\alpha = 1.27 \pm 0.04$) with $R = 14.7$ on the first images was found; see Figure 6. All BOOTES data are included in Jelínek et al. [12]; see Figure 5.

GRB 090313 (GRB by Swift, No Prompt X-Rays [40]). An optical afterglow peaked at $R \sim 15.6$. Extensive optical + infrared follow-up was carried out. The first GRB to be observed

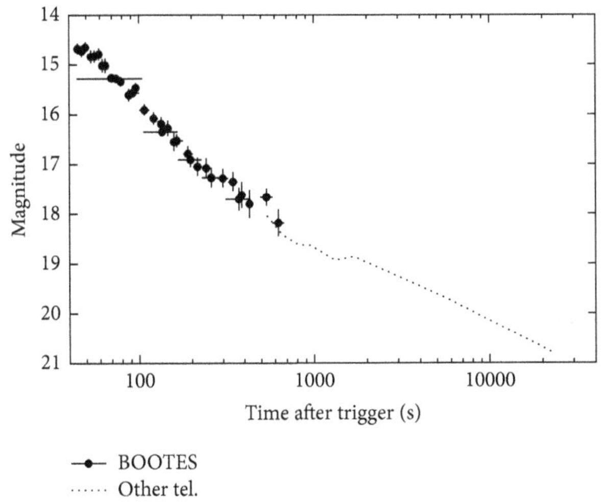

FIGURE 6: The optical light curve of GRB 080605 [12]; the dotted line is behaviour observed by Rumyantsev and Pozanenko [39] and Zafar et al. [38].

TABLE 8: GRB 090813: observing log of BOOTES-1B.

ΔT [h]	exp [s]	mag	dmag	Filter
0.175	10 × 10	17.9	0.3	Clear

was detected by X-Shooter. Also it was detected by various observatories in radio. Redshift is $z = 3.375$ [41, 42].

The GRB happened during daytime for BOOTES-1B and it was followed up manually. Due to the proximity of the moon and limitations of then-new CCD camera driver, many 2 s exposures were taken to be combined later. The optical afterglow was detected with magnitude ~18.3 ± 0.4 on a 635 × 2 s (=21 min) exposure with the midtime 11.96 h after the GRB trigger.

GRB 090813. A long GRB by *Swift*, suspected of being higher-z, observed also by *Konus*-Wind and *Fermi*-GBM [43]. Optical counterpart was observed by the 1.23 m telescope at Calar Alto with a magnitude of $I = 17.0$ [44].

BOOTES-1B started observation 53 s after the GRB, taking 10 s unfiltered exposures. The optical transient was weakly detected on a combined image of 10 × 10 s whose exposure mean time was 630 s after the burst. The optical counterpart was found having $R = 17.9 \pm 0.3$. Given that the previous and subsequent images did not show any OT detection, we might speculate about the optical emission peaking at about this time. Also the brightness is much weaker than what might be expected from the detection by Gorosabel et al. [44], supporting the high-redshift origin; see Table 8.

GRB 100418A. A weak long burst was detected by *Swift* [45] with a peculiar, late-peaking optical afterglow with $z = 0.6239$ [46]. Also it was detected in radio [47].

The first image of the GRB location was taken by BOOTES-2 at 21:50 UT (40 min after the GRB trigger). The rising optical afterglow was detected for the first time on an image obtained as a sum of 23 images, with an exposure midtime 107 minutes after the GRB trigger. The optical emission peaked at magnitude $R = 18.7$ another hour later, at an image with the midtime 163 min after the trigger. A slow decay followed, which permitted us to detect the optical counterpart until 8 days after the GRB.

TABLE 9: GRB 100418A: observing log of BOOTES-2.

ΔT [h]	exp [s]	mag	dmag	Filter
1.78	1638	19.785	0.215	Clear
2.09	597	19.127	0.127	Clear
2.55	534	18.774	0.087	Clear
2.72	656	18.668	0.073	Clear
3.10	239	18.706	0.106	Clear
3.43	238	18.759	0.189	Clear
4.70	3908	19.067	0.108	Clear
6.19	4328	18.897	0.115	Clear
7.39	551	18.493	0.078	Clear
77.3	14830	20.475	0.202	Clear
125.6	12482	20.970	0.208	Clear

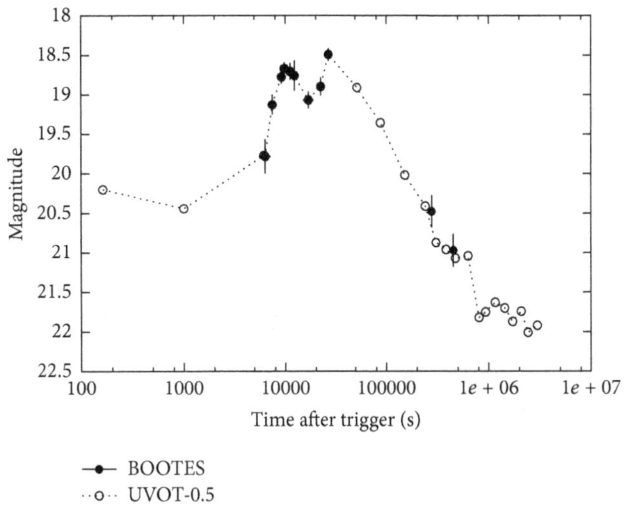

FIGURE 7: The bizarre optical light curve of GRB 100418A. Combination of BOOTES and UVOT data [45]. UVOT points were shifted by an arbitrary constant.

Because of a mount problem, many images were lost (pointed somewhere else) and the potential of the telescope was not fully used. Eventually, after combining images when appropriate, 11 photometric points were obtained. A rising part of the optical afterglow was seen that way; see Table 9 and Figure 7.

GRB 100901A (A Long Burst from Swift). Bright, slowly decaying optical afterglow was discovered by UVOT. Redshift is $z = 1.408$. It was detected also by SMA at 345 GHz [48–50].

The burst happened in daytime in Spain and the position became available only almost ten hours later after the sunset. The afterglow was still well detected with magnitude $R \simeq 17.5$ at the beginning. BOOTES-2 had some problems with CCD cooling, and some images were useless. The afterglow was

Table 10: GRB 100901A: observing log of BOOTES-2.

ΔT [h]	exp [s]	mag	dmag	Filter
10.202	268	17.52	0.08	R
10.719	415	17.61	0.07	R
11.230	354	17.67	0.09	R
11.734	238	17.99	0.16	R
12.346	730	17.78	0.13	R
12.980	759	17.68	0.12	R
13.239	759	17.82	0.16	R
13.971	997	18.21	0.12	R
14.611	1101	18.32	0.14	R
33.791	4012	19.35	0.19	R

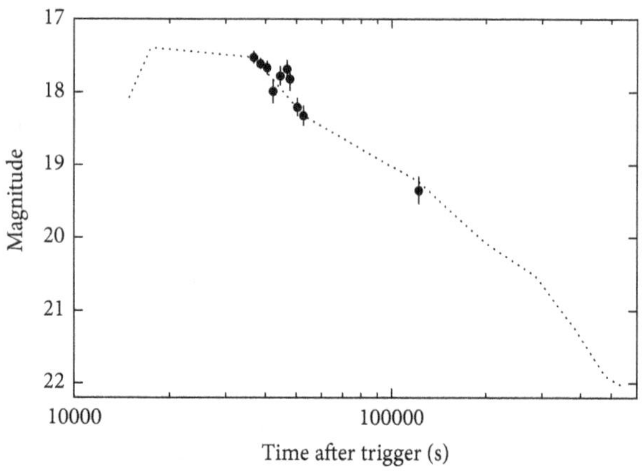

Figure 8: The optical light curve of GRB 100901A. The dotted line representing burst behaviour is based on observations by Gorbovskoy et al. [49], Kann et al. [51], and Rumyantsev et al. [52].

detected also the following night with $R = 19.35$; see Table 10 and Figure 8.

GRB 101112A. An *INTEGRAL*-localized burst [53] was also detected by *Fermi*-GBM [54], *Konus*-Wind [55], and *Swift*-XRT [56]. Optical afterglow was discovered independently by BOOTES-2 and Liverpool Telescope [57]. It was detected also in radio [58].

BOOTES-2 reacted to the GRB 101112A and started to observe 47 s after the GRB. A set of 3 s exposures was taken, but due to technical problems with the mount a significant amount of observing time was lost. An optical afterglow was discovered and reported [59]. The optical light curve exhibited first a decay, then a sudden rise to a peak at about 800 s after the trigger, and finally a surprisingly fast decay with $\alpha \simeq -4$. This behaviour seemed more like an optical flare than a "proper" GRB afterglow, but there does not seem to be contemporaneous high-energy data to make a firm statement; see Table 11 and Figure 9.

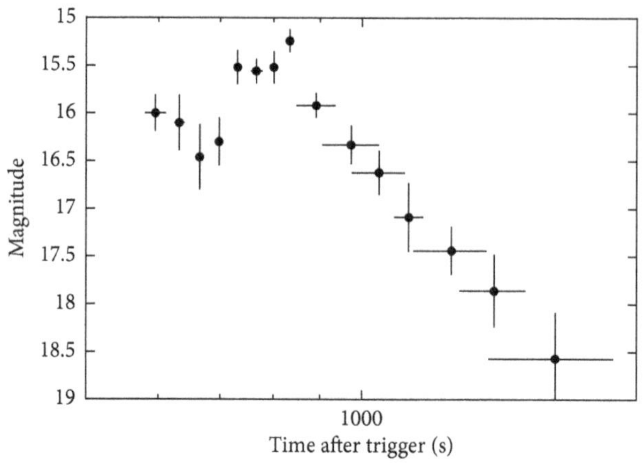

Figure 9: The optical light curve of GRB 101112A.

Table 11: GRB 101112A: observing log of BOOTES-2.

ΔT [s]	exp [s]	mag	dmag	Filter
595.0	16	16.00	0.19	r'
631.8	8	16.10	0.29	r'
664.9	7	16.46	0.34	r'
697.8	7	16.30	0.25	r'
731.0	7	15.52	0.18	r'
766.1	11	15.56	0.13	r'
800.9	7	15.52	0.17	r'
833.8	7	15.24	0.12	r'
891.2	44	15.92	0.13	r'
973.7	69	16.33	0.20	r'
1044.0	69	16.62	0.23	r'
1124.2	41	17.09	0.36	r'
1252.7	115	17.44	0.25	r'
1393.8	116	17.86	0.38	r'
1629.5	255	18.57	0.48	r'

GRB 110205A (A Very Long and Bright Burst by Swift). Detected also by *Konus*-Wind and *Suzaku*-WAM, optical afterglow peaked at $R \sim 14.0$, with extensive multiwavelength follow-up; $z = 2.22$ "*Textbook burst*" [13, 60].

BOOTES-1B reacted automatically to the *Swift* trigger. First 10 s unfiltered exposure was obtained 102 s after the beginning of the GRB (with $T_{90} = 257$ s), that is, while the gamma-ray emission was still taking place. After taking 18 images, the observatory triggered on a false alarm from the rain detector, which caused the observation to be stopped for 20 minutes. After resuming the observation, 3 × 30 s images were obtained and another false alert struck over. This alert was remotely overridden by Kubánek, so that all 20 minutes was not lost. From then on, the observation continued until sunrise. The afterglow is well detected in the images until 2.2 hours after the GRB. 16 photometric points from combined images were eventually published.

BOOTES-2 started observations 15 min after the trigger, clearly detecting the afterglow in R-band until 3.2 hours after the burst. 13 photometric points were obtained. The delay was

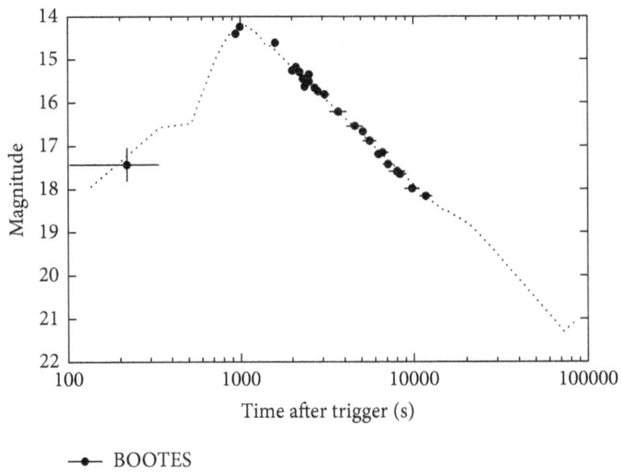

FIGURE 10: The optical light curve of GRB 110205A.

TABLE 12: GRB 110213A: observing log of BOOTES-1B.

ΔT [h]	exp [s]	mag	dmag	Filter
15.5	100 × 30	18.29	0.30	Clear

caused by technical problems. BOOTES observations of this GRB are included in Zheng et al. [13]; see Figure 10.

GRB 110213A. A bright burst was detected by *Swift*; it was detected also by *Konus*-Wind and *Fermi*-GBM. Optical afterglow is $R \sim 14.6$, with extensive follow-up [61].

BOOTES-1B started to observe 15 hours after the GRB (the position was below horizon at the time of the trigger) and continued for an hour; eventually, 100 × 30 s unfiltered images were combined; the OT brightness calibrated against USNO-A2 is 18.3 ± 0.3 at the exposure midtime of 15.5 h after the GRB trigger; see Table 12.

GRB 120326A (A Swift-Detected Burst). Afterglow was discovered by Tarot [62]. It is long-lived optical emission; redshift is $z = 1.78$ by GTC. It was detected also by *Fermi*-GBM and *Suzaku*-WAM (see [63] and the references therein).

At BOOTES-1B the mount failed, because of the serial port communication failure. After a manual recovery, 40 minutes after the GRB, images were taken in hope for a detection, but the counterpart with the brightness of $R \sim 19.5$ was detected only at about 2σ level.

GRB 120327A. A bright burst by *Swift* with an afterglow is discovered by UVOT [64]. Redshift is $z = 2.813$ [65]. Extensive optical follow-up was carried out.

BOOTES-1B reacted in 41 min (similar failure as the day before: the mount failed, because of the serial port communication failure), obtaining a series of 20 s exposures. These images were combined to get 600 s effective exposures and permitted detection of the afterglow on six such images. The brightness was decaying from $R = 17.5$ to $R = 18.6$; see Table 13.

TABLE 13: GRB 120327A: observing log of BOOTES-1B.

ΔT [h]	exp [s]	mag	dmag	Filter
0.955	654	17.50	0.12	Clear
1.140	674	17.65	0.12	Clear
1.337	748	17.82	0.13	Clear
1.533	660	18.24	0.21	Clear
1.718	673	18.17	0.21	Clear
1.905	656	18.59	0.29	Clear

TABLE 14: GRB 121024A: observing log of BOOTES-1B.

ΔT [h]	exp [s]	mag	dmag	Filter
0.900	1200	18.2	0.5	Clear

All-sky camera at BOOTES-1 (CASSANDRA1) covered the event in real time and detected nothing down to $R \sim 7.5$ *(Zanioni et al. in prep.).*

GRB 121001A. A bright and long *Swift*-detected GRB was originally designated as possibly galactic [66]. Afterglow was discovered by Andreev et al. [67].

BOOTES-2 observed this trigger starting 32 min after the trigger. An optical afterglow is detected in I-band with $I \sim 19.7$ (Vega) for a sum of images between 20:49 and 21:52 UT [68].

GRB 121024A. It is a bright *Swift*-detected GRB with a bright optical afterglow [69, 70]. It was detected also in radio [71]. Redshift is $z = 2.298$ by Tanvir et al. [72].

BOOTES-1B observed the optical afterglow of GRB 121024A. The observations started 40 minutes after the GRB trigger. The sum of 20 minutes of unfiltered images with a mean integration time 54 minutes after the GRB shows a weak detection of the optical afterglow with magnitude $R = 18.2 \pm 0.5$ [73]; see Table 14.

GRB 130418A. It is a bright and long burst with a well-detected optical afterglow somewhat peculiarly detected after a slew by *Swift* [74]. Observation by *Konus*-Wind showed that the burst started already 218 s before *Swift* triggered [75]. Redshift is $z = 1.218$ by de Ugarte Postigo et al. [76].

BOOTES-2 obtained a large set of unfiltered, r'-band and i'-band images starting 1.5 h after the trigger. The optical afterglow is well detected in the images. The light curve is steadily decaying with the power-law index of $\alpha = -0.93 \pm 0.06$, with the exception of the beginning, where there is a possible flaring with peak about 0.25 mag brighter than the steady power-law; see Table 15 and Figure 11.

GRB 130505A. A bright and intense GRB with a 14 mag optical afterglow was detected by *Swift* [77]. Redshift is $z = 2.27$ as reported by Tanvir et al. [78].

BOOTES-2 obtained the first image of this GRB 11.94 h after the trigger. A set of 60 s exposures was obtained. Combining the first hour of images taken, we clearly detect the optical afterglow, and using the calibration provided by Kann et al. [79], we measure $R_C = 19.26 \pm 0.06$; see Table 16.

Table 15: GRB 130418A: observing log of BOOTES-1B and BOOTES-2.

ΔT [h]	exp [s]	mag	dmag	Filter
1.514	3×15 s	17.09	0.08	Clear
1.529	3×15 s	16.95	0.07	Clear
1.544	3×15 s	16.90	0.06	Clear
1.558	3×15 s	16.62	0.07	Clear
1.573	3×15 s	17.03	0.07	Clear
1.590	4×15 s	16.92	0.06	Clear
1.610	4×15 s	17.04	0.07	Clear
1.749	7×15 s	17.22	0.05	Clear
1.865	60 s	16.92	0.18	r'
1.884	4×15 s	17.34	0.09	Clear
2.054	7×15 s	17.45	0.07	Clear
2.089	7×15 s	17.46	0.06	Clear
2.209	6×15 s	17.47	0.07	Clear
2.326	6×15 s	17.56	0.08	Clear
2.444	6×15 s	17.71	0.09	Clear
2.562	6×15 s	17.68	0.08	Clear
2.798	22×60 s	17.40	0.04	i'
3.061	15×60 s	17.90	0.09	r'
3.333	15×60 s	17.98	0.09	r'
3.604	15×60 s	17.90	0.09	r'
3.866	15×60 s	18.05	0.11	r'
4.130	15×60 s	18.53	0.19	r'
4.449	20×60 s	18.42	0.14	r'
4.808	20×60 s	18.61	0.23	r'

Table 16: GRB 130505A: observing log of BOOTES-2.

ΔT [h]	exp [s]	mag	dmag	Filter
12.488	51×60 s	19.26	0.06	Clear

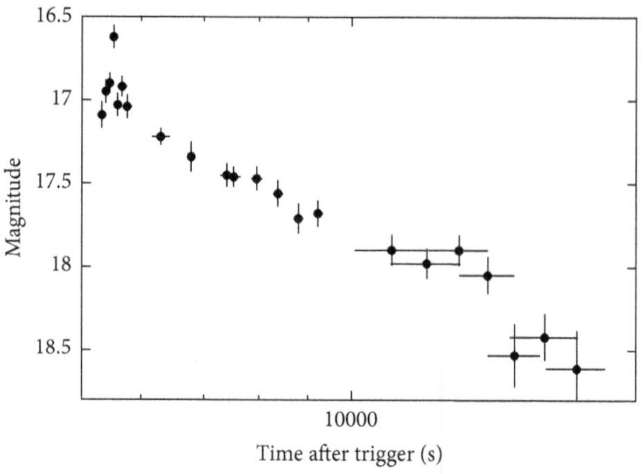

Figure 11: The optical light curve of GRB 130418A.

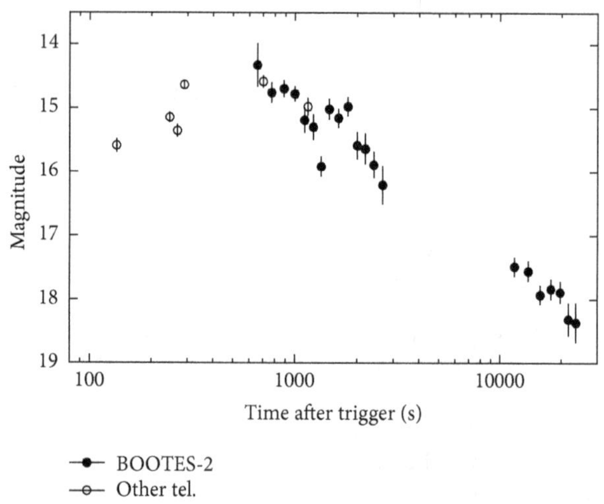

Figure 12: The optical light curve of GRB 130606A. i'-band points were shifted 2.4 mag up to match with the z'-band points.

GRB 130606A. A high-redshift GRB was detected by *Swift* [80], optical afterglow was discovered by BOOTES-2, and redshift is $z = 5.9$ by GTC [81].

BOOTES-2 reaction to this GRB alert was actually a failure; the system did not respond as well as it should and it had to be manually overridden to perform the observations. The first image has therefore been taken as late as 13 minutes after the trigger. These observations led to a discovery of a bright afterglow not seen by *Swift*-UVOT and prompted spectroscopic observations by 10.4 m GTC, which show redshift of this event to be $z = 5.9135$. Overall, 14 photometric points in i'-band and 7 in z'-band were obtained [81]; see Figure 12.

3. Summary

Eleven years of BOOTES-1B and BOOTES-2 GRB follow-up history are summarised in the textual and tabular form. Each GRB is given a short introductory paragraph as a reminder of the basic optical properties of the event. Although we do not discuss the properties in other wavelengths, we try to include a comprehensive reference of literature relevant to each burst. One by one, we show all the successful follow-ups that these telescopes have performed during the first ten years of the *Swift* era and the transition of the BOOTES network to the RTS-2 [14] observatory control system, first installed at BOOTES-2 in 2003 and made definitive during the summer of 2004.

The BOOTES telescopes, in spite of their moderate apertures ($\lesssim$60 cm), have proven to detect a significant number of afterglows—together over 20, contributing to the understanding of the early GRB phase.

Competing Interests

The authors declare that they have no competing interests.

Acknowledgments

The authors appreciate the auspices of INTA, IHSM-UMA/CSIC, and UMA as well as the financial support by the Junta de Andaluca and the Spanish Ministry of Economy and Competitiveness through the Research Projects P07-TIC-03094, P12-TIC2839, AYA 2009-14000-C03-01, AYA 2010-39727-C03-01, and AYA-2015-71718-R. Martin Jelínek was supported by the postdoctoral fellowship of the Czech Academy of Sciences. This study was carried out in the framework of the Unidad Asociada IAA-CSIC at the Group of Planetary Science of ETSI-UPV/EHU. This work was supported by the Ikerbasque Foundation for Science. The Czech CVUT FEL team acknowledges the support by GA CR Grant 13-33324S.

References

[1] R. W. Klebesadel, I. B. Strong, and R. A. Olson, "Observations of gamma-ray bursts of cosmic origin," *The Astrophysical Journal*, vol. 182, pp. L85–L88, 1973.

[2] E. Costa, F. Frontera, J. Heise et al., "Discovery of an X-ray afterglow associated with the γ-ray burst of 28 February 1997," *Nature*, vol. 387, no. 6635, pp. 783–785, 1997.

[3] J. van Paradijs, P. J. Groot, T. Galama et al., "Transient optical emission from the error box of the γ-ray burst of 28 february 1997," *Nature*, vol. 386, no. 6626, pp. 686–689, 1997.

[4] J. L. Racusin, S. V. Karpov, M. Sokolowski et al., "Broadband observations of the naked-eye big γ-ray burst GRB 080319B," *Nature*, vol. 455, pp. 183–188, 2008.

[5] A. J. Castro-Tirado, M. Jelínek, T. J. Mateo Sanguino, and A. De Ugarte Postigo, "BOOTES: a stereoscopic robotic ground support facility," *Astronomische Nachrichten*, vol. 325, article 679, 2004.

[6] M. Jelínek, A. J. Castro-Tirado, A. D. U. Postigo et al., "Four years of real-time GRB followup by BOOTES-1B (2005–2008)," *Advances in Astronomy*, vol. 2010, Article ID 432172, 10 pages, 2010.

[7] O. Rabaza, M. Jelínek, A. J. Castro-Tirado et al., "Compact low resolution spectrograph, an imaging and long slit spectrograph for robotic telescopes," *Review of Scientific Instruments*, vol. 84, no. 11, Article ID 114501, 2013.

[8] L. Resmi, K. Misra, G. Jóhannesson et al., "Comprehensive multiwavelength modelling of the afterglow of GRB 050525A," *Monthly Notices of the Royal Astronomical Society*, vol. 427, no. 1, pp. 288–297, 2012.

[9] J. Sollerman, J. P. U. Fynbo, J. Gorosabel et al., "The nature of the X-ray flash of August 24 2005," *A&A*, vol. 466, no. 3, pp. 839–846, 2007.

[10] J. B. Haislip, M. C. Nysewander, D. E. Reichart et al., "A photometric redshift of $z = 6.39 \pm 0.12$ for GRB 050904," *Nature*, vol. 440, pp. 181–183, 2006.

[11] M. Jelínek, J. Gorosabel, A. J. Castro-Tirado et al., "BOOTES observation of GRB080603B," *Acta Polytechnica*, vol. 52, Article ID 010000, 2012.

[12] M. Jelínek, E. Gómez Gauna, and A. J. Castro-Tirado, "Photometric observations of GRB 080605 by Bootes-1b and Bootes-2," in *Gamma-Ray Bursts: 15 Years of GRB Afterglows*, A. J. Castro-Tirado, J. Gorosabel, and I. H. Park, Eds., vol. 61, pp. 475–477, EAS Publications Series, 2013.

[13] W. Zheng, R. F. Shen, T. Sakamoto et al., "Panchromatic observations of the textbook GRB 110205A: constraining physical mechanisms of prompt emission and afterglow," *The Astrophysical Journal*, vol. 751, no. 2, Article ID 90, 21 pages, 2012.

[14] P. Kubánek, M. Jelínek, M. Nekola et al., "RTS2—remote telescope system, 2nd version," in *Proceedings of the Gamma-Ray Bursts: 30 Years of Discovery*, vol. 727 of *AIP Conference Proceedings*, p. 753, Sante Fe, NM, USA, 2004.

[15] A. J. Blustin, D. Band, S. Barthelmy et al., "*Swift* panchromatic observations of the bright gamma-ray burst GRB 050525a," *The Astrophysical Journal*, vol. 637, no. 2, p. 901, 2006.

[16] M. Della Valle, D. Malesani, J. S. Bloom et al., "Hypernova signatures in the late rebrightening of GRB 050525A," *The Astrophysical Journal Letters*, vol. 642, no. 2, pp. L103–L106, 2006.

[17] A. de Ugarte Postigo, M. Jelínek, J. Gorosabel et al., "GRB 050525A: bootes simultaneous optical observations," GCN Circular 3480, 2005.

[18] J. Norris, L. Barbier, D. Burrows et al., "GRB 050822C: swift detection of a bright burst," GCN Circular 4013, 2005.

[19] H. Krimm, L. Barbier, S. Barthelmy et al., "GCN circular," Tech. Rep. 4020, 2005.

[20] G. Crew, G. Ricker, J.-L. Atteia et al., "HETE fregate observations of GRB 050922C," GCN Circular 4021, 2005.

[21] P. Jakobsson, J. P. U. Fynbo, D. Paraficz et al., "GRB 050922C: refined redshift," GCN Circular 4029, 2005.

[22] N. Mirabal, J. P. Halpern, S. Tonnesen et al., "GRB 051109A: a shallow optical afterglow decay," in *Proceedings of the American Astronomical Society Meeting*, vol. 207, American Astronomical Society, 2006, Abstract 210.02.

[23] E. Fenimore, L. Angelini, L. Barbier et al., "GRB 051109: Swift-BAT refined analysis," GCN Circular 4217, 2005.

[24] R. Quimby, D. Fox, P. Hoeich, B. Roman, and J. C. Wheeler, "GRB 051109: HET optical spectrum and absorption redshift," GCN Circular 4221, 2005.

[25] M. Jelínek, A. de Ugarte Postigo, A. J. Castro-Tirado et al., "GRB 051109a: bootes R & I-band detection of the early afterglow," GCN Circular 4227, 2005.

[26] N. Mirabal, J. Halpern, S. Tonnesen, J. Eastman, and J. Prieto, 2005, http://user.astro.columbia.edu/~jules/grb/051109a/.

[27] F. Yuan, E. S. Rykoff, B. E. Schaefer et al., *Prompt Optical Observations of GRB 080330 and GRB 080413A*, vol. 1065 of *American Institute of Physics Conference Series*, American Institute of Physics, 2008, Edited by Y.-F. Huang, Z.-G. Dai, & B. Zhang.

[28] J. Mao, C. Guidorzi, C. Markwardt et al., "Swift Observation of GRB 080330," GCN Report 132, 2008.

[29] F. E. Marshall, S. D. Barthelmy, D. N. Burrows et al., "Final swift observations of GRB 080413A," GCN Report 129, 2008.

[30] P. Kubánek, M. Jelínek, J. Gorosabel et al., "GCN circular," Tech. Rep. 7603, 2008.

[31] C. Guidorzi, M. Stamatikos, W. Landsman et al., "Swift Observations of GRB 080430," GCN Report 139, 2008.

[32] J. Aleksić, H. Anderhub, L. A. Antonelli et al., "MAGIC observation of the GRB 080430 afterglow," *Astronomy & Astrophysics*, vol. 517, article A5, 2010.

[33] M. Jelínek, P. Kubánek, J. Gorosabel et al., "A decade of GRB follow-up by BOOTES in Spain," 9 Circular 7648, 2008.

[34] A. Rau, A. V. Kienlin, K. Hurley, and G. G. Lichti, "The 1st INTEGRAL SPI-ACS gamma-ray burst catalogue," *Astronomy and Astrophysics*, vol. 438, no. 3, pp. 1175–1183, 2005.

[35] V. Mangano, A. Parsons, T. Sakamoto et al., "Swift observation of GRB 080603B," GCN Report 144, 2008.

[36] B. Sbarufatti, A. Parsons, T. Sakamoto et al., "Swift observation of GRB 080605," *GCN Report*, vol. 142, 2008.

[37] T. Kruhler, J. P. U. Fynbo, S. Geier et al., "The metal-enriched host of an energetic γ-ray burst at $z \approx 1.6$," *Astronomy & Astrophysics*, vol. 546, article A8, 2012.

[38] T. Zafar, D. Watson, Á. Elíasdóttir et al., "The properties of the 2175 å extinction feature discovered in GRB afterglows," *The Astrophysical Journal*, vol. 753, no. 1, p. 82, 2012.

[39] V. Rumyantsev and A. Pozanenko, "GRB 080605: optical observations," GRB Coordinates Network 7857, 2008.

[40] J. Mao, R. Margutti, T. Sakamoto et al., "Swift observations of GRB 090313," GCN Report 204, 2009.

[41] A. de Ugarte Postigo, P. Goldoni, C. C. Thone et al., "GRB 090313: X-shooter's first shot at a gamma-ray burst," *A&A*, vol. 513, article A42, 2010.

[42] A. Melandri, S. Kobayashi, C. G. Mundell et al., "GRB 090313 and the origin of optical peaks in γ-ray burst light curves: implications for lorentz factors and radio flares," *The Astrophysical Journal*, vol. 723, no. 2, p. 1331, 2010.

[43] J. R. Cummings, A. P. Beardmore, and P. Schady, "Swift observations of GRB 090813," GCN Report 240, 2009.

[44] J. Gorosabel, V. Terron, M. Fernandez et al., "GRB 090813: optical candidate from 1.23 m CAHA telescope," GCN Circular 9782, 2009.

[45] F. E. Marshall, L. A. Antonelli, D. N. Burrows et al., "The late peaking afterglow of GRB 100418A," *The Astrophysical Journal*, vol. 727, no. 2, p. 132, 2011.

[46] A. de Ugarte Postigo, C. C. Thone, P. Goldoni, and J. P. U. Fynbo, "Time resolved spectroscopy of GRB 100418A and its host galaxy with X-shooter," *Astronomische Nachrichten*, vol. 332, no. 3, pp. 297–298, 2011.

[47] A. Moin, P. Chandra, J. C. A. Miller-Jones et al., "Radio observations of GRB 100418a: test of an energy injection model explaining long-lasting grb afterglows," *The Astrophysical Journal*, vol. 779, no. 2, p. 105, 2013.

[48] S. Immler, T. Sakamoto, K. L. Page et al., "Swift observations of GRB 100901A," GCN Report 304, 2010.

[49] E. S. Gorbovskoy, G. V. Lipunova, V. M. Lipunov et al., "Prompt, early and afterglow optical observations of five γ-ray bursts: GRB 100901A, GRB 100902A, GRB 100905A, GRB 100906A and GRB 101020A," *MNRAS*, vol. 421, no. 3, pp. 1874–1890, 2012.

[50] O. E. Hartoog, K. Wiersema, P. M. Vreeswijk et al., "The host-galaxy response to the afterglow of GRB 100901A," *Monthly Notices of the Royal Astronomical Society*, vol. 430, no. 4, pp. 2739–2754, 2013.

[51] D. A. Kann, U. Laux, and B. Stecklum, "GRB 100901A: TLS observations, SDSS calibration, decay slope," GCN Circular 11236, 2010.

[52] V. Rumyantsev, D. Shakhovkoy, and A. Pozanenko, "GRB 100901A: CrAO optical observation," GCN Circular 11255, 2010.

[53] D. Gotz, S. Mereghetti, A. Paizis et al., "GRB 101112A: a long GRB detected by INTEGRAL," GCN Circular 11396, 2010.

[54] A. Goldstein, "GRB 101112A: Fermi GBM detection," GCN Circular 11403, 2010.

[55] S. Golenetskii, R. Aptekar, D. Frederiks et al., "Konus-wind observation of GRB 101112A," GCN Circular 11400, 2010.

[56] P. A. Evans and H. A. Krimm, "GRB 101112A—XRT source detection/analysis," GCN Circular 11399, 2010.

[57] C. Guidorzi, R. J. Smith, C. G. Mundell et al., "GRB101112A: Liverpool telescope afterglow candidate," GCN Circular 11397, 2010.

[58] P. Chandra, D. A. Frail, and S. B. Cenko, "Possible detection of INTEGRAL burst GRB 101112A by the EVLA," GCN Circular 11404, 2010.

[59] A. de Ugarte Postigo, P. Kubánek, J. C. Tello et al., "GRB 101112A: BOOTES-2/TELMA optical afterglow candidate," GCN Circular 11398, 2010.

[60] B. Gendre, J. L. Atteia, M. Boer et al., "GRB 110205A: anatomy of a long gamma-ray burst," *The Astrophysical Journal*, vol. 748, no. 1, Article ID 59, 2012.

[61] V. D'Elia, G. Stratta, N. P. M. Kuin et al., "Swift observation of GRB 110213A," GCN Report 323, 2011.

[62] A. Klotz, B. Gendre, M. Boer, and J. L. Atteia, "GRB 120326A: TAROT calern observatory afterglow optical detection," GCN Circular 13107, 2012.

[63] M. H. Siegel, N. P. M. Kuin, S. Holland et al., "Swift observations of GRB 120326A," GCN Report 409, 2013.

[64] B. Sbarufatti, S. D. Barthelmy, N. Gehrels et al., "GRB 120327A: swift detection of a burst with an optical counterpart," GCN Circular 13123, 2012.

[65] V. D'Elia, "VLT/X-shooter absorption spectroscopy of the GRB 120327a afterglow," in *EAS Publications Series*, A. J. Castro-Tirado, J. Gorosabel, and I. H. Park, Eds., vol. 61 of *EAS Publications Series*, pp. 247–249, 2013.

[66] V. D'Elia, J. R. Cummings, M. Stamatikos et al., "Swift observations of GRB 121001A," GCN Report 392, 2012.

[67] M. Andreev, A. Sergeev, and A. Pozanenko, "GRB 121001A: possible optical counterpart," GCN Circular 13833, 2012.

[68] J. C. Tello, R. Gimeno, J. Gorosabel et al., "Swift trigger 535026: optical decay confirmation with IAC80 and BOOTES-2/TELMA," GCN Circular 13835, 2012.

[69] C. Pagani, S. D. Barthelmy, W. H. Baumgartner et al., "GRB 121024A: swift detection of a burst with an optical counterpart," GCN Circular 13886, 2012.

[70] A. Klotz, B. Gendre, M. Boer, and J. L. Atteia, "GRB 121024A: TAROT calern observatory optical detection of a bright counterpart," GCN Circular 13887, 2012.

[71] T. Laskar, A. Zauderer, and E. Berger, "GRB 121024A: EVLA detection," GCN Circular 13903, 2012.

[72] N. R. Tanvir, J. P. U. Fynbo, A. Melandri et al., "GRB 121024A: VLT/X-shooter redshift," GCN Circular 13890, 2012.

[73] M. Jelínek, A. J. Castro-Tirado, and J. Gorosabel, "GRB 121024A: BOOTES-1B optical detection," GCN Circular 13888, 2012.

[74] M. de Pasquale, W. H. Baumgartner, A. P. Beardmore et al., "GRB 130418A: swift detection of a burst with an optical counterpart," GCN Circular 14377, 2013.

[75] S. Golenetskii, R. Aptekar, D. Frederiks et al., "Konus-wind observation of GRB 130418A," GCN Circular 14417, 2013.

[76] A. de Ugarte Postigo, C. C. Thoene, J. Gorosabel et al., "GRB 130418A: redshift from 10.4 m GTC," GCN Circular 14380, 2013.

[77] J. K. Cannizzo, S. D. Barthelmy, J. R. Cummings, A. Melandri, and M. de Pasquale, "Swift observations of GRB 130505A," GCN Report 429, 2013.

[78] N. R. Tanvir, A. J. Levan, T. Matulonis, and A. B. Smith, "GRB 130505A—Gemini-N/GMOS redshift determination," GCN Circular 14567, 2013.

[79] D. A. Kann, B. Stecklum, and F. Ludwig, "GRB 130505A: tautenburg afterglow observations," GCN Circular 14593, 2013.

[80] T. N. Ukwatta, M. Stamatikos, A. Maselli et al., "Swift observations of GRB 130606A," GCN Report 444, 2013.

7

Caravan-Submm, Black Hole Imager in the Andes

Makoto Miyoshi,[1] Takashi Kasuga,[2] Jose K. Ishitsuka Iba,[3] Tomoharu Oka,[4] Mamoru Sekido,[5] Kazuhiro Takefuji,[5] Masaaki Takahashi,[6] Hiromi Saida,[7] and Rohta Takahashi[8]

[1]*National Astronomical Observatory, 2-21-1 Osawa, Mitaka, Tokyo 181-8588, Japan*
[2]*Department of Advanced Sciences, Faculty of Science and Engineering, Hosei University, 3-7-2 Kajino, Koganei, Tokyo 184-8584, Japan*
[3]*Geophysical Institute of Peru, Carretera Ing. Alberto A. Giesecke M. Km 15, Huachac, Peru*
[4]*Department of Physics, Institute of Science and Technology, Keio University, 3-14-1 Hiyoshi, Yokohama, Kanagawa 223-8522, Japan*
[5]*Kashima Space Technology Center, National Institute of Information and Communications Technology, 893-1 Hirai, Kashima, Ibaraki 314-8501, Japan*
[6]*Department of Physics and Astronomy, Aichi University of Education, Kariya 448-8542, Japan*
[7]*Department of Physics, Daido University, Minami-ku, Nagoya 457-8530, Japan*
[8]*National Institute of Technology, Tomakomai College, 443 Nishikioka, Tomakomai, Hokkaido 059-1275, Japan*

Correspondence should be addressed to Makoto Miyoshi; makoto.miyoshi@nao.ac.jp

Academic Editor: Roberto Turolla

Imaging a black hole horizon as a shadow at the center of black hole accretion disk is another method to prove/check Einstein's general relativity at strong gravitational fields. Such black hole imaging is expected to be achievable using a submillimeter wavelength VLBI (very long baseline interferometer) technique. Here, we introduce a Japanese black hole imaging project, Caravan-submm undertaken in the Andes.

1. Introduction

1.1. Black Hole Shadow, Evidence of a Black Hole. The detection of gravitational waves from merging a pair of black holes is the final verification of Einstein's general relativity [1]. The detection of these also opens a new field so called black hole astronomy. Also, imaging a black hole horizon as a shadow at the center of a black hole accretion disk is an independent method to prove/check Einstein's general relativity at strong gravitational fields. Because the vicinity of the horizon is a "strong" gravitational field, observations of behaviors of electromagnetic waves around a black hole must be the best laboratory for general relativity and theories of gravity. The black hole itself is entirely pitch black, but around a black hole, matter that falls onto the black hole forms a very hot disk, where strong electromagnetic waves are emitted. Black hole horizons can be observed as shadows at the centers of such bright and hot accretion disks (Figure 1).

Views of black holes in such situations have been theoretically investigated by many theorists [2–6]. From the images of the black hole shadows, the physical parameters of black holes, that is, mass, spin, and charge, can be determined without degeneracies between these parameters (Figure 2). In other words, precise measurement of the shadow shape gives us complete information of the black hole.

In this paper we introduce that imaging a black hole as a shadow is on the verge of being accomplished after several issues are resolved. First we explain the best target source is Sagittarius A* (SgrA*), the Galactic Center massive black hole in Section 1.2. Second we explain several points that

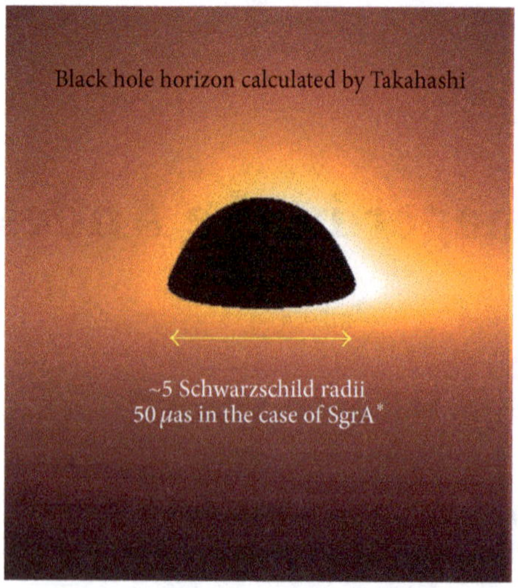

FIGURE 1: Image of black hole shadow calculated by Takahashi [7]. The black hole shadow size in SgrA* is about 50 μas in diameter.

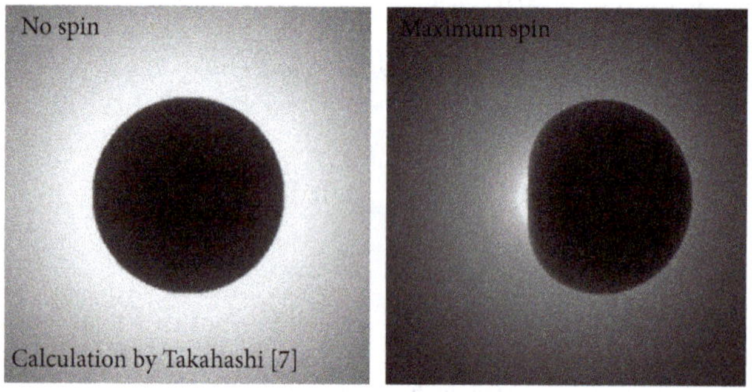

FIGURE 2: Effect of black hole rotation (spin effect). The shape of black hole horizon is distorted by the frame dragging. By measuring the aspect ratio we can investigate the spin of black hole Takahashi [7].

should be settled in Section 1.3. Then in Sections 2 and 3 we explain our proposed plan, Caravan-submm, a black hole imager in the Andes will help settle these issues.

*1.2. The Leading Target Source, Sagittarius A**. Considering the feasibility of imaging the black hole horizon, the apparent angular size of the Schwarzschild radius is the first critical factor to consider. It is almost impossible to gain an understanding of the sizes and figures of stellar black holes through observations. In the case of a stellar black hole located at 1 pc (1 pc = 3.086×10^{13} km, or 3.26 lightyears) with 1 $M_\odot$, the Schwarzschild radius, 2.95×10^3 m, corresponds to only 0.02 μ arc seconds (μas) in angular size (shadow size is 0.1 μas). This angular size is too small to distinguish with spatial resolution of present or near future telescopes.

Because the Schwarzschild radius is proportional to the mass of black hole and also because the apparent angular size is inversely proportional to the distance, super massive black holes at relatively short distance are much more promising to observe. Among them, the most promising target source is Sagittarius A* (SgrA*), the Galactic Center massive black hole (GCBH) with mass $M_{GC} \sim 4 \times 10^6\ M_\odot$ [11]. Because of both the short distance from the earth 8 kpc and the huge mass, the apparent angular size of its Schwarzschild radius is 10 μas. Further the mass of SgrA* is most precisely measured from motions of stars closely orbiting around SgrA*. Using the lower limit on the mass and the measured size ~1 AU (astronomical unit, 1.5×10^8 km) at 86 GHz, the inferred density exceeds a $6.5 \times 10^{21}\ M_\odot pc^{-3}$; therefore, SgrA* cannot be a kind of star cluster [12]. Today SgrA* is the most convincing object for a black hole. The black hole horizon or black hole shadow of ~50 μas in diameter can be observed using the VLBI technique [8, 13–15] (the next promising source is the super massive black hole at the center of M 87; the black hole mass is estimated to be $(6.6 \pm 0.4) \times 10^9\ M_\odot$ by Gebhardt et al. [16]. Assuming the distance to be 16 Mpc, the apparent angular size of its Schwarzschild radius is 8 μas. The prominent jet of M87 may hinder the viewing of the black hole shadow).

1.3. Present Status of Viewing the Black Hole Shadow. With the expectation of observing the surroundings of a black hole, the structure of SgrA* has been investigated using VLBI. However, it is now well known that the observed image of SgrA* is totally broadened and obscured due to the scattering effect of circumnuclear plasma at lower-frequency observations [12, 17–19]. This scattering effect occurs not along the long path to the sun on the galactic plane but mainly in the region within 150 pc from the Galactic Center [20]. The same kinds of plasma regions probably exist in all Galactic Centers. Therefore, the situation will be the same for observing black holes at any active galactic nuclei.

Namely, how to cope with the scattering effect of surrounding plasma is the second factor to be considered. From multiple-frequency VLBI observations of SgrA* mainly using the VLBA it has emerged that the scattering effect shows a λ^2 dependence where λ is an observing frequency. The intrinsic structure of SgrA* cannot be observed at lower frequencies, but at the higher observing frequencies, the scattering effect becomes smaller. It is expected that the intrinsic figure of SgrA* will appear at submillimeter wavelength VLBI observations [13].

The first VLBI observation of SgrA* above 200 GHz was performed by Krichbaum et al. [9]; they estimated the size of SgrA* to be 110 μas from 1100 km baseline observations. After a decade, American research group (MIT, Haystack Observatory) succeeded in obtaining the first fringe detection using multistations from SgrA* observations at 230 GHz [15]. They used a three-station VLBI array consisting of the 10 m Submillimeter Telescope Observatory (SMT) on Mt. Graham in Arizona, one 10 m element of the Combined Array for Research in Millimeter-Wave Astronomy (CARMA) in Eastern California, and the 15 m JCMT in Hawaii. They found fringes from projected baselines of $\sim 3.5 \times 10^9\ \lambda$ (4550 km) with a fringe spacing of 60 μas. Though the data were not sufficient to make a synthesis image but were well fitted by a Gaussian brightness distribution with a full width half max size of 46.0 μas. Taking the still remaining scattering effect into account, the intrinsic size of SgrA* at 230 GHz was estimated to be 40 μas. They further achieved investigation of the magnetic field of SgrA* from their continued EHT (Event Horizon Telescope) project [21] but have still not obtained a sufficient data set to make a synthesis image of the SgrA* black hole.

There are two major reasons why imaging the SgrA* black hole is still difficult. One is the insufficient u-v coverage (amount of sampled data) to make a high quality synthesis image. Radio interferometers including VLBIs sample spatial Fourier components of the brightness distribution of observing source (in radio astronomy, we call the sampled spatial Fourier component visibility). Synthesis image of the source is obtained by inverse Fourier transform of the sampled data. Naturally, if the amount of sampling data is insufficient, it is difficult to obtain a high quality synthesis image. From the simulations Miyoshi et al. [8] found that suitable arrays for imaging the SgrA* black hole shadow at 230 GHz require more than 10 stations located mainly in the southern hemisphere extending to the 8000 km region. The addition of stations in the northern hemisphere improves the image further. The present number of available radio telescopes is insufficient for SgrA* VLBI imaging at 230 GHz or above.

The other main reason lies in data calibrations. Even at relatively lower frequencies of 43 or 86 GHz, calibrating VLBI data of SgrA* was known to be very difficult. From the experiences using the VLBA, Bower et al. [22] noted the following: (1) VLBI arrays were located in the northern hemisphere and lacked north-south spatial resolutions for SgrA* ($\delta = -30^\circ$) to measure even the minor axis size of elliptical Gaussian shape assumed for SgrA*. (2) Observations of SgrA* with such existing VLBI arrays had to be performed at low elevations and it is difficult to calibrate visibility sufficiently. Therefore, even at lower frequencies, it is why we had to abandon synthesis imaging of SgrA* and to use a kind of model fitting to the data. Shen et al. [12] and Bower et al. [17] used the closure phase that is automatically free from systematic errors (sum of three visibility phases found from three baselines forming a triangle). Assuming elliptical Gaussian shapes as SgrA* structure; they fitted to closure quantities and estimated the size and the position angle of elongation of SgrA*. The main origin of such data errors is the atmospheric time variations that bring phase and amplitude variation of obtained spatial Fourier components of the SgrA* structure. The atmospheric time variations bring larger and more rapid fluctuations into data at higher observing frequencies. For imaging black holes we must cope with such errors and establish complete data calibration method.

1.4. Points at Issue. Here we summarize the issues and countermeasures in order to achieve imaging of black hole horizon. The first point is the apparent angular size of the Schwarzschild radius of the black hole. Because SgrA* has the largest one, it is the leading candidate for viewing of a black hole horizon. The second point is the observing frequency. The higher frequencies around 230 GHz or above are required. This is not because of obtaining higher spatial resolutions but because of avoiding the scattering effect of plasma around (massive) black holes. The third is imaging capability of VLBI network we use. To obtain sufficient data for high quality synthesis imaging of the SgrA* black hole horizon, we need much VLBI stations as well as 10 stations mainly in the southern hemisphere. At present, particularly shorter baselines around from 1000 to 2000 km in length are missing.

As long as we perform observations on the ground, obtained data suffer the effects caused by the atmosphere. As for present data calibrations at millimeter/submillimeter VLBI on SgrA*, we have to give up making an effort for precise calibrations and are forced to use model fitting with closure quantities. However, without synthesis imaging using correctly calibrated data, it is almost impossible to find real structures of the source, and in particular, unexpected structures are never detected. Establishing a correct calibration method is a key for obtaining real images of black hole horizons.

As previously mentioned, we need new VLBI stations to sample sufficient data set. This means we need a budget for station construction. In order to achieve the construction of new VLBI stations immediately, cost reduction for construction is necessary. Because SgrA* is an object located in the southern sky, to observe SgrA* with fine observing condition, the southern hemisphere is desirable place for new VLBI stations. In addition, a better site for submillimeter wavelength observations is in the higher mountains where the amount of atmospheric water vapor is lower. It is in such a place that we can receive radio waves at submillimeter wavelength from the universe with low noise. The Andes is a suitable location. In the Andes there are other submillimeter telescopes including the ALMA, meaning it is also suitable for obtaining shorter baselines with new VLBI stations. In the next section, we introduce our proposed plan Caravan-submm in the Andes.

2. Caravan-Submm in the Andes

Japan has a dedicated plan for black hole imaging, named "Caravan-submm," which is a project to construct a millimeter/submillimeter wavelength VLBI network in the Andes. In 2012, we put forward the plan, Caravan-submm to the Astronomy and Astrophysics subcommittee of Science Council of Japan (SCJ), where proposed future plans (middle size projects) were discussed openly (http://www.scj.go.jp/ja/member/iinkai/kiroku/3-140912.pdf. This report is written in Japanese only).

The new VLBI network contains at least its own two fixed VLBI stations and one mobile VLBI station. We plan to construct the fixed stations at around Huancayo Observatory, Geophysical Institute of Peru (IGP), and at the Mount Chacaltaya Laboratory in Bolivia. Huancayo Observatory is at Huancayo Province in Peru (12.0°S, 75.29°W, 3370 m in altitude), on the equator of the Earth's magnetic field. The observatory is famous for observation of geomagnetism. In collaboration with the National Astronomical Observatory of Japan (NAOJ), the director, Professor Ishitsuka, promotes radio astronomy with the Sicaya 32 m radio telescope.

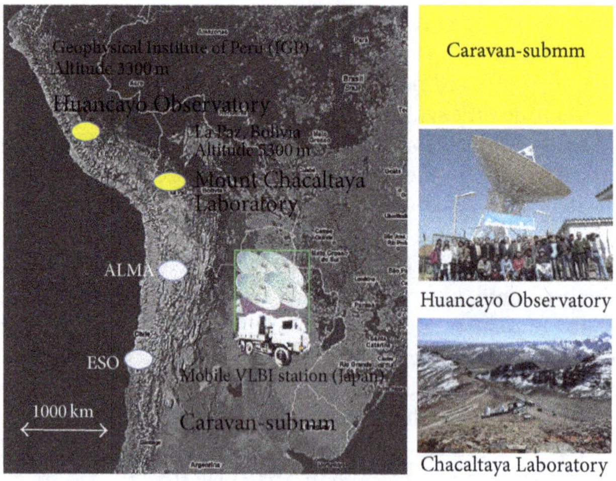

FIGURE 3: Caravan-submm in the Andes. The mobile radio telescope Caravan-submm moves around Andes.

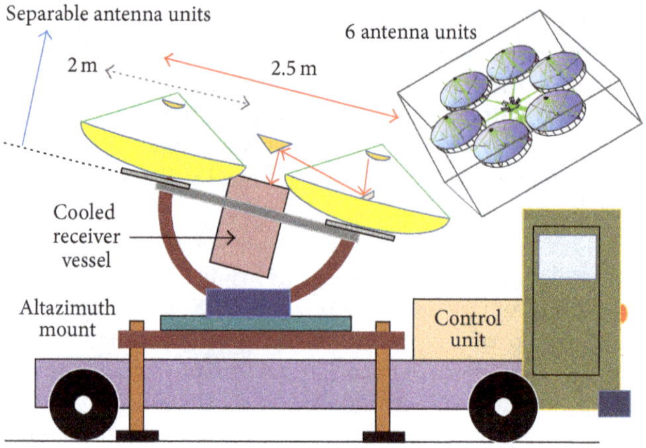

FIGURE 4: Caravan-submm, mobile VLBI station.

The Mount Chacaltaya Laboratory is on the top of Mt. Chacaltaya, near La Paz in Bolivia, and is operated in collaboration with worldwide institutes (16.35°S, 68.13°W, 5270 m in altitude). It is at the Chacaltaya Laboratory that the Yukawa's predicted π-meson was detected in 1947. The observatory's collaboration with Japan, namely, with the Institute for Cosmic Ray Research (ICRR) at the University of Tokyo, has been ongoing since 1962 (Figure 3).

2.1. Caravan-Submm, Mobile VLBI Station. For imaging black holes, as mentioned in the previous section, one of the key points is the u-v coverage and how many spatial Fourier components of the observing source can be sampled. At present, it is essentially important to obtain shorter baselines around from 1000 to 2000 km in length for the black hole imaging of SgrA*. In order to sample sufficient VLBI data for good imaging, till now, we have made efforts to construct as many new VLBI stations as possible. However, instead, Caravan-submm includes one mobile VLBI station. By changing observational positions among Andes mountains, we repeat VLBI observations of SgrA* and sample various kinds of VLBI data which allow us to make high quality images with excellent cost performance. The shape of the black hole horizon shows a constant figure because the shape is defined only by the black hole space-time. Therefore it is no matter to take the picture with longer exposure time. We intend to make the mobile station first, prior to the constructions of fixed stations. The required baseline lengths ranging from 1000 to 2000 km can be realized not only by our own fixed stations but also by collaborating with nearby ALMA (an open use radio telescope) and/or other telescopes in the Andes.

A VLBI mobile station or transportable station is a famous method in Japan because such geodetic VLBI experiments have been performed since the mid-1980s [23]. Also the Geospatial Information Authority of Japan (GSI) performed geodetic VLBI project using VLBI mobile stations [24, 25]. Using the same method, we sample VLBI data effectively and intend to attain higher quality images at submillimeter VLBI observations than those only fixed stations can do.

For transportation, the Caravan-submm mobile station will contain at least three trucks: one for antennas, another for mount, and the other for the electric power supply, VLBI recording system, and hydrogen maser. However, we intend to obtain the power supply for telescope operations from commercial base (Figure 4). The Peruvian and Bolivian Andes are developed regions with many cities and towns, meaning that we can easily obtain a commercially based electric power supply. Main roads also have been well developed throughout the Andes, which means the Caravan-submm mobile station can quite easily be transported.

The difference is the observing frequency between the mobile station for geodetic VLBI yesteryear and our new Caravan mobile station. We have to receive 230 GHz band while the old geodetic VLBI utilized SX bands (2 GHz and 8 GHz) for positional measurements. The higher the observing frequency we use, the more difficult the maintaining of coherency of observing system becomes. However, recent VLBI experiment at 230 GHz in Japan provided us with technical evidence for the Caravan mobile station. At Nobeyama radio observatory in April 2015, using two radio telescopes with independent frequency standards, VLBI fringe at 230 GHz was detected from a tentative set-up of VLBI system. It was the first VLBI fringe detection at 230 GHz in Japan, which is the world tie record for the highest frequency VLBI [26]. This proves that the idea of Caravan-submm, mobile submillimeter telescope is technically feasible. The used VLBI system for that experiment was a very temporary one, composed of instruments the participants brought together in haste. This situation was very similar to that of the Caravan-submm mobile observing system.

2.2. Multimirror System for the Mobile Station. We must further elaborate ideas to realize a mobile 230 GHz radio telescope, Caravan-submm mobile station. Today, following

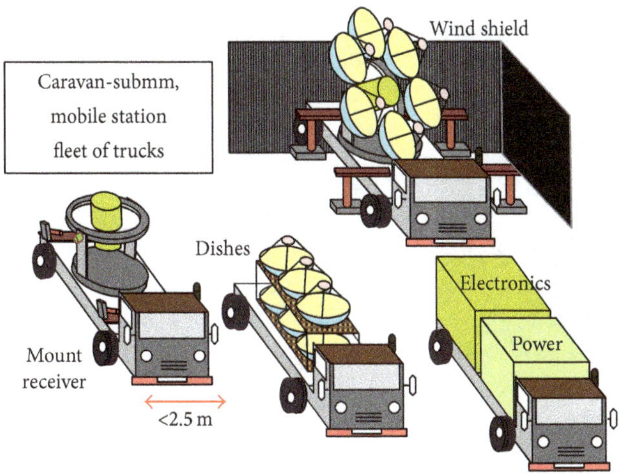

FIGURE 5: Caravan-submm, mobile VLBI station fleet of trucks.

FIGURE 6: Japanese metal spinning method (Kitajima Shibori Seisakusho Co., Ltd.).

the ALMA construction, a submillimeter telescope is not so technically difficult to construct. However it is also true that a simple application of ALMA experience is not enough to realize the mobile radio telescope for such high frequency observations. In case of ALMA 12 m telescope, to obtain a large aperture for high frequency receiving, panels on rigid structure with actuator are adopted and successful performance is attained. But the weight of the ALMA telescope means it is too heavy to easily relocate and also complicated operations are required. The ALMA telescope is designed to fulfill universal functions, and as a result it is expensive, while our concept of the Caravan-submm mobile station is quite different from that of ALMA. The Caravan-submm mobile station must move around the Andes region. Its telescope must be transportable, and it must be easy to assemble and disassemble. To fulfill the requirements we adopt a multimirror system for our mobile station. Using a couple of lightweight dishes 2 m in diameter, we get a total aperture area corresponding to that of 4 m single dish. Use of lightweight small dishes brings us two advantages. The first is that individual small dishes are so light that we can easily transport them using the usual vehicles and can assemble them by hand and disassemble them after observations. Second, we can attain cost reduction when we use several small dishes instead of making a single dish with the same collecting area. In general, the cost of making a single dish increases more rapidly than is proportional to the square of its diameter. This is why we plan to use a multimirror system for the Caravan-submm mobile telescope (Figure 5).

2.3. Antenna by Metal Spinning Method. In order to achieve the cost reduction in making high precision antennas for receiving 230 GHz, we have examined the accuracy of the parabola surfaces made using the Japanese metal spinning method (Figure 6) and found that an accuracy as well as 60 μm r.m.s. is easily obtained. Recently we confirmed that surface accuracies about 15 μm r.m.s. were achieved by adding an annealing process in spinning processing (Figure 7).

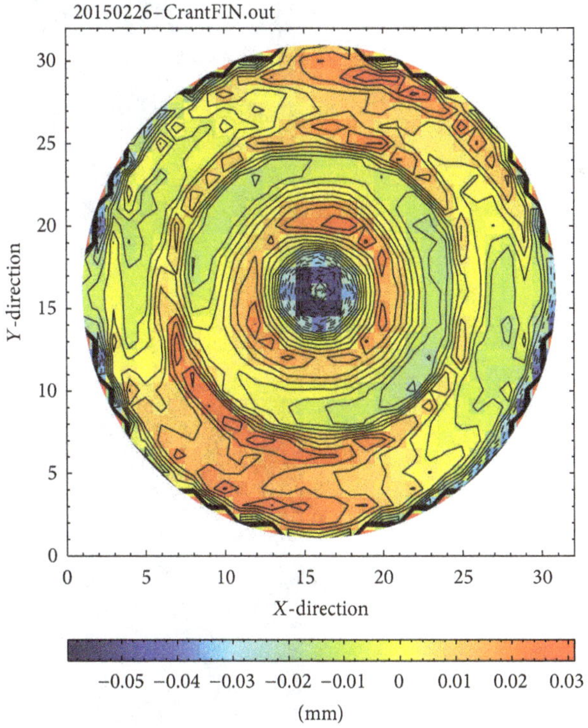

FIGURE 7: Surface accuracy about 15 μm r.m.s. was achieved using a new 30 cm mold (surface accuracy: 4 μm r.m.s.) by metal spinning method with adding an annealing process.

Now we have hopeful outlooks that cost reduced millimeter/submillimeter antennas as large as 2 m in diameter can be produced using the Japanese metal spinning method.

2.4. Site Survey. To observe with such high frequencies on the ground, high mountains such as the Andes are the best sites where the transparency of the millimeter/submillimeter wavelengths in the atmosphere is good. We performed a site survey in the Andes with the support of the ICRR. In order to

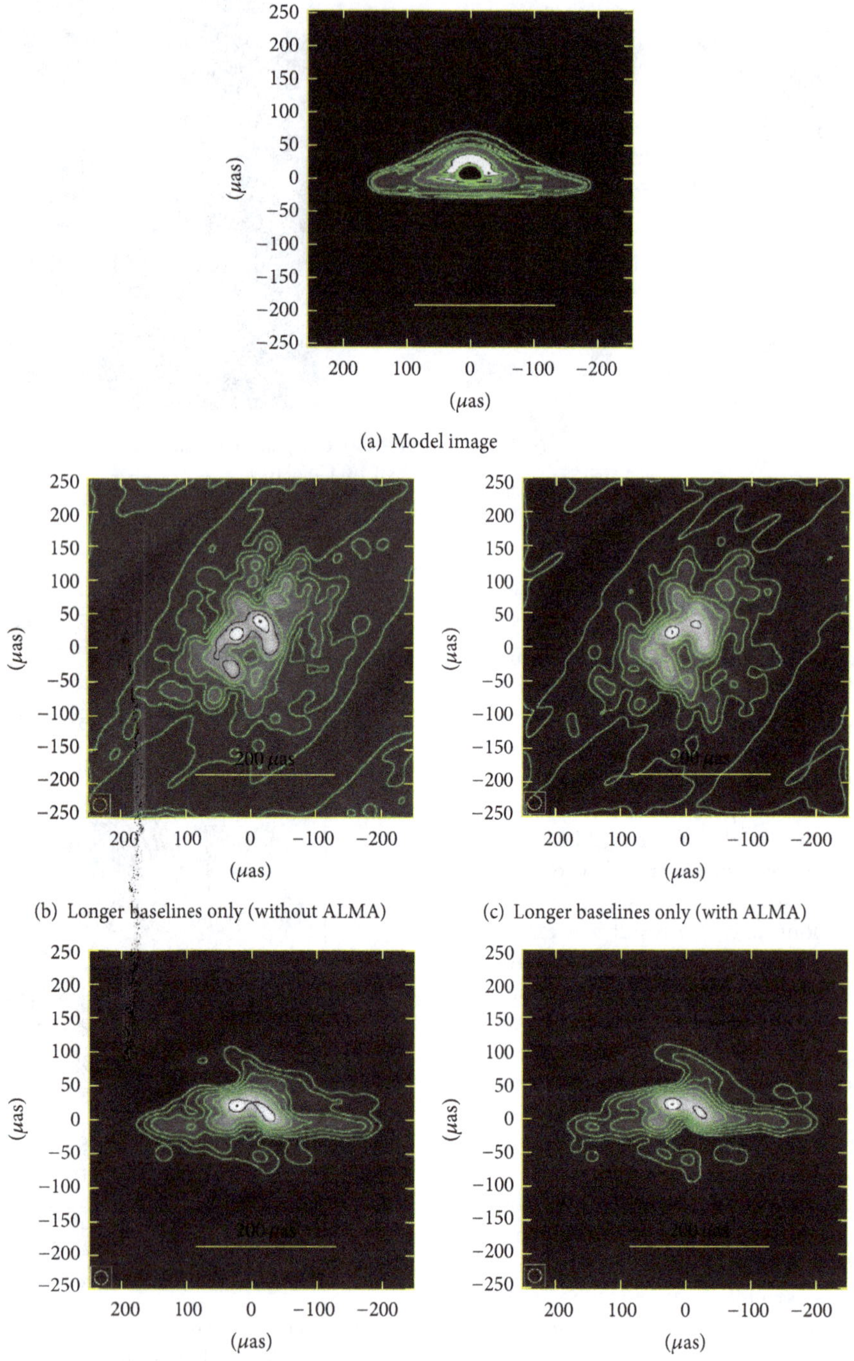

(a) Model image

(b) Longer baselines only (without ALMA)

(c) Longer baselines only (with ALMA)

(d) Addition of Caravan (without ALMA)

(e) Addition of Caravan (with ALMA)

FIGURE 8: Imaging simulation for u-v coverage. (a) Model image of edge-on view RIAF accretion disk with 50 μas black hole shadow at the center. Total flux density is 3 Jy. (b) Result from only global longer baselines without phased-ALMA. (c) Result from only global longer baselines with phased-ALMA. (d) Result from combination of global longer baselines plus Caravan-submm array without phased-ALMA. (e) Result from combination of global longer baselines plus Caravan-submm array with phased-ALMA. Here we assumed 1024 MHz recording bandwidth.

confirm the transparency at millimeter/submillimeter radio waves, measurement of the water vapor contents in the atmosphere should be done. We measured the water vapor contents at Huancayo Observatory (IGP) in Peru, Mount Chacaltaya Laboratory (Universidad Mayor de San Andres) in Bolivia, and other places. We found the precipitable water vapor amounts in the rainy reason (February to March in 2015) were at least twice as high compared with those in the dry season (June to July in 2012). This means that the best season for millimeter/submillimeter VLBI observations is limited even in the high Andes.

2.5. System Operation. The Japanese Caravan-submm team will not only perform our observations in the Andes alone, but also take part in the worldwide observations. As a first step after constructing the Caravan-submm mobile station, we begin observations with other telescopes in the Andes. By changing observational positions of the Caravan-submm mobile station, we sample data from various baselines. Checking the fringe amplitude-baseline length relations, we investigate the first null point, from which position we can estimate the horizon size and the mass of black hole assuming simple image models (concerning the details of the method we explain in the next section). It is highly possible that at this stage we will obtain crucial evidence of a black hole horizon. We will take part in the global millimeter/submillimeter VLBI observation as soon as technically possible. From global millimeter/submillimeter VLBI observations of SgrA*, we can expect to obtain reliable imaging results of the black hole horizon.

2.6. Performance of Caravan-Submm. We performed further imaging simulations with changing array configurations and model images. Here we assumed a long baseline array contains seven stations, SMA (Hawaii, US), LLAMA (Argentina), LMT (Mexico), SMT (Arizona, US), CARMA (California, US), Plateau de Bure interferometer (Spain), and IRAM (Pico Veleta, Spain), and with/without phased-ALMA, while the Caravan-submm includes the two fixed stations at Huancayo, Peru, and Mt. Chacaltaya, Bolivia, with 13 different observing positions for the mobile station. Figure 8 shows a simulation result for SgrA*. First of all, resultant images have little difference between with and without the phased-ALMA. This is because SgrA* is so bright at 230 GHz that we can detect fringes of SgrA* with high signal to noise ratio even without ALMA. The essential role of phased-ALMA as a VLBI station lies in increasing the number of detectable sources with millimeter to submillimeter VLBI. Second, there is strong possibility that we fail to obtain a correct image with only global longer baselines like the present EHT array. Needless to say, it is also difficult to obtain a correct image with only shorter baselines like Caravan-submm array. It is from the combination of global longer baselines and Andes shorter baselines that we can obtain reliable images of SgrA*. This conclusion is entirely consistent with the common sense in radio interferometry that image quality depends on how the *u-v* plain can be covered.

3. Data Analysis and Calibration

In this section we introduce methods of data analysis and data calibrations. First, we mention a model fitting method using the relation between amplitude and projected baseline length that was frequently used in the old days and then two new methods we developed. The Slit Modulation Imaging (SMI) method is developed for detecting shorter time variations

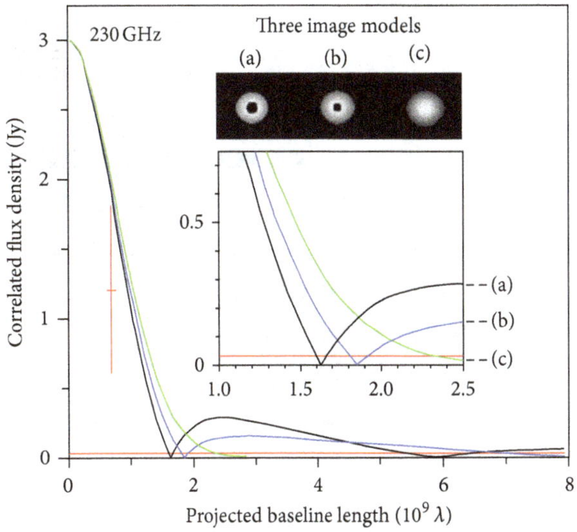

FIGURE 9: From Miyoshi et al. [8]: the visibility amplitudes of three image models as function of projected baseline: (a) the case of $M_{BH} = 3.7 \times 10^6\ M_\odot$, (b) the case of $M_{BH} = 2.6 \times 10^6\ M_\odot$, and (c) the case with no black hole or the scattering effect is still dominant. The functions of (a) and (b) have null value points that indicate the existence of the central black shadow. The 3σ noise level of present engineering performance is shown by the red horizontal line. The red point with error bar is the visibility amplitude measured by Krichbaum et al. [9].

such as quasi periodic oscillation (QPO). The verification method of calibrations (VERICA) is the one to investigate the existence of residuals of systematic errors in calibrated data.

3.1. A Simple Analysis Using Amplitude u-v Distance Relation. When an array has limited coverage in the *u-v* plane, instead of synthesis imaging, visibility analysis has frequently been performed in order to estimate the shape and size of the observed sources. In the early days of radio interferometers such methods were mainly used. Figure 9 shows visibility amplitude curves of three simple image models, (a) a simple Gaussian brightness without shadow, (b) a Gaussian with a shadow of 30 μas ($M_{BH} = 2.6 \times 10^6\ M_\odot$), and (c) a Gaussian with a shadow of 45 μas ($M_{BH} = 3.7 \times 10^6\ M_\odot$). For simplicity here we used point-symmetric images. A Gaussian brightness distribution also shows a Gaussian curve in the visibility amplitudes. If the shadow exists, the visibility function has null value points at some projected baseline length. The null value positions change with the size of the shadow. From the visibility amplitude function, we can distinguish whether the shadow exists or not. Further, because the null value points move according to the shadow size, we can estimate the shadow size, which also means we can measure the mass of the black hole from the null value positions. For measuring the correlated flux densities with *u-v* distance, a small array composed of a few stations is sufficient. It is certain that the null value points will appear in visibility amplitude and projected baseline diagrams with other types of images. One of the typical ones is double sources like core and jet structures, which are frequently observed from other AGNs.

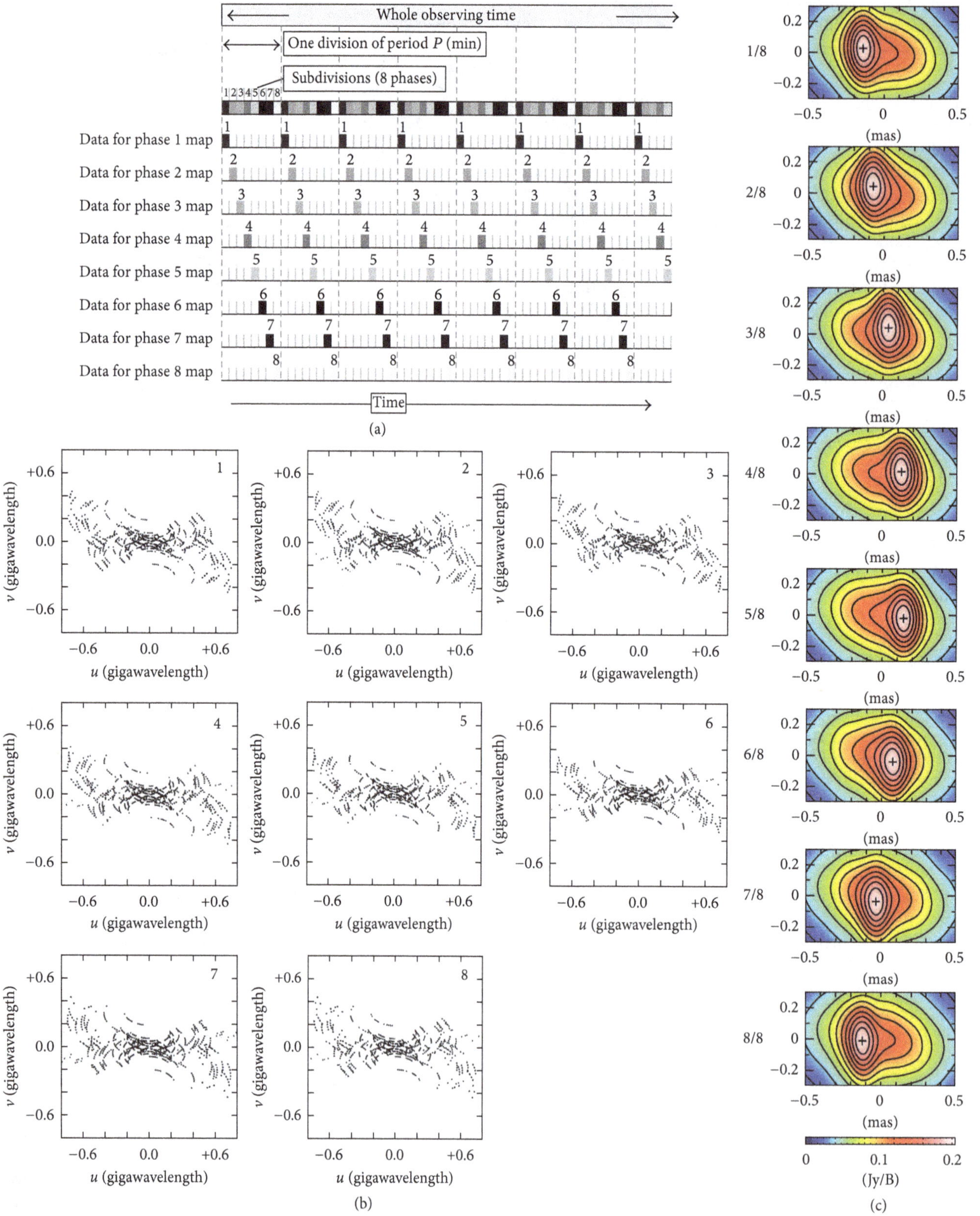

FIGURE 10: The SMI method. (a) Schematic diagram of data-sampling pattern in the SMI method. (b) u-v coverage of the 8-phase SMI maps for SgrA* ($\delta = -30°$) at 43 GHz by VLBA. $N_{div} = 8$ and $P_{trial} = 16.8$ min. (c) A simulative result of the SMI method for model of moving point on fixed Gaussian at $\delta = -30°$. The map area is 1 mas $(e - w) \times 0.5$ mas $(n - s)$. Contour levels are every 10% of the peak. Cross mark indicates the intensity peak position corresponding to the moving point position [10].

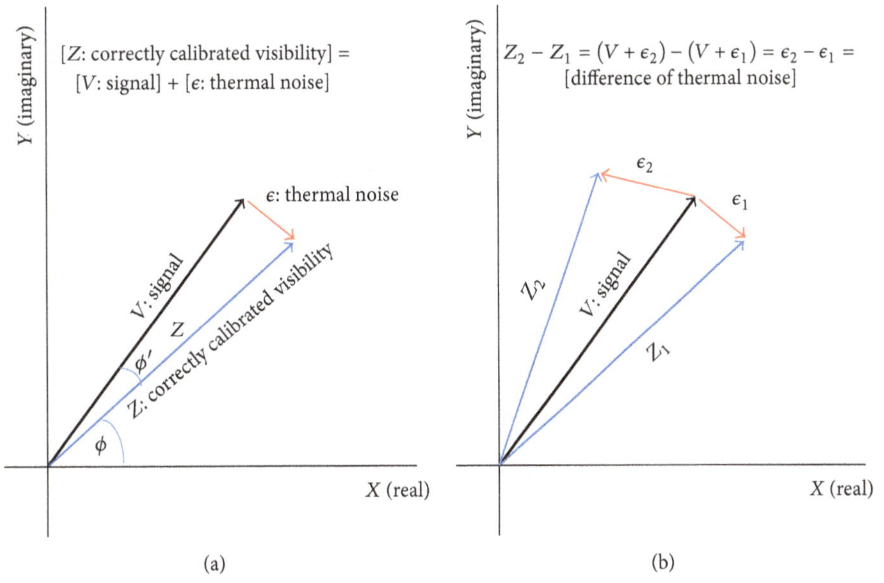

FIGURE 11: Relations of signal V, thermal noise ϵ, and correctly calibrated visibility Z. If the calibration is perfectly performed, the differential visibility shows thermal noise feature. While the calibration is insufficient, namely, systematic errors still remain, the differential visibility shows different distributions from thermal ones.

In the case of SgrA*, there is no convincing result showing separated structures from previous VLBI observations. The VLBI image of SgrA* always shows a single structure. When an eruption happens at SgrA*, the double structure will appear. But the structure is a tentative one and will soon disappear. It will be easy to distinguish the reason for the existence of null value points in visibility amplitude whether from core and jet structures or from the darkness of a black hole shadow. Even if the SgrA* structure is not simple like the three models used here, we can limit the structure model from estimated mass and spin of the black hole in SgrA* and the accretion disk model and shape of black hole shadow from ray-tracing calculations. Reliable image-model fitting to observed visibilities will certainly be possible to detect the black hole shadow in SgrA*. In the case of SgrA* the first null point will appear at projected baseline length ranging from 1000 to 2000 km. This is the reason why such shorter baselines of Caravan-submm are the key for detecting black hole horizon of SgrA* [14].

3.2. Slit Modulation Imaging (SMI) Method. The SMI method is for detecting shorter periodic structural changes of observed source from interferometric data. It is highly effective in detecting periodic change patterns whose period is shorter than the observing time span, from interferometric data [10]. The essence of the method lies in the sampling pattern of visibility data in time series. Figure 10(a) shows a schematic diagram of data selection from an observation time span. First we divide the whole observation time into several divisions with a trial period, P_{trial}, and each division is divided into subdivisions (n phase segments). Figure 10(a) shows the case of $N_{\text{div}} = 8$. The nth phase SMI map is produced from the visibilities of the nth phase segments of all the divisions. Figure 10(b) shows the resultant u-v coverage of each phase SMI map. Thus, the SMI maps have very similar u-v coverage, almost free from the influence of u-v coverage differences. Unlike the analysis of time series of snapshot maps, there is no need to discuss the effect of different u-v coverage. Figure 10(c) shows a simulation example that SMI method detects a periodic structure change: a 0.1 Jy point source oscillates with period $P_{\text{change}} = 16.8$ min on a fixed elliptical Gaussian brightness distribution shape of SgrA* at 43 GHz. By changing P_{trial} we can learn the periodic structure change of the source if it exists. As demonstrated, we can detect periodic structure change patterns with shorter periods than the observing time span.

3.3. VERICA (VERIfication Method of CAlibrations) Method. Verification of applied calibration onto visibility is important but has not been established well so far. We found such a method named VERICA recently. Because the source structure is unknown to us in general, we cannot predict how the signal V behaves. However, if we can subtract the V from the obtained visibility Z and if systematic errors are removed by calibrations, the residual shows a pure thermal noise feature, which can be estimated from measured system temperatures and antenna performance. Differential visibilities between different video channels recorded at the same time with the same baseline (and each frequency is set to be nearly equal) correspond to such quantities. If the individual calibrations applied to the visibilities are precisely appropriate, the difference should statistically show a differential thermal noise distribution (Figure 11). Therefore, by checking distributions of differential visibilities, we can verify whether the applied calibrations onto visibility are proper or not without knowledge of observed source structure. From probability density functions of signal and thermal noise [27], we calculated its differential probability density

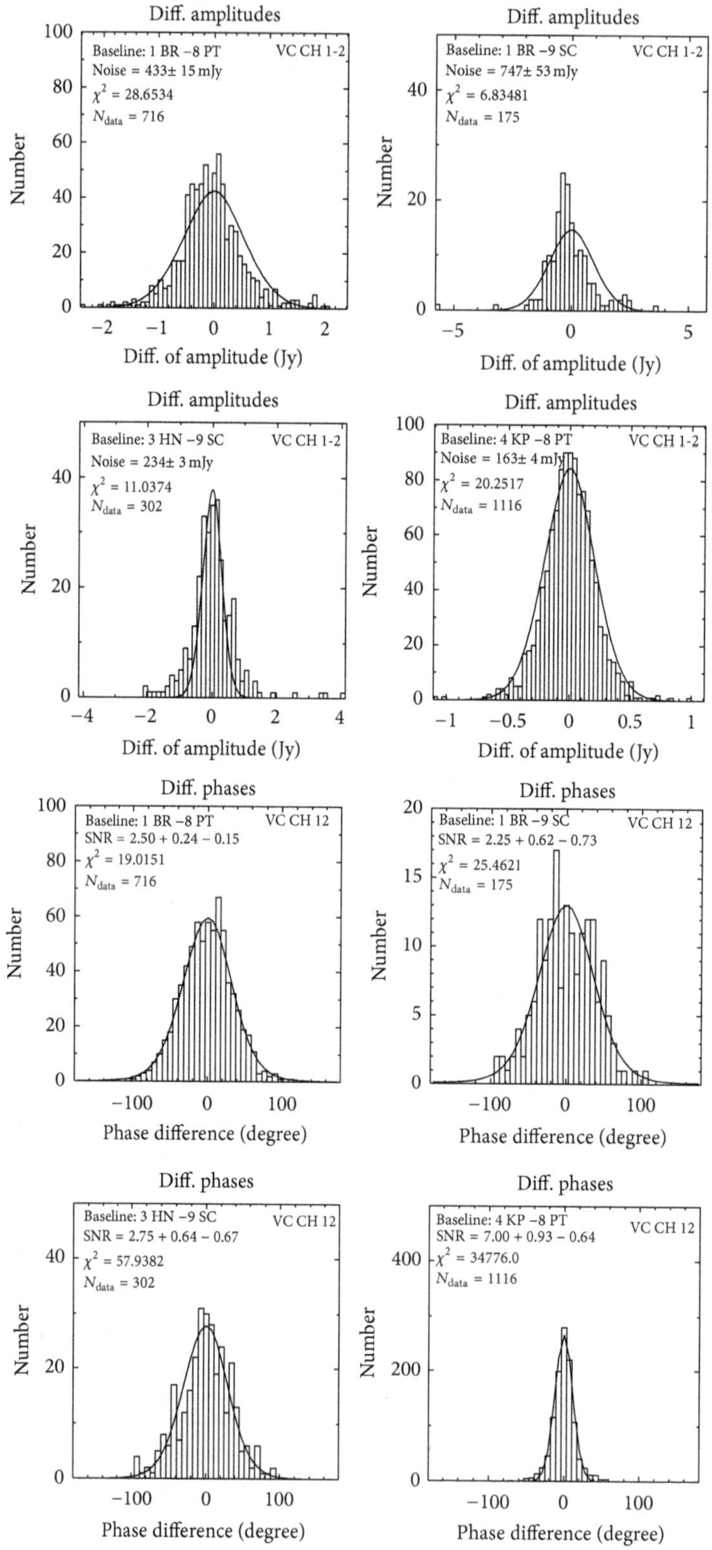

FIGURE 12: Examples of differential visibility distributions: from several baselines, BR-PT, BR-SC, HN-SC, and KP-PT. Every visibility is 10 sec integrations from VLBA 43 GHz observations of SgrA* (code: BS131B).

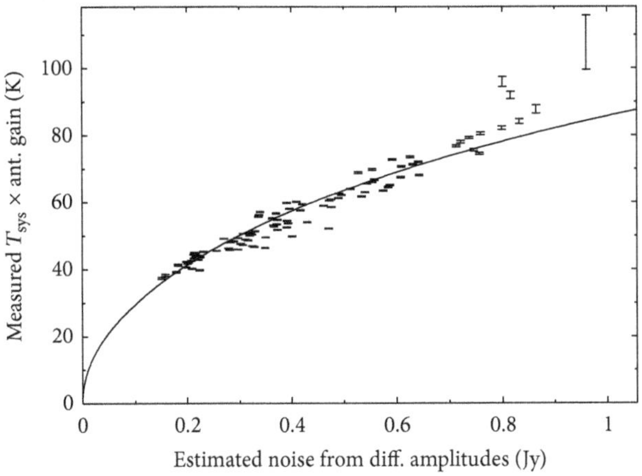

FIGURE 13: Estimated noise from fitting to differential visibility amplitude distribution and measured T_{sys} × gain. In this case, the calibrated data shows expected thermal noise features.

functions and developed this new method, VERICA [28]. Figure 12 shows examples of differential phase and amplitude distributions, respectively, from real VLBA 43 GHz data of SgrA* [29]. Real distributions are fitted well to theoretical ones. From differential amplitude fitting, we can estimate the noise levels. We found the estimated noises are consistent with measured system noise and antenna gains (Figure 13). This means systematic errors are well removed by applied calibrations. We can judge whether the applied calibrations are proper or not by VERICA even when the observed source shows time variation during observations. This is why the differential visibility contains no component of source structure. At millimeter to submillimeter wave VLBI observations of SgrA*, we will encounter such situations in future, but using the VERICA method we can find good calibration solutions to be applied and obtain correct synthesis images. (Note: here each video channel is calibrated independently by Fringe-search and self-calibration. Thus the degrees of residual systematic errors are different in each data item after individual calibration. The differential visibility distributions between individually calibrated data show deviations from thermal noise ones if residual systematic errors still exist.)

Competing Interests

The authors declare that they have no competing interests.

Acknowledgments

The authors would like to thank Professor M. Tsuboi at Institute of Space and Astronautical Science, Japan Aerospace Exploration Agency, for technical discussion about multimirror systems. This work was supported in part by Grand-in-Aid for Scientific Research of JSPS (25610044 and 16K05302, Rohta Takahashi, and 23654071, Makoto Miyoshi and Takashi Kasuga).

References

[1] B. P. Abbott, R. Abbott, T. D. Abbott et al., "Observation of gravitational waves from a binary black hole merger," *Physical Review Letters*, vol. 116, Article ID 061102, 2016.

[2] C. T. Cunningham and J. M. Bardeen, "The optical appearance of a star orbiting an extreme KERR black hole," *Astrophysical Journal*, vol. 173, p. L137, 1972.

[3] J. M. Bardeen and C. T. Cunningham, "The optical appearance of a star orbiting an extreme Kerr black hole," *Astrophysical Journal*, vol. 183, pp. 237–264, 1973.

[4] C. T. Cunningham, "The effects of redshifts and focusing on the spectrum of an accretion disk around a Kerr black hole," *Astrophysical Journal*, vol. 202, pp. 788–802, 1975.

[5] J.-P. Luminet, "Image of a spherical black hole with thin accretion disk," *Astronomy and Astrophysics*, vol. 75, no. 1-2, pp. 228–235, 1979.

[6] J. Fukue and T. Yokoyama, "Color photographs of an accretion disk around a black hole," *Publications of the Astronomical Society of Japan*, vol. 40, no. 1, pp. 15–24, 1988.

[7] R. Takahashi, "Black hole shadows of charged spinning black holes," *Publications of the Astronomical Society of Japan*, vol. 57, no. 2, pp. 273–277, 2005.

[8] M. Miyoshi, S. Kameno, J. K. Ishitsuka, Z.-Q. Shen, R. Takahashi, and S. Horiuchi, "An approach detecting the event horizon of Sgr A*," *Publication of the Astronomical Society of Japan*, vol. 10, pp. 15–23, 2007, http://www.nao.ac.jp/about-naoj/reports/publications-naoj.html, http://www.nao.ac.jp/contents/about-naoj/reports/publications-naoj/10-1234-2.pdf.

[9] T. P. Krichbaum, D. A. Graham, A. Witzel et al., "VLBI observations of the galactic center source SGR A* at 86 GHz and 215 GHz," *Astronomy and Astrophysics*, vol. 335, pp. L106–L110, 1998.

[10] M. Miyoshi, "Slit-modulation imaging method for detecting any periodic structural change in interferometric data," *Publications of the Astronomical Society of Japan*, vol. 60, no. 6, pp. 1371–1386, 2008.

[11] A. M. Ghez, S. Salim, N. N. Weinberg et al., "Measuring distance and properties of the milky way's central supermassive black hole with stellar orbits," *The Astrophysical Journal*, vol. 689, no. 2, pp. 1044–1062, 2008.

[12] Z.-Q. Shen, K. Y. Lo, M.-C. Liang, P. T. P. Ho, and J.-H. Zhao, "A size of ~1 AU for the radio source Sgr A* at the centre of the Milky Way," *Nature*, vol. 438, no. 7064, pp. 62–64, 2005.

[13] H. Falcke, F. Melia, and E. Agol, "Viewing the shadow of the black hole at the Galactic center," *The Astrophysical Journal*, vol. 528, no. 1, pp. L13–L16, 2000.

[14] M. Miyoshi, J. K. Ishitsuka, S. Kameno, Z.-Q. Shen, and S. Horiuchi, "Direct imaging of the massive black hole, SgrA*," *Progress of Theoretical Physics Supplement*, vol. 155, pp. 186–189, 2004.

[15] S. S. Doeleman, J. Weintroub, A. E. E. Rogers et al., "Event-horizon-scale structure in the supermassive black hole candidate at the Galactic Centre," *Nature*, vol. 455, pp. 78–80, 2008.

[16] K. Gebhardt, J. Adams, D. Richstone et al., "The black hole mass in M87 from GEMINI/NIFS adaptive optics observations," *The Astrophysical Journal*, vol. 729, no. 2, article 119, 13 pages, 2011.

[17] G. C. Bower, H. Falcke, R. M. Herrnstein, J.-H. Zhao, W. M. Goss, and D. C. Backer, "Detection of the intrinsic size of sagittarius A* through closure amplitude imaging," *Science*, vol. 304, no. 5671, pp. 704–708, 2004.

[18] S. S. Doeleman, Z.-Q. Shen, A. E. E. Rogers et al., "Structure of Sagittarius A* at 86 GHz using VLBI closure quantities," *Astronomical Journal*, vol. 121, no. 5, pp. 2610–2617, 2001.

[19] K. Y. Lo, Z.-Q. Shen, J.-H. Zhao, and P. T. P. Ho, "Intrinsic size of Sagittarius A*: 72 Schwarzschild radii," *The Astrophysical Journal*, vol. 508, no. 1, pp. L61–L64, 1998.

[20] G. C. Bower, D. C. Backer, and R. A. Sramek, "VLBA observations of astrometric reference sources in the galactic center," *The Astrophysical Journal*, vol. 558, no. 1, pp. 127–132, 2001.

[21] M. D. Johnson, V. L. Fish, S. S. Doeleman et al., "Resolved magnetic-field structure and variability near the event horizon of Sagittarius A*," *Science*, vol. 350, no. 6265, pp. 1242–1245, 2015.

[22] G. C. Bower, H. Falcke, D. C. Backer, and M. Wright, *The Central Parsecs of the Galaxy*, vol. 186 of *ASP Conference Series*, Astronomical Society of the Pacific, 1999.

[23] R. Ichikawa, A. Ishii, H. Takiguchi et al., "Present status and outlook of compact VLBI system development for providing over 10km baseline calibration," in *Proceedings of the 8th NICT IVS TDC Symposium*, Kashima, Japan, February 2009, http://www2.nict.go.jp/aeri/sts/stmg/ivstdc/news_30/pdf/tdcnews_30.pdf.

[24] S. Matsuzaka, M. Tobita, Y. Nakahori, J. Amagai, and Y. Sugimoto, "Detection of Philippine Sea plate motion by very long baseline interferometry," *Geophysical Research Letters*, vol. 18, no. 8, pp. 1417–1419, 1991.

[25] Y. Fukuzaki, K. Wada, R. Kawabata, M. Ishimoto, and T. Wakasugi, "First geodetic result of ishioka VGOS antenna," in *Proceedings of the 14th NICT TDC Symposium*, 35, p. 1, TDC News, Kashima, Japan, June 2015, http://www2.nict.go.jp/aeri/sts/stmg/ivstdc/news_35/tdc_news35.pdf.

[26] K. Takefuji and MICE2015 Team, "First Japanese 230 GHz VLBI experiment by MICE2015 team," in *Proceedings of the 14th NICT TDC Symposium*, p. 22, Kashima, Japan, June 2015, http://www2.nict.go.jp/aeri/sts/stmg/ivstdc/news_35/tdc_news35.pdf.

[27] A. R. Thompson, J. M. Moran, and G. W. Swenson, *Interferometry and Synthesis in Radio Astronomy*, John Wiley & Sons, New York, NY, USA, 2nd edition, 2001.

[28] M. Miyoshi, "A new verification method of applied calibrations onto visibility data," in *Proceedings of the 13th NICT TDC Symposium*, p. 3, Kashima, Japan, June 2014, http://www2.nict.go.jp/aeri/sts/stmg/ivstdc/news_34/pdf/pctdc_news34.pdf.

[29] M. Miyoshi, Z.-Q. Shen, T. Oyama, R. Takahashi, and Y. Kato, "Oscillation phenomena in the disk around the massive black hole Sagittarius A," *Publication of the Astronomical Society of Japan*, vol. 63, no. 5, pp. 1093–1116, 2011.

8

Analytic Models of Brown Dwarfs and the Substellar Mass Limit

Sayantan Auddy,[1] Shantanu Basu,[1] and S. R. Valluri[1,2]

[1]*Department of Physics and Astronomy, The University of Western Ontario, London, ON, Canada N6A 3K7*
[2]*King's University College, The University of Western Ontario, London, ON, Canada N6A 2M3*

Correspondence should be addressed to Sayantan Auddy; sauddy3@uwo.ca

Academic Editor: Gary Wegner

We present the analytic theory of brown dwarf evolution and the lower mass limit of the hydrogen burning main-sequence stars and introduce some modifications to the existing models. We give an exact expression for the pressure of an ideal nonrelativistic Fermi gas at a finite temperature, therefore allowing for nonzero values of the degeneracy parameter. We review the derivation of surface luminosity using an entropy matching condition and the first-order phase transition between the molecular hydrogen in the outer envelope and the partially ionized hydrogen in the inner region. We also discuss the results of modern simulations of the plasma phase transition, which illustrate the uncertainties in determining its critical temperature. Based on the existing models and with some simple modification, we find the maximum mass for a brown dwarf to be in the range $0.064M_{\odot}$–$0.087M_{\odot}$. An analytic formula for the luminosity evolution allows us to estimate the time period of the nonsteady state (i.e., non-main-sequence) nuclear burning for substellar objects. We also calculate the evolution of very low mass stars. We estimate that ≈11% of stars take longer than 10^7 yr to reach the main sequence, and ≈5% of stars take longer than 10^8 yr.

1. Introduction

One of the most interesting avenues in the study of stellar models lies in understanding the physics of objects at the bottom of and below the hydrogen burning main-sequence stars. The main obstacle in the study of very low mass (VLM) stars and substellar objects is their low luminosity, typically of order $10^{-4}L_{\odot}$, which makes them difficult to detect. There is also a degeneracy between mass and age for these objects, which have a luminosity that decreases with time. This makes the determination of the initial mass function (IMF) difficult in this mass regime. However, in the last two decades, there has been substantial observational evidence that supports the existence of faint substellar objects. Since the first discovery of a brown dwarf [1, 2], several similar objects were identified in young clusters [3] and Galactic fields [4] and have generated great interest among theorists and observational astronomers. The field has matured remarkably in recent years and recent summaries of the observational situation can be found in Luhman et al. [5] and Chabrier et al. [6].

In two consecutive papers, Kumar [7, 8] revolutionized the understanding of low mass objects by studying the Kelvin-Helmholtz time scale and structure of very low mass stars. He successfully estimated that stars below about $0.1M_{\odot}$ contract to a radius of about $0.1R_{\odot}$ in about 10^9 years, which was a correction to the earlier estimate of 10^{11} years. The earlier calculation was based on the understanding that low mass stars evolve horizontally in the H-R diagram and thus evolve with a low luminosity for a long period of time. However, Hayashi and Nakano [9] showed that such low mass stars remain fully convective during the pre-main-sequence evolution and are much more luminous than the previously accepted model based on radiative equilibrium. Kumar's analysis showed that, for a critical mass of $0.09M_{\odot}$, the time scale has a maximum value that decreases on either side. Although this crude model neglected any nuclear reactions, it did give a very close estimate of the time scale. The second paper [8] gave a more detailed insight into the structure of the interior of low mass stars. This model was based on the nonrelativistic degeneracy of electrons in the stellar interior. Kumar's extensive numerical analysis for a particular abundance of hydrogen, helium, and other chemical compositions yielded a limiting mass below which the central temperature and density are never high enough to fuse hydrogen. A more exact analysis required a detailed understanding of the atmosphere and surface luminosity of such contracting stars.

The next major breakthrough in theoretical understanding came from the work of Hayashi and Nakano [9], who studied the pre-main-sequence evolution of low mass stars in the degenerate limit. Although it was predicted that there exist low mass objects that cannot fuse hydrogen, the internal structure of these objects remained a mystery. A complete theory demanded a better understanding of the physical mechanisms which govern the evolution of these objects. It became essential to develop a complete equation of state (EOS).

D'Antona and Mazzitelli [10] used numerical simulations to study the evolution of VLM stars and brown dwarfs for Population I chemical composition ($Y = 0.25$, $Z = 0.02$) and different opacities. Their model showed that for the same central condition (nuclear output) an increasing opacity reduces the surface luminosity. Thus, a lower opacity causes a greater surface luminosity and subsequent cooling of the object. Their results implied that the hydrogen burning minimum mass is $M = 0.08M_{\odot}$ for opacities considered in their model. Furthermore, they showed that objects with mass close to $M = 0.08M_{\odot}$ spend more than a billion years at a luminosity of $\sim 10^{-5} L_{\odot}$.

Burrows et al. [11] modelled the structure of stars in the mass range $0.03M_{\odot}$–$0.2M_{\odot}$. They used a detailed numerical model to study the effects of varying opacity, helium fraction, and the mixing length parameter and compared their results with the existing data. Their important modification was that they considered thermonuclear burning at temperatures and densities relevant for low masses. A detailed analysis of the equation of state was performed in order to study the thermodynamics of the deep interior, which contained a combination of pressure-ionized hydrogen, helium nuclei, and degenerate electrons. This analysis clearly expressed the transition from brown dwarfs to very low mass stars. These two families are connected by a steep luminosity jump of two orders of magnitude for masses in the range of $0.07M_{\odot}$–$0.09M_{\odot}$.

Saumon and Chabrier [12] proposed a new EOS for fluid hydrogen that, in particular, connects the low density limit of molecular and atomic hydrogen to the high density fully pressure-ionized plasma. They used the consistent free energy model but with the added prediction of a first-order "plasma phase transition" (PPT) [12] in the intermediate regime of the molecular and the metallic hydrogen. As an application of this EOS, they modelled the evolution of a hydrogen and helium mixture in the interior of Jupiter, Saturn, and a brown dwarf [13, 14]. They adopted a compositional interpolation between the pure hydrogen EOS and a pure helium EOS to obtain a H/He mixed EOS. This was based on the additive volume rule for an extensive variable [15] and allowed calculations of the H/He EOS for any mixing ratio of hydrogen and helium. Their analysis suggested that the cooling of a brown dwarf with a PPT proceeds much more slowly than in previous models [11].

Stevenson [16] presented a detailed theoretical review of brown dwarfs. His simplified EOS related pressure and density for degenerate electrons and for ions in the ideal gas approximation. Although corrections due to Coulomb pressure and exchange pressure are of physical relevance, they together contribute less than 10% in comparison to the other dominant term in the pressure-density relationship for massive brown dwarfs ($M \geq 0.04M_{\odot}$). The theoretical analysis gave a very good understanding of the behavior of the central temperature T_c as a function of radius and degeneracy parameter ψ. Stevenson [16] discussed the thermal properties of the interior of brown dwarfs and provided an approximate expression for the entropy in the interior and in the atmosphere of a brown dwarf. He also derived an expression for the effective temperature as a function of mass.

A method to use the surface lithium abundance as a test for brown dwarf candidates was proposed by Rebolo et al. [17]. Lithium fusion occurs at a temperature of about 2.5×10^6 K, which is easily attainable in the interior of the low mass stars. However, brown dwarfs below the mass of $0.065M_{\odot}$ never develop this core temperature. They will then have the same lithium abundance as the interstellar medium independent of their age. However, for objects slightly more massive than $0.065M_{\odot}$, the core temperature can eventually reach 3×10^6 K. They deplete lithium in the core and the entire lithium content gets exhausted rapidly due to the convection. This causes significant change in the observable photospheric spectra. Thus, lithium can act as a brown dwarf diagnostic [18] as well as a good age detector [19].

Following this, an extensive review on the analytic model of brown dwarfs was presented by Burrows and Liebert [20]. They presented an elaborate discussion on the atmosphere and the interior of brown dwarfs and the lower edge of the hydrogen burning main sequence. Based on the convective nature of these low mass objects, they modelled them as polytropes of order $n = 1.5$. Once again, the atmospheric model was approximated based on a matching entropy condition of the plasma phase transition between molecular hydrogen at low density and ionized hydrogen at high density. The polytropic approximation enabled the calculation of the nuclear burning luminosity within the core adiabatic density profile [21]. While the luminosity did diminish with time in the substellar limit, the model did show that brown dwarfs can undergo hydrogen burning for a substantial period of time before it eventually ceases. The critical mass deduced from this model did indeed match that obtained from more sophisticated numerical calculations [11].

In this work, we give a general outline of the analytic model of the structure and the evolution of brown dwarfs. We advance some aspects of the existing analytic model by introducing a modification to the equation of state. We also discuss some of the unresolved problems like estimation of the surface temperature and the existence of PPT in the brown dwarf environment. Our paper is organized as follows. In Section 2, we discuss the derivation of a more accurate equation of state for a partially degenerate Fermi gas. We incorporate a finite temperature correction to the expression for the Fermi pressure to give a more general solution to the Fermi integral. In Section 3, we discuss the scaling laws for various thermodynamic quantities for an analytic polytrope model of index $n = 1.5$. In Section 4, we discuss the derivation of the equations [20] connecting the photospheric (surface) temperature with density, where the entropies at the interior and the exterior are matched using the first-order phase transition. In the spirit of an analytic model, we derive simplified analytic expressions for the specific entropies above and below the PPT. We also highlight the need to seek alternate methods given current concerns about the relevance of the PPT in BD interiors. We discuss the nuclear burning rates for low mass objects in Section 5 and determine the nuclear luminosity L_N [21]. In Section 6, we estimate the range of minimum mass required for stable sustainable nuclear burning. In Section 7, we discuss a cooling model and examine the evolution of photospheric properties over time. In Section 8, we estimate the number fraction of stars that enter the main sequence after more than a million years. In the concluding section, we discuss further possibilities for an improved and generalized theoretical model of brown dwarfs.

2. Equation of State

In main-sequence stars, the thermal pressure due to nuclear burning balances the gravitational pressure and the star can sustain a large radius and nondegenerate interior for a long period of time. However, substellar objects like brown dwarfs fail to have a stable hydrogen burning sequence and instead derive their stability from electron degeneracy pressure. A simple but accurate model needs to have a good equation of state that incorporates the degeneracy effect and the ideal gas behavior at a relative higher temperature. Burrows and Liebert [20] give a pressure law that applies to both extremes but has a poor connection in the intermediate zone. Stevenson [16] also gives an empirical relation for the pressure that does include an approximate correction term to connect the two extremes. Here, in order to obtain a more accurate analytic expression for the pressure, we integrate the Fermi-Dirac integral exactly using the polylogarithm functions $\mathrm{Li}_s(x)$. The most general expression for the pressure is

$$P_F = g_s \int_0^\infty \frac{4\pi p^2}{(2\pi\hbar)^3} dp \left(\frac{1}{e^{\beta(\epsilon-\mu)} + b} \right) \left(\frac{1}{3} p \frac{d\epsilon}{dp} \right) \quad (1)$$

[22], where $b = 1$ for the Fermi gas and $\epsilon(p) = \sqrt{p^2c^2 + m^2c^4} - mc^2$ and the other variables are the standard constants. For substellar objects, the electrons are mainly nonrelativistic due to the relatively low temperature and density. In the nonrelativistic limit, that is, $m^2c^4 \gg p^2c^2$, the energy density reduces to $\epsilon(p) \simeq p^2/2m$. Now, rewriting (1) in terms of the energy density gives

$$P_F = a \int_0^\infty \frac{\epsilon^{3/2} d\epsilon}{e^{\beta(\epsilon-\mu)} + 1}, \quad (2)$$

where $a = (2/3)(4\pi(2m)^{3/2}/(2\pi\hbar)^3)$, $\beta = (k_B T)^{-1}$, and we have taken $g_s = 2$. In the limit $T \to 0$, for all $\epsilon < \mu$, the argument of the exponential is negative and hence the exponential goes to zero as $\beta \to \infty$. Thus, the integral reduces to the Fermi pressure at zero temperature. However, in a physical situation at finite temperature, the integral can be solved analytically using the polylogs. The details of the exact derivation for a general Fermi integral are shown in Appendix A. The expression for the pressure of a degenerate Fermi gas at finite temperature is

$$\begin{aligned} P_F \simeq{}& a\frac{2}{5}\mu^{5/2} - \frac{1}{8}a\beta^{-1}\mu^{3/2}\ln\left(1 + e^{-\beta\mu}\right) \\ &+ \frac{3}{2}a\beta^{-2}\mu^{1/2}\frac{\pi^2}{6} + \frac{3}{4}a\beta^{-2}\mu^{1/2}\mathrm{Li}_2\left(-e^{-\beta\mu}\right)\cdots. \end{aligned} \quad (3)$$

The above expression for pressure is the most general analytic relation for the pressure of a degenerate Fermi gas at a finite temperature. The first term is the zero temperature pressure and the subsequent terms are the corrections due to the finite temperature of the gas and include Li_s, the polylogarithm functions of different orders s. The expression is terminated after the fourth term as the polylogs fall off exponentially as the gas becomes more and more degenerate. Equation (3) is a natural extension of the first-order Sommerfeld correction [23].

The central temperature of VLM stars and brown dwarfs is of the same order as the electron Fermi temperature and thus the degeneracy parameter ψ is defined as

$$\psi = \frac{k_B T}{\mu_F} = \frac{2m_e k_B T}{(3\pi^2\hbar^3)^{2/3}} \left[\frac{\mu_e}{\rho N_A}\right]^{2/3}, \quad (4)$$

where μ_F is the electron Fermi energy in the degenerate limit and $1/\mu_e = X + Y/2$ is the number of baryons per electron and X and Y are the mass fractions of hydrogen and helium, respectively. Other constants have their standard meaning.

Rewriting (3) in terms of the degeneracy parameter ψ and retaining terms only up to second order, we arrive (for $\mu = \mu_F$) at

$$\begin{aligned} P_F ={}& \frac{2}{5}aA^{5/2}\left[\frac{\rho}{\mu_e}\right]^{5/3}\left[1 - \frac{5}{16}\psi\ln\left(1 + e^{-1/\psi}\right)\right. \\ &\left. + \frac{15}{8}\psi^2\left\{\frac{\pi^2}{3} + \mathrm{Li}_2\left(-e^{-1/\psi}\right)\right\}\right], \end{aligned} \quad (5)$$

where $A = (3\pi^2\hbar^3 N_A)^{2/3}/2m_e$ is a constant. However, the interior of a brown dwarf is also composed of ionized hydrogen and helium. The total pressure is a combined effect of both electrons and ions; that is, $P = P_F + P_{\text{ion}}$, where P_F is the Fermi pressure for an ideal nonrelativistic gas at a finite temperature. The pressure due to ions for an ionized hydrogen gas can be approximated as

$$P_{\text{ion}} = \frac{k\rho T}{\mu_1 m_H}. \quad (6)$$

Therefore, the final equation of state for the combined pressure is

$$\begin{aligned} P ={}& \frac{2}{5}aA^{5/2}\left[\frac{\rho}{\mu_e}\right]^{5/3}\left[1 - \frac{5}{16}\psi\ln\left(1 + e^{-\beta\mu}\right)\right. \\ &\left. + \frac{15}{8}\psi^2\left\{\frac{\pi^2}{3} + \mathrm{Li}_2\left(-e^{-1/\psi}\right)\right\} + \alpha\psi\right], \end{aligned} \quad (7)$$

where $\alpha = 5\mu_e/2\mu_1$ and μ_1 is the mean molecular weight for helium and ionized hydrogen mixture and is expressed as

$$\frac{1}{\mu_1} = \left((1 + x_{\mathrm{H}^+}) X + \frac{Y}{4} \right), \tag{8}$$

where x_{H^+} is the ionization fraction of hydrogen. It should be noted that x_{H^+} changes as one moves from the core (completely ionized) to the surface which is mainly composed of molecular hydrogen and helium.

There are several corrections to the EOS that can be considered. The Coulomb pressure and the exchange pressure (see (13) in Stevenson [16]) are two important corrections to (7). However, as stated earlier, they are less important for more massive brown dwarfs. Hubbard [24] presents the contribution due to the electron correlation pressure, which depends on the logarithm of r_e, the mean distance between electrons. Stolzmann and Blocker [25] present an analytic formulation of the EOS for fully ionized matter to study the thermodynamic properties of stellar interiors. They show that the inclusion of both electron and the ionic correlation pressure results in a ~10% correction to the EOS. Furthermore, Gericke et al. [26] state that the main volume of the brown dwarfs and the interior of giant gas planets are in a warm dense matter state, where correlation energy, effective ionization energy, and the electron Fermi energy are of the same order of magnitude. Thus, the interiors of these objects effectively form a strongly correlated quantum system. Becker et al. [27] give an EOS for hydrogen and helium covering a wide range of densities and temperatures. They extend their ab initio EOS to the strongly correlated quantum regime and connect it with the data derived using other methods for the neighboring regions of the ρ-T plane. These simulations are within the framework of density functional theory molecular dynamics (DFT-MD) and give a detailed description of the internal structure of brown dwarfs and giant planets. This leads to a 2.5%–5% correction in the mass-radius relation.

The study of the EOS of brown dwarfs will help in understanding degenerate bodies in the thermodynamic regime that is not so close to the high pressure limit of a fully degenerate Fermi gas. In this context, the Mie-Grueneisen equation of state is of relevance to test the validity of the assumption that the Grueneisen parameter $\gamma = (\partial \log T/\partial \log \rho)_s$ is independent of the temperature T [28] at a constant volume V. The brown dwarf regime is in a way more interesting than the white dwarf regime since it is not so close to the limit of a fully degenerate Fermi gas. In Appendix C, we have provided analytic expressions for two parameters that are of particular relevance for the brown dwarfs: the specific heat (C_v or C_p) and the Grueneisen parameter.

3. An Analytic Model for Brown Dwarfs

In this section, we derive some of the essential thermodynamic properties of a polytropic gas sphere based on the discussion in Chandrasekhar [29]. As is evident from (7), the P-ρ relation for a brown dwarf is a polytrope

$$P = K\rho^{(1+1/n)}, \tag{9}$$

where the index $n = 3/2$. K is a constant depending on the composition and degeneracy and can be expressed (from (7)) as

$$K = C\mu_e^{-5/3} (1 + \gamma + \alpha\psi), \tag{10}$$

where for a simplified presentation we represent the correction terms as

$$\gamma = -\frac{5}{16}\psi \ln\left(1 + e^{-\beta\mu}\right) + \frac{15}{8}\psi^2 \left\{ \frac{\pi^2}{3} + \mathrm{Li}_2\left(-e^{-1/\psi}\right) \right\} \tag{11}$$

and C (on using the values of natural constants, we get $C = 10^{13}\,\mathrm{cm}^4\,\mathrm{g}^{-2/3}\,\mathrm{s}^{-2}) = (2/5)aA^{5/2}$ is a constant. The solution to the Lane-Emden equation subject to the zero pressure outer boundary condition can be used to arrive at useful results for R, ρ_c, and P_c for the polytropic equation of state (see (7)). The radius can be expressed as

$$R = 2.3573 \frac{K}{GM^{1/3}} \tag{12}$$

[29]. On substituting (10) for K, the radius for a brown dwarf can be expressed as the function of degeneracy and mass:

$$R = 2.80858 \times 10^9 \left(\frac{M_\odot}{M} \right)^{1/3} \mu_e^{-5/3} (1 + \gamma + \alpha\psi)\ \mathrm{cm}. \tag{13}$$

Similarly, the expressions for the central density ρ_c and central pressure are given by the relations $\rho_c = \delta_n (3M/4\pi R^3)$ and $P_c = W_n (GM^2/R^4)$, where the constant $\delta_n = 5.991$ and $W_n = 0.77$ for the polytrope of $n = 1.5$ [29]. On substituting the expression for R (13) in these relations, we get

$$\rho_c = 1.28412 \times 10^5 \left(\frac{M}{M_\odot} \right)^2 \frac{\mu_e^5}{(1 + \gamma + \alpha\psi)^3}\ \mathrm{g/cm^3}, \tag{14}$$

$$P_c = 3.26763 \times 10^9 \left(\frac{M}{M_\odot} \right)^{10/3} \frac{\mu_e^{20/3}}{(1 + \gamma + \alpha\psi)^4}\ \mathrm{Mbar}. \tag{15}$$

These are the scaling laws of the density and pressure in the interior core of a brown dwarf. Interestingly, these vary with the degeneracy parameter ψ that is a function of time. Thus, a very simple polytropic model can yield the time evolution of the internal thermodynamical conditions of a brown dwarf. From the definition of the degeneracy parameter in (4) and using (14), the central temperature can be expressed as a function of ψ:

$$T_c = 7.68097 \times 10^8\,\mathrm{K} \left(\frac{M}{M_\odot} \right)^{4/3} \frac{\psi \mu_e^{8/3}}{(1 + \gamma + \alpha\psi)^2}. \tag{16}$$

The central temperature has a maximum for a certain value of ψ, and it increases for greater values of M. Further, using (13), we have shown the variation of central temperature T_c as a function of radius R. T_c increases as the object contracts under the influence of gravity. It peaks at a certain R and then cools over time. The maximum peak temperature increases for heavier objects and also depends on the extent

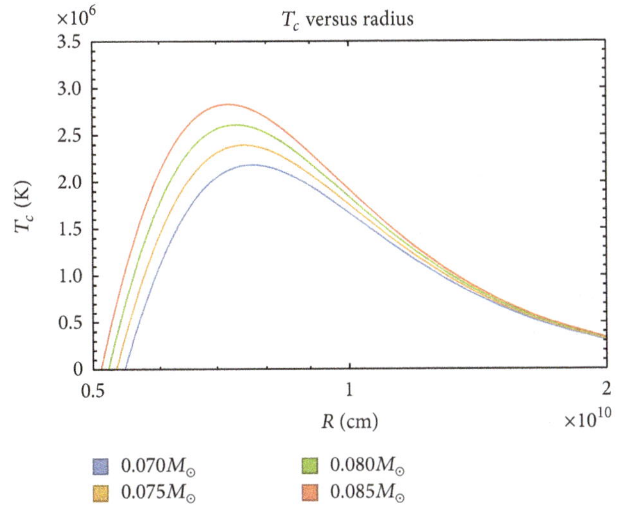

FIGURE 1: The variation of T_c versus radius R for different masses.

of ionization of hydrogen and helium. Figure 1 shows the variation of the central temperature as a function of radius for different mass ranges. If the critical temperature for thermonuclear reactions is around 3×10^6 K, we can roughly estimate the critical mass for the main sequence as $\sim 0.085M_\odot$. This is similar to the estimated critical mass ($\sim 0.084M_\odot$) for the main sequence (see Figure 1 in Stevenson [16]). However, it should be noted that the estimate of minimum mass is very sensitive to the mean molecular mass μ_1. In Figure 1, we have used $\mu_1 = 1.23$ for fully neutral gas (similar to $\overline{A} \sim 1.24$ for cosmic mixture as used in Stevenson [16]). Depending upon the value of μ_1, the minimum mass may vary significantly. For example, if we consider a fully ionized gas, that is, $\mu_1 = 0.59$, it yields a minimum mass of $0.12M_\odot$.

4. Surface Properties

In this section, we discuss a very simple but crude model which is broadly based on the phase transition proposed by Chabrier et al. [14] and the isentropic nature of the interior of brown dwarfs. The development of a theoretical model for studying the variation of surface luminosity over time for low mass stars (LMS) and brown dwarfs is a great challenge. There is no stable phase of nuclear burning for brown dwarfs and the luminosity gradually decreases with time. This leads to an age-mass degeneracy in observational determinations. Our lack of knowledge in understanding the physics of the interior of brown dwarfs restricts the development of a comprehensive model. However, extensive simulations were done on the molecular-metallic transition of hydrogen for LMS and planets [13, 14]. Chabrier et al.'s [14] model predicts a first-order transition for the metallization of hydrogen at a pressure of ~1 Mbar and critical temperature of ~15300 K. Such pressure and temperature values are appropriate for giant planets and brown dwarfs. Modern numerical simulations [30, 31] do confirm the existence of such phase transitions at the same pressure range but predict a much different range of temperature ~2000 K–3000 K. This new temperature regime is certainly too low for brown dwarfs. Although the pressure estimate is relatively well established in these numerical simulations, the phase transition temperature is still a matter of continuing investigation [27]. Having noted these caveats, we present the existing model for the surface temperature, based on Chabrier et al. [14] and Burrows and Liebert [20]. We also introduce a simpler treatment of the specific entropy.

The PPT occurs over a narrow range of densities near $1.0\,\mathrm{g/cm^3}$ from a partially ionized phase ($x_{H^+} \sim 0.5$) to a neutral molecular phase ($x_{H^+} < 10^{-3}$). For massive brown dwarfs, the phase transition occurs nearer to the surface. Burrows and Liebert [20] used the following approximate analytic expressions for the specific entropy (equations (2.48) and (2.49) in Burrows and Liebert [20]) for the two phases of the PPT:

$$\sigma_1 = -1.594 \ln \frac{1}{\psi} + 12.43,$$
$$\sigma_2 = 1.032 \frac{\ln T}{\rho^{0.42}} - 2.438. \tag{17}$$

Similar expressions for the entropy at the interior and the atmosphere are given in equations (21) and (22) in Stevenson [16]. We use a simplified approach to make the origin of the above equation clear in the spirit of an analytic model. We derive analytic expressions for the entropy of the ionized and the molecular hydrogen separately for the two phases (similar to equations (2.48) and (2.49) in Burrows and Liebert [20]) and match them via the phase transition. It is assumed that the presence of helium does not affect the hydrogen PPT [14].

The region between the strongly correlated quantum regime and the ideal gas limit can be modelled with corrections to the ideal gas equation. Such correction terms can be expressed by virial coefficients (see equation (1) in Becker et al. [27]). For simplicity, we ignore such corrections in our EOS (see (7)) and consider only the contribution of electron pressure (see (5)) and the ion pressure (see (6)) of the partially ionized hydrogen (of ionization fraction x_{H^+}) and helium mixture.

The total entropy for our EOS (see (7)) is the sum of the entropies of the atomic/molecular gas and the degenerate electrons. The internal energy per gram for the monoatomic gas particles is

$$U = \frac{3}{2} \frac{k_B N_0 T}{\mu_1}. \tag{18}$$

Ideally, we can consider the total energy as a combination of kinetic energy, radiation energy, and ionization energy (B.3). But as the electron gas is degenerate, the radiation pressure is relatively unimportant. Furthermore, as shown in Appendix B, we may even neglect the contribution of the ionization energy as the gas is only partially ionized. Taking the partial derivatives of the expression for the internal energy equation (18), we can express the change in heat dQ using the first law of thermodynamics (see (B.2)). However, as the ionization fraction $x_{H^+}(\rho, T)$ is a function of density and temperature, we further use Saha's ionization equations (B.5) to get

$$\frac{dQ}{T} = -\frac{3}{2}\frac{k_B N_0}{\mu_1}\frac{dT}{T} + \frac{k_B N_0}{\mu_1}\frac{dW}{W} + \left(\frac{3}{2}\right)^2 H k_B N_0 \frac{dT}{T} + \frac{3}{2} H k_B N_0 \frac{dV}{V}, \quad (19)$$

where $W = T^3 V$, $H = x_{H^+}(1 - x_{H^+})/(2 - x_{H^+})$. A more generalized version including the radiation and the ionization terms is shown in (B.6) in Appendix B. For $TdS = dQ$, we integrate (19) to express the entropy in the interior of the brown dwarf as

$$S_1 = \frac{k_B N_0}{\mu_{1\text{mod}}} \ln \frac{T^{3/2}}{\rho_1} + C_1, \quad (20)$$

where

$$\frac{1}{\mu_{1\text{mod}}} = \left(\frac{1}{\mu_1} + \frac{3}{2}\frac{x_{H^+}(1 - x_{H^+})}{2 - x_{H^+}}\right) \quad (21)$$

and μ_1 is different for each model in this region and is calculated using x_{H^+} from Table 1. However, the contributions to entropy due to radiation and the degenerate electrons (see equation 2–145 in Clayton [32]) are negligible in the range of temperature and density applicable for brown dwarfs. Based on a similar argument, the analytic expression for the entropy of nonionized molecular hydrogen and helium mixture at the photosphere is expressed as

$$S_2 = \frac{k_B N_0}{\mu_2} \ln \frac{T^{5/2}}{\rho_2} + C_2, \quad (22)$$

where $1/\mu_2 = X/2 + Y/4$ is the mean molecular weight for the hydrogen and helium mixture. The expression for entropy S_2 is derived using the first law of thermodynamics (Appendix B) and the relation of the internal energy of diatomic molecules ($U = (5/2)(k_B N_0 T/\mu_2)$). Here, we have considered only five degrees of freedom as the temperatures are just sufficient to excite the rotational degrees of H_2 but the vibrational degrees remain dormant. It should be noted that (20) and (22) are just simplified forms of the entropy expressions presented in Burrows and Liebert [20] and Stevenson [16].

Thus, the entropy in the two phases is dominated by contributions from the ionic and molecular gas, respectively. Using the same argument as Burrows and Liebert [20] that the two regions of different temperature and density are separated by a phase transition of order one, we can estimate the surface temperature. Using the expression for the degeneracy parameter ψ from (4), we can simplify (20) to be

$$S_1 = \frac{3}{2}\frac{k_B N_0}{\mu_{1\text{mod}}} (\ln \psi + 12.7065) + C_1. \quad (23)$$

Furthermore, the jump of entropy

$$\Delta\sigma = \frac{S_2 - S_1}{k_B N_0} \quad (24)$$

(see Table 1) for the phase transition at each point of the coexistence curve of PPT [12] is used to estimate the relation $|C_1 - C_2|$ between the two constants in (20) and (22). For $T = T_{\text{eff}}$ and $\rho_2 = \rho_e$ in (22), we can use (23) and (22) in (24) and the value of $|C_1 - C_2|$ to obtain a wide range of possible values of surface temperature T_{eff} in terms of the degeneracy parameter and photospheric density ρ_e:

$$T_{\text{eff}} = b_1 \times 10^6 \rho_e^{0.4} \psi^{\nu}\,\text{K}. \quad (25)$$

The values of the parameters b_1 and ν for different models are shown in Table 1. According to Chabrier's model, the critical temperature $\sim 1.53 \times 10^4$ K and critical density $\sim 0.35\,\text{g cm}^{-3}$ mark the end of the phase transition with $\Delta\sigma = \Delta S/k_B N = 0$. In the following discussion, we briefly summarize the steps from Burrows and Liebert [20]. We replace equation (2.50) in Burrows and Liebert [20] by (25) to estimate the surface luminosity. As an example, we select a particular phase transition point (model D) and show the derivation of surface luminosity using hydrostatic equilibrium and ideal gas approximation. The photosphere of a brown dwarf is located at approximately the $\tau = 2/3$ surface, where

$$\tau = \int_r^{\infty} \kappa_R \rho\, dr \quad (26)$$

is the optical depth. Using the general equation for hydrostatic equilibrium, $dP = -(GM/r^2)\rho dr$, and (26), the photospheric pressure can be expressed as

$$P_e = \frac{2}{3}\frac{GM}{\kappa_R R^2}, \quad (27)$$

where κ_R is the Rosseland mean opacity and the other variables have their standard meanings. Furthermore, our EOS (see (7)) in the approximation of negligible degeneracy pressure near the photosphere gives the photospheric pressure as

$$P_e = \frac{\rho_e N_A k_B T_{\text{eff}}}{\mu_2}. \quad (28)$$

Now, using the expression for radius R (13) in (27), we can calculate the external pressure P_e as a function of M and ψ:

$$P_e = \frac{11.2193\,\text{bar}}{\kappa_R}\left(\frac{M}{M_\odot}\right)^{5/3} \frac{\mu_e^{10/3}}{(1 + \gamma + \alpha\psi)^2}. \quad (29)$$

On using (29) in (28) and substituting T_{eff} for model D with $b_1 = 2.00$ and $\nu = 1.60$ from Table 1, the effective density ρ_e can be expressed as a function of M and ψ:

$$\rho_e^{1.40} = \frac{6.89811}{\kappa_R N_A k_B}\left(\frac{M}{M_\odot}\right)^{5/3} \frac{\mu_e^{10/3}\mu_2\,\text{g/cm}^3}{(1 + \gamma + \alpha\psi)^2 \psi^{1.58}}. \quad (30)$$

Substituting the expression for ρ_e from (30) in (25), we derive the expression for effective temperature for model D as a function of M and ψ:

$$T_{\text{eff}} = \frac{2.57881 \times 10^4\,\text{K}}{\kappa_R^{0.2856}}\left(\frac{M}{M_\odot}\right)^{0.4764} \frac{\psi^{1.1456}}{(1 + \gamma + \alpha\psi)^{0.5712}}. \quad (31)$$

Similarly, the surface temperature can be evaluated for all the other models. Since the procedure is the same for all the models in Table 1, we just show one calculation. For this range of surface temperatures, the Stefan-Boltzmann law, $L = 4\pi R^2 \sigma T_{\rm eff}^4$, yields a set of possible values of the surface luminosity L as a function of the degeneracy parameter ψ. The luminosity for model D using (13) and (31) is

$$L = \frac{0.41470 \times L_\odot}{\kappa_R^{1.1424}} \left(\frac{M}{M_\odot} \right)^{1.239} \frac{\psi^{4.5797}}{(1+\gamma+\alpha\psi)^{0.2848}}, \quad (32)$$

where σ is the Stefan-Boltzmann constant. Substellar objects below the main-sequence mass gradually evolve towards complete degeneracy and a state of stable equilibrium as their luminosity decreases over time. In the following sections, we show that the degeneracy parameter ψ is a function of time and it evolves towards $\psi = 0$ over the lifetime of brown dwarfs. This gives us an estimate of the luminosity at different epochs of time.

4.1. Validity of PPT in Brown Dwarfs. There is a distinction between the temperature-driven PPT with a critical point at ~0.5 Mbar and between 10000 K and 20000 K as predicted by the chemical models [33] and the pressure-driven transition from an insulating molecular liquid to a metallic liquid with a critical point below 2000 K at pressures between 1 and 3 Mbar. The latter is predicted, for example, by Lorenzen et al. [34], Mazzola et al. [35], and Morales et al. [31] based on the ab initio simulations. Lorenzen et al. [34] rule out the presence of PPT above 10000 K and give an estimate of the critical points for the transition at $T_c = (1400 \pm 100)$ K, $P_c = 1.32 \pm 0.1$ Mbar, and $\rho_c = 0.79 \pm 0.05\,{\rm g/cm^3}$. Similarly, Morales et al. [31] estimated the critical point of the transition at a temperature near 2000 K and pressure near 1.2 Mbar. Signatures of pressure-driven PPT in a cold regime below 2000 K are obtained by Knudson et al. [36]. Figure 1 in Knudson et al. [36] shows the melting line (black) as well as the different predictions for the coexistence lines for the first-order transition (green curves). Brown dwarf interior temperatures are far above these estimates for a first-order transition from the insulating to the metallic system. The same is true for Jupiter. Of course, a continuous transition may be possible in Jupiter and brown dwarfs, but a first-order transition may not be possible. Thus, the determination of the range of temperature of this transition provides a much needed benchmark for the theory of the standard models for the internal structures of the gas-giant planets and low mass stars.

5. Nuclear Processes

VLM stars and brown dwarfs contract during their evolution due to gravitational collapse. The core temperature increases and the contraction is halted by either the degeneracy pressure of the electrons or the onset of the nuclear burning, whichever comes first. In the first case, the brown dwarf continues to lose energy through radiation and cools down with time without any further compression. However, massive brown dwarfs or stars at the edge of the main sequence can burn hydrogen for a very long time before they either cease nuclear burning or settle into a steady state main sequence. The thermonuclear reactions suitable for the brown dwarfs and VLM stars are

$$p + p \longrightarrow d + e^+ + \nu, \quad (33)$$

$$p + d \longrightarrow {}^3{\rm He} + \gamma. \quad (34)$$

As the central temperature is not high enough to overcome the Coulomb barrier of the ^{3}He–^{3}He reaction and the *p-p* chain is truncated, ^{4}He is not produced. Most of the thermonuclear energy is produced from the burning of the primordial deuterium (see (34)). The energy generation rates of the above processes are given as

$$\dot{\epsilon}_{pp} = 2.5 \times 10^6 \left[\frac{\rho X^2}{T_6^{2/3}} \right] e^{-33.8/T_6^{1/3}}\ {\rm erg/g \cdot s},$$
$$\dot{\epsilon}_{pd} = 1.4 \times 10^{24} \left[\frac{\rho X Y_d}{T_6^{2/3}} \right] e^{-37.2/T_6^{1/3}}\ {\rm erg/g \cdot s} \quad (35)$$

[37]. However, one can fit the thermonuclear rates to a power law in T and ρ in terms of the central temperature (T_c) and density (ρ_c) as in Fowler and Hoyle [21]:

$$\dot{\epsilon}_n = \dot{\epsilon}_c \left[\frac{T}{T_c} \right]^s \left[\frac{\rho}{\rho_c} \right]^{u-1}, \quad (36)$$

where $u \simeq 2.28$ and $s = 6.31$ are constants that depend on the core conditions [20]. To obtain the luminosity due to the nuclear burning $L_N = \int \dot{\epsilon}_n dm$, we use the power law form for the nuclear burning rate $\dot{\epsilon}_n$ (see (36)), and making the polytropic approximation $\rho = \rho_c \theta^n$ and setting $T/T_c = (\rho/\rho_c)^{2/3}$, we obtain

$$L_N = \int \dot{\epsilon}_n dm = 4\pi a^3 \dot{\epsilon}_c \rho_c \int \theta^{n(u+2s/3)} \zeta^2 d\zeta, \quad (37)$$

where $r = a\zeta$ [29]. Inserting (14) and (16) in (37) yields the final expression for luminosity as

$$L_N = 7.33 \times 10^{16} L_\odot \left(\frac{M}{M_\odot} \right)^{11.977} \frac{\psi^{6.0316}}{(1+\gamma+\alpha\psi)^{16.466}}. \quad (38)$$

6. Estimate of the Minimum Mass

In this section, we estimate the minimum main-sequence mass by comparing the surface luminosity (32) with the luminosity (L_N) due to nuclear burning at the core of LMS and brown dwarfs. Instead of just quoting one value as the critical mass, we have presented a range of values depending on the various phase transition points listed in Table 1. This will give us a range of values for the minimum critical mass that is sufficient to ignite hydrogen burning. Model H marks the end of the phase transition and gives a lower limit of the critical mass. However, we calculate the mass limit for model B and model D only. Equating L_N of (38) with L of (32) gives us

$$\frac{M}{M_\odot} = \frac{0.02440}{\kappa_R^{0.106}} \frac{(1+\gamma+\alpha\psi)^{1.507}}{\psi^{0.1617}} = F(\psi), \quad (39)$$

TABLE 1: Effective temperature for different phase transition points[a].

Model	$\log T$ (K)	P (Mbar)	ρ_1 (g cm^{-3})	ρ_2 (g cm^{-3})	$\Delta\sigma$	$2x_{H^+}$	b_1	ν
A	3.70	2.14	0.75	0.92	0.62	0.48	2.87	1.58
B	3.78	1.95	0.70	0.88	0.59	0.50	2.70	1.59
C	3.86	1.62	0.64	0.80	0.54	0.50	2.26	1.59
D	3.94	1.39	0.58	0.74	0.51	0.51	2.00	1.60
E	4.02	1.13	0.51	0.65	0.46	0.52	1.68	1.61
F	4.10	0.895	0.43	0.55	0.42	0.50	1.29	1.59
G	4.18	0.631	0.35	0.38	0.14	0.33	0.60	1.44
H	4.185	0.614	0.35	0.35	0.00	0.18	0.40	1.30

[a]The phase transition points are taken from Chabrier et al. [14]. This gives the possible range of surface temperature depending on the phase transition points. For different values of temperature and density at which the phase transition takes place, the effective surface temperature is calculated using (25).

where $\alpha = 2.32$ for $\mu_e = 1.143$ and $\mu_1 = 1.24$, which are the number of baryons per electron and the mean molecular weight of neutral ($x_{H^+} = 0$) hydrogen and helium, respectively. These mass densities are evaluated for hydrogen and helium mass fractions of $X = 0.75$ and $Y = 0.25$, respectively.

The right hand side of (39) has a minimum at a certain value of ψ. This gives the lowest mass at which (39) has a solution and this corresponds to the boundary of brown dwarfs and VLM stars. The minimum of $F(\psi)$ is at $\psi_{\min} = 0.042$. Substituting this in (39) and for $\kappa_R = 0.01\ \text{cm}^2/\text{g}$, the minimum mass (model D) is

$$M = 0.078 M_\odot. \tag{40}$$

A similar analysis for model B gives the value of minimum mass of $M = 0.085 M_\odot$ for $\psi_{\min} = 0.042$. For the other models, the minimum main-sequence mass is in the range of 0.064–$0.087 M_\odot$.

The solution is relatively independent of the mean molecular weight μ_1. For example, using partially ionized gas; that is, $\mu_1 = 0.84$ in α; the minimum stellar mass increases by only ~5%.

7. A Cooling Model

A simple cooling model for a brown dwarf is presented in both Burrows and Liebert [20] and Stevenson [16]. In this section, we review some of these steps using our more exact EOS (7) and represent the evolution of the brown dwarfs over time. Using the first and the second law of thermodynamics, the time varying energy equation for a contracting star is expressed as

$$\frac{dE}{dt} + P\frac{dV}{dt} = T\frac{dS}{dt} = \dot{\epsilon} - \frac{\partial L}{\partial M}, \tag{41}$$

where S is the entropy per unit mass and the other symbols have their standard meaning. The energy generation term $\dot{\epsilon}$ is ignored. On integrating over mass, we get

$$\frac{d\sigma}{dt}\left[\int N_A k_B T\, dM\right] = -L, \tag{42}$$

where L is the surface luminosity and $\sigma = S/k_B N_A$. Now, replacing T in terms of the degeneracy parameter ψ in (4) and using the polytropic relation $P = K\rho^{5/3}$, we arrive at

$$\frac{d\sigma}{dt}\frac{N_A k_B \psi}{\mu_e^{2/3}}\int P\, dV = -L. \tag{43}$$

Using the standard expression, $\int P\, dV = (2/7)(GM^2/R)$, for polytropes of $n = 1.5$, the integral in (42) reduces to

$$\int N_A k_B T\, dM = \frac{6.73857 \times 10^{49}\psi\mu_e^{8/3}}{(1+\gamma+\alpha\psi)^2}\left(\frac{M}{M_\odot}\right)^{7/3}. \tag{44}$$

The variation of the entropy with time (42) can be expressed as the rate of change of degeneracy over time. As the star collapses, the gas in the interior becomes more and more degenerate and finally the degeneracy pressure halts further contraction. A completely degenerate star ($\psi = 0$) becomes static and cools with time. Thus, by substituting the time variation of entropy, using (23), that is,

$$\frac{d\sigma}{dt} = \frac{1.5}{\mu_{1\text{mod}}}\frac{1}{\psi}\frac{d\psi}{dt}, \tag{45}$$

and (44) into the energy equation (42) and using the luminosity expression for model D (32), we obtain an evolutionary equation for ψ:

$$\frac{d\psi}{dt} = \frac{9.4486 \times 10^{-18}}{\kappa_R^{1.1424}}\left(\frac{M_\odot}{M}\right)^{1.094}(1+\gamma+\alpha\psi)^{1.715} \cdot \psi^{4.5797}. \tag{46}$$

This is a nonlinear differential equation of ψ for model D, and an exact solution can only be obtained numerically. However, we use some very simple and physical approximations to solve this differential equation to yield a simple relation of ψ as a function of time and mass M. As we are trying to estimate the critical mass for hydrogen burning, it is safe to ignore the early evolution of VLM stars and brown dwarfs. Thus, we will solve the differential equation (46) with the assumption $\psi \ll 1$. Thus, we can drop the term $(1+\gamma+\alpha\psi)^{1.715}$ as it is almost

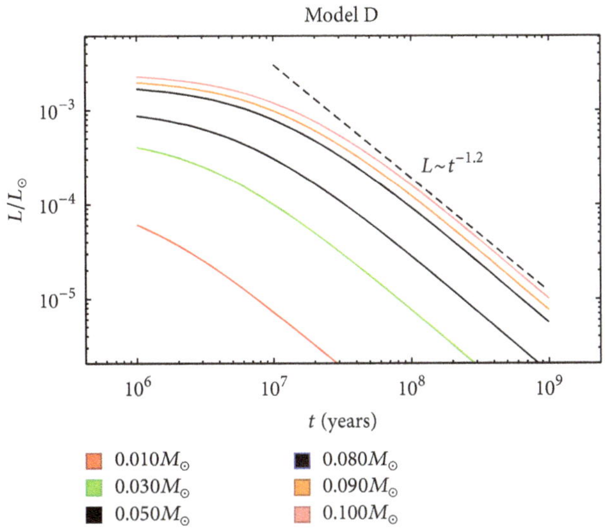

FIGURE 2: The variation of $L/L_\odot$ over time t for different masses.

unity in the range $0 < \psi < 0.1$ and integrate (46) in the above limit to obtain

$$\psi = \left(317.8 + 2.053 \times 10^{-6} \left(\frac{M_\odot}{M} \right)^{1.094} \frac{t}{\mathrm{yr}} \right)^{-0.2794}. \quad (47)$$

Similarly, we can solve for ψ for all other models and obtain the evolution of degeneracy over time. We use this expression of ψ, for model D, and can express luminosity as a function of time t and mass M. The time evolution of luminosity (model D) is represented in Figure 2. It is evident that such low mass objects continue to have low luminosity for millions of years before they gradually start to cool. For $t > 10^7$ yr, the luminosity declines as a function of time, $L \simeq t^{-1.2}$, as shown in Figure 2 (dashed black line). A simplified expression of the variation of luminosity after 10^7 yr for model D is

$$L \simeq L_\odot \left(\frac{M}{M_\odot} \right)^{2.63} \left(\frac{t}{10^7\,\mathrm{yr}} \right)^{-1.2}. \quad (48)$$

Our luminosity model is consistent with the simulation results of the present day stellar evolution code Modules for Experiments in Stellar Astrophysics (MESA) (see Figure 17 in Paxton et al. [38]). Paxton et al. [38] use a one-dimensional stellar evolution module, MESA star, to study evolutionary phases of brown dwarfs, pre-main-sequence stars, and LMS.

In Figures 3 and 4, the ratio L_N/L is plotted against time for different masses in the substellar regime for models B and D, respectively. As evident for both models, there is a nonsteady state of substantial nuclear burning for millions of years for substellar objects. For a critical mass of $0.085M_\odot$ (model B) and $0.078M_\odot$ (model D), the ratio L_N/L approaches 1 in about a few billion years and marks the beginning of main-sequence nuclear burning (note that the time to reach the main sequence will increase if we use a partially ionized gas; for example, it becomes $\approx$10 Gyr if we use $\mu_1 = 0.84$). Stars with greater mass reach a steady state where the thermal energy balances the gravitational collapse. However, as our model does not consider any feedback from hydrogen fusion, the curves do not stabilize to a steady state main-sequence regime, in which L_N/L remains 1 until nuclear burning stops. Interestingly, the ratio L_N/L is close to unity for many objects below the main-sequence transition mass. This suggests that they burn nuclear fuel for a part of their evolutionary cycle but do not have enough mass to sustain a steady state. Note that the results of Becker et al. [27] discussed in Section 2 would affect the luminosity L by less than a few percent. For example, if the radius R increases (or decreases) by 2.5% for a constant value of mass M, K in (12) increases by 2.5%, and T_{eff} (31) decreases by 1.5%, therefore L decreases by ~1%.

7.1. Brown Dwarfs as Clocks. Interestingly, the cooling properties of brown dwarfs (Figure 2) can be calibrated to serve as an astronomical clock. As the electron degeneracy pressure puts a lower limit to the size of the dwarf, it cools slowly and radiates its internal energy. The luminosity of a brown dwarf is the most directly accessible observable quantity. As luminosity is a time variable, one can get important information on the age of a brown dwarf depending on its mass and the cooling rate. As evident from Figure 2, given mass and the luminosity, one can roughly identify the age of the dwarfs. However, it is still a challenge to estimate the mass of a brown dwarf. An essential part of the solution is to find brown dwarfs in a binary system where one can get an accurate estimate of the mass and then compare its luminosity against available models. Newly discovered brown dwarfs in eclipsing binaries [39, 40] can provide a data set of directly measured mass and radii. This can yield an empirical mass-radius relation that also tests the prediction of the theoretical models. Furthermore, lithium in brown dwarfs has been used as a clock to obtain the ages of young open clusters as originally suggested by Martin et al. [41] and Basri et al. [18] and most recently applied to the Pleiades by Dahm [42]. Massive brown dwarfs ($M > 0.065M_\odot$) deplete their lithium on a longer time scale, but VLM stars and objects above the hydrogen burning limit fuse lithium on a much shorter time scale [43]. A limitation of our model is that it does not include rotation. But the mechanical equilibrium in our models may not be significantly affected by this. For example, in model D, using (47) in (13), we find the radius of a 10^7-year-old brown dwarf of mass $0.075M_\odot$ to be $\sim 8 \times 10^9$ cm. This implies that for a median observed rotational period $(2\pi/\omega)$ of one day [44] the ratio of magnitude of the rotational energy ($\sim MR^2\omega^2$) to gravitational energy ($\sim GM^2/R$) for a $0.075M_\odot$ brown dwarf is $\sim 10^{-4}$. However, convective mixing and the consequent lithium abundance have a strong connection to the rotation rate of brown dwarfs and pre-main-sequence stars [45]. Fast rotators are lithium-rich compared to their slow rotating counterparts, indicating a connection between the lithium content and the spin rate of young pre-main-sequence stars and brown dwarfs [46]. Rapid rotation reduces the convective mixing, resulting in a higher lithium abundance in fast rotating pre-main-sequence stars. Thus, the rotational evolution of a brown dwarf can potentially be used as a clock as discussed in Scholz et al. [44].

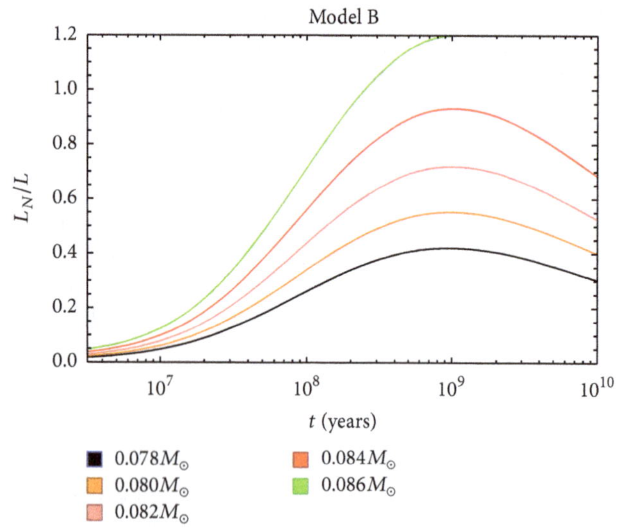

Figure 3: The variation of L_N/L over time t for different masses for model B in Table 1.

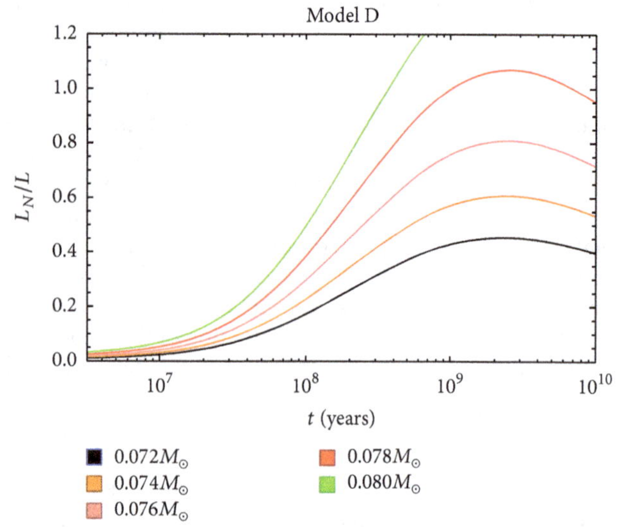

Figure 4: The variation of L_N/L over time t for different masses for model D in Table 1.

8. Low Mass Stars That Reach the Main Sequence

In Figure 5, we plot the time required by low mass stars to reach the main sequence. The curves represent the steady state limit where $L_N/L = 1$ for four different models. It is interesting to note that objects of masses at the critical mass boundary between brown dwarfs and main-sequence stars, for example, $0.078M_\odot$ for model D, reach the steady state in about 2.5 Gyr. This suggests that objects just below the critical mass undergo nuclear burning for an extended period of time but fail to enter the main sequence. Furthermore, stars in the mass range $0.078M_\odot$–$0.086M_\odot$ for model D take more than 10^8 yr to reach the main sequence. Depending on the phase transition points for different models, these numbers vary but the fact that stars close to the minimum mass limit can take an

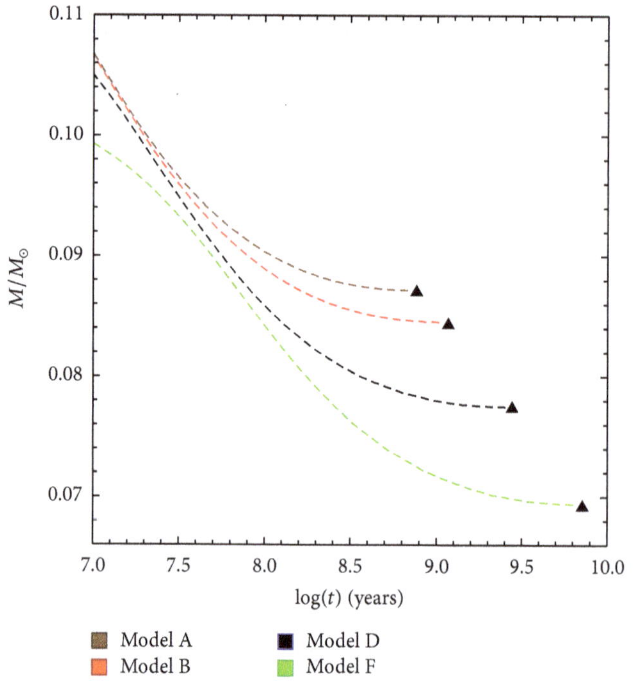

Figure 5: The time needed to reach the main sequence, that is, $L_N/L = 1$, for objects of different masses. Four colored dashed lines represent different models as labeled. The triangles mark the main-sequence critical mass for the respective models.

extended amount of time to reach the main sequence means that young stellar clusters may contain a significant fraction of objects that are still in a phase of decreasing luminosity and behave like brown dwarfs. These objects will ultimately settle into an extremely low luminosity main sequence. Here, we estimate the fraction of stars that take more than a specified time to reach the main sequence by using the modified log-normal power law (MLP) probability distribution function of Basu et al. [47]. Their cumulative distribution function is

$$F(m) = \frac{1}{2}\operatorname{erfc}\left(-\frac{\ln(m)-\mu_0}{\sqrt{2}\sigma_0}\right) - \frac{1}{2}\exp(\alpha\mu_0)\, m^{-\alpha}\operatorname{erfc}\left(\frac{\alpha\sigma_0}{\sqrt{2}} - \frac{\ln(m)-\mu_0}{\sqrt{2}\sigma_0}\right). \quad (49)$$

We use the best fit MLP parameters corresponding to Chabrier's [48] initial mass function (IMF) as obtained by Basu et al. [47], where $\mu_0 = -2.404$, $\sigma_0 = 1.044$, and $\alpha = 1.396$. We then use the cumulative function to calculate the fraction of stars taking more than either 10^7 yr, 10^8 yr, or 10^9 yr to reach the main sequence. Table 2 contains our results for models A to H. It turns out that about 0.2% of stars take more than 10^9 yr to reach the main sequence and about 4% take longer than 10^8 yr and about 12% take longer than 10^7 yr, for model D. Some of these objects will end up on an extremely low luminosity main sequence, and a sample of luminosity values when an object just above the substellar

TABLE 2: Minimum mass and fractions of stars.

Model	M_{min} [a]	M_9 [b]	N_9 [c]	M_8 [b]	N_8 [c]	M_7 [b]	N_7 [c]
A[d]	0.087	—	—	0.090	0.014	0.107	0.085
B	0.085	0.085	0.000	0.089	0.020	0.107	0.095
C	0.081	0.081	0.001	0.087	0.029	0.106	0.108
D	0.078	0.078	0.002	0.086	0.038	0.105	0.118
E	0.073	0.075	0.006	0.085	0.052	0.103	0.127
F	0.069	0.072	0.013	0.084	0.069	0.099	0.129
G	0.064	0.068	0.018	0.082	0.081	0.089	0.109
H	0.064	0.067	0.014	0.080	0.073	0.085	0.093

[a] M_{min} is the minimum mass to reach the main sequence.
[b] M_9, M_8, and M_7 are the masses up to which the stars take at least 10^9 yr, 10^8 yr, and 10^7 yr, respectively, to reach the main sequence.
[c] N_9, N_8, and N_7 are the number fraction of stars reaching the main sequence in more than 10^9 yr, 10^8 yr, and 10^7 yr, respectively.
[d] Note that, for model A, low mass stars reach the main sequence in $<10^9$ yr.

TABLE 3: Luminosity at the main sequence.

$M/M_\odot$	10^8 (yr)	$L/L_\odot$ [a]
0.080	3.66	1.90×10^{-5}
0.085	1.15	9.12×10^{-5}
0.090	0.56	2.33×10^{-4}
0.095	0.31	4.67×10^{-4}

[a] The luminosity of low mass stars in model D when they enter the main sequence.

limit achieves $L_N/L = 1$, that is, reaches the main sequence, is given in Table 3.

9. Discussion and Conclusions

This paper presents a simple analytic model of substellar objects. A focus of the paper was to revisit both the development and the shortcomings of the theoretical understanding of the physics governing the evolution of low mass stars and substellar objects over the last 50 years. We have also made some modifications to the existing models to better explain the physics using analytic forms. Although observational constraints hinder our understanding, a simple analytic model can answer many questions. We have summarized the method of determining the minimum mass for sustained hydrogen burning. Objects in the mass range $0.064M_\odot$–$0.087M_\odot$ mark this critical boundary between brown dwarfs and the main-sequence stars.

We have derived a general equation of state using polylogarithm functions [49] to obtain the P-ρ relation in the interior of brown dwarfs. The inclusion of the finite temperature correction gives us a much more complete and sophisticated analytic expression of the Fermi pressure (3). The application of this relation can extend to other branches of physics, especially for semiconductor and thermoelectric materials [50].

The estimate of the surface luminosity is a challenge given our limitations in understanding the physics inside such low mass objects. Also, it is still an open question if a phase transition actually occurs in a brown dwarf. The results of modern day simulations [30, 31] do raise doubts about the relevance of phase transitions in the brown dwarf scenario. We are not aware of well defined analytic models that have a unique way of estimating the surface luminosity apart from using the PPT technique as given in Burrows and Liebert [20]. In this work, rather than considering a single value for the phase transition point, we have used the entire range of temperatures from the phase transition coexistence table [14]. These are within the uncertainty range of the critical temperature of the PPT as proposed by the recent simulations [30]. Thus, considering the large uncertainties involved in such models, this range of values of the minimum mass is much more acceptable than a single distinct transition mass. However, the next step forward is to develop an analytic model for surface temperature that is independent of the PPT.

We estimate that ≈5% of stars take more than 10^8 yr to reach the main sequence, and ≈11% of stars take more than 10^7 yr to reach the main sequence (Table 2). The stars in these categories have mass very close to the minimum hydrogen burning limit and will eventually settle into an extremely low luminosity main sequence with $L/L_\odot$ in the range $\approx 10^{-5}$–10^{-4}. The very low luminosity non-main-sequence hydrogen burning in substellar objects and the pre-main-sequence nuclear burning in very low mass stars are very interesting to study further, and our simplified model can certainly be improved in its ability to estimate the time evolution.

Appendix

A. Pressure Integral

The general Fermi integral can be written as

$$\begin{aligned} F_n &= a\int_0^\infty \frac{\epsilon^n d\epsilon}{e^{\beta(\epsilon-\mu)}+1}, \\ &= a\int_0^\mu \frac{\epsilon^n d\epsilon}{e^{\beta(\epsilon-\mu)}+1} + a\int_\mu^\infty \frac{\epsilon^n d\epsilon}{e^{\beta(\epsilon-\mu)}+1}, \\ &= a\int_0^\mu \epsilon^n d\epsilon - a\int_0^\mu \epsilon^n d\epsilon + a\int_0^\mu \frac{\epsilon^n d\epsilon}{e^{\beta(\epsilon-\mu)}+1} \\ &\quad + a\int_\mu^\infty \frac{\epsilon^n d\epsilon}{e^{\beta(\epsilon-\mu)}+1}, \end{aligned}$$

$$= a\int_0^{\mu} \epsilon^n d\epsilon - a\int_0^{\mu} \frac{\epsilon^n d\epsilon}{e^{-\beta(\epsilon-\mu)}+1} + a\int_{\mu}^{\infty} \frac{\epsilon^n d\epsilon}{e^{\beta(\epsilon-\mu)}+1}. \quad (A.1)$$

On substituting $x = -\beta(\epsilon - \mu)$ in the second term and $x = \beta(\epsilon - \mu)$ in the third term, we arrive at

$$F_n = a\int_0^{\mu} \epsilon^n d\epsilon - \frac{a}{\beta}\int_0^{\beta\mu} \frac{(\mu - x/\beta)^n}{e^x+1} dx + \frac{a}{\beta}\int_0^{\infty} \frac{(\mu + x/\beta)^n}{e^x+1} dx. \quad (A.2)$$

The numerator in the above integrals can be expanded as

$$\left(\mu \pm \frac{x}{\beta}\right)^n \simeq \mu^n \pm n\mu^{n-1}\left(\frac{x}{\beta}\right) + \frac{n(n-1)}{2}\mu^{n-2}\left(\frac{x}{\beta}\right)^2 + \cdots. \quad (A.3)$$

Substituting this in the integral for the pressure, we can proceed as follows:

$$\begin{aligned} F_n &= a\int_0^{\mu} \epsilon^n d\epsilon + \frac{a}{\beta}\mu^n \left\{\int_0^{\infty} \frac{dx}{e^x+1} - \int_0^{\beta\mu} \frac{dx}{e^x+1}\right\} \\ &+ n\frac{a}{\beta^2}\mu^{n-1}\left\{\int_0^{\infty} \frac{x\,dx}{e^x+1} + \int_0^{\beta\mu} \frac{x\,dx}{e^x+1}\right\} \\ &- \frac{n(n-1)}{2}\frac{a}{\beta^3}\mu^{n-2}\left\{\int_0^{\beta\mu} \frac{x^2dx}{e^x+1} - \int_0^{\infty} \frac{x^2dx}{e^x+1}\right\}. \end{aligned} \quad (A.4)$$

Substituting $n = 3/2$ for the Fermi pressure (2) and evaluating these integrals using the polylogs [49], we arrive at a simplified form

$$\begin{aligned} P_F &\simeq a\frac{2}{5}\mu^{5/2} - \frac{1}{8}a\beta^{-1}\mu^{3/2}\ln\left(1+e^{-\beta\mu}\right) \\ &+ \frac{3}{2}\frac{\pi^2}{6}a\beta^{-2}\mu^{1/2} + \frac{3}{4}a\beta^{-2}\mu^{1/2}\mathrm{Li}_2\left(-e^{-\beta\mu}\right) \\ &- \frac{3}{4}a\beta^{-3}\mu^{-1/2}\mathrm{Li}_3\left(-e^{-\beta\mu}\right)\cdots. \end{aligned} \quad (A.5)$$

Similarly, for $n = 1/2$, the expression for the number density can be obtained as

$$\begin{aligned} \rho &\simeq a\frac{2}{3}\mu^{3/2} + \frac{3}{8}a\beta^{-1}\mu^{1/2}\ln\left(1+e^{-\beta\mu}\right) \\ &+ \frac{\pi^2}{12}a\beta^{-2}\mu^{-1/2} + \frac{3}{4}a\beta^{-2}\mu^{-1/2}\mathrm{Li}_2\left(-e^{-\beta\mu}\right) \\ &+ \frac{1}{4}a\beta^{-3}\mu^{-1/2}\mathrm{Li}_3\left(-e^{-\beta\mu}\right)\cdots. \end{aligned} \quad (A.6)$$

B. Surface Properties

The surface properties of a brown dwarf can be analyzed by studying the phase diagram of hydrogen in the interior and the photospheric region. The total pressure inside a stellar or substellar object can be represented as

$$P = P_g + P_r, \quad (B.1)$$

where P_g is the gas pressure due to the adiabatic ideal gas and P_r is the radiation pressure. At a temperature comparable to that of the envelope surrounding the interior of a substellar object, hydrogen is partially ionized and the helium gas is mostly molecular. For a quasistatic change, the first law of thermodynamics yields

$$dQ = \left(\frac{\partial U}{\partial T}\right)_V dT + \left(\frac{\partial U}{\partial V}\right)_T dV + PdV \quad (B.2)$$

[32]. The most general expression for the internal energy of a monatomic gas is

$$U = \frac{3}{2}\frac{k_B N_0 T}{\mu_1} + aT^4V + x\chi N_0, \quad (B.3)$$

where we have considered the energy due to photon radiation and gas ionization. Here, χ is the ionization energy and x is the ionization fraction of hydrogen. Taking the partial derivatives of (B.3) and using the second law of thermodynamics $TdS = dQ$, we arrive at

$$\begin{aligned} dS &= -\frac{3}{2}\frac{k_B N_0}{\mu_1}\frac{dT}{T} + \frac{k_B N_0}{\mu_1}\frac{dW}{W} + \frac{4}{3}adW \\ &+ N_0\left(\chi + \frac{3}{2}k_B T\right)\left(\frac{\partial x}{\partial T}\frac{dT}{T} + \frac{\partial x}{\partial V}\frac{dV}{T}\right), \end{aligned} \quad (B.4)$$

where $W = T^3V$. The ionization fraction is a function of density and temperature, $x(\rho, T)$. Using the Saha equation, we obtain

$$\begin{aligned} \left(\frac{\partial x}{\partial T}\right)_V &= \frac{x(1-x)}{2-x}\frac{1}{T}\left(\frac{3}{2} + \frac{\chi}{k_B T}\right), \\ \left(\frac{\partial x}{\partial V}\right)_T &= \frac{x(1-x)}{2-x}\frac{1}{V}. \end{aligned} \quad (B.5)$$

Using the above relations, we can simplify (B.4) to be

$$\begin{aligned} \frac{dS}{k_B N_0} &= -\frac{3}{2\mu_1}\frac{dT}{T} + \frac{dW}{\mu_1 W} + \left(\frac{3}{2}\right)^2\frac{HdT}{T} + \frac{3}{2}\frac{HdV}{V} \\ &+ \frac{4adW}{3k_B N_0} + \chi\left(\frac{dV}{TV} + 3\frac{dT}{T^2} + \frac{\chi}{k_B}\frac{dT}{T^3}\right), \end{aligned} \quad (B.6)$$

where $H = x(1-x)/(2-x)$. This expression is the same as (19) except for the final two terms due to radiation and ionization energy, respectively. We can ignore the final term and retain terms of linear order in T. On integrating and simplifying the above expression, we get the entropy for the partially ionized hydrogen and helium gas:

$$S_1 = k_B N_0\left(\frac{1}{\mu_1} + \frac{3}{2}\frac{x(1-x)}{2-x}\right)\ln\frac{T^{3/2}}{\rho} + \frac{4}{3}aT^3V + C_1. \quad (B.7)$$

At low temperature ($T < 4000$ K) and low pressure, hydrogen is predominantly molecular and fluid. Repeating the above derivation for the nonionized diatomic hydrogen gas with energy $U = (5/2)(k_B T N_0/\mu_2)$ and molecular helium, we can arrive at an expression for the entropy:

$$S_2 = \frac{k_B N_0}{\mu_2} \ln \frac{T^{5/2}}{\rho} + \frac{4}{3} a T^3 V + C_2. \quad \text{(B.8)}$$

In the above expressions for entropy, ρ and T are the density and the temperature, respectively. Other variables are as described in the text.

C. Thermal Properties

We discuss some of the more accurate expressions for the important thermal properties of a degenerate system.

C.1. Fermi Energy. Using (A.6) for the number density, we can write the general expression for the chemical potential μ in terms of the Fermi energy μ_F at $T = 0$ [51]. Considering only the first three terms (A.6) and for $\rho = (2/3)\mu_F^{3/2}$, we find

$$\mu \simeq \mu_F - \frac{\pi^2}{12} \frac{1}{(\beta\mu_F)^2} - \frac{3}{8} \frac{1}{\beta\mu_F} \ln\left(1 + \exp^{-\beta\mu_F}\right). \quad \text{(C.1)}$$

The second and the third terms are the correction factor C to the zero temperature Fermi energy. For $0.03 < 1/\mu_F\beta < 0.20$, the correction factor C will be in the range $\sim 8 \times 10^{-4} < C < 4 \times 10^{-2}$.

C.2. Specific Heat. In the nondegenerate completely ionized limit, the specific heat $C_v \sim 3k_B N/2$. At finite temperatures, the value of the specific heat is less than the limiting value; that is, $C_v < 3Nk_B/2$. The specific heat of the ideal Fermi gas decreases monotonically. At low but finite temperatures,

$$\frac{C_v}{N} \simeq \frac{\pi^2}{2} \frac{k_B T}{\mu_F}. \quad \text{(C.2)}$$

A detailed analysis in the calculation of the specific heat shows that the numerical coefficients in the expansion approached a limiting value of 2.

C.3. Grueneisen Parameter. Applying the condition of constant entropy to (20) leads to the condition

$$T = C\rho^{2/3}, \quad \text{(C.3)}$$

where C is a constant. The Grueneisen parameter γ is given by the expression

$$\gamma = \left(\frac{\partial \log T}{\partial \log \rho}\right)_s. \quad \text{(C.4)}$$

Using (C.3) in the above expression, we estimate the value of the Grueneisen parameter γ to be $2/3$. This value is in approximate accord with Stevenson [16], who indicated that $\gamma \simeq 0.6$ in dense Coulomb plasma when obtained from computer simulations.

C.4. Ionic Correlation. Ionic correlation is an important contribution as considered by Stolzmann and Blocker [25], Becker et al. [27], Hubbard et al. [52], and Gericke et al. [26] to name but a few. Stolzmann and Blocker [25] use the method of Pade's approximations to provide explicit expressions for the fully ionized plasma of the Helmholtz free energy and pressure. They have considered the nonideal effects of different correlations such as the electron-electron, ion-ion, electron-ion, and exchange contribution for a wide range of values of the Coulomb coupling parameter Γ, which is the ratio of the Coulomb to thermal energy:

$$\Gamma = \frac{e^2}{k_B T} \left(\frac{4\pi n_e}{3}\right)^{1/3} = \frac{\Gamma_{\text{ion}}}{\langle z^{5/3} \rangle}. \quad \text{(C.5)}$$

Here, n_e stands for the electron density and $\langle z^{5/3} \rangle$ is the charge average, given as

$$\langle z^{5/3} \rangle = \frac{\sum n_i Z_i^{5/3}}{\sum n_i}, \quad \text{(C.6)}$$

for ions of different species i. The greatest effect is in the relative pressure contribution P_{ii}/P_{ideal} of the ionic correlation term for hydrogen at $T = 10^5$ K, estimated to be ~ -0.1 at $\rho \sim 10^3$ g/cm^3 and a minimum of ~ -0.2 at $\rho \sim 10$ g/cm^3.

Competing Interests

The authors declare that they have no competing interests regarding the publication of this paper.

Acknowledgments

The authors thank Dr. Andrea Becker and Dr. Gilles Chabrier for their valuable comments and input during the preparation of this article. They also thank Anushrut Sharma for contributing to the initial stages of this project. Shantanu Basu and S. R. Valluri were supported by Discovery Grant from the Natural Sciences and Engineering Research Council (NSERC) of Canada. S. R. Valluri also thanks King's University College for its continued support for his research endeavours.

References

[1] B. R. Oppenheimer, S. R. Kulkarni, K. Matthews, and T. Nakajima, "Infrared spectrum of the cool brown dwarf GI 229B," *Science*, vol. 270, no. 5241, pp. 1478–1479, 1995.

[2] R. Rebolo, M. R. Zapatero Osorio, and E. L. Martín, "Discovery of a brown dwarf in the Pleiades star cluster," *Nature*, vol. 377, no. 6545, pp. 129–131, 1995.

[3] E. L. Martín, R. Rebolo, and M. R. Zapatero-Osorio, "Spectroscopy of new substellar candidates in the pleiades: toward a spectral sequence for young brown dwarfs," *Astrophysical Journal*, vol. 469, no. 2, pp. 706–714, 1996.

[4] M. T. Ruiz, S. K. Leggett, and F. Allard, "Kelu-1: a free-floating brown dwarf in the solar neighborhood," *Astrophysical Journal*, vol. 491, no. 2, pp. L107–L110, 1997.

[5] K. L. Luhman, V. Joergens, C. Lada, J. Muzerolle, I. Pascucci, and R. White, "The formation of brown dwarfs: observations,"

in *Protostars and Planets V*, B. Reipurth, D. Jewitt, and K. Keil, Eds., The University of Arizona Press, 2007.

[6] G. Chabrier, A. Johansen, M. Janson, and R. Rafikov, "Giant planet and brown dwarf formation," https://arxiv.org/abs/1401.7559.

[7] S. S. Kumar, "The structure of stars of very low mass," *The Astrophysical Journal*, vol. 137, p. 1121, 1963.

[8] S. S. Kumar, "The Helmholtz-Kelvin time scale for stars of very low mass," *The Astrophysical Journal*, vol. 137, p. 1126, 1963.

[9] C. Hayashi and T. Nakano, "Evolution of stars of small masses in the pre-main-sequence stages," *Progress of Theoretical Physics*, vol. 30, no. 4, pp. 460–474, 1963.

[10] F. D'Antona and I. Mazzitelli, "Evolution of very low mass stars and brown dwarfs. I—the minimum main-sequence mass and luminosity," *Astrophysical Journal*, vol. 296, pp. 502–513, 1985.

[11] A. Burrows, W. B. Hubbard, and J. I. Lunine, "Theoretical models of very low mass stars and brown dwarfs," *The Astrophysical Journal*, vol. 345, pp. 939–958, 1989.

[12] D. Saumon and G. Chabrier, "Fluid hydrogen at high density: the plasma phase transition," *Physical Review Letters*, vol. 62, no. 20, pp. 2397–2400, 1989.

[13] G. Chabrier and H. M. van Horn, "The role of the molecular-metallic transition of hydrogen in the evolution of Jupiter, Saturn, and brown dwarfs," *The Astrophysical Journal*, vol. 391, no. 2, pp. 827–831, 1992.

[14] G. Chabrier, D. Saumon, W. B. Hubbard, and J. I. Lunine, "The molecular-metallic transition of hydrogen and the structure of Jupiter and Saturn," *The Astrophysical Journal*, vol. 391, no. 2, pp. 817–826, 1992.

[15] G. Fontaine, H. C. Graboske, and H. M. van Horn, "Equations of state for stellar partial ionization zones," *The Astrophysical Journal Supplement Series*, vol. 35, p. 293, 1977.

[16] D. J. Stevenson, "The search for brown dwarfs," *Annual Review of Astronomy and Astrophysics*, vol. 29, no. 1, pp. 163–193, 1991.

[17] R. Rebolo, E. L. Martin, and A. Magazzu, "Spectroscopy of a brown dwarf candidate in the Alpha Persei open cluster," *Astrophysical Journal*, vol. 389, pp. L83–L86, 1992.

[18] G. Basri, W. G. Marcy, and R. J. Graham, "Lithium in brown dwarf candidates: the mass and age of the faintest pleiades stars," *Astrophysical Journal*, vol. 458, p. 600, 1996.

[19] J. Stauffer, G. Schultz, and D. J. Kirkpatrick, "Keck spectra of pleiades brown dwarf candidates and a precise determination of the lithium depletion edge in the pleiades," *The Astrophysical Journal*, vol. 499, no. 2, p. L199, 1998.

[20] A. Burrows and J. Liebert, "The science of brown dwarfs," *Reviews of Modern Physics*, vol. 65, no. 2, pp. 301–336, 1993.

[21] W. A. Fowler and F. Hoyle, "Neutrino processes and pair formation in massive stars and supernovae," *Astrophysical Journal Supplement*, vol. 9, p. 201, 1964.

[22] T. Padmanabhan, *Theoretical Astrophysics, Volume 1: Astrophysical Processes*, Cambridge University Press, Cambridge, UK, 1999.

[23] A. Sommerfeld, "Zur Elektronentheorie der Metalle auf Grund der Fermischen Statistik. I. Teil: Allgemeines, Strömungs- und Austrittsvorgänge," *Zeitschrift für Physik*, vol. 47, no. 1-2, pp. 1–32, 1928.

[24] W. B. Hubbard, *Planetary Interiors*, Van Nostrand Reinhold, New York, NY, USA, 1984.

[25] W. Stolzmann and T. Blocker, "Thermodynamical properties of stellar matter. I. Equation of state for stellar interiors," *Astronomy & Astrophysics*, vol. 314, pp. 1024–1040, 1996.

[26] O. D. Gericke, K. Wünsch, A. Grinenko, and J. Vorberger, "Structural properties of warm dense matter," *Journal of Physics: Conference Series*, vol. 220, no. 1, Article ID 012001, 2010.

[27] A. Becker, W. Lorenzen, J. J. Fortney, N. Nettelmann, M. Schöttler, and R. Redmer, "AB initio equations of state for hydrogen (H-REOS.3) and helium (H-REOS.3) and their implications for the interior of brown dwarfs," *The Astrophysical Journal Supplement Series*, vol. 215, no. 2, p. 21, 2014.

[28] O. L. Anderson, "The Grüneisen ratio for the last 30 years," *Geophysical Journal International*, vol. 143, no. 2, pp. 279–294, 2000.

[29] S. Chandrasekhar, *An Introduction to the Study of Stellar Structure*, The University of Chicago Press, Chicago, Ill, USA, 1939.

[30] J. Yang, Ch. L. Tian, F. Sh. Liu, K. H. Yuan, H. M. Zhong, and F. Xiao, "A new evidence of first-order phase transition for hydrogen at 3000 K," *Europhysics Letters*, vol. 109, no. 3, Article ID 36003, 2015.

[31] M. A. Morales, C. Pierleoni, E. Schwegler, and D. M. Ceperley, "Evidence for a first order liquid-liquid transition in high-pressure hydrogen from ab initio simulations," *Proceedings of the National Academy of Sciences of the United States of America*, vol. 107, no. 29, pp. 12799–12803, 2010.

[32] D. D. Clayton, *Principles of Stellar Evolution and Nucleosynthesis*, McGraw-Hill, New York, NY, USA, 1968.

[33] D. Saumon, G. Chabrier, and H. M. Van Horn, "An equation of state for low-mass stars and giant planets," *Astrophysical Journal, Supplement Series*, vol. 99, no. 2, pp. 713–741, 1995.

[34] W. Lorenzen, B. Holst, and R. Redmer, "First-order liquid-liquid phase transition in dense hydrogen," *Physical Review B*, vol. 82, no. 19, Article ID 195107, 6 pages, 2010.

[35] G. Mazzola, S. Yunoki, and S. Sorella, "Unexpectedly high pressure for molecular dissociation in liquid hydrogen by electronic simulation," *Nature Communications*, vol. 5, article 3487, 2014.

[36] M. D. Knudson, M. P. Desjarlais, A. Becker et al., "Direct observation of an abrupt insulator-to-metal transition in dense liquid deuterium," *Science*, vol. 348, no. 6242, pp. 1455–1460, 2015.

[37] W. A. Fowler, G. R. Caughlan, and B. A. Zimmerman, "Thermonuclear reaction rates, II," *Annual Review of Astronomy & Astrophysics*, vol. 13, pp. 69–112, 1975.

[38] B. Paxton, L. Bildsten, A. Dotter, F. Herwig, P. Lesaffre, and F. Timmes, "Modules for Experiments in Stellar Astrophysics (MESA)," *The Astrophysical Journal Supplement*, vol. 192, no. 1, 3 pages, 2011.

[39] K. G. Stassun, R. D. Mathieu, and J. A. Valenti, "Discovery of two young brown dwarfs in an eclipsing binary system," *Nature*, vol. 440, no. 7082, pp. 311–314, 2006.

[40] T. J. David, L. A. Hillenbrand, A. M. Cody, J. M. Carpenter, and A. W. Howard, "K2 discovery of young eclipsing binaries in upper scorpius: direct mass and radius determinations for the lowest mass stars and initial characterization of an eclipsing brown dwarf binary," *The Astrophysical Journal*, vol. 816, no. 1, article 21, 27 pages, 2016.

[41] E. L. Martin, R. Rebolo, and T. Magazzu, "Constraints to the masses of brown dwarf candidates form the lithium test," *The Astrophysical Journal*, vol. 436, no. 1, pp. 262–269, 1994.

[42] E. S. Dahm, "Reexamining the lithium depletion boundary in the pleiades and the inferred age of the cluster," *The Astrophysical Journal*, vol. 813, no. 2, article 108, 2015.

[43] A. Magazzù, E. L. Martin, and R. Rebolo, "A spectroscopic test for substellar objects," *Astrophysical Journal*, vol. 404, no. 1, pp. L17–L20, 1993.

[44] A. Scholz, V. Kostov, R. Jayawardhana, and K. Mužić, "Rotation periods of young brown dwarfs: k2 survey in upper scorpius," *The Astrophysical Journal*, vol. 809, article L29, 2015.

[45] E. L. Martín and A. Claret, "Stellar models with rotation: an exploratory application to pre-main sequence lithium depletion," *Astronomy & Astrophysics*, vol. 306, no. 2, pp. 408–416, 1996.

[46] J. Bouvier, C. A. Lanzafame, L. Venuti et al., "The Gaia-ESO Survey: a lithium-rotation connection at 5 Myr?" *Astronomy & Astrophysics*, vol. 590, article A78, 2016.

[47] S. Basu, M. Gil, and S. Auddy, "The MLP distribution: a modified lognormal power-law model for the stellar initial mass function," *Monthly Notices of the Royal Astronomical Society*, vol. 449, no. 3, pp. 2413–2420, 2015.

[48] G. Chabrier, "The initial mass function: from Salpeter 1955 to 2005," in *The Initial Mass Function 50 Years Later*, E. Corbelli, F. Palla, and H. Zinnecker, Eds., vol. 327 of *Astrophysics and Space Science Library*, pp. 41–50, Springer, Dordrecht, Netherlands, 2005.

[49] J. Tanguay, M. Gil, D. J. Jeffrey, and S. R. Valluri, "D-dimensional Bose gases and the Lambert W function," *Journal of Mathematical Physics*, vol. 51, no. 12, Article ID 123303, 2010.

[50] M. Molli, K. Venkataramaniah, and S. R. Valluri, "The polylogarithm and the Lambert W functions in thermoelectrics," *Canadian Journal of Physics*, vol. 89, no. 11, pp. 1171–1178, 2011.

[51] R. P. Feynman, *Statistical Mechanics: A Set of Lectures*, Benjamin/Cummings Publishing, Reading, Mass, USA, 1972.

[52] W. B. Hubbard and H. E. Dewitt, "Statistical mechanics of light elements at high pressure. VII—a perturbative free energy for arbitrary mixtures of H and He," *Astrophysical Journal*, vol. 290, pp. 388–393, 1985.

9

The Observer's Guide to the Gamma-Ray Burst Supernova Connection

Zach Cano,[1,2] Shan-Qin Wang,[3,4] Zi-Gao Dai,[3,4] and Xue-Feng Wu[5,6]

[1] *Centre for Astrophysics and Cosmology, Science Institute, University of Iceland, Dunhagi 5, 107 Reykjavik, Iceland*
[2] *Instituto de Astrofísica de Andalucía (IAA-CSIC), Glorieta de la Astronomía s/n, 18008 Granada, Spain*
[3] *School of Astronomy and Space Science, Nanjing University, Nanjing 210093, China*
[4] *Key Laboratory of Modern Astronomy and Astrophysics (Nanjing University), Ministry of Education, Nanjing, China*
[5] *Purple Mountain Observatory, Chinese Academy of Sciences, Nanjing 210008, China*
[6] *Joint Center for Particle, Nuclear Physics and Cosmology, Nanjing University-Purple Mountain Observatory, Nanjing 210008, China*

Correspondence should be addressed to Zach Cano; zewcano@gmail.com

Academic Editor: Josep M. Trigo-Rodríguez

We present a detailed report of the connection between long-duration gamma-ray bursts (GRBs) and their accompanying supernovae (SNe). The discussion presented here places emphasis on how observations, and the modelling of observations, have constrained what we know about GRB-SNe. We discuss their photometric and spectroscopic properties, their role as cosmological probes, including their measured luminosity–decline relationships, and how they can be used to measure the Hubble constant. We present a statistical summary of their bolometric properties and use this to determine the properties of the "average" GRB-SN. We discuss their geometry and consider the various physical processes that are thought to power the luminosity of GRB-SNe and whether differences exist between GRB-SNe and the SNe associated with ultra-long-duration GRBs. We discuss how observations of their environments further constrain the physical properties of their progenitor stars and give a brief overview of the current theoretical paradigms of their central engines. We then present an overview of the radioactively powered transients that have been photometrically associated with short-duration GRBs, and we conclude by discussing what additional research is needed to further our understanding of GRB-SNe, in particular the role of binary-formation channels and the connection of GRB-SNe with superluminous SNe.

1. Introduction

Observations have proved the massive-star origins of long-duration GRBs (LGRBs) beyond any reasonable doubt. The temporal and spatial connection between GRB 980425 and broad-lined type Ic (IcBL) SN 1998bw offered the first clues to their nature [1, 2] (Figure 1). The close proximity of this event ($z = 0.00866$; ~40 Mpc), which is still the closest GRB to date, resulted in it becoming one of the most, if not *the* most, scrutinized GRB-SN in history. It was shown that SN 1998bw had a very large kinetic energy (see Section 4 and Table 3) of ~2–5 × 10^{52} erg, which led it to being referred to as a hypernova [3]. However, given several peculiarities of its γ-ray properties, including its underluminous γ-ray luminosity ($L_{\gamma,\mathrm{iso}} \sim 5 \times 10^{46}$ erg s^{-1}), it was doubted whether this event was truly representative of the general LGRB population. This uncertainty persisted for almost five years until the spectroscopic association between cosmological/high-luminosity GRB 030329 ($L_{\gamma,\mathrm{iso}} \sim 8 \times 10^{50}$ erg s^{-1}) and SN 2003dh [4–6]. GRB 030329 had an exceptionally bright optical afterglow (AG; see Figures 2 and 3), and a careful decomposition of the photometric and spectroscopic observations was required in order to isolate the SN features from the dominant AG light [7] (see Section 2.1 and Figure 4). As was seen for SN 1998bw, SN 2003dh was a type IcBL SN, and its kinetic energy was in excess of 10^{52} erg, showing that it too was a hypernova.

The launch of the *Swift* satellite [8] dramatically changed the way we studied GRBs and the GRB-SN association, and the number of events detected by this mission has helped increase the GRB-SN sample size by a factor of three since

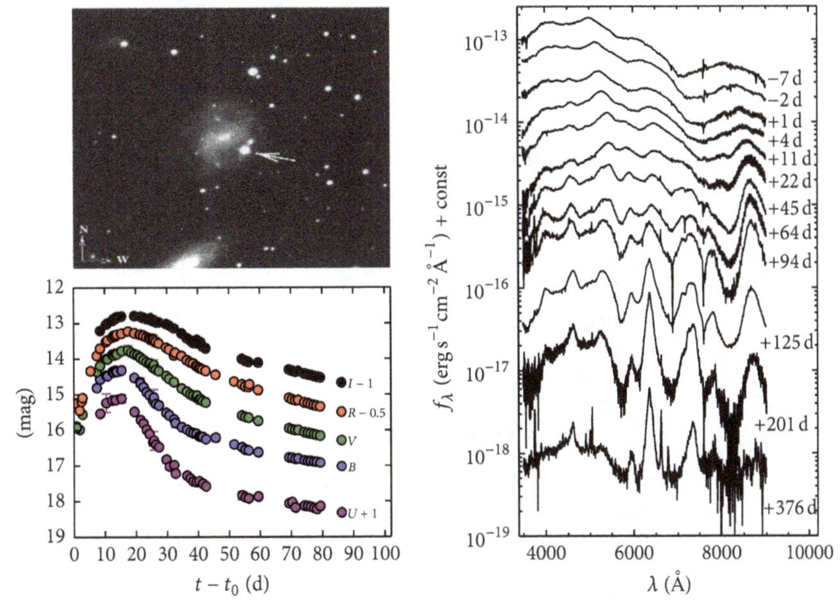

FIGURE 1: GRB 980425/SN 1998bw: the archetype GRB-SN. Host image (ESO 184-G82) is from [1], where the position of the optical transient is clearly visible. Optical light curves are from [9] and spectra from [2].

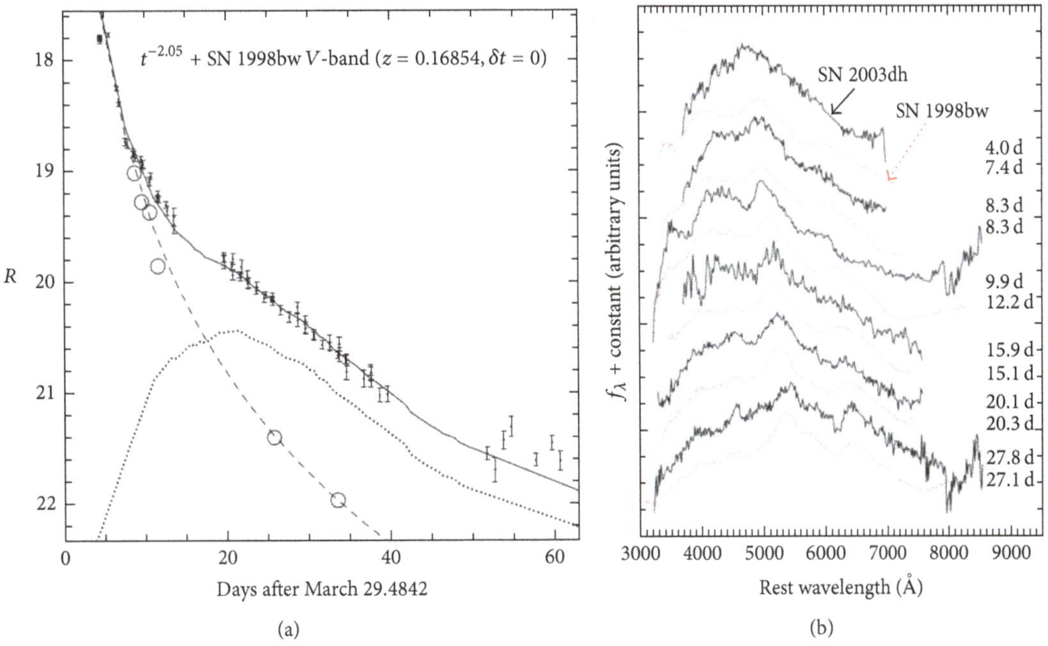

FIGURE 2: (a) The photometric (R-band) evolution of GRB 030329/SN 2003dh, from [6]; (b) the spectral evolution of GRB 030329/SN 2003dh, as compared with that of SN 1998bw, from [4].

the pre-*Swift* era. This includes, among many others, the well-studied events GRB 060218/SN 2006aj, GRB 100316D/SN 2010bh, GRB 111209A/SN 2011kl, GRB 120422A/SN 2012bz, GRB 130427A/SN 2013cq, and GRB 130702A/SN 2013dx. A full list of the references to these well-studied spectroscopic GRB-SN associations is found in Table 4.

This review paper represents a continuation of other review articles presented to date, including the seminal work by Woosley and Bloom (2006) [11]. As such, we have focused the majority of the content on achievements made in the 10 years since [11] was published. In this review and many others [12–19], thorough historical accounts of the development of the gamma-ray burst supernova (GRB-SN) connection are presented, and we encourage the reader to consult the detailed presentation given in section one of [11] for further details. In Tables 2 and 3 we present the most comprehensive database yet compiled of the observational and physical properties of the GRB prompt emission and

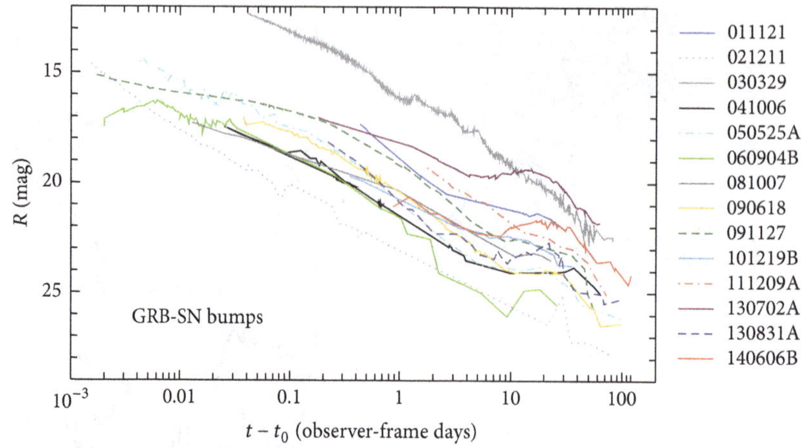

Figure 3: A mosaic of GRB-SNe (AG + SN). Clear SN bumps are observed for all events except SN 2003dh, for which the SN's properties had to be carefully decomposed from photometric and spectroscopic observations [7]. The lack of an unambiguous SN bump in this case is not surprising given the brightness of its AG relative to the other GRB-SN in the plot: SN 2013dx was at a comparable redshift ($z = 0.145$, compared with $z = 0.1685$ for 2003dh), but its AG was much fainter (2–5 mag) at a given moment in time. The redshift range probed in this mosaic spans almost an order of magnitude ($0.145 < z < 1.006$) and shows the variation in peak observed magnitude for GRB-SNe. It is important to remember that given the large span of distances probed here, observer-frame R-band samples a wide range of rest-frame SEDs (from U-band to V-band).

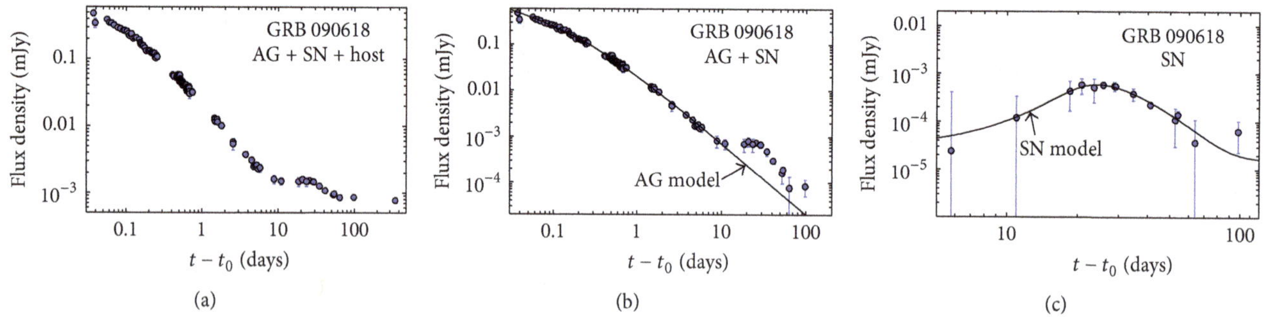

Figure 4: An example decomposition of the optical (R-band) light curve of GRB 090618 [10]. (a) For a given GRB-SN event, the single-filter monochromatic flux is attributed as arising from three sources: the AG, the SN, and a constant source of flux from the host galaxy. (b) Once the observations have been dereddened, the host flux is removed, either via the image-subtraction technique or by being mathematically subtracted away. At this point a mathematical model composed of one or more power laws punctuated by break-times is fit to the early light curve to determine the temporal behaviour of the AG. (c) Once the AG model has been determined, it is subtracted from the observations leaving just light from the SN.

GRB-SNe, respectively, which consists of 46 GRB-SNe. It is the interpretation of these data which forms a substantial contribution to this review. We have adopted the grading scheme devised by [17] to assign a significance of the GRB-SN association to each event, where A is strong spectroscopic evidence, B is a clear light curve bump as well as some spectroscopic evidence resembling a GRB-SN, C is a clear bump consistent with other GRB-SNe at the spectroscopic redshift of the GRB, D is a bump, but the inferred SN properties are not fully consistent with other GRB-SNe or the bump was not well sampled or there is no spectroscopic redshift of the GRB, and E is a bump, either of low significance or inconsistent with other GRB-SNe. This is found in Table 3.

Throughout this article we use a ΛCDM cosmology constrained by [20] of $H_0 = 67.3\ \mathrm{km\,s^{-1}\,Mpc^{-1}}$, $\Omega_M = 0.315$, $\Omega_\Lambda = 0.685$. All published data, where applicable, have been renormalized to this cosmological model. Foreground extinctions were calculated using the dust extinction maps of [21, 22]. Unless stated otherwise, errors are statistical only. Nomenclature is as follows: σ denotes the standard deviation of a sample, whereas the root-mean square of a sample is expressed as RMS. A symbol with an overplotted bar denotes an average value. LGRB and SGRB are long- and short-duration GRBs, respectively, while a GRB-SN is implicitly understood to be associated with an LGRB. The term t_0 refers to the time that a given GRB was detected by a GRB satellite.

2. Observational Properties

2.1. Photometric Properties. The observer-frame, optical light curves (LCs) of GRBs span more than 8–10 magnitudes at a given observer-frame postexplosion epoch (see, e.g., Figure 1 in [23]). Similarly, if we inspect the observer-frame R-band LCs of GRB-SNe (redshift range $0.145 < z < 1.006$) shown in Figure 3, they too span a similar range at a given epoch. Indeed, the peak SN brightness during the SN "bump" phase ranges from $R = 19.5$ for GRB 130702A (the brightest GRB-SN bump observed to date) to $R = 25$ for GRB 021211.

For a typical GRB-SN, there are three components of flux being measured: (1) the afterglow (AG), which is associated with the GRB event, (2) the SN, and (3) the constant source of flux coming from the host galaxy. A great deal of information can be obtained from modelling each component, but, for the SN component to be analysed, it needs to be *decomposed* from the optical/NIR LCs (Figure 4). To achieve this task, the temporal behaviour of the AG, the constant source of flux from the host galaxy, and the line of sight extinction, including foreground extinction arising from different sight-lines through the Milky Way (MW) [21, 22], and extinction local to the event itself [10, 23–26], in a given filter need to be modelled and quantified. The host contribution can be considered either by removing it via the image-subtraction technique [27–29], by simple flux-subtraction [30–32], or by including it as an additional component in the fitting routine [33–36]. The AG component is modelled using either a single or a set of broken power laws (SPL/BPL; [37]). This phenomenological approach is rooted in theory however, as standard GRB theory states that the light powering the AG is synchrotron in origin and therefore follows a power law behaviour in both time and frequency ($f_\nu \propto (t - t_0)^{-\alpha}\nu^{-\beta}$, where the respective decay and energy spectral indices are α and β).

Once the SN LC has been obtained, traditionally it is compared to a template supernova, that is, SN 1998bw, where the relative brightness (k) and width (also known as a stretch factor, s) are determined. Such an approach has been used extensively over the years [10, 18, 31–34, 38–43]. Another approach to determining the SN's properties is to fit a phenomenological model to the resultant SN LC [10, 42–44], such as the Bazin function [45], in order to determine the magnitude/flux at peak SN light, the time it takes to rise and fade from peak, and the width of the LC, such as the Δm_{15} parameter (in a given filter, the amount a SN fades in magnitudes from peak light to 15 days later). All published values of these observables are presented in Table 3.

2.2. Spectroscopic Properties. Optical and NIR spectra, of varying levels of quality due to their large cosmological distances, have been obtained for more than a dozen GRB-SNe. Those of the highest quality show broad observation lines of O I, Ca II, Si II, and Fe II near maximum light. The line velocities of two specific transitions (Si II λ6355 and Fe II λ5169; Figure 6) indicate that near maximum light the ejecta that contain these elements move at velocities of order 20,000–40,000 km s^{-1} (Fe II λ5169) and about 15,000–25,000 km s^{-1} (Si II λ6355). The weighted mean absorption velocities at peak V-band light of a sample of SNe IcBL that included GRB-SNe were found to be 23,800 ± 9500 km s^{-1} (Fe II λ5169) by [46] (see as well Table 3). SNe IcBL (including and excluding GRB-SNe) have Fe II λ5169 widths that are ~9,000 km s^{-1} broader than SNe Ic, while

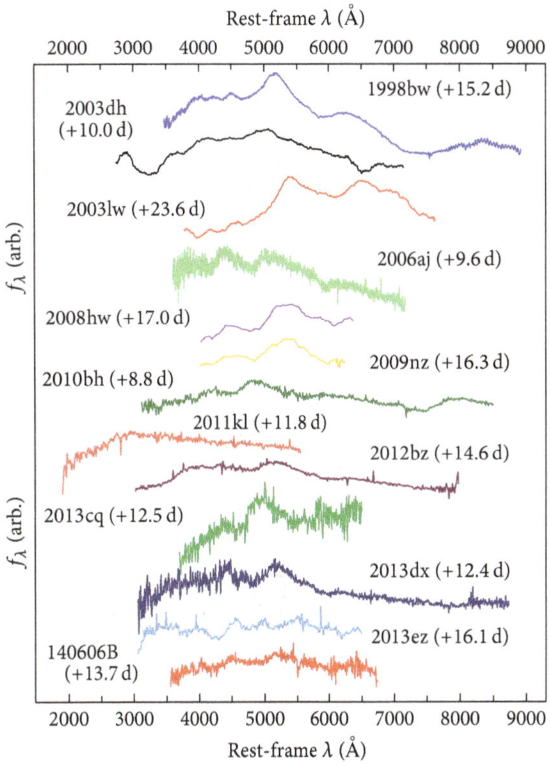

FIGURE 5: Peak/near-peak spectra of GRB-SNe. The spectra have been arbitrarily shifted in flux for comparison purposes and to exaggerate their main features, and host emission lines have been manually removed. The spectra of SNe 2012bz, 2013cq, and 2013dx have been Kaiser smoothed [31] in order to suppress noise. Most of the spectra are characterized by broad absorption features, while such features are conspicuously absent in the spectra of SN 2013ez and SN 2011kl.

GRB-SNe appear to be, on average, about ~6,000 km s^{-1} more rapid than SNe IcBL at peak light [46]. Si II λ6355 appears to have a tighter grouping of velocities than Fe II λ5169, though SN 2010bh is a notable outlier, being roughly 15,000 to 20,000 km s^{-1} more rapid than the other GRB-SNe. SN 2013ez is also a notable outlier due to its low line velocity (4000–6000 km s^{-1}), and inspection of its spectrum (Figure 5) reveals fewer broad features than other GRB-SNe, where it more closely resembles type Ic SNe rather than type IcBL [31]. Nevertheless, this relative grouping of line velocities may indicate similar density structure(s) in the ejecta of these SN, which in turn could indicate some general similarities in their preexplosion progenitor configurations. For comparison, [46] found that the dispersion of peak SNe Ic Fe II λ5169 line velocities is tighter than those measured for GRB-SNe and SNe IcBL not associated with GRBs (σ = 1500, 9500, 2700 km s^{-1}, resp.). This suggests that GRB-SNe and SNe IcBL have more diversity in their spectral velocities, and in turn their density structures, than SNe Ic. Finally, [46] found no differences in the spectra of *ll*GRB-SNe relative to high-luminosity GRB-SNe.

During the nebular phase of SN 1998bw (one of only a few GRB-SNe that has been spectroscopically observed

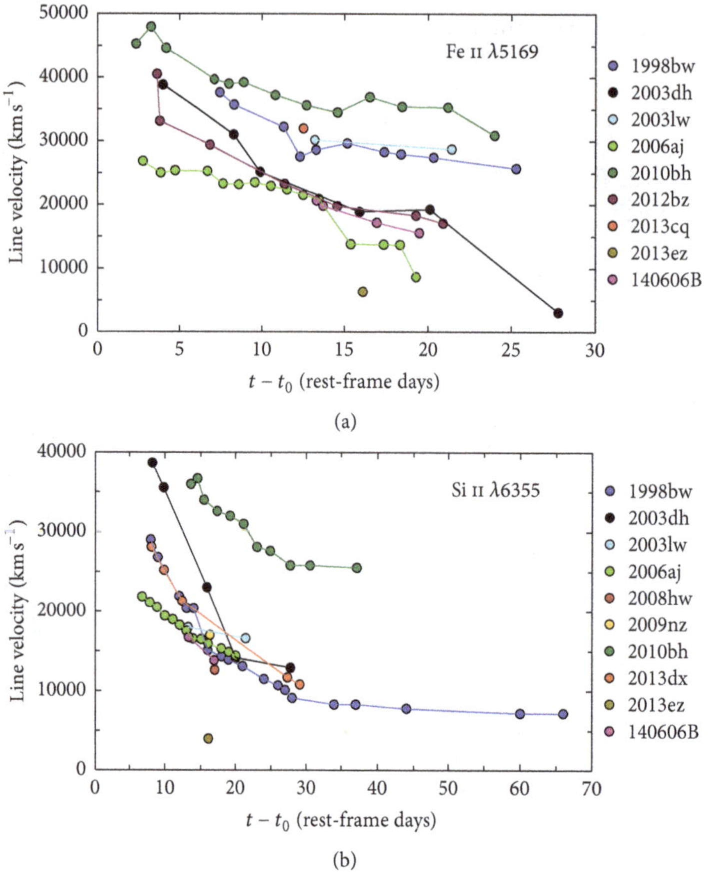

FIGURE 6: Measured line velocities of a sample of GRB-SNe. See Table 4 for their respective references.

during this phase due to its close proximity; see as well Section 5), observed lines include [O I] λ5577, λλ6300,6364; O [II] λ7322; Ca II λλ3934,3963, λλ7291,7324; Mg I] λ4570; Na I λλ5890,5896; [Fe II] λ4244, λ4276, λ4416, λ4458, λ4814, λ4890, λ5169, λ5261, λ5273, λ5333, λ7155, λ7172, λ7388, λ7452; [Fe III] λ5270; Co II λ7541; C I] λ8727 [47]. Nebular [O I] λλ6300,6364 was also observed for nearby GRB-SNe 2006aj [48] and 100316D [49], though in the latter case strong lines from the underlying HII are considerably more dominant. For SN 2006aj, [Ni II λ7380] was tentatively detected [48], which, given the short half-life of ^{56}Ni, implies the existence of roughly 0.05 $M_\odot$ of ^{58}Ni. Such a large amount of stable neutron-rich Ni strongly indicates the formation of a neutron star [48]. Moreover, the absence of [Ca II] lines for SN 2006aj also supported the lower kinetic energy of this event relative to other GRB-SNe, which is likely less than that attributed to a hypernova.

3. Phenomenological Classifications of GRB-SNe

Replicating previous works [19, 41], in this review, we divided GRB-SNe into the following subclasses based primarily on their isotropic γ-ray luminosity $L_{\gamma,\mathrm{iso}}$:

(i) *ll*GRB-SNe: GRB-SNe associated with low-luminosity GRBs ($L_{\gamma,\mathrm{iso}} < 10^{48.5}$ erg s^{-1}).

(ii) INT-GRB-SNe: GRB-SNe associated with intermediate-luminosity GRBs ($10^{48.5} < L_{\gamma,\mathrm{iso}} < 10^{49.5}$ erg s^{-1}). (Not to be confused with intermediate-duration GRBs, i.e., those with durations of 2–5 s [50–52].)

(iii) GRB-SNe: GRB-SNe associated with high-luminosity GRBs ($L_{\gamma,\mathrm{iso}} > 10^{49.5}$ erg s^{-1}).

(iv) ULGRB-SNe: ultra-long-duration GRB-SNe, which are classified according to the exceptionally long duration of their γ-ray emission (~10^4 seconds [53, 54]) rather than on their γ-ray luminosities.

Historically, the term X-ray flash (XRF) was used throughout the literature, which has slowly been replaced with the idiom of "low-luminosity." Strictly speaking, the definition of an XRF [55] arises from the detection of soft, X-ray rich events detected by the Wide Field Camera on *BeppoSax* in the energy range 2–25 keV. Here we make no distinction based on the detection of a given satellite and instrumentation, where the "*ll*" nomenclature refers only to the magnitude of a given GRB's $L_{\gamma,\mathrm{iso}}$.

The luminosity, energetics, and shape of the γ-ray pulse of a given GRB can reveal clues to the origin of its high-energy emission and thus its emission process. Of particular importance is whether the γ-rays emitted by *ll*GRBs arise from the same mechanism as high-luminosity GRBs (i.e., from a jet) or whether from a relativistic shock breakout

(SBO) [30, 56–60] (see as well Section 9). It was demonstrated by [61, 62] that a key observable of *ll*GRBs are their single-peaked, smooth, nonvariable γ-ray LCs compared to the more erratic γ-ray LCs of jetted-GRBs, which become softer over time. It was shown by [60] that an SBO is likely present in all LGRB events, but for any realistic configuration the energy in the SBO pulse is lower by many orders of magnitude compared to those observed in the GRB prompt emission ($E_{\mathrm{SBO}} = 10^{44}$–$10^{47}$ erg, for reasonable estimates of the ejecta mass and progenitor radii). These low energies (compared with $E_{\gamma,\mathrm{iso}}$) suggest that relativistic SBOs are not likely to be detected at redshifts exceeding $z \approx 0.1$. In cases where they are detectable, the SBO may be in the form of a short pulse of photons with energies >1 MeV. Inspection of the E_p values in Table 2 shows that only a few events have photons with peak γ-ray energies close to this value: GRB 140606B has $E_p \approx$ 800 keV [32]; however suspected *ll*GRBs 060218 and 100316D only have E_p = 5 keV and 30 keV, respectively. It should be noted that while the SBO model of [60] successfully explains the observed properties (namely, the energetics, temperature, and duration of the prompt emission) of GRBs 980425, 031203, 060218, and 100316D, their SBO origins are still widely debated [63, 64], with no firm consensus yet achieved.

Thermal, black body (BB) components in UV and X-ray spectra have been detected for several events, including GRB 060218 (X-ray: kT = 0.17 keV, time averaged from first 10,000 s, [58]); GRB 100316D (X-ray: kT = 0.14 keV, time averaged from 144–737 s, [65]); GRB 090618 (X-ray: kT = 0.3–1 keV up to first 2500 s, [66]); GRB 101219B (X-ray: kT = 0.2 keV, [67]); and GRB 120422A (UV: kT = 16 eV at observer-frame $t - t_0$ = 0.054 d, [41]). A sample of LGRBs with associated SNe was analysed by [68] who found that thermal components were present in many events, which could possibly be attributed to thermal emission arising from a cocoon that surrounds the jet [69] or perhaps associated with SBO emission. Reference [67] analysed a larger sample of LGRBs and found that, for several events, a model that included a BB contribution provided better fits than absorbed power laws. Reference [70] found that, in their sample of 28 LGRBs, eight had evidence of thermal emission in their X-ray spectra, indicating such emission may be somewhat prevalent. However, the large inferred BB temperatures (kT ranging from 0.16 keV for 060218 to 3.2 keV for 061007, with an average of ≈1 keV) indicates that the origin of the thermal emission may not be a SBO. Moreover, the large superluminal expansions inferred for the thermal components instead hint at a connection with late photospheric emission. In comparison, some studies indicate a SBO temperature of ~ 1 keV [71], while [60, 72–74] showed that for a short while the region behind the shock is out of thermal equilibrium, and temperatures can reach as high as ~50 keV.

The radius of the fitted BB component offers additional clues. References [58, 59] derived a BB radius of 5–8×10^{12} cm for GRB 060218; [65] found $\approx 8 \times 10^{11}$ cm for GRB 100316D; [41] found $\approx 7 \times 10^{13}$ cm for GRB 120422A; and [75] derived a radius of $\approx 9 \times 10^{13}$ cm for GRB 140606B. The radii inferred for GRBs 060218, 120422A, and 140606B are commensurate with the radii of red supergiants (200–1500 $R_\odot$), while that measured for GRB 100316D is similar to that of the radius of a

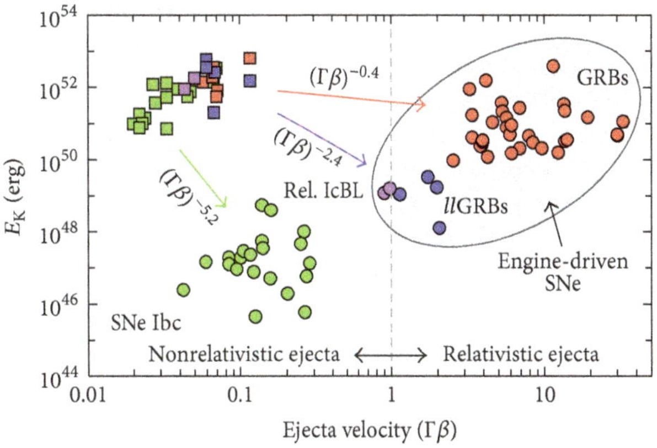

FIGURE 7: The positions of GRBs, SNe Ibc, and GRB-SNe in the E_K-$\Gamma\beta$ plane [32, 78–81]. Ordinary SNe Ibc are shown in green, *ll*GRBs in blue, relativistic SNe IcBL in purple, and jetted-GRBs in red. Squares are used for the slow-moving SN ejecta, while circles represent the kinetic energy and velocity of the nonthermal radio-emitting ejecta associated with these events (e.g., the GRB jet). The velocities were computed for $t - t_0$ = 1 day (rest-frame), where the value $\Gamma\beta$ = 1 denotes the division between relativistic and nonrelativistic ejecta. The solid lines correspond to (green) ejecta kinetic energy profiles of a purely hydrodynamical explosion $E_K \propto (\Gamma\beta)^{-5.2}$ [57, 82, 83]; (blue/purple dashed) explosions powered by a short-lived central engine (SBO-GRBs and relativistic IcBL SNe 2009bb and 2012ap: $E_K \propto (\Gamma\beta)^{-2.4}$); (red) those arising from a long-lived central engine (i.e., jetted-GRBs; $E_K \propto (\Gamma\beta)^{-0.4}$ [84]). Modified, with permission, from Margutti et al. [78, 81].

blue supergiant (≲25 $R_\odot$). These radii, which are much larger than those expected for Wolf-Rayet (WR) stars (of order a few solar radius to a few tens of solar radii), were explained by these authors by the presence of a massive, dense stellar wind surrounding the progenitor star, where the thermal radiation is observed once the shock, which is driven into the wind, reaches a radius where the wind becomes optically thin. An alternative explanation for the large BB radii was presented by [76] (see as well [77]), where the breakout occurs in an extended (R = 100 $R_\odot$) low-mass (0.01 $M_\odot$) envelope surrounding the preexplosion progenitor star. The origin of envelope is likely material stripped just prior to explosion, and such an envelope is missing for high-luminosity GRB-SNe [77].

For a given GRB-SN event there are both relativistic and nonrelativistic ejecta, where the former is responsible for producing the prompt emission, and the latter is associated with the SN itself. The average mass between the two components is large: the ejecta mass of a GRB-SN is of order 2–8 $M_\odot$, while that in the jet that produces the γ-rays is of order 10^{-6} $M_\odot$, based on arguments for very low baryon loading [88]. A GRB jet decelerates very rapidly, within a few days, because the very low-mass ejecta is rapidly swept up into the comparatively larger mass of the surrounding CSM. Conversely, SNe have much heavier ejecta and can be in free-expansion for many years or even centuries. Measuring the amount of kinetic energy associated with each ejecta component can offer additional clues to the explosion mechanisms operating in these events. Figure 7 shows the position of SNe

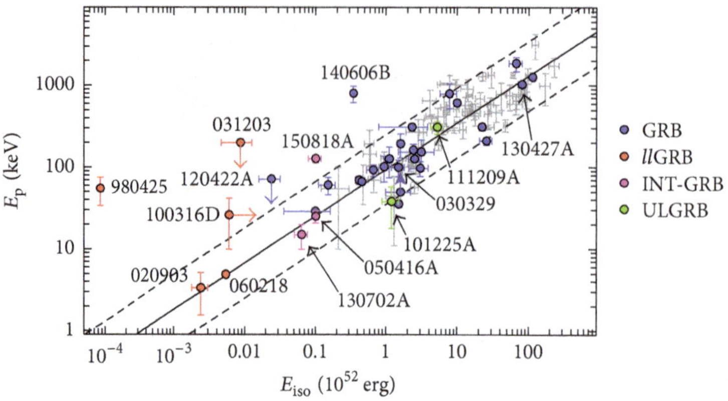

FIGURE 8: Properties of the prompt emission for different classes of GRBs in the $E_{\gamma,\mathrm{iso}}$-E_p plane [85]. Data from [85–87] are shown in grey along with their best fit to a single power law (index of $\alpha = 0.57$) and the 2σ uncertainty in their fit. Notable events that do not appear to follow the Amati relation include *ll*GRBs (980425 and 031203), INT-GRBs (150818A), and high-luminosity GRB 140606B. Both ULGRBs are consistent with the Amati relation, so are GRBs 030329 and 130427A, while GRB 120422A and *ll*GRB 100316D are marginally consistent.

Ibc (green), GRBs (red), *ll*GRBs (blue), and relativistic SNe IcBL (purple) in the E_K-$\Gamma\beta$ plane [32, 78–81], where $\beta = v/c$ (not to be confused with the spectral PL index of synchrotron radiation) and Γ is the bulk Lorentz factor. Squares indicate slow-moving SN ejecta, while circles represent the kinetic energy and velocity of the nonthermal radio-emitting ejecta associated with these events (e.g., the jet in GRBs). The velocities were computed for $t-t_0 = 1$ day (rest-frame), where the value $\Gamma\beta = 1$ denotes the division between relativistic and nonrelativistic ejecta. The solid lines show the ejecta kinetic energy profiles of a purely hydrodynamical explosion (green) $E_\mathrm{K} \propto (\Gamma\beta)^{-5.2}$ [57, 82, 83]; explosions powered by a short-lived central engine (blue/purple dashed), SBO-GRBs and relativistic IcBL SNe 2009bb and 2012ap: $E_\mathrm{K} \propto (\Gamma\beta)^{-2.4}$; and those arising from a long-lived central engine (red), that is, jetted-GRBs: $E_\mathrm{K} \propto (\Gamma\beta)^{-0.4}$ [84].

It is seen that *ll*GRBs and high-luminosity GRBs span a wide range of engine energetics, as indicated by the range of PL indices seen in Figure 7. The two relativistic SNe IcBL considered in this review (SNe 2009bb and 2012ap), which are also thought to be engine-driven SNe [79, 81, 89], occur at the lower-end of central engine energetics. Modelling of GRB 060218 [90] showed that $\sim 10^{48}$ erg of energy was coupled to the mildly relativistic ejecta ($\Gamma \sim 2$). Reference [78] showed the presence of a very weak central engine for GRB 100316D, where $\sim 10^{49}$ erg of energy was coupled to mildly relativistic ($\Gamma = 1.5$–2), quasi-spherical ejecta. It was shown by [79] that $\geq 10^{49}$ erg was associated with the relativistic ($v = 0.9c$), radio-emitting ejecta of SN 2009bb. These authors also showed that, unlike GRB jets, the ejecta was in free-expansion, which implied it was baryon loaded. For SN 2012ap, [89] estimated there was $\sim 1.6 \times 10^{49}$ erg of energy associated with the mildly relativistic ($0.7c$) radio-emitting ejecta. The weak X-ray emission of SN 2012ap [81] implied no late-time activity of its central engine, which led these authors to suggest that relativistic SNe IcBL represent the weakest engine-driven explosions, where the jet is unable to successfully break out of the progenitor. *ll*GRBs then represent events where the jet does not or just barely escapes into space. Note that [91] calculated an estimate to the dividing line between SBO-GRBs and jet-GRBs, finding that for γ-ray luminosities above 10^{48} erg s^{-1} a jet-GRB may be possible.

Next, the distribution of T_{90} (the time over which a burst emits from 5% of its total measured counts to 95%) as measured by the various GRB satellites can be used to infer additional physical properties of the GRB jet duration and progenitor radii. A basic assertion of the collapsar model is that the duration of the GRB prompt phase (where T_{90} is used as a proxy) is the difference of the time that the central engine operates minus the time it takes for the jet to break out of the star: $T_{90} \sim t_\mathrm{engine} - t_\mathrm{breakout}$. A direct consequence of this premise is that there should be a plateau in the distribution of T_{90} for GRBs produced by collapsars when $T_{90} < t_\mathrm{breakout}$ [92]. Moreover, the value of T_{90} found at the upper limit of the plateaus seen for three satellites (*BATSE*, *Swift*, and *Fermi*) was approximately the same ($T_{90} \sim 20$–30 s), which is interpreted as the typical breakout time of the jet. This short breakout time suggests that the progenitor star at the time of explosion is quite compact ($\sim 5\,R_\odot$ [93]). Reference [94] then used these distributions to calculate the probability that a given GRB arises from a collapsar or not based on its T_{90} and hardness ratio. Note however that T_{90} might not always be the best indicator of the engine on-time. For example, [91] showed that while GRB 120422A had $T_{90} = 5$ s, the actual duration of the jet was actually 86 s, as constrained by modelling of the curvature effect. Though see [95] who state that curvature radiation is not from a central engine that is still on but from electrons that were off-axis and hence had a lower Lorentz factor and which are received over a time interval that is long compared to the duration of the burst.

Finally, Figure 8 shows the properties of the prompt emission for the various GRB-SN subclasses in the $E_{\gamma,\mathrm{iso}}$-E_p plane, that is, the Amati relation [85]. Data from [85–87] are shown in grey along with their best fit to a single power law (index of $\alpha = 0.57$) and the 2σ uncertainty in their fit. Several events do not appear to follow the Amati relation, including *ll*GRBs 980425 and 031203, INT-GRB 150818A, and

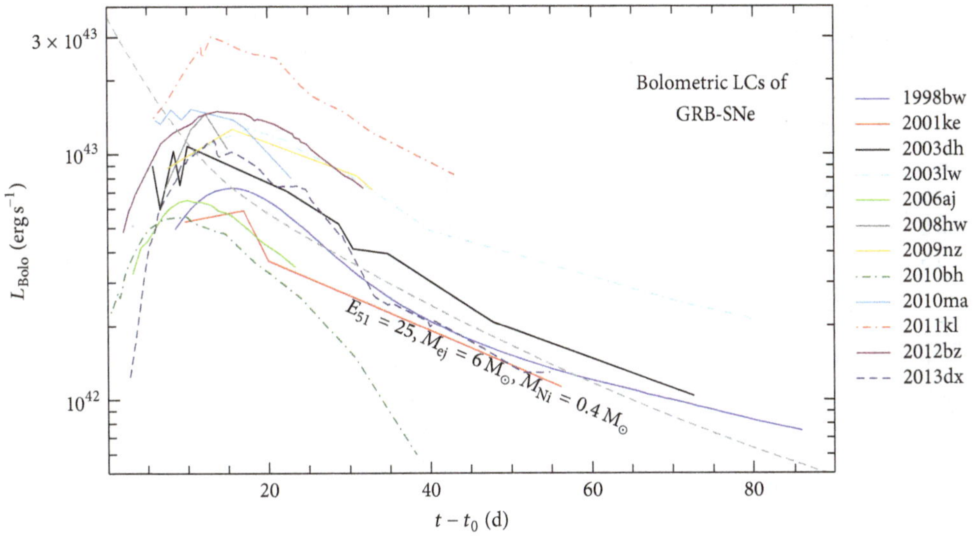

FIGURE 9: Bolometric LCs of a sample of GRB-SNe. Times are given in the rest-frame. The average peak luminosity of all GRB-SNe except SN 2011kl is $\overline{L}_p = 1.0 \times 10^{43}$ erg s^{-1}, with a standard deviation of 0.36×10^{43} erg s^{-1}. The peak luminosity of SN 2011kl is $L_p = 2.9 \times 10^{43}$ erg s^{-1}, which makes it more than 5σ more luminous than the average GRB-SN. The average peak time of the entire sample is $t_p = 13.2$ d, with a standard deviation of 2.6 d. If SN 2011kl is excluded from the sample, this changes to 13.0 d. Plotted for reference is an analytical model that considers the luminosity produced by the average GRB-SN ($E_K = 25 \times 10^{51}$ erg, $M_{ej} = 6\,M_\odot$, and $M_{Ni} = 0.4\,M_\odot$).

high-luminosity GRB 140606B. Both ULGRBs are consistent with the Amati relation, so are GRBs 030329 and 130427A, while GRB 120422A and *ll*GRB 100316D are marginally consistent. It was once supposed that the placement of a GRB in the $E_{\gamma,\rm iso}$-E_K plane could be a discriminant of GRB's origins, where it is seen that SGRBs also do not follow the Amati relation. However, over the years many authors have closely scrutinized the Amati relation, with opinions swinging back and forth as to whether it reflects a physical origin or is simply due to selection effects [96–103]. To date, no consensus has yet been reached.

4. Physical Properties: Observational Constraints

4.1. Bolometrics. The bolometric LCs of a sample of 12 GRB-SNe, which includes *ll*GRB-SNe and ULGRB-SN 2011kl, are shown in Figure 9. The Bazin function was fit to the GRB-SN bolometric LCs in order to determine their peak luminosity (L_p), the time of peak luminosity (t_p), and the amount the bolometric LC fades from peak to 15 days later (Δm_{15}). (*NB.* that SNe 2001ke, 2008hw, and 2009nz were excluded from the fitting and the subsequent calculated averages, as their bolometric LCs contained too few points to be fit with the Bazin function, which has four free parameters. As such, their luminosities and peak times were approximated by eye and are not included in the average GRB-SN properties presented here.) These values are presented in Table 3.

The average peak luminosity of the GRB-SN sample, excluding SN 2011kl, is $\overline{L}_p = 1.0 \times 10^{43}$ erg s^{-1}, with a standard deviation of $\sigma_{L_p} = 0.4 \times 10^{43}$ erg s^{-1}. The peak luminosities of SNe 2003dh and 2013dx are $\approx 1 \times 10^{43}$ erg s^{-1}, meaning that they are perhaps better representatives of a typical GRB-SN than the archetype SN 1998bw ($L_p = 7 \times 10^{42}$ erg s^{-1}). The peak luminosity of SN 2011kl is $L_p = 2.9 \times 10^{43}$ erg s^{-1}, which makes it more than 5σ more luminous than the average GRB-SN. This is not, however, as bright as superluminous supernovae (SLSNe), whose luminosities exceed $>7 \times 10^{43}$ erg s^{-1} [104]. This makes SN 2011kl an intermediate SN event between GRB-SNe and SLSNe and perhaps warrants a classification of a "superluminous GRB-SNe" (SLGRB-SN); however, in this chapter we will stick with the nomenclature ULGRB-SN. When SN 2011kl is included in the sample, $\overline{L}_p = 1.2 \times 10^{43}$ erg s^{-1}, with $\sigma_{\overline{L}_p} = 0.7 \times 10^{43}$ erg s^{-1}. Even using this average value, SN 2011kl is still 2.5σ more luminous than the average GRB-SN.

The average peak time, when SN 2011kl is and is not included in the sample, is $t_p = 13.2$ d ($\sigma_{t_p} = 2.6$ d) and $t_p = 13.0$ d ($\sigma_{t_p} = 2.7$ d), respectively. Similarly, $\Delta m_{15} = 0.7$ mag ($\sigma_{\Delta m_{15}} = 0.1$ mag) and 0.8 mag ($\sigma_{\Delta m_{15}} = 0.1$ mag), respectively. As such, the inclusion/exclusion of SN 2011kl has little effect on these derived values. The fact that SN 2011kl peaks at a similar time as the average GRB-SN, but does so at a much larger luminosity, strongly suggests that ULGRB-SNe do not belong to the same class of standardizable candles as GRB-SNe. This can be readily explained in that SN 2011kl is powered by emission from a magnetar central engine [36, 105–107], whereas GRB-SNe, including *ll*GRB-SNe, are powered by radioactive heating [106]. Whether ULGRB-SNe represent the same set of standardizable candles as type I SLSNe [108, 109], which are also thought to be powered by a magnetar central engine, their own subset, or perhaps none at all, requires additional well-monitored events.

Over the years, and since the discovery of SN 1998bw, the bolometric properties (kinetic energy, E_K, ejecta mass, M_{ej},

Table 1: Average Bolometric properties of GRB-SNe.

Type*					$(M_\odot)$			$(M_\odot)$						(d, rest)			(mag)			(km s^{-1})	
	N	$E_K^\dagger$	σ	N	M_{ej}	σ	N	M_{Ni}	σ	N	$L_p^\ddagger$	σ	N	t_p	σ	N	Δm_{15}	σ	N	v_{ph}	σ
GRB	19	26.0	18.3	19	5.8	4.0	20	0.38	0.13	2	1.26	0.35	2	12.28	0.67	2	0.85	0.21	6	18400	9700
INT	1	8.2	—	1	3.1	—	1	0.37	—	1	1.08	—	1	12.94	—	1	0.85	—	1	21300	—
LL	6	27.8	19.6	6	6.5	4.0	6	0.35	0.19	5	0.94	0.41	5	13.22	3.53	5	0.75	0.12	4	22800	8200
ULGRB	2	18.8	18.7	2	6.1	2.9	1	0.41	—	1	2.91	—	1	14.80	—	1	0.78	—	1	21000	—
Rel IcBL	2	13.5	6.4	2	3.4	1.0	2	0.16	0.05	2	0.35	0.50	2	12.78	0.84	2	0.90	0.21	2	14000	1400
GRB ALL	28	25.2	17.9	28	5.9	3.8	28	0.37	0.20	9	1.24	0.71	9	13.16	2.61	9	0.79	0.12	12	20300	8100
GRB ALL**	27	25.9	17.9	27	5.9	3.9	28	0.37	0.20	8	1.03	0.36	8	12.95	2.72	8	0.79	0.13	11	20200	8500
Ib	19	3.3	2.6	19	4.7	2.8	12	0.21	0.22	—	—	—	—	—	—	—	—	—	11	8000	1700
Ic	13	3.3	3.3	13	4.6	4.5	7	0.23	0.19	—	—	—	—	—	—	—	—	—	10	8500	1800

*Classifications (Section 3): *ll*GRBs: GRB-SNe associated with low-luminosity GRBs ($L_{\gamma,iso}$ < $10^{48.5}$ erg s^{-1}); INT-GRBs: GRB-SNe associated with intermediate-luminosity GRBs ($10^{48.5}$ < $L_{\gamma,iso}$ < $10^{49.5}$ erg s^{-1}); GRBs: GRB-SNe associated with high-luminosity GRBs ($L_{\gamma,iso}$ > $10^{49.5}$ erg s^{-1}); ULGRBs: GRB-SNe associated with ultra-long-duration GRBs (see Section 3).
**Excluding SN 2011kl.
†Units: 10^{51} erg.
‡Units: 10^{43} erg s^{-1}.
Note: average bolometric properties of SNe Ib and Ic are from [18].

and nickel mass, M_{Ni}) of the best-observed GRB-SNe have been determined by sophisticated numerical simulations (hydrodynamical models coupled with radiative transfer, RT, codes) [3, 7, 36, 47, 48, 107, 110–119] and analytical modelling [10, 18, 31, 32, 40, 41, 43, 106, 120–122]. A summary of the derived bolometric properties for individual GRB-SNe is presented in Table 3, while a summary of the average bolometric properties, broken down by GRB-SN subtype and compared against other subtypes of SNe Ibc, is shown in Table 1. It should be noted that the values presented have been derived over different wavelength ranges: some are observer-frame *BVRI*, while others include UV, *U*-band, and NIR contributions. Further discussion on the effects of including additional filters when constructing a bolometric LC of a given SN can be found in [30, 41, 49, 120, 123], who show that including NIR flux leads to brighter bolometric LCs that decay slower at later times and including UV flux leads to an increase in luminosity at earlier times (during the first couple of weeks, rest-frame) when the UV contribution is nonnegligible.

From this sample of $N = 28$ GRB-SNe we can say that the average GRB-SN (grey-dashed line in Figure 9) has a kinetic energy of $E_K = 2.5 \times 10^{52}$ erg ($\sigma_{E_K} = 1.8 \times 10^{52}$ erg), an ejecta mass of $M_{ej} = 6\,M_\odot$ ($\sigma_{M_{ej}} = 4\,M_\odot$), a nickel mass of $M_{Ni} = 0.4\,M_\odot$ ($\sigma_{M_{Ni}} = 0.2\,M_\odot$), and a peak photospheric velocity of $v_{ph} = 20{,}000$ km s^{-1} ($\sigma_{v_{ph}} = 8{,}000$ km s^{-1}). Here we have assumed that the line velocities of various transitions, namely, Fe II λ5169 and Si II λ6355, are suitable proxies for the photospheric velocities. An in-depth discussion of this assumption and its various caveats can be found in [46]. It has a peak luminosity of $L_p = 1 \times 10^{43}$ erg s^{-1} ($\sigma_{L_p} = 0.4 \times 10^{43}$ erg s^{-1}), reaches peak bolometric light in $t_p = 13$ days ($\sigma_{t_p} = 2.7$ days), and has $\Delta m_{15} = 0.8$ mag ($\sigma_{\Delta m_{15}} = 0.1$ mag). There are no statistical differences in the average bolometric properties, rise times, and decay rates, between the different GRB-SN subtypes, and excluding ULGRB-SN 2011kl, there are no differences in their peak luminosities. As found in previous studies [18, 120], relativistic SNe IcBL are roughly half as energetic as GRB-SNe and contain approximately half as much ejecta mass and nickel content therein. However, we are comparing GRB-SNe against a sample of two relativistic SNe IcBL, meaning we should not draw any firm conclusions as of yet.

There are a few caveats to keep in mind when interpreting these results. The first is the comparison of bolometric properties derived for SNe observed over different filter/wavelength ranges, as discussed above. Secondly, for a GRB-SN to be observed there are several stringent requirements [11], including AGs that fade at a reliably determined rate (e.g., for GRB 030329 the complex AG behaviour led to a range of 1 mag in the peak brightness of accompanying SN 2003dh [4–6]; thus in this case the reported peak brightness was strongly model-dependent), have a host galaxy that can be readily quantified, and are to be at a relatively low redshift ($z \leq 1$ for current 10 m class ground telescopes and *HST*). Moreover, the modelling techniques used to estimate the bolometric properties contain their own caveats and limitations. For example, the analytical Arnett model [208] contains assumptions such as spherical symmetry, homogeneous ejecta distribution, homologous expansion, and a central location for the radioactive elements [18].

4.2. What Powers a GRB-SN? Observations of GRB-SNe can act as a powerful discriminant of the different theoretical models proposed to produce them. The analysis presented in the previous section made the assumption that GRB-SNe are powered by radioactive heating. In this scenario, it is assumed that during the initial core-collapse (see Section 9 for further discussion), roughly $0.1\,M_\odot$ or so of nickel can be created via explosive nucleosynthesis [209] if the stellar material has nearly equal amounts of neutrons and protons (such as silicon and oxygen), and approximately 10^{52} erg of energy is focused

TABLE 2: GRB-SN master Table 1: γ-ray properties.

GRB	SN	Type	z	T_{90} (s)	(10^{52} erg) $E_{\gamma,\mathrm{iso}}$	(keV) E_{p}	(erg s^{-1}) $L^{\dagger}_{\mathrm{iso}}$
970228		GRB	0.695	56	1.6 (0.12)	195 (64)	4.84×10^{50}
980326		GRB			0.48 (0.09)	935 (36)	
980425	1998bw	*ll*GRB	0.00867	18	0.000086 (0.000002)	55 (21)	4.80×10^{46}
990712		GRB	0.4331	19	0.67 (0.13)	93 (15)	5.05×10^{50}
991208		GRB	0.7063	60	22.3 (1.8)	313 (31)	6.34×10^{51}
000911		GRB	1.0585	500	67 (14)	1859 (371)	2.75×10^{51}
011121	2001ke	GRB	0.362	47	7.8 (2.1)	793 (265)	2.26×10^{51}
020305							
020405		GRB	0.68986	40	10 (0.9)	612 (10)	4.22×10^{51}
020410				>1600			
020903		*ll*GRB	0.2506	3.3	0.0011 (0.0006)	3.37 (1.79)	4.20×10^{48}
021211	2002lt	GRB	1.004	2.8	1.12 (0.13)	127 (52)	8.02×10^{51}
030329	2003dh	GRB	0.16867	22.76	1.5 (0.3)	100 (23)	7.70×10^{50}
030723						<0.023	
030725							
031203	2003lw	*ll*GRB	0.10536	37	0.0086 (0.004)	<200	2.55×10^{48}
040924		GRB	0.858	2.39	0.95 (0.09)	102 (35)	7.38×10^{51}
041006		GRB	0.716	18	3 (0.9)	98 (20)	2.86×10^{51}
050416A		INT	0.6528	2.4	0.1 (0.01)	25.1 (4.2)	6.89×10^{50}
050525A	2005nc	GRB	0.606	8.84	2.5 (0.43)	127 (10)	4.54×10^{51}
050824		GRB	0.8281	25	$0.041 < E < 0.34$	$11 < E < 32$	
060218	2006aj	*ll*GRB	0.03342	2100	0.0053 (0.0003)	4.9 (0.3)	2.60×10^{46}
060729		GRB	0.5428	115	1.6 (0.6)	>50	2.14×10^{50}
060904B		GRB	0.7029	192	2.4 (0.2)	163 (31)	2.12×10^{50}
070419A		INT	0.9705	116	≈0.16		2.71×10^{49}
080319B		GRB	0.9371	124.86	114 (9)	1261 (65)	1.76×10^{52}
081007	2008hw	GRB	0.5295	9.01	0.15 (0.04)	61 (15)	2.54×10^{50}
090618		GRB	0.54	113.34	25.7 (5)	211 (22)	3.49×10^{51}
091127	2009nz	GRB	0.49044	7.42	1.5 (0.2)	35.5 (1.5)	3.01×10^{51}
100316D	2010bh	*ll*GRB	0.0592	1300	>0.0059	26 (16)	4.80×10^{46}
100418A		GRB	0.6239	8	0.0990 (0.0630)	29 (2)	2.00×10^{50}
101219B	2010ma	GRB	0.55185	51	0.42 (0.05)	70 (8)	1.27×10^{50}
101225A		ULGRB	0.847	7000	1.2 (0.3)	38 (20)	3.16×10^{48}
111209A	2011kl	ULGRB	0.67702	10000	58.2 (7.3)	520 (89)	9.76×10^{49}
111211A			0.478				
111228A		GRB	0.71627	101.2	4.2 (0.6)	58.4 (6.9)	7.12×10^{50}
120422A	2012bz	GRB	0.28253	5.4	0.024 (0.008)	<72	5.70×10^{49}
120714B	2012eb	INT	0.3984	159	0.0594 (0.0195)	101.4 (155.7)	5.22×10^{48}
120729A		GRB	0.8	71.5	2.3 (1.5)	310.6 (31.6)	5.79×10^{50}
130215A	2013ez	GRB	0.597	65.7	3.1 (1.6)	155 (63)	7.53×10^{50}
130427A	2013cq	GRB	0.3399	163	81 (10)	1028 (50)	6.65×10^{51}
130702A	2013dx	INT	0.145	58.881	0.064 (0.013)	15 (5)	1.24×10^{49}
130831A	2013fu	GRB	0.479	32.5	0.46 (0.02)	67 (4)	2.09×10^{50}
140606B		GRB	0.384	22.78	0.347 (0.02)	801 (182)	2.10×10^{50}
150518A			0.256				
150818A		INT	0.282	123.3	0.1 (0.02)	128 (13)	1.03×10^{49}

$^{\dagger}L_{\mathrm{iso}} = E_{\mathrm{iso}}(1+z)/T_{90}$.
$^{\ddagger}\gamma$-ray properties calculated by [85] for a redshift range of $0.9 \le z \le 1.1$.
*ll*GRB: GRB-SN associated with a low-luminosity GRB ($L_{\gamma,\mathrm{iso}} < 10^{48.5}$ erg s^{-1}); INT: GRB-SN associated with an intermediate-luminosity GRB ($10^{48.5} < L_{\gamma,\mathrm{iso}} < 10^{49.5}$ erg s^{-1}); GRB: GRB-SN associated with a high-luminosity GRB ($L_{\gamma,\mathrm{iso}} > 10^{49.5}$ erg s^{-1}); ULGRB: GRB-SN associated with an ultra-long-duration GRB (see Section 3).

TABLE 3: GRB-SN master Table 2: SN properties.

GRB	SN	Type	z	S?	Grade	(mag) M_V^*	(mag) $\Delta m_{15,V}$	(d) $t_{V,p}^*$	(erg s^{-1}) $L_{p,Bol}$	(d) $t_{p,rest}$	(mag) $\Delta m_{15,Bol}$	(10^{51} erg) E_K	($M_\odot$) M_{ej}	($M_\odot$) M_{Ni}	(km s^{-1}) v_{ph}	$\bar{k}$	$\bar{s}$	Filters
970228		GRB	0.695		C													
980326		GRB			D													
980425	1998bw	*ll*GRB	0.00866	S	A	−19.29 ± 0.08	0.75 ± 0.02	16.09 ± 0.18	7.33 × 10^{42}	15.16	0.80	20–30	6–10	0.3–0.6	18000	1	1	
990712		GRB	0.4331		C							$26.1^{+24.6}_{-15.0}$	$6.6^{+3.5}_{-2.9}$	0.14 ± 0.04		0.36 ± 0.05	1.10 ± 0.20	*R*
991208		GRB	0.7063		E							$38.7^{+44.6}_{-26.0}$	$9.7^{+6.8}_{-5.6}$	0.96 ± 0.48		2.11 ± 0.58	1.10 ± 0.20	*R*
000911		GRB	1.0585		E													
011121	2001ke	GRB	0.362	S	B				~5.9 × 10^{42}	~17		$17.7^{+8.8}_{-6.4}$	4.4 ± 0.8	0.35 ± 0.01		1.13 ± 0.23	0.84 ± 0.17	*BV**
020305					E													
020405		GRB	0.68986		C							$8.9^{+5.4}_{-3.8}$	$2.2^{+0.6}_{-0.5}$	0.23 ± 0.02		0.82 ± 0.14	0.62 ± 0.03	*R*
020410					D													
020903		*ll*GRB	0.2506	S	B							$28.9^{+32.2}_{-18.9}$	$7.3^{+4.9}_{-4.0}$	0.25 ± 0.13		0.61 ± 0.19	0.98 ± 0.02	*R*
021211	2002lt	GRB	1.004	S	B							$28.5^{+45.0}_{-13.0}$	$7.2^{+7.4}_{-6.0}$	0.16 ± 0.14		0.40 ± 0.19	0.98 ± 0.26	*R*
030329	2003dh	GRB	0.16867	S	A	−19.39 ± 0.14	0.90 ± 0.50	10.74 ± 2.57	1.01 × 10^{43}	12.75	0.70	20–50	5–10	0.4–0.6	20000	1.28 ± 0.28	0.87 ± 0.18	*UBV**
030723					D													
030725					E													
031203	2003lw	*ll*GRB	0.10536	S	A	−19.90 ± 0.16	0.64 ± 0.10	19.94 ± 1.48	1.26 × 10^{43}	17.33	0.62	60.0 ± 15	13.0 ± 4.0	0.55 ± 0.20	18000	1.65 ± 0.36	1.10 ± 0.24	*VRI**
040924		GRB	0.858		C													
041006		GRB	0.716		C							$76.4^{+39.8}_{-28.7}$	$19.2^{+3.9}_{-3.6}$	0.69 ± 0.07		1.16 ± 0.06	1.47 ± 0.04	*R*
050416A		INT	0.6528		D													
050525A	2005nc	GRB	0.606	S	B	−18.59 ± 0.31	1.17 ± 0.88	11.08 ± 3.37				$18.9^{+10.7}_{-7.5}$	$4.8^{+1.1}_{-1.0}$	0.24 ± 0.02		0.69 ± 0.03	0.83 ± 0.03	*R*
050824		GRB	0.8281		E							$5.7^{+9.3}_{-3.7}$	$1.4^{+1.6}_{-0.6}$	0.26 ± 0.17		1.05 ± 0.42	0.52 ± 0.14	*R*
060218	2006aj	*ll*GRB	0.03342	S	A	−18.85 ± 0.08	1.08 ± 0.06	9.96 ± 0.18	6.47 × 10^{42}	10.42	0.83	1.0 ± 0.5	2.0 ± 0.5	0.20 ± 0.10	20000	0.58 ± 0.13	0.67 ± 0.14	*UBVR**
060729		GRB	0.5428		D							$24.4^{+14.3}_{-9.9}$	$6.1^{+1.6}_{-1.4}$	0.36 ± 0.05		0.94 ± 0.10	0.92 ± 0.04	*RI*
060904B		GRB	0.7029		C							$9.9^{+5.1}_{-3.7}$	2.5 ± 0.5	0.12 ± 0.01		0.42 ± 0.02	0.65 ± 0.01	*R*
070419A		GRB	0.9705		D													
080319B		GRB	0.9371		C							$22.7^{+19.1}_{-11.9}$	$5.7^{+2.6}_{-2.2}$	0.86 ± 0.45		2.30 ± 0.90	0.89 ± 0.10	*I*
081007	2008hw	GRB	0.5295	S	B				~1.4 × 10^{43}	~12		19.0 ± 15.0	2.3 ± 1.0	0.39 ± 0.08	12600	0.71 ± 0.10	0.85 ± 0.11	*riz*
090618		GRB	0.54		C	−19.34 ± 0.13	0.65 ± 0.17	17.54 ± 1.64				$36.5^{+20.0}_{-14.2}$	$9.2^{+2.1}_{-1.9}$	0.37 ± 0.03		1.11 ± 0.22	0.98 ± 0.20	*B**
091127	2009nz	GRB	0.49044	S	B				~1.2 × 10^{43}	~15	~0.5	13.5 ± 0.4	4.7 ± 0.1	0.33 ± 0.01	17000	0.89 ± 0.01	0.88 ± 0.01	*I*
100316D	2010bh	*ll*GRB	0.0592	S	A	−18.89 ± 0.10	1.10 ± 0.05	8.76 ± 0.37	5.67 × 10^{42}	8.76	0.89	15.4 ± 1.4	2.5 ± 0.2	0.12 ± 0.02	35000	0.53 ± 0.15	0.53 ± 0.11	*VRI**
100418A		INT	0.6239		D/E													
101219B	2010ma	GRB	0.55185	S	A/B				1.5 × 10^{43}	11.80	0.99	10.0 ± 6.0	1.3 ± 0.5	0.43 ± 0.03		1.16 ± 0.63	0.76 ± 0.10	*griz*
101225A		ULGRB	0.847		D							32.0 ± 16.0	8.1 ± 1.5	0.41 ± 0.03		0.96 ± 0.05	1.02 ± 0.03	*i*
111209A	2011kl	ULGRB	0.67702		A/B				2.91 × 10^{43}	14.80	0.78	20–90	3–5		21000	1.81 ± 0.19	1.08 ± 0.11	*iz*
111211A			0.478	S	B/C													
111228A			0.71627		E													
120422A	2012bz	*ll*GRB	0.28253	S	A	−19.50 ± 0.03	0.73 ± 0.06	14.20 ± 0.34	1.48 × 10^{43}	14.45	0.62	25.5 ± 2.1	6.1 ± 0.5	0.57 ± 0.07	20500	1.13 ± 0.25	0.93 ± 0.19	*BV**
120714B	2012eb		0.3984	S	B													
120729A		GRB	0.8		D/E									0.42 ± 0.11		1.02 ± 0.26	1‡	*ri*
130215A	2013ez	GRB	0.597	S	B									0.25–0.30	6000	0.6–0.75	1‡	*ri*
130427A	2013cq	GRB	0.3399	S	B							64.0 ± 7.0	6.3 ± 0.7	0.28 ± 0.02	35000	0.85 ± 0.03	0.77 ± 0.03	*r*
130702A	2013dx	INT	0.145	S	A	−19.26 ± 0.24	1.05 ± 0.05	13.86 ± 0.70	1.08 × 10^{43}	12.94	0.85	8.2 ± 0.4	3.1 ± 0.1	0.37 ± 0.01	21300	0.98 ± 0.07	0.78 ± 0.05	*griz*
130831A	2013fu	GRB	0.479	S	A/B							18.7 ± 9.0	4.7 ± 0.8	0.30 ± 0.07		0.95 ± 0.19	0.82 ± 0.19	*B**
140606B		GRB	0.384	S	A/B							19.0 ± 11.0	4.8 ± 1.9	0.42 ± 0.17	19800	1.04 ± 0.24	0.81 ± 0.13	*V**
150518A			0.256		C/D													
150818A		INT	0.282	S	B													
—	2009bb	Rel IcBL	0.009987	S	—	−18.61 ± 0.28	1.13 ± 0.04	13.37 ± 0.32				18.0 ± 8.0	4.1 ± 1.9	0.19 ± 0.03	15000	0.60 ± 0.05	0.73 ± 0.07	*BVRI**
—	2012ap	Rel IcBL	0.012141	S		−18.76 ± 0.33	0.92 ± 0.08	14.43 ± 0.19				9.0 ± 3.0	2.7 ± 0.5	0.12 ± 0.02	13000	1.10 ± 0.23	0.82 ± 0.09	*BVRI**

S denotes one or more spectra of the SN were obtained.

Grades are from Hjorth and Bloom (2012): A: strong spectroscopic evidence. B: a clear light curve bump as well as some spectroscopic evidence resembling a GRB-SN. C: a clear bump consistent with other GRB-SNe at the spectroscopic redshift of the GRB. D: a bump, but the inferred SN properties are not fully consistent with other GRB-SNe or the bump was not well sampled or there is no spectroscopic redshift of the GRB. E: a bump, either of low significance or inconsistent with other GRB-SNe.

$*$ denotes exact, K-corrected rest-frame filter observable.

‡Values fixed during fit.

$\bar{k}$ and $\bar{s}$ denote the filter-averaged luminosity (k) and stretch (s) factors relative to SN 1998bw.

Table 4: References.

GRB	References(s)
970228	[124, 125]
980326	[126, 127]
980425	[1–3, 9, 18, 42, 44, 47, 110, 111, 120, 128]
990712	[18, 129]
991208	[18, 130]
000911	[131, 132]
011121	[18, 42, 44, 133–136]
020305	[137]
020405	[18, 138]
020410	[139, 140]
020903	[18, 141–143]
021211	[18, 144, 145]
030329	[4–7, 18, 30, 42, 44, 112, 128, 146, 147]
030723	[148, 149]
030725	[150]
031203	[18, 42, 44, 113, 128, 151–156]
040924	[90, 157]
041006	[18, 39, 90, 158]
050416A	[159, 160]
050525A	[18, 44, 128, 161]
050824	[18, 35, 162, 163]
060218	[18, 33, 42, 44, 48, 58, 90, 113, 120, 128, 156, 164–168]
060729	[10, 18, 169, 170]
060904B	[18, 171]
070419A	[172, 173]
080319B	[174–176]
081007	[177–181]
090618	[10, 18, 42, 128]
091127	[18, 181–183]
100316D	[18, 30, 42, 44, 49, 65, 78, 120, 121, 128, 184]
100418A	[185–187]
101219B	[181, 188]
101225A	[34, 53], *here*
111209A	[36, 53, 105–107, 189], *here*
111211A	[190]
111228A	[191]
120422A	[18, 41, 44, 91, 128, 192]
120714B	[193, 194]
120729A	[31]
130215A	[31]
130427A	[40, 195–197]
130702A	[43, 44, 198]
130831A	[31, 42, 199]
140606B	[32, 75]
150518A	[200]
150818A	[201–203]
SN 2009bb	[18, 42, 44, 79, 120, 204, 205]
SN 2012ap	[76, 89, 206, 207], *here*

into ~1% of the star, which occurs in the region between the newly formed compact object and 4×10^9 cm [210]. In this scenario, temperatures in excess of 4×10^9 K can be attained. However, the precise amount of ^{56}Ni that is generated is quite uncertain and depends greatly on how much the star has expanded (or collapsed), prior to energy deposition. The radioactive nickel decays into cobalt with a half-life of 6.077 d and then cobalt into iron with a half-life of 77.236 d: $^{56}_{28}$Ni $\rightarrow$ $^{56}_{27}$Co $\rightarrow$ $^{56}_{26}$Fe [208, 211]. Given its short half-life, the synthesized nickel must be generated during the explosion itself and not long before core-collapse. Gamma-rays that are emitted during the different radioactive decay processes are thermalised in the optically thick SN ejecta, which heat the ejecta that in turn radiates this energy at longer wavelengths (optical and NIR). This physical process is expected to power other types of SNe, including all type I SNe (Ia, Ib, Ic, and type Ic SLSNe) and the radioactive tail of type IIP SNe.

Observationally, there are hints that suggest that the best-observed GRB-SNe are powered, at least in part [212], by radioactive heating. At late times, the decay of ^{56}Co leads to an exponential decline in the nebular-phase bolometric LC of type I SNe. An example of this is the grey-dashed line in Figure 9, which is an analytical model [122, 213] that considers the luminosity produced by a fiducial SN with a kinetic energy of $E_K = 25 \times 10^{51}$ erg, an ejecta mass of M_{ej} = 6 $M_\odot$, and a nickel mass of M_{Ni} = 0.4 $M_\odot$ (e.g., the "average" GRB-SN). Such a model and others of this ilk assume full trapping of the emitted γ-rays and thermalised energy. For comparison, the late-time LC of SN 1998bw appears to fade more rapidly than this, presumably because some of the γ-rays escape directly into space without depositing energy into the expanding ejecta. At times later than 500 d [9, 214], the observed flattening seen in the LC can be interpreted in terms of both more of the energy and γ-rays being retained in the ejecta, and more energy input from the radioactive decay of species in addition to cobalt.

In the collapsar model, there are additional physical processes that can lead to the creation of greater masses of radioactive nickel. One potential source of ^{56}Ni arises from the wind emitted by the accretion disk surrounding the newly formed black hole (BH). According to the numerical simulations of [215], the amount of generated nickel depends on the accretion rate as well as the viscosity of the inflow. In theory, at least, the only upper bound on the amount of nickel that can be synthesized by the disk wind is the mass of material that is accreted. In an analytical approach, [216] demonstrated that enough ^{56}Ni can be synthesized (in order to match observations of GRB-SNe), over the course of a few tens of seconds, in the convective accretion flow arising from the initial circularization of the infalling envelope around the BH.

In the millisecond magnetar model, it is more difficult to produce a sufficient amount of ^{56}Ni via energy injection from a central engine. Some simulations suggest that only a few hundredths of a solar mass of nickel can be synthesized in the magnetar model [217]. However it may be possible to generate more nickel by tapping into the initial rotational energy of the magnetar via magnetic stresses, thus enhancing the shocks induced by the collision of the energetic wind emanated by the magnetar with material already processed by the SN shock [218, 219]. Another route would be via a shock

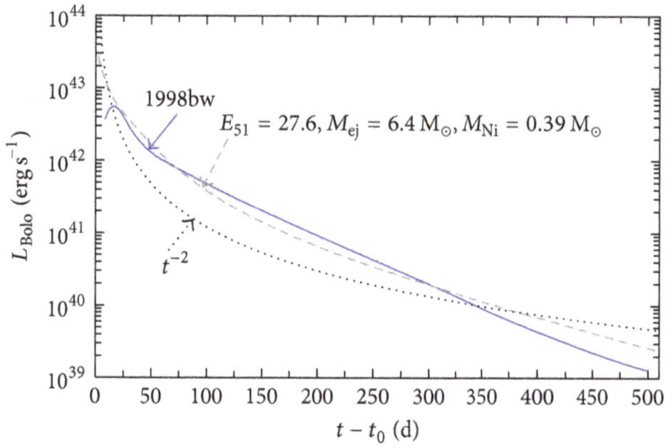

FIGURE 10: Late-time bolometric LC of SN 1998bw in filters *BVRI*. Two analytical models have been plotted to match the peak luminosity: (1) a single-zone analytical model for a fiducial SN that is powered by radioactive heating, where $E_K = 27.6 \times 10^{51}$ erg, an ejecta mass of $M_{ej} = 6.4\,M_\odot$, and a nickel mass of $M_{Ni} = 0.39\,M_\odot$, and (2) a t^{-2} curve, which is the expected decay rate for luminosity powered by a magnetar central engine. At late times the decay rate of model (1) provides a much better fit than the t^{-2} decay, which grossly overpredicts the bolometric luminosity at times later than 400 d. This is one line of observational evidence that GRB-SNe are powered by radioactive heating and not via dipole-extracted radiation from a magnetar central engine (i.e., a magnetar-driven SN).

wave driven into the ejecta by the magnetar itself, which for certain values of P and B could generate the required nickel masses [220]. However, in this scenario an isotropic-equivalent energy input rate of more than 10^{52} erg is required, and the subsequent procurement of additional nickel mass via explosive nucleosynthesis will inevitably lead to a more rapid spin-down of the magnetar central engine, rendering it unable to produce energy input during the AG phase. It is also worth considering that if a magnetar (and the subsequent GRB) is formed via the accretion-induced collapse of a white dwarf star, or perhaps the merger of two white dwarfs, there is no explosive nucleosynthesis and thus a very low ^{56}Ni yield [221].

The uncertainties unpinning both models mean that neither can be ruled out at this time, though perhaps the collapsar model offers a slightly easier route for producing the necessary quantity of nickel needed to explain the observed luminosities of GRB-SNe. But what if GRB-SNe are not powered by radioactive heating, but instead via another mechanism? Could, instead, GRB-SNe be powered by a magnetar central engine [223, 224], as has been proposed for some type I SLSNe [225–227]? A prediction of the magnetar-driven SN model is that at late times the bolometric LC should decay as t^{-2} [106, 217, 223–225, 228, 229]. Plotted in Figure 10 is the bolometric LC of SN 1998bw to $t - t_0 = 500$ d. Overplotted are two analytical models: (1) a single-zone analytical model for a fiducial SN that is powered by radioactive heating, where $E_K = 27.6 \times 10^{51}$ erg, an ejecta mass of $M_{ej} = 6.4\,M_\odot$, and a nickel mass of $M_{Ni} = 0.39\,M_\odot$, and (2) a t^{-2} curve (i.e., the decay rate expected for luminosity powered by a magnetar central engine). Both have been fitted to the bolometric LC of SN 1998bw to match its peak luminosity. At late times the decay rate of the radioactive-heated analytical LC provides a much better fit than the t^{-2} decay, which grossly overpredicts the bolometric luminosity at times later than 400 d. The difference between observations and the radioactive decay model can be attributed to incomplete trapping of γ-rays produced during the radioactive decay process.

Further evidence against the magnetar model are the observed line velocities as a function of time. In 1D analytical magnetar models [228, 229], a mass shell forms due to the expanding magnetar bubble. This feature of the 1D models has the implication that the observed line velocities will have a flat, plateau-like evolution. Inspection of Figure 6 reveals that this is indeed not the case for all the GRB-SNe of which there are time-series spectra. This particularly applies in the measured Si II λ6355 velocities, where all appear to decrease from a maximum value early on, rather than maintaining a flat evolution throughout. This is a second line of evidence that rules against magnetar heating in GRB-SNe.

However, it appears that not all GRB-SNe subtypes are powered by radioactive heating. Several investigations have provided compelling evidence that ULGRB 111209A/SN 2011kl was powered instead by a magnetar central engine. Reference [36] showed that SN 2011kl could not be powered entirely (or at all) by radioactive heating. Their argument was based primarily on the fact that the inferred ejecta mass ($3.2 \pm 0.5\,M_\odot$), determined via fitting the Arnett model [208] to their constructed bolometric LC, was too low for the amount of nickel needed to explain the observed bolometric luminosity ($1.0 \pm 0.1\,M_\odot$). The ratio of $M_{Ni}/M_{ej} = 0.3$ was much larger than that inferred for the general GRB-SN population ($M_{Ni}/M_{ej} \approx 0.07$; [18]), which rules against radioactive heating powering SN 2011kl. Secondly, the shape and relative brightness of an optical spectrum obtained of SN 2011kl just after peak SN light ($t - t_0 = 20$ d, rest-frame) was entirely unlike the spectra observed for GRB-SNe (Figure 5), including SN 1998bw [2]. Instead, the spectrum more closely resembled those of SLSNe in its shape, including the sharp cut-off at wavelengths bluewards of 3000 Å. Several authors [36, 105–107, 230] modelled different phases of the entire ULGRB event to determine the ejecta mass (M_{ej}), initial spin period (P), and the initial magnetic field strength (B), with some general consensus among the derived values: $M_{ej} = 3$–$5\,M_\odot$ (for various values of the assumed grey opacity), $P = 2$–11 ms, and $B = 0.4$–2×10^{15} G. Note that some models assumed additional heating from some nucleosynthesised nickel ($0.2\,M_\odot$ [105, 107]), while [106] assumed that energy injection from the magnetar central engine was solely responsible for powering the entire event. The general consensus of all the modelling approaches is that SN 2011kl was not powered entirely by radioactive heating, and additional energy, likely arising from a magnetar central engine, was needed to explain the observations of this enigmatic event.

5. Geometry

Measuring the geometry of GRB-SNe can lead to additional understanding of their explosion mechanism(s) and the role and degree of nickel mixing within the ejecta. A starting

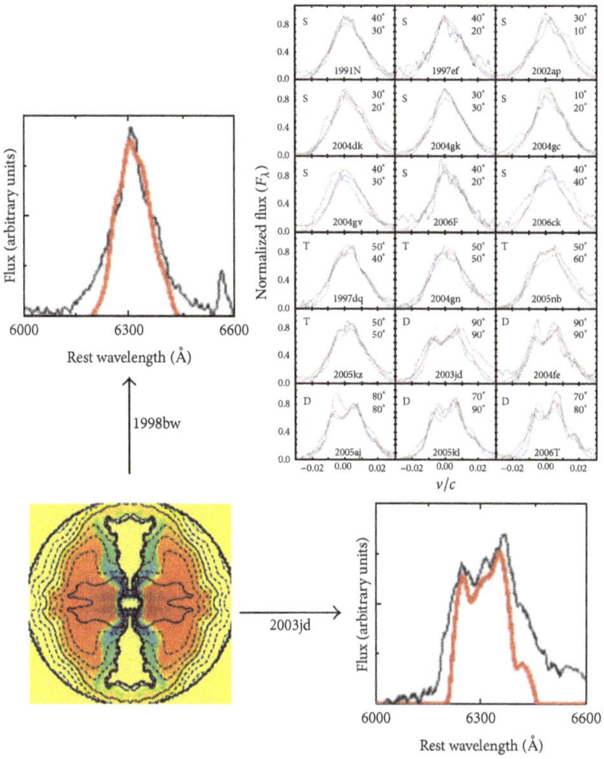

FIGURE 11: Observed [O I] $\lambda\lambda$6300,6364 emission-line profiles for a sample of SNe Ibc. Top right: emission lines classified into characteristic profiles (from [222]): single-peaked (S), transition (T), and double-peaked (D). Model predictions from a bipolar model (red curves) and a less aspherical model (blue), for different viewing directions are shown (directions denoted by the red and blue text). All other panels: nebular line profiles observed for an aspherical explosion model for different viewing angles (from [113]). The figure shows the properties of the explosion model: Fe (coloured in green and blue) is ejected near the jet direction and oxygen (red) in a torus-like structure near the equatorial plane. Synthetic [O I] $\lambda\lambda$6300,6364 emission-line profiles are compared with the spectra of SN 1998bw (top left) and SN 2003jd (bottom right).

point is to understand the geometry of GRB-SNe relative to other types of stripped-envelope core-collapse SNe (CCSNe) and ascertain whether any differences exist. In this section we will recap the results of photometric, spectroscopic, and polarimetric/spectropolarimetry observations of SNe Ibc. The collective conclusion of these studies is that asphericity appears to be ubiquitous to *all* SNe Ibc.

5.1. Non-GRB-SNe Ibc

5.1.1. Spectroscopy. The best way to investigate the inner ejecta geometry of a given SN is through late-time spectroscopy, as done by [110, 111, 222, 231–235]. At ≥200 d after the explosion, expansion makes the density of the ejecta so low that it becomes optically thin, thus allowing optical photons produced anywhere in the ejecta to escape without interacting with the gas. At these epochs the SN spectrum is nebular, showing emission lines mostly of forbidden transitions. Because the expansion velocity is proportional to the radius of any point in the ejecta, the Doppler shift indicates where the photon was emitted: those emitted from the near side of the ejecta are detected at a shorter (blueshifted) wavelengths, while those from the far side of the ejecta are detected at a longer (redshifted) wavelength. The late-time nebular emission profiles thus probe the geometry and the distribution of the emitting gas within the SN ejecta [236, 237]. Importantly for SNe Ibc, nebular spectra allow the observer to look directly into the oxygen core.

One of the strongest emission lines is the [O I] $\lambda\lambda$6300,6364 doublet, which behaves like a single transition if the lines are sufficiently broad (≥0.01c) because the red component is weaker than the blue one by a factor of three; see Figure 11. The appearance of this line can then be used to infer the approximate ejecta geometry: (1) a radially expanding spherical shell of gas produces a square-topped profile; (2) and a filled uniform sphere, where ^{56}Ni is confined in a central high-density region with an inner hole that is surrounded by a low-density O-rich region [238], produces a parabolic profile. These authors also considered a third scenario: (3) a bipolar model [215, 238, 239] characterized by a low-density ^{56}Ni-rich region located near the jet axis, where the jets convert stellar material (mostly O) into Fe-peak elements. The [O I] profile in the bipolar model depends on both the degree of asphericity and the viewing angle. If a bipolar SN explosion is viewed from a direction close to the jet axis, the O-rich material in the equatorial region expands in a direction perpendicular to the line of sight, and the [O I] emission profile is observed to be sharp and single-peaked.

On the other hand, for a near-equatorial view, the profile is broader and double-peaked. It is important to note that a double-peaked profile cannot be accounted for in the spherical model. Furthermore, the separation of the blueshifted and redshifted peaks, which represent the forward and rear portions of an expanding torus of O-rich material, suggests that the two peaks actually originate from the two lines of the doublet from a single emitting source on the front of the SN moving towards the observer. Double emission peaks seen in asymmetric profiles with separations larger or smaller than the doublet spacing do not share this problem. The high incidence of ≈64 Å separation between emission peaks of symmetric profiles plus the lack of redshifted emission peaks in asymmetric profiles suggests that emission from the rear of the SN may be suppressed. This implies that the double-peaked [O I] $\lambda\lambda$6300,6364 line profiles of SNe Ibc are not necessarily signatures of emission from a torus. The underlying cause of the observed predominance of blueshifted emission peaks is unclear but may be due to internal scattering or dust obscuration of emission from far side ejecta [235]. These models are for single-star progenitors, and they do not consider the effects that binary interactions or merger might impart to the observed geometry of the SN ejecta [240].

References [222, 232, 234] found that all SNe Ibc and IIb are aspherical explosions. The degree of asphericity varies in severity, but all studies concluded that most SNe Ibc are not as extremely aspherical as GRB-SNe (specifically SN 1998bw). Interestingly [234] found that, for some SNe Ibc, the [O I] line exhibits a variety of shifted secondary peaks or shoulders, interpreted as blobs of matter ejected at high velocity and possibly accompanied by neutron-star kicks to assure momentum conservation. The interpretation of massive blobs in the SN ejecta is expected to be the signature of very one-sided explosions.

Some notable and relevant nebular spectra analyses include SNe IcBL 2003jd [231], 2009bb [204], and 2012ap [206]. Reference [231] interprets their double-peaked [O I] $\lambda\lambda$6300,6364 nebular lines of SN 2003jd as an indication of an aspherical axisymmetric explosion viewed from near the equatorial plane, and directly perpendicular to the jet axis, and suggested that this asphericity could be caused by an off-axis GRB jet. Reference [204] obtained moderately noisy nebular spectra of SN 2009bb, which nevertheless displayed strong nebular lines of [O I] $\lambda\lambda$6300,6364 and [Ca II] $\lambda\lambda$7291,7324 that had all single-peaked profiles. In their derived synthetic spectra, a single velocity provided a good fit to these lines, thus implying that the ejecta is not overly aspherical. The nebular spectra (>200 d) of SN 2012ap [206] had an asymmetric double-peaked [O I] $\lambda\lambda$6300,6364 emission profile that was attributed to either absorption in the supernova interior or a toroidal ejecta geometry.

5.1.2. Polarimetry. Further enlightening clues to the geometry of SNe Ibc have arisen via polarimetric and spectropolarimetric observations (see [242] for an extensive review and Figure 12). When light scatters through the expanding debris of a SN, it retains information about the orientation of the scattering layers. Since it is not possible to spatially resolve extragalactic SNe through direct imaging, polarization is a powerful tool to determine the morphology of the ejecta. Spectropolarimetry measures both the overall shape of the emitting region and the shape of regions composed of particular chemical elements. Collectively, numerous polarimetric data have provided overwhelming evidence that all CCSNe are intrinsically three-dimensional phenomena with significant departures from spherical symmetry, and they routinely show evidence for strong alignment of the ejecta in single well-defined directions, suggestive of a jet-like flow. As discussed in [242], many of these CCSNe often show a rotation of the position angle with time of 30–40° that is indicative of a jet of material emerging at an angle with respect to the rotational axis of the inner layers. Another recent investigation by [241] showed that all SNe Ibc show nonzero polarization at the wavelength of strong lines. More importantly, they demonstrated that five of the six SNe Ibc they investigated had a "loop" in their Stokes Q-U diagram (where Q is the radiance linearly polarized in the direction parallel or perpendicular to the reference plane and U is the radiance linearly polarized in the directions 45° to the reference plane), which indicates that a nonaxisymmetric, three-dimensional ion distribution is ubiquitous for SNe Ibc ejecta.

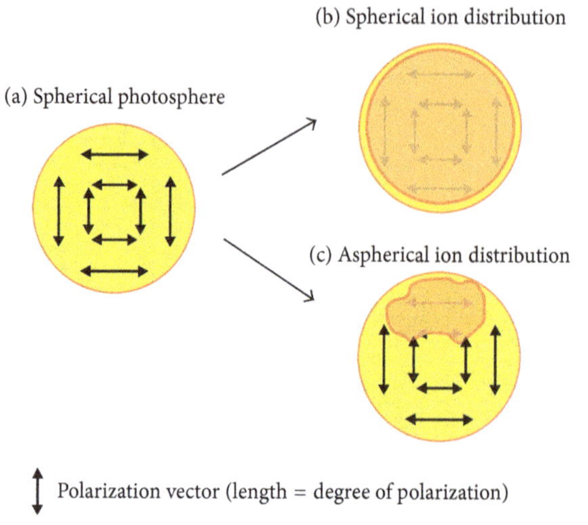

FIGURE 12: Schematic illustration of polarization in the SN ejecta. (a) When the photosphere is spherical, polarization is canceled out, and no polarization is expected. At the wavelength of a line, polarization produced by the electron scattering is depolarized by the line transition. (b) When the ion distribution is spherical, the remaining polarization is canceled, and no polarization is expected. (c) When the ion distribution is not spherical, the cancelation becomes incomplete, and line polarization could be detected (figure and caption taken from [241]).

The results of [242] suggest that the mechanism that drives CCSNe must produce energy and momentum aspherically from the start, either induced from the preexplosion progenitor star (i.e., rotation and/or magnetic fields) or perhaps arising from the newly formed neutron star (NS) [243–246]. In any case, it appears that the asphericity is permanently frozen into the expanding matter. Collimated outflows might be caused by magnetohydrodynamic jets, as is perhaps the case for GRB-SNe [210, 215, 247, 248], from accretion flows around the central neutron star, via asymmetric neutrino emission, from magnetoacoustic flux, jittering jets (jets that have their launching direction rapidly change [249]), or by some combination of those mechanisms. Another alternative idea, perhaps intimately related, is that material could be ejected in clumps that block the photosphere in different ways in different lines. It may be that jet-like flows induce clumping so that these effects occur simultaneously. Alternatively, the results of [241] suggest that the global asymmetry of SNe Ibc ejecta may rather arise from convection and preexisting asymmetries in the stellar progenitor before and during the time of core-collapse (e.g., [250, 251]), rather than induced by two-dimensional jet-like asphericity.

In addition to the above analyses, there are more clues which show that asphericity is quite ubiquitous in CCSN ejecta. A jet model was proposed for type Ic SN 2002ap [252], where the jet was buried in the ejecta and did not bore through the oxygen mantle. The lack of Fe polarization suggests that a nickel jet had not penetrated all the way to the surface. For CCSNe, we know that pulsars are somehow kicked at birth in a manner that requires a departure from both spherical and up/down symmetry [253]. The spatial distribution of various elements, including ^{44}Ti in supernova remnants [254], is also consistent with an aspherical explosion, arising from the development of low-mode convective instabilities (e.g., standing accretion shock instabilities [255]) that can produce aspherical or bipolar explosions in CCSNe. The anisotropies inferred by the oxygen distribution instead suggest that large-scale (plume-like) mixing is present, rather than small-scale (Rayleigh-Taylor) mixing, in supernova remnants. Additionally, the Cassiopeia A supernova remnant shows signs of a jet and counterjet that have punched holes in the expanding shell of debris [256], and there are examples of other asymmetric supernova remnants [257, 258] and remnants with indications of being jet-driven explosions or possessing jet-like features [259, 260].

5.1.3. Role of Mixing in the Ejecta. The analytical modelling of late-time (>50–100 d) bolometric LCs of SNe Ibc also implies a departure from spherical symmetry (or perhaps a range grey optical opacities [261]). Modelling performed by [122] showed that the late-time bolometric LC behaviour of a sample of three SNe Ic and IcBL (SNe 1998bw, 1997ef, and 2002ap) was better described by a two-component model (two concentric shells that approximated the behaviour of a high-velocity jet and a dense inner core/torus) than spherical models. Their modelling also showed that there was a large degree of nickel mixing throughout the ejecta. A similar result was inferred by [262] for a sample of SNe Ibc, who showed that the outflow of SNe Ib is thoroughly mixed. Helium lines arise via nonthermal excitation and nonlocal thermodynamic equilibrium [263–266]. High-energy γ-rays produced during the radioactive decay of nickel, cobalt, and iron Compton scatter with free and bound electrons, ultimately producing high-energy electrons that deposit their energy in the ejecta through heating, excitation, and ionization.

To address the question of whether the lack of helium absorption lines for SNe Ic was due to a lack of this element in the ejecta or that the helium was located at large distances from the decaying nickel [266–268], [262] showed that the ejecta of type SN Ic 2007gr was also thoroughly mixed, meaning that the lack of helium lines in this event could not be attributed to poor mixing. A similar conclusion was reached by [46] who demonstrated that He lines cannot be "smeared out" in the spectra of SNe IcBL, that is, blended so much that they disappear; instead He really must be absent in the ejecta (see as well [269]). A prediction of RT models [270] is if the lack of mixing is the only discriminant between SNe Ib and Ic, then well-mixed SNe Ib should have higher ejecta velocities than the less well-mixed SNe Ic. The investigation by [271] tested this prediction with a very large sample of SNe Ibc spectra, finding the opposite to be true: SNe Ic have higher ejecta velocities than SNe Ib, implying that the lack of He lines in the former cannot be attributed entirely to poor mixing in the ejecta. Next, [272] showed that for a sample of SNe Ibc, SNe Ib, Ic, and IcBL have faster rising LCs than SNe Ib, implying that the ejecta in these events are probably well mixed. The collective conclusion of these observational investigations states that the lack of helium features in SNe Ic spectra cannot be attributed to poor mixing but rather the absence of this element in the ejecta, which agrees with the conclusion of [268] that no more than 0.06–0.14 $M_{\odot}$ of He can be "hidden" in the ejecta of SNe Ic.

5.2. GRB-SNe. The key result presented in the previous sections is that all CCSNe possess a degree of asphericity: either two-dimensional [242] asymmetries where most CCSNe possess a jet or three-dimensional asymmetries [241]. Taken at face value, if all CCSNe possess two-dimensional axisymmetric geometry, then the observation of the 30–40° rotation of the position angle with time is suggestive of a jet of material emerging at an angle with respect to the rotational axis of the inner layers. This observation differs to that expected for GRB-SNe, where the jet angle is expected to be along or very near to the rotation axis of the preexplosion progenitor star. If jets are almost ubiquitous in CCSNe, but they are usually at an angle to the rotational axis, does this suggest that GRB-SNe are different because the jet emerges along, or very near to, the rotational axis? If so, then something is required to maintain that collimation: that is, more rapid rotation of GRB-SN progenitors and/or strong collimation provided by magnetic fields [248]. Moreover, is the difference between *ll*GRBs and high-luminosity GRBs due to less collimation in the former? In turn, perhaps more SNe Ibc arise from central engine that is currently accounted for, but for whatever reason the jets very quickly lose their collimation, perhaps to underenergetic or very short-lived central engines, and deposit their energy in the interior of the star, where perhaps a combination of jets and a neutrino-driven explosion mechanism is responsible for the observed SN. Note that this supposition is also consistent with the study of [273] who looked for off-axis radio emission from GRBs pointed away from Earth, finding <10% of all SNe Ibc are associated with GRBs pointed away from our line of sight. In this scenario, no imprint of the jet in the non-GRB-SNe Ibc is imparted to the ejecta. Nevertheless, the results of [241] need to be kept in mind when considering this speculative scenario, where the asymmetries in SNe Ibc may not be axisymmetric, but instead may be intrinsically three-dimensional.

More observations are sorely needed of nearby GRB-SNe to help address this outstanding question. To date only two GRB-SNe have occurred at close enough distances that reasonable quality nebular spectra have been obtained: SN 1998bw (~40 Mpc) and SN 2006aj (~150 Mpc). Even SN 2010bh was too distant (~270 Mpc) for the nebular emission lines to be reasonably modelled [49]. In the following section we will present a brief summary of the results of spectroscopic and polarimetric analyses of these two GRB-SNe.

References [47, 110, 111] investigated the nebular spectra of SN 1998bw, which exhibited properties that could not be explained with spherical symmetry. Instead, a model with high-velocity Fe-rich material ejected along the jet axis, and lower-velocity oxygen torus perpendicular to the jet axis, was proposed. From this geometry a strong viewing-angle dependence of nebular line profiles was obtained [110]. Reference [47] noted that the [Fe II] lines were unusually strong for a SN Ic and that lines of different elements have different widths, indicating different expansion velocities, where iron appeared to expand more rapidly than oxygen (i.e., a rapid Fe/Ni-jet and a slower moving O-torus). The [O I] nebular lines declined more slowly than the [Fe II] ones, signalling deposition of γ-rays in a slowly moving O-dominated region. These facts suggest that the explosion was aspherical. The absence of [Fe III] nebular lines can be understood if the ejecta are significantly clumped. Reference [111] noted that their models show an initial large degree (~4 depending on model parameters) of boosting luminosity along the polar/jet direction relative to the equatorial plane, which decreased as the SN approached peak light. After the peak, the factor of the luminosity boost remains almost constant (~1.2) until the supernova entered the nebular phase. This behaviour was attributed to an aspherical ^{56}Ni distribution in the earlier phase and to the disk-like inner low-velocity structure in the later phase.

Early polarization measurements of ≈0.5%, possibly decreasing with time, were detected for SN 1998bw [2, 274], which imply the presence of aspherical ejecta, with an axis ratio of about 2 : 1 [115]. In contrast, radio emission of GRB 980425/SN 1998bw showed no evidence for polarization [56], which suggested that the mildly relativistic ejecta were not highly asymmetric, at least in projection. However it should be noted that internal Faraday dispersion in the ejecta can suppress radio polarization. As mentioned in the previous section, modelling of the late-time bolometric LC of SN 1998bw [9, 117, 214, 275] showed that some degree of asymmetry in the explosion is required to explain its decay behaviour (see as well Figure 10).

For SN 2006aj, the [Fe II] lines were much weaker than those observed for SN 1998bw, which supports its lower luminosity relative to the archetype GRB-SN [113]. Most of the nebular lines had similar widths, and their profiles indicated that no major asymmetries were present in the ejecta at velocities below 8000 km s^{-1}. The modelling results of [48] implied a 1.3 $M_\odot$ oxygen core that was produced by a mildly asymmetric explosion. The mildly peaked [O I] $\lambda\lambda$6300,6364 profile showed an enhancement of the material density at velocities less than <3000 km s^{-1}, which also indicated an asymmetric explosion. If SN 2006aj was a jetted SN explosion, the jet was wider than in SN 1998bw (intrinsically or due to stronger lateral expansion [238]), since the signature is seen only in the innermost part. Linear polarization was detected by [276] between three and 39 days after explosion, which implied the evolution of an asymmetric SN expansion. Reference [277] concluded that their polarization measurements were not very well constrained, and considering the low polarization observed of 6000–6500 Å, the global asymmetry was ≤15%.

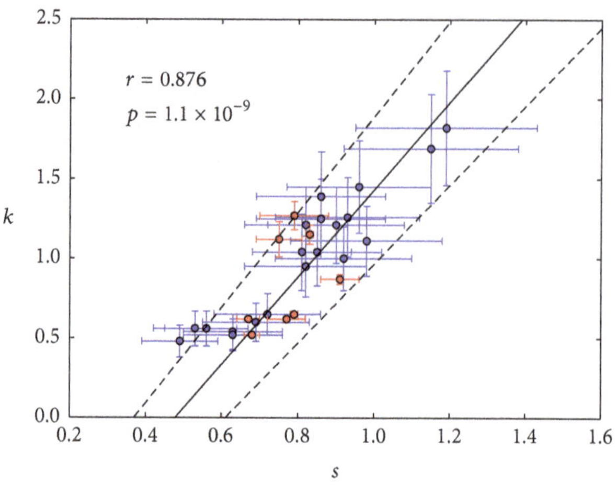

Figure 13: Luminosity (k)-stretch (s) relation for relativistic type IcBL SNe [42]. For all filters from $UBVRI$, and combinations thereof, GRB-SNe are shown in blue, and the two known relativistic type IcBL SNe (2009bb and 2012ap) are shown in red. A bootstrap analysis was performed to fit a straight line to the dataset to find the slope (m) and y-intercept (b), which used Monte-Carlo sampling and $N = 10{,}000$ simulations. The best-fitting values are $m = 2.72 \pm 0.26$ and $b = -1.29 \pm 0.20$. The correlation coefficient is $r = 0.876$, and the two-point probability of a chance correlation is $p = 1.1 \times 10^{-9}$. This shows that the k-s relationship is statistically significant at the 0.001 significance level.

6. GRB-SNe as Cosmological Probes

6.1. Luminosity-Stretch/Decline Relationships. In 2014, [42, 44, 128] (see as well [41]) demonstrated, using entirely different approaches, that GRB-SNe (which included *ll*GRB-SNe, INT-GRB-SNe, and high-luminosity GRB-SNe) have a luminosity–decline relationship that is perfectly akin to that measured for type Ia SNe [279]. All approaches investigated decomposed GRB-SN LCs (see Section 2.1). In [42], a template SN LC (1998bw) was created in filters $BVRI(1+z)$ as it would appear at the redshift of the given GRB-SN. A spline function ($g(x)$) was then fit to the template LC, and the relative brightness (k) and width (s) were determined (i.e., $f(x) = k \times g(x/s)$) for each GRB-SN in each rest-frame filter. These were then plotted, and a straight line was fit to the data, where the slope and intercept were constrained via a bootstrap fitting analysis that used Monte-Carlo sampling. An example of the k-s relation is shown in Figure 13, where GRB-SNe are shown in blue points, and the two relativistic SNe IcBL (2009bb and 2012ap) are shown in red. This relation

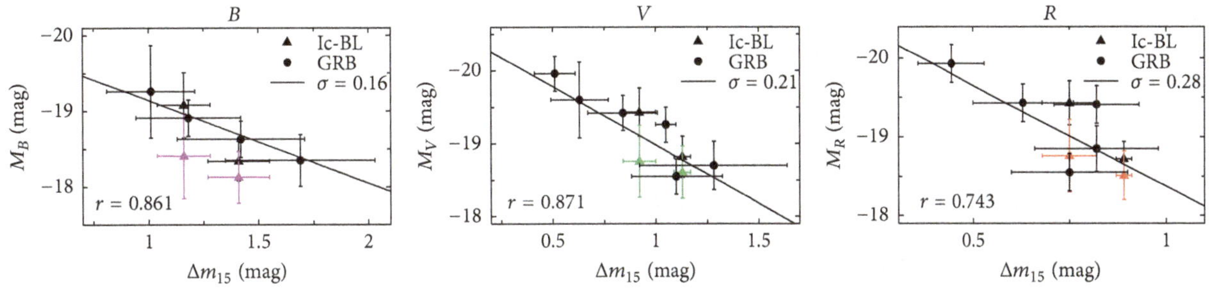

FIGURE 14: Luminosity–decline relationships of relativistic SNe IcBL (GRB-SNe: filled circles; SNe IcBL: filled triangles) in filters B (purple), V (green), and R (red), from [44]. Solid black lines and points correspond to absolute magnitudes calculated for luminosity distances, while coloured points and lines correspond to absolute magnitudes calculated for those events where independent distance measurements have been made to the SN's host galaxy. The correlation coefficient for each dataset is shown (in black and in their respective colours) as well as the best-fitting luminosity–decline relationship determined using a bootstrap method and the corresponding rms (σ) of the fitted model. It is seen that statistically significant correlations are present for both the GRB-SNe and combined GRB-SN and SN IcBL samples.

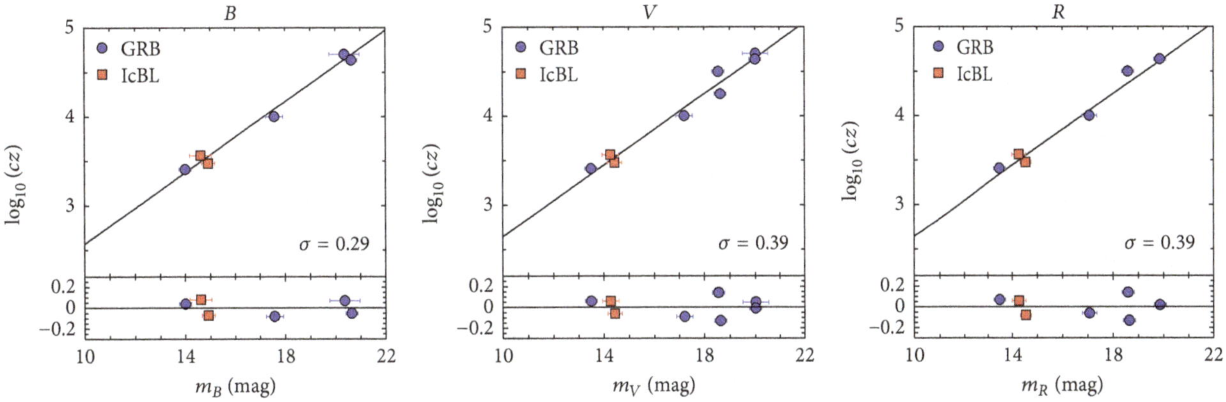

FIGURE 15: Hubble diagrams of relativistic SNe IcBL in filters BVR, from [44]. GRB-SNe are shown in blue and SNe IcBL (SNe 2009bb and 2012ap) in red. Plotted in each subplot are the uncorrected magnitudes of each subtype and the fitted Hubble ridge line as determined using a bootstrap method. Also plotted are the rms values (σ) and residuals of the magnitudes about the ridge line. In the B-band, the amount of scatter in the combined SNe IcBL sample is the same as that for SNe Ia up to $z = 0.2$ [44, 278], which is $\sigma \approx 0.3$ mag.

shows that GRB-SNe with larger k values also have larger s values; that is, brighter GRB-SNe fade slower. The statistical significance of the fit is shown as Pearson's correlation coefficient, where $r = 0.876$, and the two-point probability of a chance correlation is $p = 1.1 \times 10^{-9}$, which clearly shows that the relationship is significant at more than $p < 0.001$ significance level. This implies that not only are GRB-SNe standardizable candles, but all relativistic type IcBL SNe are.

The result in [42] clearly superseded the results of [30, 33, 39] who searched for correlations in the observer-frame R-band LCs of a sample of GRB-SNe, concluding that none was present. However, the method used in [32, 42, 44] had one key difference to previous methods: they considered precise, K-corrected rest-frame filters. Instead, previous approaches were all sampling different portions of the rest-frame spectral energy distribution (SED), which removed any trace of the k-s relationship. While such a correlation implies that, like SNe Ia, there is a relationship between the brightness of a given GRB-SN and how fast it fades, where brighter GRB-SNe fade slower, this relationship is not very useful if GRB-SNe want to be used for cosmological research: the template LCs of SN 1998bw are created for a specific cosmological model and are therefore model-dependent. Instead, the luminosity–decline relationship presented by [44, 128] relates the same observables as those used in SN Ia-cosmology research: their peak absolute magnitude and Δm_{15} in a given filter. Reference [128] considered rest-frame V-band only, while [44] considered rest-frame BVR. Figure 14 shows the relationships from the latter paper, where the two relativistic SNe IcBL are included in the sample. The amount of RMS scatter increases from blue to red filters and is only statistically significant in B and V (at the $p = 0.02$ level).

6.2. Constraining Cosmological Parameters. Once the luminosity–decline relationship was identified, the logical next step is to use GRB-SNe to constrain cosmological models, in an attempt to determine the rate of universal expansion in the local universe (the Hubble constant, H_0) and perhaps even the mass and energy budget of the cosmos. In a textbook example of how to use any standard(izable) candle to measure H_0 in a Hubble diagram of low-redshift objects (typically $z \ll 1$), [44] followed the procedure outlined in numerous SNe Ia-cosmology papers [280–288]. Figure 15 shows Hubble diagrams of relativistic SNe IcBL in filters BVR (GRB-SNe in blue, relativistic SNe IcBL in red) for redshifts less than $z = 0.2$. The amount of RMS scatter (shown as σ) is less in the B-filter, ≈ 0.3 mag, and about 0.4 mag in the redder V and R filters. Compared with the sample of SNe Ia in [278] over the same redshift range, it is seen that SNe Ia in the B-band also have a scatter in their Hubble diagram of 0.3 mag. Moreover, when the large SNe Ia sample ($N = 318$) was decreased to the same sample size of the relativistic SNe IcBL sample, the same amount of scatter was measured, meaning that GRB-SNe and SNe IcBL are as accurate as SNe Ia when used as cosmological probes.

A key observable needed to measure H_0 is independent distance measurements to one or more of the objects being used. However, to date no independent distance has yet been determined for a GRB or GRB-SN. However, relativistic SNe IcBL 2009bb and 2012ap were included in the same sample as the GRB-SNe, which was justified by [44] because both are subtypes of engine-driven SNe (Figure 7), and indeed they also follow the same luminosity–stretch (Figure 13) and luminosity–decline (Figure 14) relationship as GRB-SNe. Thus, one can use the independent distance measurements to their host galaxies (Tully-Fisher distances) and use them as probes of the local Hubble flow to provide a model-independent estimate of H_0. Reference [44] constrained a weighted-average value of $H_{0,w} = 82.5 \pm 8.1\,\mathrm{km\,s^{-1}\,Mpc^{-1}}$. This value is 1σ greater than that obtained using SNe Ia and 2σ larger than that determined by Planck. This difference can be attributed to large peculiar motions of the host galaxies of the two SNe IcBL, which are members of galaxy groups. Interestingly, when the same authors used a sample of SNe Ib, Ic, and IIb, they found an average value of H_0 that had a standard deviation of order 20–40 $\mathrm{km\,s^{-1}\,Mpc^{-1}}$, which demonstrates that these SNe are poor cosmological candles. In a separate analysis, [289] used their sample of GRB-SNe, which did not include non-GRB-SNe IcBL but instead covered a larger redshift range (up to $z = 0.6$), to derive the mass and energy budget of the universe, finding loosely constrained values of $\Omega_M = 0.58^{+0.22}_{-0.25}$ and $\Omega_\Lambda = 0.42^{+0.25}_{-0.22}$.

6.3. Physics of the Luminosity-Decline Relationship. A physical explanation for why GRB-SNe are standardizable candles is not immediately obvious. If the luminosity of GRB-SNe (excluding SN 2011kl) is powered by radioactive heating (see Section 4.2), then more nickel production leads to brighter SNe. So far however, no correlation has been found between the bolometric properties of GRB-SNe and the properties (E_{iso} and T_{90}) of the accompanying γ-ray emission [11, 32, 43]. To a first order, this is at odds with the simplest predictions of the collapsar model, which suggests that more energy input by a central engine should lead to increased nickel production and more relativistic ejecta. However, γ-ray energetics are a poor proxy of the total energy associated with the central engine, so the absence of a correlation is perhaps not surprising. Moreover, as pointed out by [11], one expects large variations in the masses and rotation rates of the pre-explosion progenitor stars, especially when metallicity effects are factored in. Different stellar rotation rates will result in different rotation rates imparted to their cores, leading to different amounts of material being accreted and ultimately

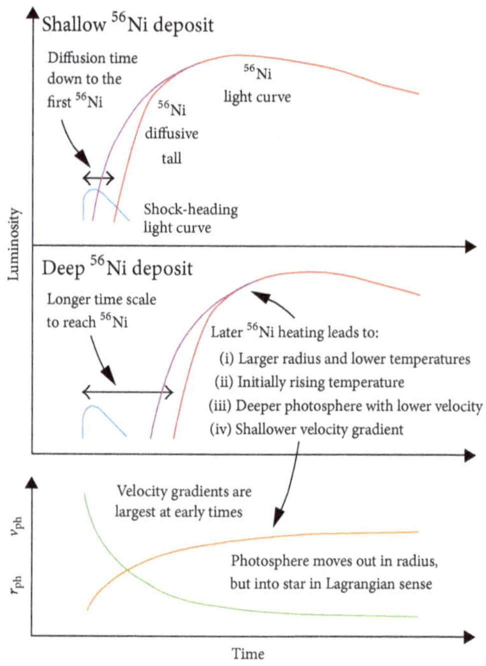

Figure 16: The effect of different degrees of nickel mixing in the ejecta of SNe Ibc on their observed LCs, from [290]. Top and middle panels: how the relative positions of the shock-heating contribution (blue curves), ^{56}Ni diffusive tail contribution (purple curves), and the ^{56}Ni contribution (red curves) to the observed LC can differ depending on the depth and amount of mixing of the ^{56}Ni. The total observed LC is the sum of these three components. When the ^{56}Ni is located deep in the ejecta (middle panel) and the shock-heating light curve (blue curve) is below the detection limits, there can be a significant dark phase between the time of explosion and the moment of first detection. Bottom: temporal evolution of the photospheric radius (orange curve) and velocity (green curve). Depending on the position of the ^{56}Ni LC, different photospheric radii, velocities, and velocity gradients will be present during the rising LC.

resulting in a variation of the final BH masses. Along with variations in the stellar density, all of these factors will result in a range of nickel masses being produced. Moreover, even if the same amount of nickel is produced in each event, SNe that expand at a slower rate will be fainter because their LCs will peak later after which more of the nickel has decayed and suffered adiabatic degradation. Additionally, the location of the nickel in the ejecta will also result in different looking LCs, where nickel that is located deeper in the ejecta takes longer to diffuse out of the optically thick ejecta, leading to later peak times (Figure 16). If the degree of mixing in the early SN is heterogeneous for GRB-SNe, a range of rise times is expected, along with a large variation in the velocity gradients and photospheric radii. However, inspection of Figure 6 shows that, if we naively take a single transition as a proxy of the photospheric velocity, the distribution of say Si II λ6355 shows that the velocity gradient of most GRB-SNe has a similar evolution, though the range of velocities of the Fe II λ5169 transition implies that they still have a wide range of velocities at a given epoch. This similar behaviour might suggest a similar degree of nickel mixing in the SN ejecta.

Nevertheless, it appears from independent studies using different approaches that GRB-SNe *are* standardizable candles. Whether this observation implies similarities in the physical properties of the central engine driving the explosion, or the SNe themselves, is uncertain. For the most part, it is expected that most GRB-SNe are viewed close to the jet axis [42], which also appears to apply to SN 2009bb [204], meaning we are observing SNe more or less with the same approximate geometry. The fact that GRB-SNe are standard*izable* and have a range of brightness implies that different amounts of nickel are being generated. A naive conclusion to be drawn is that the observed luminosity–decline relationships suggest that a correlation exists between the strength and energetics produced by the central engine and the resultant nucleosynthetic yields of ^{56}Ni. Moreover, the lack a luminosity–decline relationship for SNe Ibc [42, 44] implies that the explosion and nucleosynthesis mechanism(s) are not correlated.

In the context of SNe Ia, which are, of course, also standardizable candles, their LCs are also powered solely by the radioactive decay of nickel and cobalt, the amount of which determines the LC's peak brightness and width. The width also depends on the photon diffusion time, which in turn depends on the physical distribution of the nickel in the ejecta, as well as the mean opacity of the ejecta. In general, the opacity increases with increasing temperature and ionization [291], thus implying that more nickel present in the ejecta leads to larger diffusion times. This directly implies that fainter SNe Ia fade faster than brighter SNe Ia, thus satisfying the luminosity–decline relation [279]. This is not the only effect however, as the distribution of nickel in the ejecta also affects how the LC evolves, where nickel located further out has a faster bolometric LC decline. Additionally, following maximum *B*-band light, SNe Ia colours are increasingly affected by the development of Fe ɪɪ and Co ɪɪ lines that blanket/suppress the blue *B*-band light. Dimmer SNe are thus cooler, and the onset of Fe ɪɪɪ → Fe ɪɪ recombination occurs quicker than in brighter SNe Ia, resulting in a more rapid evolution to redder colours [292]. Therefore the faster *B*-band decline rate of dimmer SNe Ia reflects their faster ionization evolution and provides additional clues as to why fainter SNe Ia fade more rapidly. Thus, as the LCs of GRB-SNe are also powered by radioactive decay, the physics that govern SNe Ia also govern those of GRB-SNe and may go some way to explaining why GRB-SNe are also standardizable candles.

7. Host Environments

Direct observations of the SNe that accompany LGRBs, and their subtypes, provide a rich range of clues as to the physical properties of their preexplosion progenitor stars. LGRBs represent a rare endpoint of stellar evolution, and their production and subsequent properties are likely to be a consequence of environmental factors. As such, many in-depth investigations of their host environments, both their global/galaxy-wide properties and, where possible, host-resolved environmental conditions, have been performed. Indeed, the information gained from this myriad of investigations warrants their own reviews, and the gathered nuances of these studies are beyond the scope of this GRB-SN review. Instead, in this section we highlight what we regard as the most important developments in this branch of GRB phenomenology that have directly furthered our understanding of the GRB-SN connection. For further insight, we refer the reader to excellent reviews and seminal studies by, among others, [293–296], and references therein.

7.1. Global Properties. With the advent of X-ray localizations of GRB AGs came the ability to study the type of galaxies that LGRBs occur in. Over the years, evidence mounted that LGRBs appeared to prefer low-luminosity, low-mass, blue, star-forming galaxies that have higher specific star-formation rates (SFRs) than the typical field galaxy [293, 297–308]. Visual inspection of optical *HST* imaging of LGRB host galaxies [307, 309, 310] showed a high fraction of merging/interacting systems: 30% showed clear signs of interaction, and another 30% showed irregular and asymmetric structure, which may be the result of recent mergers. The position of a GRB within its host also provided additional clues: both [311], who examined the offsets of LGRBs from their host nuclei (see as well [312, 313]), and [309] demonstrated that, within their hosts, LGRBs were more likely to be localized in the brightest UV regions of the galaxy, which are associated with concentrated populations of young massive stars.

At the same time, several early studies were converging towards the idea that LGRBs favoured subsolar, low-metallicity (Z) host/environments [314–317]. As the progenitors of LGRBs are massive stars with short lifetimes (of order a few million years), they are not expected to travel far from their birth in H ɪɪ regions, and the measured metallicity of the associated H ɪɪ region at the site of an LGRB can be used as a proxy of the natal metallicity. Reference [305] found that the metallicities of half a dozen low-redshift ($z < 0.3$) LGRB hosts were lower than their equally luminous counterparts in the local star-forming galaxy population and proposed that LGRB formation was limited by a strong metallicity threshold. This was based on the observation that LGRB hosts were placed below the standard L-Z relation for star-forming galaxies, where galaxies with higher masses, and therefore luminosities, generally have higher metallicities [318–321]. A metallicity cut-off for LGRB formation was also proposed by [322]. Reference [323] demonstrated that nearby LGRB host galaxies had systematically lower metallicities than the host galaxies of nearby ($z < 0.14$) SNe IcBL. Reference [205] showed that most LGRB host galaxies fall below the general L-Z relation for star-forming galaxies and are statistically distinct to the host galaxies of SNe Ibc and the larger star-forming galaxy population. LGRB hosts followed their own mass-metallicity relation out to $z \sim 1$ that is offset from the general mass-metallicity relation for star-forming galaxies by an average of 0.4 ± 0.2 dex in metallicity. This marks LGRB hosts as distinct from the host galaxies of SNe Ibc and reinforced the idea that LGRB host galaxies are not representative of the general galaxy population [303, 324, 325].

For the better part of a decade, this general picture became the status-quo for the assumed host properties of LGRBs: blue, low-luminosity, low-mass, star-forming galaxies with low metal content. However, more recently this previously quite uniform picture of GRB hosts became somewhat more diverse: several metal-rich GRB hosts were discovered [205, 326, 327], which revealed a population of red, high-mass, high-luminosity hosts that were mostly associated with dust-extinguished afterglows [328–331]. Next, the offset of GRB-selected galaxies towards lower metallicities in the mass-metallicity relation [205] could, for example, be partially explained with the dependence of the metallicity of star-forming galaxies on SFR [332, 333]. Moreover, it was shown that LGRBs do not exclusively form in low-metallicity environments [328, 331, 334, 335], where the results of [296] are an excellent example of this notion. Analysing the largest sample of LGRB-selected host spectra yet considered (up to $z = 3.5$), they found that a fraction of LGRBs occur in hosts that contain super-solar ($Z > Z_{\odot}$) metal content (<20% at $z = 1$). This shows that while some LGRBs can be found in high-Z galaxies, this fraction is significantly less than the fraction of star-forming regions in similar galaxies, indicating GRBs are actually quite scarce in high-metallicity hosts. They found a range of host metallicities of 12 + log(O/H) = 7.9 to 9.0, with a median of 8.5. Reference [296] therefore concluded that GRB host properties at lower redshift ($z < 1$-2) are driven by a given LGRB's preference to occur in lower-metallicity galaxies without fully avoiding metal-rich ones and that one or more mechanism(s) may operate to quench GRB formation at the very highest metallicities. This result supported similar conclusions from numerous other recent studies [205, 326, 331, 335–340] which show that LGRBs seem to prefer environments of lower metallicity, with possibly no strict cut-off in the upper limit of metal content (though see [341]).

Another revealing observation was made by [342] who showed that low-z SNe IcBL and $z < 1.2$ LGRBs (i.e., core-collapse explosions in which a significant fraction of the ejecta moves at velocities exceeding 20,000–30,000 km s^{-1}) preferentially occur in host galaxies of high stellar-mass and star-formation densities when compared with SDSS galaxies of similar mass ($z < 0.2$). Moreover, these hosts are compact for their stellar masses and SFRs compared with SDSS field galaxies. More importantly, [342] showed that the hosts of low-z SNe IcBL and $z < 1.2$ LGRBs have high gas velocity dispersions for their stellar masses. It was shown that core-collapse SNe (types Ibc and II) showed no such preferences. It appears that only SLSNe occur in more extreme environments than GRB-SNe and relativistic SNe IcBL: [343] showed that SLSNe occur in extreme emission-line galaxies, which are on average more extreme than those of LGRBs and that type I SLSNe may result from the very first stars exploding in a starburst, even earlier than LGRBs. Finally, [342] concluded that the preference for SNe IcBL and LGRBs for galaxies with high stellar-mass densities and star-formation densities may be just as important as their preference for low-metallicity environments.

The result of [342] is the latest in a long line of investigations that suggest that LGRBs are useful probes of high-z star formation. This result stems from a long-debated question of whether LGRBs may be good tracers of the universal star-formation rate over all of cosmic history [293, 296, 311, 331, 335, 344–348]. Reference [296] showed that there is an increase in the (median) SFR of their sample of LGRB host galaxies at increasing redshift, where they found 0.6 $M_{\odot}\,yr^{-1}$ at $z \approx 0.6$ to 15.0 $M_{\odot}\,yr^{-1}$ at $z \approx 2.0$. Moreover, these authors suggest that by $z \sim 3$ GRB hosts will probe a large fraction of the total star formation. In absence of further secondary environmental factors, GRB hosts would then provide an extensive picture of high-redshift star-forming galaxies. However, the connection between LGRBs and low-metallicity galaxies may hinder their utility as unbiased tracers of star formation [305, 317, 323, 349], though if LGRBs do occur in galaxies of all types, as suggested above, then they may be only mildly biased tracers of star formation [350].

7.2. Immediate Environments. Most LGRB host galaxies are too distant for astronomers to discern their spatially resolved properties. These limitations are important to consider when extrapolating LGRB progenitor properties from the global host properties, as it may be possible that the location of a given LGRB may differ to that of the host itself. Where spatially resolved studies have been performed, such as for GRB 980425 [351–354], GRB 060505 [355], GRB 100316D [356], and GRB 120422A [41, 357], it was found that, in at least two of these cases, the metallicity and SFR of other H II regions in their hosts had comparable properties as those associated with the LGRB location (within 3σ). In these studies the host galaxies had a minimal metallicity gradient [355], and there were multiple low-metallicity locations within the host galaxies, where in some cases the location of the LGRB was in that of the lowest metallicity [357]. These studies suggest that, in general, the host-wide metallicity measurement can be used as a first-order approximation of the LGRB site.

Next, the line ratios of [Ne III] to [O II] suggest that H II regions associated with LGRBs are especially hot [358], which may indicate a preference for the hosts of LGRBs to produce very massive stars. Absorption line spectroscopy has revealed some fine-structure lines (e.g., Fe [II]), which could indicate the presence of absorption occurring in fast-moving winds emanated by WR stars (i.e., stars that are highly stripped of their outer layers of hydrogen and helium). The distances implied by variable fine-structure transitions (e.g., their large equivalent widths imply large distances to avoid photoionization) show that the absorption occurs at distances of order tens to hundreds of parsecs from the GRB itself [359–362], which makes sense given that the dust and surrounding stellar material around a GRB is completely obliterated by the explosion. Such absorption must arise from nearby WR stars whose winds dissect the line of sight between the GRB and Earth.

The type of environment in which a given LGRB occurs is also of interest: is it a constant interstellar medium (ISM) or a wind-like medium? Do the progenitors of LGRB carve out large wind-blown bubbles [363, 364], as has been observed for galactic WR stars [365, 366]? Using a statistical approach to the modelling of GRB AGs, [367] demonstrated that the majority of GRBs (L- and SGRBs) in their sample (18/27) were compatible with a constant ISM, and only six showed evidence of a wind profile at late times. They concluded that, observationally, ISM profiles appear to dominate and that most GRB progenitors likely have relatively small wind termination-shock radii, where a variable mass-loss history, binarity of a dense ISM, and a weak wind can bring the wind termination shock radius closer to the star [368, 369]. A smaller group of progenitors, however, seem to be characterized by significantly more extended wind regions [367]. In this study, the AG is assumed to be powered by the standard forward-shock model, which has been shown to not always be the best physical description for all LGRBs observed in nature [195].

Finally, it appears that LGRBs generally occur in environments that possess strong ionization fields, which likely arise from hot, luminous massive stars in the vicinity of LGRBs. Reference [296] showed that the GRB hosts in their sample occupied a different phase-space than SDSS galaxies in the Baldwin-Phillips-Terlevich (BPT) diagram [370]: they are predominantly above the ridge line that denotes the highest density of local star-forming field galaxies. A similar offset was also observed for galaxies hosting type I SLSNe [343]. This offset is often attributed to harder ionization fields, higher ionization parameters, or changes in the ISM properties [371–373]. This result is consistent with the hypothesis that the difference in the location in the BPT diagram between GRB hosts and $z \sim 0$ star-forming galaxies is caused by an increase in the ionization fraction; that is, for a given metallicity a larger percentage of the total oxygen abundance is present at higher ionization states at higher redshifts. This could be caused by a harder ionization field originating from hot O-type stars [373] that emit a large number of photons capable of ionizing oxygen into [O III].

7.3. Implications for Progenitor Stars. Before LGRBs were conclusively associated with the core-collapse of a massive star, their massive-star origins were indirectly inferred. If LGRBs were instead associated with the merger of binary compact objects, two "kicks" arising from two SN explosions would imply a long delay before coalescence and likely lead to GRBs occurring at large distances from star-forming regions [374–377]. With subarcsecond localization came observations that showed LGRBs, on average, were offset from the apparent galactic centre by roughly 1 kpc [311], which did not agree with a compact object binary-merger scenario. Further statistical studies showed a strong correlation between the location of LGRBs and the regions of bluest light in their host galaxies [309, 310], which implied an association with massive-star formation. Thus result was furthered by [378] that showed that LGRBs and type Ic SNe have similar locations in their host galaxies, providing additional indirect evidence of LGRBs and massive stars.

The general consensus that LGRBs occur, on average, in metal-poor galaxies (or location within more metal-rich hosts), aligned well with theoretical expectations that LGRB formation has a strong dependence on metallicity. In theoretical models [215, 379–383], the progenitors of LGRBs need to be able to lose their outer layers of hydrogen and helium (as these transitions are not observed spectroscopically), but do so in a manner that does not remove angular momentum from the core (to then power the GRB). At high metallicities, high-mass loss rates will decrease the surface rotation velocities of massive stars and, due to coupling between the outer envelopes and the core, will rob the latter of angular momentum and hence the required rapid rotation to produce a GRB. In quasi-chemically homogeneous models [382–384], rapid rotation creates a quasi-homogeneous internal structure, whereby the onion-like structure retained by non- or slowly rotating massive stars is effectively smeared out, and the recycling of material from the outer layers to the core results in the loss of hydrogen and helium in the star because it is fused in the core. Intriguingly, quasi-chemically homogeneous stars do appear to exist in nature. The FLAMES survey [385] observed over 100 O- and B-type stars in the Large Magellanic Cloud (LMC) and the Milky Way galaxy and showed the presence of a group of rapidly rotating stars that were enriched with nitrogen at their surfaces. The presence of nitrogen at the surface could only be due to rotationally triggered internal transport processes that brought nuclear processed material, in this case nitrogen, from the core to the stellar surface. Observations of metal-poor O-type stars in the LMC by [386, 387] show the signature of CNO cycle-processed material at their surfaces, while modelling of the spectra of galactic and extragalactic oxygen-sequence WR stars shows very low surface He mass fractions, thus making them plausible single-star progenitors of SNe Ic [388].

However, other observations of Local Group massive-star populations have revealed that the WR population actually decreases strongly at lower metallicities, particularly the carbon- and oxygen-rich subtypes [389], suggesting that these proposed progenitors may be extremely rare in LGRB host environments. Moreover, the results of [390], based on the analysis of two LGRBs, suggest that some LGRBs may be associated with progenitors that suffer a great degree of mass loss before exploding and hence a great deal of core angular momentum. Moreover, the association of some LGRBs with super-solar metallicity environments also contradicts the predictions of the collapsar model. However, other recent models have considered alternative evolutionary scenarios whereby LGRB progenitors can lose a great deal of mass before exploding, but still retain enough angular momentum to power a GRB [391–393]. Such models consider the complex connection between surface and core angular momentum loss and show that single stars arising from a wide range of metal content can actually produce a GRB. Moreover, the effects of anisotropic stellar winds need to also be considered [205]. Polar mass loss removes considerably less angular momentum than equatorial mass loss [394], which provides the means for the progenitor to lose mass but sustain a high rotation rate. Alternatively, episodic mass loss, as has been observed for luminous blue variable stars may also offer another means of providing a way to lose mass but retain core angular momentum.

8. Kilonovae Associated with SGRBs

To date, the amount of direct and indirect evidence for the massive-star origins of LGRBs is quite comprehensive and thoroughly beyond any conceivable doubt. The same however cannot be stated about the progenitors of SGRBs. For many years, since the discovery that there are two general classes of GRBs [399, 400], general expectations were that they arose from different physical scenarios, where SGRBs are thought to occur via the merger of a binary compact object system

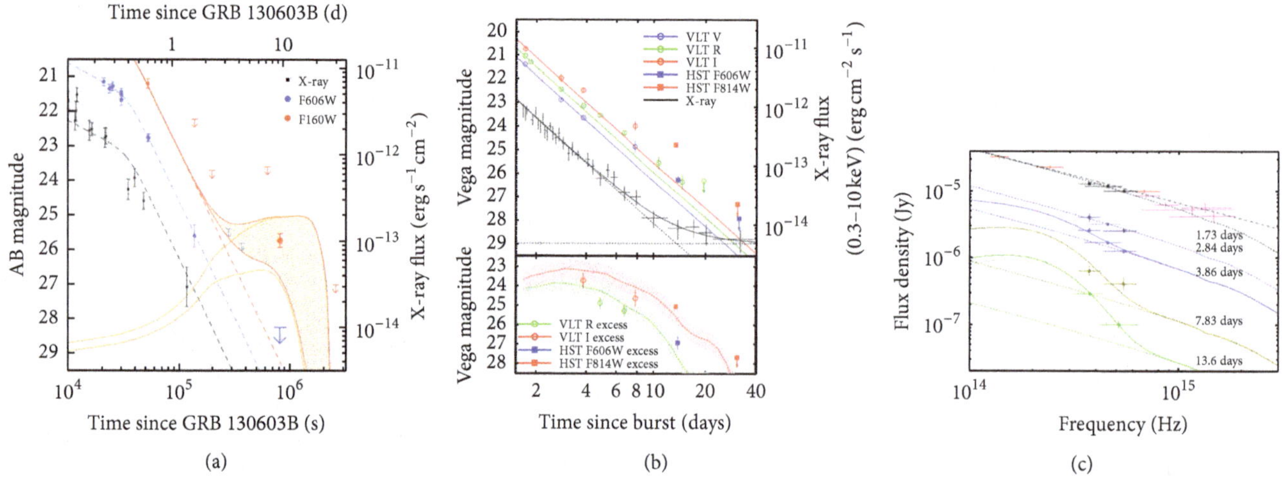

FIGURE 17: Observations of KNe associated with SGRBs: (a) GRB 130603B, from [395]. The decomposed optical and NIR LCs show an excess of flux in the NIR ($F160W$) filter, which is consistent with theoretical predictions of light coming from a KN. (b) GRB 060614, from [396]. Multiband LCs show an excess in the optical LCs (R and I), which once the AG light is removed, the resultant KN LCs match those from hydrodynamic simulations of a BH-NS merger (ejecta velocity of ~0.2c and an ejecta mass of 0.1 $M_\odot$ [397]). (c) SEDs of the multiband observations of GRB 060614, also from [396]. The early SEDs are well described by a power law spectrum, which implies synchrotron radiation. However, at later epochs the SEDs are better described by thermal, black body spectra, with peak temperatures of ~2700 K, which are in good agreement with theoretical expectations [398].

containing at least one neutron star (i.e., NS-NS or NS-BH). Circumstantial evidence for the compact object merger origins of SGRBs [401, 402] includes their locations in elliptical galaxies, the lack of associated supernovae (as observed for LGRBs) [403–406], the distribution of explosion-site offsets relative to their host galaxies (0.5–75 kpc away, median of 5 kpc [407, 408]), and a weak spatial correlation of SGRB locations with star-formation activity within their host galaxies.

The compact coalescence scenario predicts SGRB AGs at longer wavelengths [401, 402, 409–411], which have been observed [412]. As well as the expected AG emission, emission from a SN-like transient was also predicted [413–417], which have been referred to as a "kilonova" (KN), "merger-nova," or "macronova" (see the recent review by [418]), where we have adopted the former terminology in this review.

The KN prediction is a natural consequence of the unavoidable decompression of NS material, where a compact binary coalescence provides excellent conditions for the rapid-neutron capture process (r-process [409, 419–422]). The neutron capture process occurs very quickly and is completed in less than a second, and it leaves behind a broad distribution of radioactive nuclei whose decay, once the ejected material becomes transparent, powers an electromagnetic transient in a process similar to that expected to cause GRB-SNe to shine. Hydrodynamic simulations suggest that, during a merger, mass is ejected via two mechanisms: (I) during the merger, surface layers may be tidally stripped and dynamically flung out in tidal tails; (II) following the merger, material that has accumulated into a centrifugally supported disk may be blown off in a neutrino or nuclear-driven wind. In mechanism (I), the amount of material ejected depends primarily on the mass ratio of the compact objects and the equation of state of the nuclear matter. The material is very neutron-rich ($Y_e \sim 0.1$), and the ejecta is expected to assemble into heavy ($Z > 50$) elements (including Lanthanides, $58 < Z < 70$, and Actinides, $90 < Z < 100$) via the r-process. In mechanism (II), however, neutrinos emitted by the accretion disk raise the electron fraction ($Y_e \sim 0.5$) to values where a Lanthanide-free outflow is created [423]. In both cases 10^{-4}–$10^{-1}\,M_\odot$ of ejecta is expected to be expelled. A direct observational consequence of mechanism (I) is a radioactively powered transient that resembles a SN, but which evolves over a rapid timescale (~1 week, due to less material ejected compared with a typical SN) and whose spectrum peaks at IR wavelengths. In contrast to other types of SNe, for example, SNe Ia whose optical opacity is dictated by the amount of iron-group elements present in the ejecta, r-process ejecta that is composed of Lanthanides has a much larger expansion opacity (≈100 times greater) due to the atoms/ions having a greater degree of complexity in the number of ways in which their electrons can be arranged in their valence shells (relative to iron-group elements).

There have been a handful of observational searches for KN emission: GRB 050709 [424, 425]; GRB 051221A [426]; GRB 060614 [396, 427]; GRB 070724A [428, 429]; GRB 080503 [430, 431]; GRB 080905A [432]; and GRB 130603B [395, 433]. In almost all cases null results were obtained, with the notable exceptions being GRB 130603B, GRB 060614 (see Figure 17), and GRB 050709. In these cases, the optical and NIR LCs required a careful decomposition, and once the AG components were accounted for, an excess of emission was detected. In the case of GRB 130603B, a single NIR datapoint was found to be in excess of the extrapolated AG decay, which was interpreted by [395] as arising from emission from a KN. The (observer-frame) colour term $R_{F606W} - H_{F160W} < 1.7$ mag at +0.6 d, and $R_{F606W} - H_{F160W} < 2.5$ mag at +9 days, which is inconsistent with a colour change due to FS emission and was argued to be evidence of nonsynchrotron emission arising from a possible KN. The dataset of GRB 060614 considered by [396] is more extensive than that of GRB 130603B, and KN bumps were detected in two filters (observer-frame R and I), which peaked at 4–6 d (rest-frame). The decomposed KN LCs were shown to be consistent with LCs arising from hydrodynamic simulations of a BH-NS merger, which had an ejecta velocity of ~0.2c and an ejecta mass of 0.1 $M_\odot$ [233]. The larger dataset also allowed for the construction of SEDs, which showed a clear transition from a power law spectrum at early epochs (<3 d), which appeared to transition into a thermal, black body spectrum over the next two weeks. Moreover, the inferred temperature of the black body was around 2700 K, which fitted well with theoretical expectations. However, the precise nature of GRB 060614 is still not understood, and it is still uncertain if it is a short or a long GRB.

9. Theoretical Overview

While the focus of this review is geared towards what observations tell us about the GRB-SN connection, a keen understanding of the leading theoretical models is also required. The finer intricacies of each model are presented elsewhere, and we suggest the reader to start with the comprehensive review by [434] (and references therein), which is just one of many excellent reviews of the physics of the prompt emission and AGs. As such, what is presented here is meant only as an overview of the rich and complex field of GRB phenomenology.

9.1. Central-Engine Models: Millisecond Magnetars versus Collapsars. The main consensus of all GRB models is that LGRBs and their associated SNe arise via the collapse of massive stars, albeit ones endowed with physical properties that must arise only seldom in nature, given the fact GRB-SNe are very rare. In the leading theoretical paradigms, after the core-collapse of the progenitor stars, the leftover remnant is either a NS or a BH, and under the correct conditions, both can operate as a central engine to ultimately produce an LGRB.

In reality, very few solid facts are known about the true nature of the central-engine(s) operating to produce LGRBs. Nevertheless, one of the most prevailing models of the central engines of GRBs associated with SNe is the collapsar model [210, 215, 247], where the accretion of material from a centrifugally supported disk onto a BH leads to the launch a bipolar relativistic jet, and material within the jet leads to the production of γ-ray emission. The collapsar model suggests that there is enough kinetic energy (2–5 × 10^{52} erg) in the accretion disk wind which can be used to explosively disrupt the star, as well as synthesizing ~0.5 $M_\odot$ of ^{56}Ni. In this model, the duration of the prompt emission is directly related to the stellar envelope infall time, and the jet structure is maintained either magnetically or via neutrino-annihilation-driven jets along the rotation axes. The other promising mechanism that could lead to the production of an LGRB and its hypernovae is the millisecond magnetar model [221, 435–437]. In this scenario, the compact remnant is a rapidly rotating (P ~ 1–10 ms), highly magnetized (B ~ 10^{14-15} G) NS, where the relativistic Poynting-flux jets are supported by stellar confinement [436].

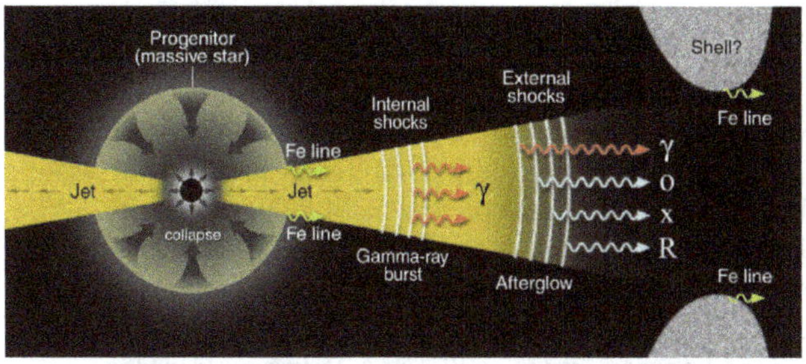

Figure 18: The death of a massive star produces a GRB (and its multiband AG) and an energetic and bright SN (from [439]).

A cartoon visualization of the formation of an LGRB, including its AG and associated SN, is shown in Figure 18. In the standard fireball model, shells of material within the jet interact to produce the initial burst of γ-rays, called the prompt emission, via internal shocks. As the jet propagates away from the explosion site, it eventually collides with the surrounding medium producing external shocks that power an AG that is visible across almost the entire electromagnetic spectrum, from X-rays to radio, and which lasts for several weeks to months. In this leptonic model, the prompt and AG radiation is synchrotron or synchrotron-self-Compton in origin [438]. It is interesting to note that this scenario is pretty much independent of the nature of the central engine; all that is required is the formation of an ultra-relativistic jet. It is generally thought that luminous GRBs with bulk Lorentz factors of order Γ_B ~ 300 must stem from ultra-relativistic collisionless jets produced by millisecond magnetars and/or collapsars. As discussed in Section 3, in order to penetrate the stellar envelope, the active timescale of the jet produced by the central engine (t_{engine}) must be longer than the penetrating timescale, where the latter is ~R/v_{jet}. Some *ll*GRBs whose Γ_B ~ 2 can also be explained by this model, but in these cases, the active timescale is likely to be slightly smaller than the penetrating timescale so that the ultra-relativistic jet from the central engine either just barely or completely fails to completely penetrate the stellar envelope.

The first class of models for the prompt emission of GRBs was the internal-shock model, where synchrotron or synchrotron self-Compton radiation was emitted by electrons that were accelerated by internal shocks [438, 440] in the form of high-energy γ-ray photons. Inverse Compton (IC) scattering and synchrotron self-Compton (SSC) scattering can enhance the seed photons and account for the very high-energy γ-ray photons measured for some GRBs. One prediction of the internal-shock model is the production of high-energy neutrinos, which to date have not been observed by neutrino detectors such as IceCube (only upper limits have been obtained so far, see the review by [441]). Although more detailed calculations performed by [442, 443] have demonstrated that the internal-shock model which includes benchmark parameters (e.g., the bulk Lorentz factor Γ_B = 300) is consistent with the upper limits obtained by IceCube, these results have posed more stringent constraints on the internal shocks model because there is a possible correlation between the bulk Lorentz factor Γ_B and the GRB luminosity [444–446]. Instead, alternative scenarios in the context of the ultra-relativistic jet model are the photospheric emission model [447–451] and the Internal-Collision-induced MAgnetic Reconnection and Turbulence (ICMART) model [452, 453]. Photospheric models assume that thermal energy stored in the jet is radiated as prompt emission at the Thomson photosphere [454–456], while ICMART models envisage that collisions between "mini-shells" in a Poynting-flux-dominated outflow distort the ordered magnetic field lines in a runaway manner, which accelerates particles that then radiate synchrotron γ-ray photons at radii of $\sim 10^{15}$-10^{16} cm [452, 453].

9.2. Shock Breakout Models. It has long been believed that when the diffusion timescale of photons at the shock-wave front is comparable to the dynamical timescale, a SBO can occur (see as well Section 3). The SBO of a CCSN can produce a brief and bright flash whose spectral energy distribution peaks in the near UV or X-ray regimes [57, 60, 62, 73, 83, 457–464]. When the SN progenitor is a red supergiant whose radius is larger than several hundred $R_{\odot}$, the SBO is nonrelativistic (Newtonian) [459–461] and the emission is dominated by optical and UV radiation, which is detectable with space telescopes [465–467]. When the explosion is energetic enough, and the progenitor is a WR star whose radius is of order a few solar radii, the SBO emission typically peaks at X-rays or soft γ-rays, with a duration of ~10–2000 s. This class of relativistic SBOs can naturally explain some LGRBs [60, 62, 463, 464]. Reference [62] demonstrated that *ll*GRB jets either fail or just barely pierce through the stellar envelope. This choked/stifled jet can also help accelerate the shock to a mildly relativistic velocities (~30,000–100,000 km s^{-1}). In the shock breakout model, the AG emission is produced when the stellar ejecta collides with the CSM, and [464] showed that the data of the afterglows of GRBs 980425, 031203, 060218, and 100316D are in good agreement with the predictions of this model.

10. Future Research

While considerable progress has been made in the field of GRB-SNe, there are still uncertainties related to several aspects of their true nature. Solidifying their role as standardizable candles and cosmological probes requires both more work and considerably more events. Indeed for GRB-SNe to be used as cosmological probes, independent distance measurements to their host galaxies need to be obtained. Sample studies of GRB-SNe are the ideal way to approach this question, and with the hopeful launch of *JWST* in the next few years, their use over larger redshift ranges than SNe Ia could make them appealing cosmological candles. Additional attention is also required to determine the physical configuration and properties of their preexplosion progenitor stars, to help address the question of whether they arise from single versus binary systems. Moreover, further ULGRB-SNe are needed to address the question of whether all are ultra-luminous compared with typical GRB-SNe, as seen for SN 2011kl, or whether this event is quite anomalous.

10.1. Role of Binarity. Throughout this review, discussions of their stellar progenitors were primarily focused on single-star candidates. However the role of binarity may prove to be one of the most important ingredients to eventually producing a GRB. Theoretically, there are strong motivations for considering a binary evolution. To date, the best theoretical stellar models find it hard to produce enough angular momentum in the core at the time of collapse to make a centrifugally supported disk, though some progress has been made [382, 383, 468]. Instead, it is possible to impart angular momentum into the core of a star through the inspiral of a companion star during a common envelope phase (CEP) [469]: that is, converting orbital angular momentum into core angular momentum. The general idea is to consider a binary system comprised of, among others, a red supergiant and a NS [217], a NS with the He core of a massive star [470], or the merger of two helium stars [471]. During the inspiral of the compact object into the secondary/companion, angular momentum is imparted to the core, which is spun up via disc accretion. During this process, the core of the secondary will increase in mass as well as gain additional angular momentum, while the inspiralling NS will also accrete gas via the Bondi-Hoyle mechanism, which can lead to the NS reaching periods of order milliseconds before it eventually merges with the secondary's core. If a merger of the NS with the core occurs, a collapsar can be created, where a GRB can be produced depending on the initial mass of the secondary, the spin of the newly formed BH, and the amount of angular momentum imparted to the BH.

For the binary model to be a viable route for LGRB formation, one or more mechanism is required to expel the outer envelopes out into space prior to explosion. Generally there are different ways for this to be achieved, either through noncontact methods such as stellar winds, through semi-contact processes such as Roche lobe overflow, or through contact mechanisms that operate during a CEP. The spin-rates of a small sample of O-type star and WR binaries indicate that Roche lobe overflow mass transfer from the WR progenitor companion may play a critical role in the evolution of WR–O-star binaries, where equatorial rotational velocities of 140–500 km s^{-1} have been measured [472]. In the CE scenario, during the inspiral, the orbital separation decreases via drag forces inside the envelope which also results in a loss of kinetic energy. Some of this energy is lost to the surrounding envelope, which heats up and expands. Over a long-enough period the entire envelope can be lost into space. Another mechanism to expel the CE arises via nuclear energy rather than orbital energy [473]. For example, during the slow merger of a massive primary that has completed helium-core burning with a 1–3 $M_{\odot}$ secondary, H-rich material from the secondary is injected into the He-burning shell of the primary. This leads to nuclear runaway and the explosive ejection of the H and He envelopes and produces a binary comprised of a CO star and the low-mass companion. Should a further merger occur, this could lead to the formation of a GRB. If GRB-SNe arise via this formation channel, then this scenario can naturally explain why GRB-SNe are all of type Ic.

A generalization of the binary-merger model is that the more massive the stars are, the more accretion will occur.

This in turn leads to more convection in the core, which results in larger magnetic fields being generated and hence more magnetic collimation for any jets that are produced. In the case of GRB-SNe versus SNe Ibc, if jets are ubiquitous, then the difference between them may be the mass of the merging stars, where lower masses imply lower magnetic fields and hence less collimation. Moreover, the mass ratio of the secondary to the primary is also important, where higher mass ratios will result in more asymmetric explosions [469].

There is a growing list of observations that show that most massive stars exist in binaries, including [474] who estimated that over 70% of all massive stars will exchange mass with a companion star, which in a third of all cases will lead to a merger of the binary system. Moreover, closely orbiting binaries are more common at lower metallicities [475], where the progenitors of GRBs are normally found (though see [476] who showed that the close binary frequency of WRs is not metallicity dependent). Additional support for the notion that the progenitors of SNe Ibc are massive stars in binary systems has come from [477] who argued that, for a standard initial-mass function, the observed abundances of the different types of CCSNe are not consistent with expectations of single-star evolution. Progenitor nondetection of 10 SNe Ibc strongly indicates that single massive WR stars cannot be their solitary progenitor channel [478]. Reference [479] derived a 15% probability that all SNe Ibc arise from single-star WR progenitors. The large gas velocity dispersions measured for the host galaxies of GRBs by [342] may imply the efficient formation of tight massive binary progenitor systems in dense star-forming regions. Rotationally supported galaxies that are more compact and have dense mass configurations are expected to have higher velocity dispersions. Observations of extra galactic star clusters show evidence that bound-cluster formation efficiency increases with star-formation density [480, 481]. Binaries may form more frequently in bound clusters, and they evolve to become more tightly bound through dynamical interactions with other members of the cluster. Alternatively, if the progenitors of GRBs are actually single stars, but which are more massive than those that produce SNe Ibc and II, a top-heavy initial-mass function (IMF) in dense, highly star-forming regions can also explain their observations. A similar conclusion was made by [343], who suggested that if the progenitors of SLSNe are single stars, the extreme emission-line galaxies in which they occur may indicate a bottom-light IMF in these systems. However, observations of low-mass stars in elliptical galaxies that are thought to have undergone high star-formation densities in their star-forming epochs instead suggest that the IMF is bottom heavy [482, 483].

A major hurdle therefore is finding ways to provide observational evidence to distinguish between single and binary progenitors. One such indication may be idea that the progenitors of GRB-SNe are "runaway" stars: that is, massive stars ejected from compact massive-star clusters [484, 485]. This notation was prompted by the observation that the very nearest GRB-SNe, which can be spatially resolved in their host galaxies, are offset from the nearest sites of star formation by 400–800 pc. If GRB-SNe do arise from runaway stars, the lack of obvious wind-features in AG modelling (Section 7.2) can naturally be explained: simulations [486] suggest that a high density of OB stars is required to produce the r^{-2} wind profile, in the region of 10^4-10^5 OB stars within a few tens of parsecs. This is a much larger density than has been observed in nature, where the densest known cluster is R136 (e.g., [487, 488]) which contains many of the most massive and luminous stars known, including R136a1 ($M \sim 315\,M_{\odot}$, $L \sim 8.7 \times 10^6\,L_{\odot}$). Within the central five parsecs of R136 there are 32 of the hottest known type O stars (spectral type O2.0–3.5), 40 other O stars, and 12 Wolf-Rayet stars, mostly of the extremely luminous WNh type (which are still burning hydrogen in their cores and have nitrogen at their surfaces).

For non-GRB related SNe, such as the very nearby peculiar type II SN 1987A (in the LMC, $D \sim 50$ kpc), constraining the nature of its progenitor was made possible due to a combination of a spatially resolved SN remnant and an enormously rich photometric and spectroscopic dataset compiled over a time-span of nearly three decades. These observations have shown that the most likely progenitor of SN 1987A was the merger of a binary system [240, 489], which can explain the triple-ringed structure seen in *HST* images [490], as well as explain the He-enriched outer layers of the blue supergiant progenitor [491]. It was also shown that type IIb SN 1993J (~3.5 Mpc) likely originated from a binary system via analysis of its early LC [492], hydrodynamical modelling [493], and by detection of the preexplosion progenitor star in spatially resolved *HST* images [494] and a possible companion [495]. The direct imaging revealed the progenitor was a red supergiant, where excess of UV and *B*-band flux implied the presence of a hot stellar companion, or it was embedded in an unresolved young stellar cluster. These studies are possible because of the close proximity of the SNe to our vantage point as observers on Earth. However, the nearest GRB to date is GRB 980425, which, at ~40 Mpc, means the progenitor is too distant to be direct imaged. For any progress to be made concerning single versus binary progenitors, nearby events are required that will allow either for exceptionally detailed observations to be obtained and modelled or even the remote chance of directly detecting the progenitor. For lack of better ideas, what we then require is a healthy dose of patience.

10.2. From GRB-SNe to ULGRB-SNe to SLSNe. As discussed previously, the most luminous GRB-SNe to date is SN 2011kl, which had a peak absolute magnitude of ≈−20 mag [496]. This is roughly 0.5–1.0 mag brighter than most GRB-SNe, but still one magnitude fainter than those associated with SLSNe, which peaked at ≈−21 mag [104]. Moreover, it appears that SN 2011kl is not the only object that falls in this gap between ordinary SNe and SLSNe: four objects discovered by PTF and the SNLS have similar peak absolute magnitudes and LC evolution as SN 2011kl [497]. No accompanying γ-ray emission was detected for any of these events, which begs the question of whether they are off-axis ULGRB-SNe or represent yet another type of explosion transient.

In contrast to the cases of GRB-SNe whose optical light curves appear to be mainly powered by heating arising from ^{56}Ni decay, it seems that most SLSNe cannot be explained by the simple radioactive heat deposition model. Instead the luminosity of SLSNe appears to be either driven by energy input from a magnetar [226, 227] or powered by the interaction between SN ejecta and the CSM, which is the likely mechanism for SNe IIn. Indeed, one could argue that the magnetar model is the most promising model to explain the luminosity of SLSNe-Ic. For the most luminous SNe Ic, such as SN 2010ay [498] and SN 2011kl, if the former event arose from radioactive heating, the ratio of the inferred nickel mass to the total ejecta mass was too large, implying that the radioactive heat deposition model was not a viable model. Instead it is possible that events such as this could be powered by both nickel decay and a magnetar [499]. Then, for true SLSNe-Ic, nickel heating can be ignored, and conversely for SNe Ic, including GRB-SNe, magnetar input is negligible. It is only for SNe Ic (all types) with peak absolute magnitudes that exceed ≈ -20 mag that both energy sources must be considered. Clearly more observations of luminous SNe Ic are needed to test this hypothesis.

One final point of interest is determining whether all ULGRB-SNe are superluminous compared with GRB-SNe or whether GRB 111209A/SN 2011kl is a one-off event. As stated previously, the number of GRB-SNe is very small, and only two are considered here: the aforementioned case and ULGRB 101225A. Modelling of the (observer-frame) i-band LC of the accompanying SN in the latter event showed that its brightness was not exceptional: we found $k = 0.96 \pm 0.05$ and $s = 1.02 \pm 0.03$ (Table 3), which implies that some ULGRB-SNe have luminosities that are similar to those of other GRB-SNe. Moreover, the definition of an ULGRB is important: here we have defined an ULGRB as an event that is still detected after several thousand seconds by a gamma-ray instrument. This definition is inherently detector- and redshift-dependent. Based on this definition alone, it appears that GRB 091127 is also an ULGRB; an inspection of the third version of the *Swift*/BAT catalog [500] reveals that this event was detected by BAT at more than 5000 s. In turn, accompanying SN 2009nz is also quite typical of the general GRB-SN population, with $k = 0.89 \pm 0.01$ and $s = 0.88 \pm 0.01$. However, our definition is of course limited and does not include additional facts of this situation: First, the BAT detection at the late times is very marginal, with a signal-to-noise ratio of just 4.36 (where a value of 7.0 is required in a typical image-trigger threshold). Secondly, the BAT event data value of T_{90} is only 7.42 s, whereas the BAT value in [500] is obtained in survey mode. Thus an alternative interpretation of the extended GRB emission seen in the survey data is that it is soft gamma-rays emitted by the very bright X-ray afterglow and not from the prompt emission. In summary, more unambiguous ULGRB events at redshifts lower than unity are needed in order to measure the properties of their accompanying SN and address the peculiar nature of GRB 111209A/SN 2011kl.

Competing Interests

The authors declare that there is no conflict of interests regarding the publication of this manuscript. This applies both to the scientific content of this work and to their funding.

Acknowledgments

Zach Cano thanks David Alexander "Dr Data" Kann, Steve Schulze, Maryam Modjaz, Jens Hjorth, Jochen Greiner, Elena Pian, Vicki Toy, and Antonio de Ugarte Postigo for sharing their photometric and spectroscopic data and stimulating discussions, which without a doubt led to a much improved manuscript. Some data was extracted using the Dexter applet [501], while others were downloaded from the WiseREP archive [502]. The work of Zach Cano was partially funded by a Project Grant from the Icelandic Research Fund. Shan-Qin Wang, Zi-Gao Dai, and Xue-Feng Wu are supported by the National Basic Research Program (973 Program) of China (Grant nos. 2014CB845800 and 2013CB834900) and the National Natural Science Foundation of China (Grant nos. 11573014 and 11322328). Xue-Feng Wu is partially supported by the Youth Innovation Promotion Association (2011231) and the Strategic Priority Research Program "Multi-Waveband Gravitational Wave Universe" (Grant no. XDB23000000) of the Chinese Academy of Sciences.

References

[1] T. J. Galama, P. M. Vreeswijk, J. van Paradijs et al., "An unusual supernova in the error box of the γ-ray burst of 25 April 1998," *Nature*, vol. 395, no. 6703, pp. 670–672, 1998.

[2] F. Patat, E. Cappellaro, J. Danziger et al., "The metamorphosis of SN 1998bw," *Astrophysical Journal*, vol. 555, no. 2, pp. 900–917, 2001.

[3] K. Iwamoto, P. A. Mazzali, K. Nomoto et al., "A hypernova model for the supernova associated with the γ-ray burst of 25 April 1998," *Nature*, vol. 395, no. 6703, pp. 672–674, 1998.

[4] J. Hjorth, J. Sollerman, P. Møller et al., "A very energetic supernova associated with the γ-ray burst of 29 March 2003," *Nature*, vol. 423, no. 6942, pp. 847–850, 2003.

[5] K. Z. Stanek, T. Matheson, P. M. Garnavich et al., "Spectroscopic discovery of the supernova 2003dh associated with GRB 030329," *The Astrophysical Journal*, vol. 591, no. 1, pp. L17–L20, 2003.

[6] T. Matheson, P. M. Garnavich, K. Z. Stanek et al., "Photometry and spectroscopy of GRB 030329 and its associated supernova 2003dh: the first two months," *The Astrophysical Journal*, vol. 599, no. 1, pp. 394–407, 2003.

[7] J. Deng, N. Tominaga, P. A. Mazzali, K. Maeda, and K. Nomoto, "On the light curve and spectrum of SN 2003dh separated from the optical afterglow of GRB 030329," *Astrophysical Journal*, vol. 624, no. 2, pp. 898–905, 2005.

[8] N. Gehrels, G. Chincarini, P. Giommi et al., "The swift gamma-ray burst mission," *The Astrophysical Journal*, vol. 611, no. 2, pp. 1005–1020, 2004.

[9] A. Clocchiatti, N. B. Suntzeff, R. Covarrubias, and P. Candia, "The ultimate light curve of SN 1998bw/GRB 980425," *Astronomical Journal*, vol. 141, no. 5, article 163, 2011.

[10] Z. Cano, D. Bersier, C. Guidorzi et al., "A tale of two GRB-SNe at a common redshift of $z = 0.54$," *Monthly Notices of the Royal Astronomical Society*, vol. 413, no. 1, pp. 669–685, 2011.

[11] S. E. Woosley and J. S. Bloom, "The supernova-gamma-ray burst connection," *Annual Review of Astronomy and Astrophysics*, vol. 44, pp. 507–556, 2006.

[12] L. Wang and J. C. Wheeler, "The supernova-gamma-ray burst connection," *Astrophysical Journal*, vol. 504, no. 2, pp. L87–L90, 1998.

[13] K. Nomoto, N. Tominaga, M. Tanaka et al., "Diversity of the supernova—gamma-ray burst connection," *Nuovo Cimento B Serie*, vol. 121, pp. 1207–1222, 2006.

[14] E. Bissaldi, F. Calura, F. Matteucci, F. Longo, and G. Barbiellini, "The connection between gamma-ray bursts and supernovae Ib/c," *Astronomy & Astrophysics*, vol. 471, no. 2, pp. 585–597, 2007.

[15] M. D. Valle, "Supernovae and gamma-ray bursts: a decade of observations," *International Journal of Modern Physics D*, vol. 20, no. 10, pp. 1745–1754, 2011.

[16] M. Modjaz, "Stellar forensics with the supernova-GRB connection—ludwig Biermann Award Lecture 2010," *Astronomische Nachrichten*, vol. 332, no. 5, pp. 434–447, 2011.

[17] J. Hjorth and J. S. Bloom, "The gamma-ray burst—supernova connection," in *Gamma-Ray Bursts*, pp. 169–190, Cambridge University Press, Cambridge, UK, 2012.

[18] Z. Cano, "A new method for estimating the bolometric properties of Ibc supernovae," *Monthly Notices of the Royal Astronomical Society*, vol. 434, no. 2, pp. 1098–1116, 2013.

[19] J. Hjorth, "The supernova-gamma-ray burst-jet connection," *Philosophical Transactions of the Royal Society A: Mathematical, Physical and Engineering Sciences*, vol. 371, no. 1992, Article ID 20120275, 2013.

[20] P. A. R. Ade, N. Aghanim, C. Armitage-Caplan et al., "Planck 2013 results. XVI. Cosmological parameters," *A & A*, vol. 571, article A16, 2013.

[21] D. J. Schlegel, D. P. Finkbeiner, and M. Davis, "Maps of dust infrared emission for use in estimation of reddening and cosmic microwave background radiation foregrounds," *The Astrophysical Journal Letters*, vol. 500, no. 2, pp. 525–553, 1998.

[22] E. F. Schlafly and D. P. Finkbeiner, "Measuring reddening with Sloan Digital Sky Survey stellar spectra and recalibrating SFD," *Astrophysical Journal*, vol. 737, no. 2, article no. 103, 2011.

[23] D. A. Kann, S. Klose, B. Zhang et al., "The afterglows of Swift-era gamma-ray bursts. I. Comparing pre-Swift and Swift-era long/soft (type II) GRB optical afterglows," *Astrophysical Journal*, vol. 720, no. 2, pp. 1513–1558, 2010.

[24] D. A. Kann, S. Klose, and A. Zeh, "Signatures of extragalactic dust in PRE-swift GRB afterglows," *The Astrophysical Journal*, vol. 641, no. 2, pp. 993–1009, 2006.

[25] D. A. Kann, S. Klose, B. Zhang et al., "The afterglows of swift-era gamma-ray bursts. II. Type I GRB versus type II GRB optical afterglows," *The Astrophysical Journal*, vol. 734, no. 2, article 96, 2011.

[26] J. Japelj, S. Covino, A. Gomboc et al., "Spectrophotometric analysis of gamma-ray burst afterglow extinction curves with X-Shooter," *Astronomy and Astrophysics*, vol. 579, article no. A74, 2015.

[27] C. Alard and R. H. Lupton, "A method for optimal image subtraction," *Astrophysical Journal*, vol. 503, no. 1, pp. 325–331, 1998.

[28] C. Alard, "Image subtraction using a space-varying kernel," *Astronomy and Astrophysics Supplement Series*, vol. 144, no. 2, pp. 363–370, 2000.

[29] L.-G. Strolger, A. G. Riess, T. Dahlen et al., "The *hubble* higher z supernova search: supernovae to $z \approx 1.6$ and constraints on type Ia progenitor models," *The Astrophysical Journal*, vol. 613, no. 1, pp. 200–223, 2004.

[30] Z. Cano, D. Bersier, C. Guidorzi et al., "XRF 100316D/SN 2010bh and the nature of gamma-ray burst supernovae," *Astrophysical Journal*, vol. 740, no. 1, article no. 41, 2011.

[31] Z. Cano, A. De Ugarte Postigo, A. Pozanenko et al., "A trio of gamma-ray burst supernovae: GRB 120729A, GRB 130215A/SN 2013ez, and GRB 130831A/SN 2013fu," *Astronomy and Astrophysics*, vol. 568, article no. A19, 2014.

[32] Z. Cano, A. de Ugarte Postigo, D. Perley et al., "GRB 140606B/iPTF14bfu: detection of shock-breakout emission from a cosmological γ-ray burst?" *Monthly Notices of the Royal Astronomical Society*, vol. 452, no. 2, pp. 1535–1552, 2015.

[33] P. Ferrero, D. A. Kann, A. Zeh et al., "The GRB 060218/SN 2006aj event in the context of other gamma-ray burst supernovae," *Astronomy & Astrophysics*, vol. 457, no. 3, pp. 857–864, 2006.

[34] C. C. Thöne, A. De Ugarte Postigo, C. L. Fryer et al., "The unusual γ-ray burst GRB 101225A from a helium star/neutron star merger at redshift 0.33," *Nature*, vol. 480, no. 7375, pp. 72–74, 2011.

[35] J. Sollerman, J. P. U. Fynbo, J. Gorosabel et al., "The nature of the X-ray flash of August 24 2005. Photometric evidence for an on-axis z = 0.83 burst with continuous energy injection and an associated supernova?" *Astronomy and Astrophysics*, vol. 466, no. 3, pp. 839–846, 2007.

[36] J. Greiner, P. A. Mazzali, D. A. Kann et al., "A very luminous magnetar-powered supernova associated with an ultra-long γ-ray burst," *Nature*, vol. 523, no. 7559, pp. 189–192, 2015.

[37] K. Beuermann, F. V. Hessman, K. Reinsch et al., "VLT observations of GRB 990510 and its environment," *Astronomy and Astrophysics*, vol. 352, no. 1, pp. L26–L30, 1999.

[38] A. Zeh, S. Klose, and D. H. Hartmann, "A systematic analysis of supernova light in gamma-ray burst afterglows," *Astrophysical Journal*, vol. 609, no. 2, pp. 952–961, 2004.

[39] K. Z. Stanek, P. M. Garnavich, P. A. Nutzman et al., "Deep photometry of grb 041006 afterglow: hypernova bump at redshift $z = 0.716$," *Astrophysical Journal*, vol. 626, no. 1, pp. L5–L9, 2005.

[40] D. Xu, A. De Ugarte Postigo, G. Leloudas et al., "Discovery of the broad-lined type IC SN 2013CQ associated with the very energetic GRB 130427A," *Astrophysical Journal*, vol. 776, no. 2, article no. 98, 2013.

[41] S. Schulze, D. Malesani, A. Cucchiara et al., "GRB 120422A/SN 2012bz: bridging the gap between low- and high-luminosity gamma-ray bursts," *Astronomy & Astrophysics*, vol. 566, article A102, 2014.

[42] Z. Cano, "Gamma-ray burst supernovae as standardizable candles," *Astrophysical Journal*, vol. 794, no. 2, article no. 121, 2014.

[43] V. L. Toy, S. B. Cenko, J. M. Silverman et al., "Optical and near-infrared observations of SN 2013dx associated with GRB 130702A," *Astrophysical Journal*, vol. 818, article 79, 2016.

[44] Z. Cano, P. Jakobsson, and O. Pall Geirsson, "Hubble diagrams of relativistic broad-lined type Ic supernovae," https://arxiv.org/abs/1409.3570.

[45] G. Bazin, V. Ruhlmann-Kleider, N. Palanque-Delabrouille et al., "Photometric selection of type Ia supernovae in the supernova legacy survey," *Astronomy and Astrophysics*, vol. 534, article A43, 2011.

[46] M. Modjaz, Y. Q. Liu, F. B. Bianco, and O. Graur, "The Spectral SN-GRB connection: systematic spectral comparisons between Type Ic Supernovae, and broad-lined Type Ic supernovae with

and without gamma-ray bursts," *The Astrophysical Journal*, vol. 832, no. 2, p. 108, 2015.

[47] P. A. Mazzali, K. Nomoto, F. Patat, and K. Maeda, "The nebular spectra of the hypernova SN 1998bw and evidence for asymmetry," *The Astrophysical Journal*, vol. 559, no. 2, pp. 1047–1053, 2001.

[48] K. Maeda, K. Kawabata, M. Tanaka et al., "SN 2006aj associated with XRF 060218 at late phases: nucleosynthesis signature of a neutron star-driven explosion," *Astrophysical Journal*, vol. 658, no. 1, pp. L5–L8, 2007.

[49] F. Bufano, E. Pian, J. Sollerman et al., "The highly energetic expansion of SN 2010bh associated with GRB 100316D," *The Astrophysical Journal*, vol. 753, no. 1, article 67, 2012.

[50] S. Mukherjee, E. D. Feigelson, G. Jogesh Babu, F. Murtagh, C. Fralev, and A. Raftery, "Three types of gamma-ray bursts," *Astrophysical Journal*, vol. 508, no. 1, pp. 314–327, 1998.

[51] I. Horvath, L. G. Balazs, Z. Bagoly, and P. Veres, "Classification of Swift's gamma-ray bursts," *Astronomy and Astrophysics*, vol. 489, no. 1, pp. L1–L4, 2008.

[52] A. de Ugarte Postigo, I. Horváth, P. Veres et al., "Searching for differences in Swift's intermediate GRBs," *Astronomy & Astrophysics*, vol. 525, article no. A109, 2011.

[53] A. J. Levan, N. R. Tanvir, R. L. C. Starling et al., "A new population of ultra-long duration gamma-ray bursts," *The Astrophysical Journal*, vol. 781, artile 13, 2014.

[54] A. J. Levan, "Swift discoveries of new populations of extremely long duration high energy transient," *Journal of High Energy Astrophysics*, vol. 7, pp. 44–55, 2015.

[55] J. Heise, "X-ray flashes and X-ray counterparts of gamm-ray bursts," in *Proceedings of the Gamma-Ray Burst and Afterglow Astronomy 2001: A Workshop Celebrating the First Year of the HETE Mission*, G. R. Ricker and R. K. Vanderspek, Eds., vol. 662 of *American Institute of Physics Conference Series*, pp. 229–236, April 2003.

[56] S. R. Kulkarni, D. A. Frail, M. H. Wieringa et al., "Radio emission from the unusual supernova 1998bw and its association with the γ-ray burst of 25 April 1998," *Nature*, vol. 395, no. 6703, pp. 663–669, 1998.

[57] J. C. Tan, C. D. Matzner, and C. F. McKee, "Trans-relativistic blast waves in supernovae as gamma-ray burst progenitors," *Astrophysical Journal*, vol. 551, no. 2, pp. 946–972, 2001.

[58] S. Campana, V. Mangano, A. J. Blustin et al., "The association of GRB 060218 with a supernova and the evolution of the shock wave," *Nature*, vol. 442, no. 7106, pp. 1008–1010, 2006.

[59] E. Waxman, P. Mészáros, and S. Campana, "GRB 060218: a relativistic supernova shock breakout," *Astrophysical Journal*, vol. 667, no. 1, pp. 351–357, 2007.

[60] E. Nakar and R. Sari, "Relativistic shock breakouts—a variety of gamma-ray flares: from low-luminosity gamma-ray bursts to type Ia supernovae," *Astrophysical Journal*, vol. 747, no. 2, article 88, 2012.

[61] Y. Kaneko, E. Ramirez-Ruiz, J. Granot et al., "Prompt and afterglow emission properties of gamma-ray bursts with spectroscopically identified supernovae," *The Astrophysical Journal*, vol. 654, no. 1 I, pp. 385–402, 2007.

[62] O. Bromberg, E. Nakar, and T. Piran, "Are low-luminosity gamma-ray bursts generated by relativistic jets?" *The Astrophysical Journal Letters*, vol. 739, no. 2, article L55, 2011.

[63] G. Ghisellini, G. Ghirlanda, and F. Tavecchio, "Did we observe the supernova shock breakout in GRB 060218?" *Monthly Notices of the Royal Astronomical Society: Letters*, vol. 382, no. 1, pp. L77–L81, 2007.

[64] C. M. Irwin and R. A. Chevalier, "Jet or shock breakout? The low-luminosity GRB 060218," *Monthly Notices of the Royal Astronomical Society*, vol. 460, no. 2, pp. 1680–1704, 2016.

[65] R. L. C. Starling, K. Wiersema, A. J. Levan et al., "Discovery of the nearby long, soft GRB 100316D with an associated supernova," *Monthly Notices of the Royal Astronomical Society*, vol. 411, no. 4, pp. 2792–2803, 2011.

[66] K. L. Page, R. L. C. Starling, G. Fitzpatrick et al., "GRB 090618: detection of thermal X-ray emission from a bright gamma-ray burst," *Monthly Notices of the Royal Astronomical Society*, vol. 416, no. 3, pp. 2078–2089, 2011.

[67] M. Sparre and R. L. C. Starling, "A search for thermal X-ray signatures in gamma-ray bursts—II. The swift sample," *Monthly Notices of the Royal Astronomical Society*, vol. 427, no. 4, pp. 2965–2974, 2012.

[68] R. L. C. Starling, K. L. Page, A. Pe'er, A. P. Beardmore, and J. P. Osborne, "A search for thermal X-ray signatures in gamma-ray bursts—I. Swift bursts with optical supernovae," *Monthly Notices of the Royal Astronomical Society*, vol. 427, no. 4, pp. 2950–2964, 2012.

[69] A. Pe'er, P. Mészáros, and M. J. Rees, "Radiation from an expanding cocoon as an explanation of the steep decay observed in GRB early afterglow light curves," *Astrophysical Journal*, vol. 652, no. 1 I, pp. 482–489, 2006.

[70] M. Friis and D. Watson, "Thermal emission in the early X-ray afterglows of gamma-ray bursts: following the prompt phase to late times," *The Astrophysical Journal*, vol. 771, no. 1, article 15, 2013.

[71] L.-X. Li, "Shock breakout in Type Ibc supernovae and application to GRB 060218/SN 2006aj," *Monthly Notices of the Royal Astronomical Society*, vol. 375, no. 1, pp. 240–256, 2007.

[72] T. A. Weaver, "The structure of supernova shock waves," *The Astrophysical Journal Supplement Series*, vol. 32, pp. 233–282, 1976.

[73] B. Katz, R. Budnik, and E. Waxman, "Fast radiation mediated shocks and supernova shock breakouts," *Astrophysical Journal*, vol. 716, no. 1, pp. 781–791, 2010.

[74] Y. Ohtani, A. Suzuki, and T. Shigeyama, "Generation of high-energy photons at ultra-relativistic shock breakout in supernovae," *The Astrophysical Journal*, vol. 777, no. 2, article 113, 2013.

[75] L. P. Singer, M. M. Kasliwal, S. B. Cenko et al., "The needle in the 100 deg^2 haystack: uncovering afterglows of *Fermi* GRBs with the palomar transient factory," *The Astrophysical Journal*, vol. 806, no. 1, article 52, 2015.

[76] R. Margutti, C. Guidorzi, D. Lazzati et al., "Dust in the wind: the role of recent mass loss in long gamma-ray bursts," *Astrophysical Journal*, vol. 805, no. 2, article 159, 2015.

[77] E. Nakar, "A unified picture for low-luminosity and long gamma-ray bursts based on the extended progenitor of llGRB 060218/SN 2006aj," *The Astrophysical Journal*, vol. 807, p. 172, 2015.

[78] R. Margutti, A. M. Soderberg, M. H. Wieringa et al., "The signature of the central engine in the weakest relativistic explosions: GRB 100316D," *Astrophysical Journal*, vol. 778, no. 1, article no. 18, 2013.

[79] A. M. Soderberg, S. Chakraborti, G. Pignata et al., "A relativistic type Ibc supernova without a detected γ-ray burst," *Nature*, vol. 463, no. 7280, pp. 513–515, 2010.

[80] D. Xu, D. Watson, J. Fynbo et al., "Mildly relativistic X-ray transient 080109 and SN 2008D: towards a continuum from energetic GRB/XRF to ordinary Ibc SN," in *Proceedings of the*

37th COSPAR Scientific Assembly, vol. 37 of *COSPAR Meeting*, p. 3512, 2008.

[81] R. Margutti, D. Milisavljevic, A. M. Soderberg et al., "Relativistic supernovae have shorter-lived central engines or more extended progenitors: the case of SN 2012ap," *The Astrophysical Journal*, vol. 797, article 107, 2014.

[82] A. Sakurai, "On the problem of a shock wave arriving at the edge of a gas," *Communications on Pure and Applied Mathematics*, vol. 13, pp. 353–370, 1960.

[83] C. D. Matzner and C. F. Mckee, "The expulsion of stellar envelopes in core-collapse supernovae," *Astrophysical Journal*, vol. 510, no. 1, pp. 379–403, 1999.

[84] D. Lazzati, B. J. Morsony, C. H. Blackwell, and M. C. Begelman, "Unifying the zoo of jet-driven stellar explosions," *Astrophysical Journal*, vol. 750, no. 1, article no. 68, 2012.

[85] L. Amati, F. Frontera, M. Tavani et al., "Intrinsic spectra and energetics of BeppoSAX Gamma-Ray Bursts with known redshifts," *Astronomy & Astrophysics*, vol. 390, pp. 81–89, 2002.

[86] L. Amati, M. Della Valle, F. Frontera et al., "On the consistency of peculiar GRBs 060218 and 060614 with the E p,i—E iso correlation," *Astronomy and Astrophysics*, vol. 463, no. 3, pp. 913–919, 2007.

[87] L. Amati, C. Guidorzi, F. Frontera et al., "Measuring the cosmological parameters with the $E_{p,i}$–E_{iso} correlation of gamma-ray bursts," *Monthly Notices of the Royal Astronomical Society*, vol. 391, no. 2, pp. 577–584, 2008.

[88] A. Shemi and T. Piran, "The appearance of cosmic fireballs," *Astrophysical Journal*, vol. 365, no. 2, pp. L55–L58, 1990.

[89] S. Chakraborti, A. Soderberg, L. Chomiuk et al., "A missing-link in the supernova-GRB connection: the case of SNp," *The Astrophysical Journal*, vol. 805, no. 2, p. 187, 2015.

[90] A. M. Soderberg, S. R. Kulkarni, E. Nakar et al., "Relativistic ejecta from X-ray flash XRF 060218 and the rate of cosmic explosions," *Nature*, vol. 442, no. 7106, pp. 1014–1017, 2006.

[91] B.-B. Zhang, Y.-Z. Fan, R.-F. Shen et al., "GRB 120422A: a low-luminosity gamma-ray burst driven by a central engine," *The Astrophysical Journal*, vol. 756, article 190, 2012.

[92] O. Bromberg, E. Nakar, T. Piran, and R. Sari, "An observational imprint of the Collapsar model of long gamma-ray bursts," *Astrophysical Journal*, vol. 749, no. 2, article no. 110, 2012.

[93] T. Piran, O. Bromberg, E. Nakar, and R. Sari, "The long, the short and the weak: the origin of gamma-ray bursts," *Philosophical Transactions of the Royal Society A*, vol. 371, no. 1992, 2013.

[94] O. Bromberg, E. Nakar, T. Piran, and R. Sari, "Short versus long and collapsars versus non-collapsars: a quantitative classification of gamma-ray bursts," *The Astrophysical Journal*, vol. 764, no. 2, article 179, 2013.

[95] P. Kumar and A. Panaitescu, "Afterglow emission from naked gamma-ray bursts," *The Astrophysical Journal*, vol. 541, no. 2, pp. L51–L54, 2000.

[96] E. Nakar and T. Piran, "Outliers to the peak energy-isotropic energy relation in gamma-ray bursts," *Monthly Notices of the Royal Astronomical Society: Letters*, vol. 360, no. 1, pp. L73–L76, 2005.

[97] C. Firmani, J. I. Cabrera, V. Avila-Reese et al., "Time-resolved spectral correlations of long-duration γ-ray bursts," *Monthly Notices of the Royal Astronomical Society*, vol. 393, no. 4, pp. 1209–1218, 2009.

[98] A. Shahmoradi and R. J. Nemiroff, "Hardness as a spectral peak estimator for gamma-ray bursts," *Monthly Notices of the Royal Astronomical Society*, vol. 407, no. 4, pp. 2075–2090, 2010.

[99] A. Shahmoradi and R. J. Nemiroff, "The possible impact of gamma-ray burst detector thresholds on cosmological standard candles," *Monthly Notices of the Royal Astronomical Society*, vol. 411, no. 3, pp. 1843–1856, 2011.

[100] G. Ghirlanda, L. Nava, and G. Ghisellini, "Spectral-luminosity relation within individual Fermi gamma rays bursts," *Astronomy & Astrophysics*, vol. 511, no. 1, article A43, 2010.

[101] G. Ghirlanda, "Gamma Ray Bursts Spectral-Energy correlations: recent results," in *Proceedings of the Jets at All Scales*, G. E. Romero, R. A. Sunyaev, and T. Belloni, Eds., vol. 275 of *IAU Symposium*, pp. 344–348, 2011.

[102] A. C. Collazzi, B. E. Schaefer, A. Goldstein, and R. D. Preece, "A significant problem with using the Amati relation for cosmological purposes," *The Astrophysical Journal*, vol. 747, no. 1, article 39, 2012.

[103] R. Preece, A. Goldstein, N. Bhat et al., "Which E peak?-the characteristic energy of gamma-ray burst spectra," *The Astrophysical Journal*, vol. 821, no. 1, article 12, 2016.

[104] A. Gal-Yam, "Luminous supernovae," *Science*, vol. 337, no. 6097, pp. 927–932, 2012.

[105] B. D. Metzger, B. Margalit, D. Kasen, and E. Quataert, "The diversity of transients from magnetar birth in core collapse supernovae," *Monthly Notices of the Royal Astronomical Society* , vol. 454, no. 3, pp. 3311–3316, 2015.

[106] Z. Cano, K. G. Johansson Andreas, and K. Maeda, "A self-consistent analytical magnetar model: the luminosity of γ-ray burst supernovae is powered by radioactivity," *Monthly Notices of the Royal Astronomical Society*, vol. 457, no. 3, Article ID stw122, pp. 2761–2772, 2016.

[107] M. C. Bersten, O. G. Benvenuto, M. Orellana, and K. Nomoto, "The unusual super-luminous supernovae SN 2011KL and asassn-15LH," *Astrophysical Journal Letters*, vol. 817, no. 1, article L8, 2016.

[108] C. Inserra and S. J. Smartt, "Superluminous supernovae as standardizable candles and high-redshift distance probes," *The Astrophysical Journal*, vol. 796, no. 2, p. 87, 2014.

[109] J.-J. Wei, X.-F. Wu, and F. Melia, "Testing cosmological models with type Ic super luminous supernovae," *Astronomical Journal*, vol. 149, no. 5, article no. 165, 2015.

[110] K. Maeda, T. Nakamura, K. Nomoto, P. A. Mazzali, F. Patat, and I. Hachisu, "Explosive nucleosynthesis in aspherical hypernova explosions and late-time spectra of SN 1998bw," *Astrophysical Journal*, vol. 565, no. 1, pp. 405–412, 2002.

[111] K. Maeda, "Three-dimensional simulation of gamma-ray emission from asymmetric supernovae and hypernovae," *Astrophysical Journal*, vol. 644, no. 1, pp. 385–399, 2006.

[112] S. E. Woosley and A. Heger, "The light curve of the unusual supernova SN 2003dh," https://arxiv.org/abs/astro-ph/0309165.

[113] P. A. Mazzali, J. Deng, K. Nomoto et al., "A neutron-star-driven X-ray flash associated with supernova SN 2006aj," *Nature*, vol. 442, no. 7106, pp. 1018–1020, 2006.

[114] S. E. Woosley, R. G. Eastman, and B. P. Schmidt, "Gamma-ray bursts and type Ic supernova SN 1998bw," *The Astrophysical Journal*, vol. 516, no. 2, pp. 788–796, 1999.

[115] P. Höflich, J. C. Wheeler, and L. Wang, "Aspherical explosion models for SN 1998bw/GRB 980425," *Astrophysical Journal*, vol. 521, no. 1, pp. 179–189, 1999.

[116] K. Iwamoto, "On the radio-to-X-ray light curves of SN 1998bw and GRB 980425," *Astrophysical Journal*, vol. 512, no. 1, pp. L47–L50, 1999.

[117] T. Nakamura, P. A. Mazzali, K. Nomoto, and K. Iwamoto, "Light curve and spectral models for the hypernova SN 1998bw associated with GRB 980425," *Astrophysical Journal*, vol. 550, no. 2, pp. 991–999, 2001.

[118] P. A. Mazzali, J. Deng, N. Tominaga et al., "The type Ic hypernova SN 2003dh/GRB 030329," *The Astrophysical Journal*, vol. 599, no. 2, pp. L95–L98, 2003.

[119] S. Nagataki, A. Mizuta, and K. Sato, "Explosive nucleosynthesis in GRB jets accompanied by hypernovae," *Astrophysical Journal*, vol. 647, no. 2, pp. 1255–1268, 2006.

[120] J. D. Lyman, D. Bersier, P. A. James et al., "Bolometric light curves and explosion parameters of 38 stripped-envelope core-collapse supernovae," *Monthly Notices of the Royal Astronomical Society*, vol. 457, no. 1, pp. 328–350, 2016.

[121] E. F. Olivares, J. Greiner, P. Schady et al., "The fast evolution of SN 2010bh associated with XRF 100316D," *Astronomy and Astrophysics*, vol. 539, article no. A76, 2012.

[122] K. Maeda, P. A. Mazzali, J. Deng et al., "A two-component model for the light curves of hypernovae," *The Astrophysical Journal*, vol. 593, no. 2, pp. 931–940, 2003.

[123] M. Modjaz, W. Li, N. Butler et al., "From shock breakout to peak and beyond: extensive panchromatic observations of the type Ib supernova 2008D associated with swift x-ray transient 080109," *Astrophysical Journal*, vol. 702, no. 1, pp. 226–248, 2009.

[124] D. E. Reichart, "GRB 970228 revisited: evidence for a supernova in the light curve and late spectral energy distribution of the afterglow," *Astrophysical Journal*, vol. 521, no. 2, pp. L111–L115, 1999.

[125] T. J. Galama, N. Tanvir, P. M. Vreeswijk et al., "Evidence for a supernova in reanalyzed optical and near-infrared images of GRB 970228," *The Astrophysical Journal*, vol. 536, no. 1, pp. 185–194, 2000.

[126] J. S. Bloom, S. R. Kulkarni, S. G. Djorgovski et al., "The unusual afterglow of the γ-ray burst of 26 March 1998 as evidence for a supernova connection," *Nature*, vol. 401, no. 6752, pp. 453–456, 1999.

[127] A. J. Castro-Tirado and J. Gorosabel, "Optical observations of GRB afterglows: GRB 970508 and GRB 980326 revisited," *Astronomy and Astrophysics Supplement Series*, vol. 138, no. 3, pp. 449–450, 1999.

[128] X. Li and J. Hjorth, "Light curve properties of supernovae associated with gamma-ray bursts," https://arxiv.org/abs/1407.3506.

[129] G. Björnsson, J. Hjorth, P. Jakobsson, L. Christensen, and S. Holland, "The jet and the supernova in GRB 990712," *The Astrophysical Journal*, vol. 552, no. 2, pp. L121–L124, 2001.

[130] A. J. Castro-Tirado, V. V. Sokolov, J. Gorosabel et al., "The extraordinarily bright optical afterglow of GRB 991208 and its host galaxy," *Astronomy & Astrophysics*, vol. 370, no. 2, pp. 398–406, 2001.

[131] D. Lazzati, S. Covino, G. Ghisellini et al., "The optical afterglow of GRB 000911: evidence for an associated supernova?" *Astronomy and Astrophysics*, vol. 378, no. 3, pp. 996–1002, 2001.

[132] N. Masetti, E. Palazzi, E. Pian et al., "Late-epoch optical and near-infrared observations of the GRB 000911 afterglow and its host galaxy," *Astronomy and Astrophysics*, vol. 438, no. 3, pp. 841–853, 2005.

[133] J. S. Bloom, S. R. Kulkarni, P. A. Price et al., "Detection of a supernova signature associated with GRB 011121," *Astrophysical Journal*, vol. 572, no. 1, pp. L45–L49, 2002.

[134] P. A. Price, E. Berger, D. E. Reichart et al., "GRB 011121: a massive star progenitor," *The Astrophysical Journal Letters*, vol. 572, no. 1, pp. L51–L55, 2002.

[135] P. M. Garnavich, K. Z. Stanek, L. Wyrzykowski et al., "Discovery of the low-redshift optical afterglow of GRB 011121 and its progenitor supernova sn 2001ke," *Astrophysical Journal*, vol. 582, no. 2, pp. 924–932, 2003.

[136] J. Greiner, S. Klose, M. Salvato et al., "GRB 011121: a collimated outflow into wind-blown surroundings," *Astrophysical Journal*, vol. 599, no. 2 I, pp. 1223–1237, 2003.

[137] J. Gorosabel, J. P. U. Fynbo, A. Fruchter et al., "A possible bright blue supernova in the afterglow of GRB 020305," *The Astronomy and Astrophysics*, vol. 437, no. 2, pp. 411–418, 2005.

[138] P. A. Price, S. R. Kulkarni, E. Berger et al., "Discovery of GRB 020405 and its late red bump," *The Astrophysical Journal*, vol. 589, no. 2, pp. 838–843, 2003.

[139] L. Nicastro, J. J. M. In't Zand, L. Amati et al., "Multiwavelength study of the very long GRB 020410," *Astronomy and Astrophysics*, vol. 427, no. 2, pp. 445–452, 2004.

[140] A. Levan, P. Nugent, A. Fruchter et al., "GRB 020410: a gamma-ray burst afterglow discovered by its supernova light," *Astrophysical Journal*, vol. 624, no. 2, pp. 880–888, 2005.

[141] G. Ricker, J.-L. Atteia, N. Kawai et al., "GRB020903(=H2314): an X-ray flash localized by HETE," *GRB Coordinates Network*, vol. 1530, article 1, 2002.

[142] A. M. Soderberg, S. R. Kulkarni, D. B. Fox et al., "An HST search for supernovae accompanying X-ray flashes," *Astrophysical Journal*, vol. 627, no. 2, pp. 877–887, 2005.

[143] D. Bersier, A. S. Fruchter, L.-G. Strolger et al., "Evidence for a supernova associated with the X-ray flash 020903," *Astrophysical Journal*, vol. 643, no. 1, pp. 284–291, 2006.

[144] M. D. Valle, D. Malesani, S. Benetti et al., "Evidence for supernova signatures in the spectrum of the late-time bump of the optical afterglow of GRB 021211," *The Astronomy and Astrophysics*, vol. 406, no. 2, pp. L33–L37, 2003.

[145] S. B. Pandey, G. C. Anupama, R. Sagar et al., "The optical afterglow of the not so dark GRB 021211," *Astronomy and Astrophysics*, vol. 408, no. 3, pp. L21–L24, 2003.

[146] J. S. Bloom, P. G. Van Dokkum, C. D. Bailyn, M. M. Buxton, S. R. Kulkarni, and B. P. Schmidt, "Optical-infrared ANDICAM observations of the transient associated with GRB 030329," *Astronomical Journal*, vol. 127, no. 1, pp. 252–263, 2004.

[147] Y. M. Lipkin, E. O. Ofek, A. Gal-Yam et al., "The detailed optical light curve of GRB 030329," *Astrophysical Journal*, vol. 606, no. 1, pp. 381–394, 2004.

[148] J. P. U. Fynbo, J. Sollerman, J. Hjorth et al., "On the afterglow of the X-ray flash of 2003 July 23: photometric evidence for an off-axis gamma-ray burst with an associated supernova?" *Astrophysical Journal*, vol. 609, no. 2, pp. 962–971, 2004.

[149] N. Tominaga, J. Deng, P. A. Mazzali et al., "Supernova light-curve models for the bump in the optical counterpart of X-ray flash 030723," *Astrophysical Journal*, vol. 612, no. 2, pp. L105–L108, 2004.

[150] G. Pugliese, P. Møller, J. Gorosabel et al., "The red optical afterglow of GRB 030725," *Astronomy and Astrophysics*, vol. 439, no. 2, pp. 527–532, 2005.

[151] D. Malesani, G. Tagliaferri, G. Chincarini et al., "SN 2003lw and GRB 031203: a bright supernova for a faint gamma-ray burst," *Astrophysical Journal*, vol. 609, no. 1, pp. L5–L8, 2004.

[152] A. M. Soderberg, S. R. Kulkarni, E. Berger et al., "The sub-energetic γ-ray burst GRB 031203 as a cosmic analogue to the nearby GRB 980425," *Nature*, vol. 430, no. 7000, pp. 648–650, 2004.

[153] A. Gal-Yam, D.-S. Moon, D. B. Fox et al., "The J-band light curve of SN 2003lw, associated with GRB 031203," *Astrophysical Journal*, vol. 609, no. 2, pp. L59–L62, 2004.

[154] D. Watson, J. Hjorth, A. Levan et al., "A very low luminosity X-ray flash: XMM-Newton observations of GRB 031203," *Astrophysical Journal*, vol. 605, no. 2, pp. L101–L104, 2004.

[155] B. Thomsen, J. Hjorth, D. Watson et al., "The supernova 2003lw associated with X-ray flash 031203," *Astronomy and Astrophysics*, vol. 419, no. 2, pp. L21–L25, 2004.

[156] B. E. Cobb, C. D. Bailyn, P. G. Van Dokkum, and P. Natarajan, "SN 2006aj and the nature of low-luminosity gamma-ray bursts," *Astrophysical Journal*, vol. 645, no. 2, pp. L113–L116, 2006.

[157] K. Wiersema, A. J. Van Der Horst, D. A. Kann et al., "Spectroscopy and multiband photometry of the afterglow of intermediate duration γ-ray burst GRB 040924 and its host galaxy," *Astronomy and Astrophysics*, vol. 481, no. 2, pp. 319–326, 2008.

[158] K. Misra, L. Resmi, S. B. Pandey, D. Bhattacharya, and R. Sagar, "Optical observations and multiband modelling of the afterglow of GRB 041006: evidence of a hard electron energy spectrum," *Bulletin of the Astronomical Society of India*, vol. 33, no. 4, pp. 487–497, 2005.

[159] T. Sakamoto, L. Barbier, S. D. Barthelmy et al., "Confirmation of the E^{src}_{peak}-Eiso (Amati) relation from the X-ray flash XRF 050416A observed by the Swift burst alert telescope," *Astrophysical Journal*, vol. 636, no. 2, pp. L73–L76, 2006.

[160] A. M. Soderberg, E. Nakar, S. B. Cenko et al., "A spectacular radio flare from XRF 050416a at 40 days and implications for the nature of X-ray flashes," *Astrophysical Journal*, vol. 661, no. 2 I, pp. 982–994, 2007.

[161] M. Della Valle, D. Malesani, J. S. Bloom et al., "Hypernova signatures in the late rebrightening of GRB 050525A," *Astrophysical Journal*, vol. 642, no. 2, pp. L103–L106, 2006.

[162] H. Krimm, L. Barbier, S. Barthelmy et al., "GRB 050824: BAT refined analysis of a soft weak burst," *GRB Coordinates Network, Circular Service*, vol. 3871, article 1, 2005.

[163] G. Crew, G. Ricker, J.-L. Atteia et al., "GRB050824, HETE-2 observation," GRB Coordinates Network 3890, 2005.

[164] N. Mirabal, J. P. Halpern, D. An, J. R. Thorstensen, and D. M. Terndrup, "GRB 060218/SN 2006aj: a gamma-ray burst and prompt supernova at z = 0.0335," *The Astrophysical Journal Letters*, vol. 643, no. 2, pp. L99–L102, 2006.

[165] M. Modjaz, K. Z. Stanek, P. M. Garnavich et al., "Early-time photometry and spectroscopy of the fast evolving SN 2006aj associated with GRB 060218," *Astrophysical Journal*, vol. 645, no. 1, pp. L21–L24, 2006.

[166] E. Pian, P. A. Mazzali, N. Masetti et al., "An optical supernova associated with the X-ray flash XRF 060218," *Nature*, vol. 442, pp. 1011–1013, 2006.

[167] J. Sollerman, A. O. Jaunsen, J. P. U. Fynbo et al., "Supernova 2006aj and the associated X-Ray Flash 060218," *Astronomy and Astrophysics*, vol. 454, no. 2, pp. 503–509, 2006.

[168] K. Misra, A. S. Fruchter, and P. Nugent, "Late-time HST observations of XRF 060218/SN 2006aj," in *AIP Conference Proceedings*, J. E. McEnery, J. L. Racusin, and N. Gehrels, Eds., vol. 1358 of *American Institute of Physics Conference Series*, pp. 299–302, 2011.

[169] D. Grupe, C. Gronwall, X.-Y. Wang et al., "SWIFT and *XMM-Newton* observations of the extraordinary gamma-ray burst 060729: more than 125 days of X-ray afterglow," *Astrophysical Journal*, vol. 662, no. 1 I, pp. 443–458, 2007.

[170] N. R. Butler and D. Kocevski, "X-ray hardness evolution in GRB afterglows and flares: late-time GRB activity without NH variations," *Astrophysical Journal*, vol. 663, no. 1, pp. 407–419, 2007.

[171] C. Markwardt, L. Barbier, S. D. Barthelmy et al., "GRB 060904B: Swift-BAT refined analysis," GRB Coordinates Network 5520, 2006.

[172] S. B. Cenko, S. Gezari, T. Small, D. B. Fox, and R. Chornock, "GRB 070419: Keck/LRIS absorption redshift," GRB Coordinates Network 6322, 2007.

[173] J. Hill, P. Garnavich, O. Kuhn et al., "GRB 070419A, deep LBT photometry and possible supernova detection," GRB Coordinates Network 6486, 2007.

[174] D. A. Kann, S. Schulze, and A. C. Updike, "GRB 080319B: jet break, energetics, supernova," GRB Coordinates Network 7627, 2008.

[175] J. S. Bloom, D. A. Perley, W. Li et al., "Observations of the naked-eye GRB 080319B: implications of nature's brightest explosion," *Astrophysical Journal*, vol. 691, no. 1, pp. 723–737, 2009.

[176] N. R. Tanvir, E. Rol, A. J. Levan et al., "Late-time observations of GRB 080319B: jet break, host galaxy, and accompanying supernova," *The Astrophysical Journal*, vol. 725, no. 1, pp. 625–632, 2010.

[177] A. Soderberg, E. Berger, and D. Fox, "GRB 081007: detection of a supernova," GRB Coordinates Network 8662, 2008.

[178] E. Berger, D. B. Fox, A. Cucchiara, and S. B. Cenko, "GRB 081007: Gemini-south redshift," GRB Coordinates Network 8335, 2008.

[179] M. Della Valle, S. Benetti, P. Mazzali et al., "Supernova 2008hw and GRB 081007," Central Bureau Electronic Telegrams 1602, 2008.

[180] Z.-P. Jin, S. Covino, M. Della Valle et al., "GRB 081007 and GRB 090424: the surrounding medium, outflows, and supernovae," *The Astrophysical Journal*, vol. 774, no. 2, p. 114, 2013.

[181] E. F. Olivares, J. Greiner, P. Schady et al., "Multiwavelength analysis of three supernovae associated with gamma-ray bursts observed by GROND," *Astronomy & Astrophysics*, vol. 577, article A44, 2015.

[182] B. E. Cobb, J. S. Bloom, D. A. Perley, A. N. Morgan, S. B. Cenko, and A. V. Filippenko, "Discovery of sn 2009nz associated with grb 091127," *Astrophysical Journal Letters*, vol. 718, no. 2, pp. L150–L155, 2010.

[183] E. Berger, R. Chornock, T. R. Holmes et al., "The spectroscopic classification and explosion properties of SN 2009nz associated with GRB 091127 AT z = 0.490," *Astrophysical Journal*, vol. 743, no. 2, article 204, 2011.

[184] R. Chornock, E. Berger, E. M. Levesque et al., "Spectroscopic discovery of the broad-lined type Ic supernova 2010bh associated with the low-redshift GRB 100316D," https://arxiv.org/abs/1004.2262.

[185] S. T. Holland, F. E. Marshall, M. Page, M. de Pasquale, and M. H. Siegel, "GRB 100418A: possible evidence for a supernova," GRB Coordinates Network 10661, 2010.

[186] F. E. Marshall, L. A. Antonelli, D. N. Burrows et al., "The late peaking afterglow of GRB 100418A," *Astrophysical Journal*, vol. 727, no. 2, 2011.

[187] A. De Ugarte Postigo, C. C. Thöne, P. Goldoni, and J. P. U. Fynbo, "Time resolved spectroscopy of GRB 100418A and its host galaxy with X-shooter," *Astronomische Nachrichten*, vol. 332, no. 3, pp. 297–298, 2011.

[188] M. Sparre, J. Sollerman, J. P. U. Fynbo et al., "Spectroscopic evidence for SN 2010ma associated with GRB 101219B," *The Astrophysical Journal Letters*, vol. 735, no. 1, p. L24, 2011.

[189] S. Golenetskii, R. Aptekar, E. Mazets et al., "Konus-wind observation of GRB 111209A," GRB Coordinates Network 12663, 2011.

[190] A. de Ugarte Postigo, C. C. Thoene, and J. Gorosabel, "GRB 111211A: detection of the SN with the 10.4m GTC," GRB Coordinates Network 12802, 2012.

[191] P. D'Avanzo, A. Melandri, E. Palazzi et al., "GRB 111228A: possible detection of the SN with the TNG," GRB Coordinates Network 13069, 2012.

[192] A. Melandri, E. Pian, P. Ferrero et al., "The optical SN 2012bz associated with the long GRB 120422A," *Astronomy and Astrophysics*, vol. 547, article no. A82, 2012.

[193] S. Klose, J. Greiner, J. Fynbo et al., "Supernova 2012eb = GRB 120714B," Central Bureau Electronic Telegrams 3200, 2012.

[194] J. R. Cummings, S. D. Barthelmy, W. H. Baumgartner et al., "GRB 120714B: swift-BAT refined analysis," GRB Coordinates Network 13481, 2012.

[195] M. De Pasquale, M. J. Page, D. A. Kann et al., "The 80 Ms follow-up of the X-ray afterglow of GRB 130427A challenges the standard forward shock model," *Monthly Notices of the Royal Astronomical Society*, vol. 462, no. 1, pp. 1111–1122, 2016.

[196] A. J. Levan, N. R. Tanvir, A. S. Fruchter et al., "Hubble space telescope observations of the afterglow, supernova, and host galaxy associated with the extremely bright GRB 130427A," *The Astrophysical Journal*, vol. 792, no. 2, article 115, 2014.

[197] A. Melandri, E. Pian, V. D'Elia et al., "Diversity of gamma-ray burst energetics vs. supernova homogeneity: SN 2013cq associated with GRB 130427A," *Astronomy and Astrophysics*, vol. 567, article A29, 2014.

[198] V. D'Elia, E. Pian, A. Melandri et al., "SN 2013dx associated with GRB 130702A: A detailed photometric and spectroscopic monitoring and a study of the environment," *Astronomy and Astrophysics*, vol. 577, article A116, 2015.

[199] S. Klose, A. Nicuesa Guelbenzu, T. Kruehler et al., "Supernova 2013fu = GRB 130831A," Central Bureau Electronic Telegrams 3677, 2013.

[200] A. Pozanenko, E. Mazaeva, A. Sergeev et al., "GRB 150518A: possible SN observations," GRB Coordinates Network 17903, 2015.

[201] A. de Ugarte Postigo, Z. Cano, D. A. Perley et al., "GRB 150818A: spectroscopic confirmation of the SN from GTC," GRB Coordinates Network 18213, 2015.

[202] S. Golenetskii, R. Aptekar, D. Frederiks et al., "Konus-Wind observation of GRB 150818A," GRB Coordinates Network 18198, 2015.

[203] D. M. Palmer, S. D. Barthelmy, J. R. Cummings et al., "GRB 150818A: Swift-BAT refined analysis," GRB Coordinates Network 18157, 2015.

[204] G. Pignata, M. Stritzinger, A. Soderberg et al., "SN 2009bb: a peculiar broad-lined Type Ic supernova," *Astrophysical Journal*, vol. 728, no. 1, 2011.

[205] E. M. Levesque, L. J. Kewley, E. Berger, and H. J. Zahid, "The host galaxies of gamma-ray bursts. II. A mass-metallicity relation for long-duration gamma-ray burst host galaxies," *Astronomical Journal*, vol. 140, no. 5, pp. 1557–1566, 2010.

[206] D. Milisavljevic, R. Margutti, J. T. Parrent et al., "The broad-lined type Ic SN 2012ap and the nature of relativistic supernovae lacking a gamma-ray burst detection," *The Astrophysical Journal*, vol. 799, no. 1, article 51, 2015.

[207] Z. Liu, X.-L. Zhao, F. Huang et al., "Optical observations of the broad-lined type Ic supernova SN 2012ap," *Research in Astronomy and Astrophysics*, vol. 15, no. 2, article 007, pp. 225–236, 2015.

[208] W. D. Arnett, "Type I supernovae. I—analytic solutions for the early part of the light curve," *The Astrophysical Journal*, vol. 253, pp. 785–797, 1982.

[209] K. Maeda and N. Tominaga, "Nucleosynthesis of 56Ni in wind-driven supernova explosions and constraints on the central engine of gamma-ray bursts," *Monthly Notices of the Royal Astronomical Society*, vol. 394, no. 3, pp. 1317–1324, 2009.

[210] A. I. Macfadyen, S. E. Woosley, and A. Heger, "Supernovae, jets, and collapsars," *Astrophysical Journal*, vol. 550, no. 1, pp. 410–425, 2001.

[211] S. E. Woosley and T. A. Weaver, "The physics of supernova explosions," *Annual Review of Astronomy & Astrophysics*, vol. 24, pp. 205–253, 1986.

[212] C. L. Fryer, P. A. Mazzali, J. Prochaska et al., "Constraints on Type Ib/c supernovae and gamma-ray burst progenitors," *Publications of the Astronomical Society of the Pacific*, vol. 119, no. 861, pp. 1211–1232, 2007.

[213] K. Takaki, K. S. Kawabata, M. Yamanaka et al., "A luminous and fast-expanding type Ib supernova SN 2012au," *Astrophysical Journal Letters*, vol. 772, no. 2, article no. L17, 2013.

[214] J. Sollerman, S. T. Holland, P. Challis et al., "Supernova 1998bw—the final phases," *Astronomy and Astrophysics*, vol. 386, no. 3, pp. 944–956, 2002.

[215] A. I. MacFadyen and S. E. Woosley, "Collapsars: gamma-ray bursts and explosions in 'failed supernovae'," *The Astrophysical Journal*, vol. 524, no. 1, pp. 262–289, 1999.

[216] M. Milosavljević, C. C. Lindner, R. Shen, and P. Kumar, "Supernovae powered by collapsar accretion in gamma-ray burst sources," *Astrophysical Journal*, vol. 744, no. 2, article no. 103, 2012.

[217] M. V. Barkov and S. S. Komissarov, "Recycling of neutron stars in common envelopes and hypernova explosions," *Monthly Notices of the Royal Astronomical Society*, vol. 415, no. 1, pp. 944–958, 2011.

[218] T. A. Thompson, "Assessing millisecond proto-magnetars as GRB central engines," *Revista Mexicana de Astronomia y Astrofisica*, vol. 27, pp. 80–90, 2007.

[219] T. A. Thompson, B. D. Metzger, and N. Bucciantini, "Proto-magnetars as GRB central engines: uncertainties, limitations, & particulars," in *Proceedings of the American Institute of Physics Conference Series*, N. Kawai and S. Nagataki, Eds., vol. 1279 of *American Institute of Physics Conference Series*, pp. 81–88, 2010.

[220] Y. Suwa and N. Tominaga, "How much can ^{56}Ni be synthesized by the magnetar model for long gamma-ray bursts and hypernovae?" *Monthly Notices of the Royal Astronomical Society*, vol. 451, no. 1, pp. 282–287, 2015.

[221] B. D. Metzger, T. A. Thompson, and E. Quataert, "Proto-neutron star winds with magnetic fields and rotation," *The Astrophysical Journal*, vol. 659, no. 1 I, pp. 561–579, 2007.

[222] K. Maeda, K. Kawabata, P. A. Mazzali et al., "Asphericity in supernova explosions from late-time spectroscopy," *Science*, vol. 319, no. 5867, pp. 1220–1223, 2008.

[223] J. P. Ostriker and J. E. Gunn, "Do pulsars make supernovae?" *The Astrophysical Journal*, vol. 164, pp. L95–L104, 1971.

[224] B. Zhang and P. Mészáros, "Gamma-ray burst afterglow with continuous energy injection: signature of a highly magnetized millisecond pulsar," *Astrophysical Journal*, vol. 552, no. 1, pp. L35–L38, 2001.

[225] E. Chatzopoulos, J. C. Wheeler, J. Vinko et al., "SN 2008am: a super-luminous type IIn supernova," *The Astrophysical Journal*, vol. 729, article 143, 2011.

[226] C. Inserra, S. J. Smartt, A. Jerkstrand et al., "Super-luminous type Ic supernovae: catching a magnetar by the tail," *The Astrophysical Journal*, vol. 770, no. 2, p. 128, 2013.

[227] M. Nicholl, S. J. Smartt, A. Jerkstrand et al., "Slowly fading super-luminous supernovae that are not pair-instability explosions," *Nature*, vol. 502, pp. 346–349, 2013.

[228] D. Kasen and L. Bildsten, "Supernova light curves powered by young magnetars," *Astrophysical Journal*, vol. 717, no. 1, pp. 245–249, 2010.

[229] S. E. Woosley, "Bright supernovae from magnetar birth," *Astrophysical Journal Letters*, vol. 719, no. 2, pp. L204–L207, 2010.

[230] P. A. Mazzali, M. Sullivan, E. Pian, J. Greiner, and D. A. Kann, "Spectrum formation in superluminous supernovae (Type I)," *Monthly Notices of the Royal Astronomical Society*, vol. 458, no. 4, pp. 3455–3465, 2016.

[231] P. A. Mazzali, K. S. Kawabata, K. Maeda et al., "An asymmetric energetic type Ic supernova viewed off-axis, and a link to gamma ray bursts," *Science*, vol. 308, no. 5726, pp. 1284–1287, 2005.

[232] M. Modjaz, R. P. Kirshner, S. Blondin, P. Challis, and T. Matheson, "Double-peaked oxygen lines are not rare in nebular spectra of core-collapse Supernovae," *The Astrophysical Journal Letters*, vol. 687, pp. L9–L12, 2008.

[233] M. Tanaka, N. Tominaga, K. Nomoto et al., "Type Ib supernova 2008D associated with the luminous X-ray transient 080109: an energetic explosion of a massive helium star," *The Astrophysical Journal*, vol. 692, no. 2, pp. 1131–1142, 2009.

[234] S. Taubenberger, S. Valenti, S. Benetti et al., "Nebular emission-line profiles of Type Ib/c supernovae—probing the ejecta asphericity," *Monthly Notices of the Royal Astronomical Society*, vol. 397, no. 2, pp. 677–694, 2009.

[235] D. Milisavljevic, R. A. Fesen, C. L. Gerardy, R. P. Kirshner, and P. Challis, "Doublets and double peaks: late-time [O I] $\lambda\lambda$6300, 6364 line profiles of stripped-envelope, core-collapse supernovae," *The Astrophysical Journal*, vol. 709, no. 2, pp. 1343–1355, 2010.

[236] C. Fransson and R. A. Chevalier, "Late emission from SN 1987A," *The Astrophysical Journal*, vol. 322, no. 1, pp. L15–L20, 1987.

[237] E. M. Schlegel and R. P. Kirshner, "The type Ib supernova 1984L in NGC 991," *Astronomical Journal*, vol. 98, pp. 577–589, 1989.

[238] K. Maeda and K. Nomoto, "Bipolar supernova explosions: nucleosynthesis and implications for abundances in extremely metal-poor stars," *The Astrophysical Journal*, vol. 598, no. 2, pp. 1163–1200, 2003.

[239] A. M. Khokhlov, P. A. Höflich, E. S. Oran, J. C. Wheeler, L. Wang, and A. Y. Chtchelkanova, "Jet-induced explosions of core collapse supernovae," *Astrophysical Journal*, vol. 524, no. 2, pp. L107–L110, 1999.

[240] T. Morris and P. Podsiadlowski, "The triple-ring nebula around SN 1987A: fingerprint of a binary merger," *Science*, vol. 315, no. 5815, pp. 1103–1106, 2007.

[241] M. Tanaka, K. S. Kawabata, T. Hattori et al., "Three-dimensional explosion geometry of stripped-envelope core-collapse supernovae. I. Spectropolarimetric observations," *The Astrophysical Journal*, vol. 754, no. 1, p. 63, 2012.

[242] L. Wang and J. C. Wheeler, "Spectropolarimetry of supernovae," *Annual Review of Astronomy & Astrophysics*, vol. 46, pp. 433–474, 2008.

[243] S. Akiyama, J. C. Wheeler, D. L. Meier, and I. Lichtenstadt, "The magnetorotational instability in core-collapse supernova explosions," *The Astrophysical Journal*, vol. 584, no. 2, pp. 954–970, 2003.

[244] T. A. Thompson, P. Chang, and E. Quataert, "Magnetar spin-down, hyperenergetic supernovae, and gamma-ray bursts," *Astrophysical Journal*, vol. 611, no. 1, pp. 380–393, 2004.

[245] Y. Masada, T. Sano, and H. Takabe, "Nonaxisymmetric magnetorotational instability in proto-neutron stars," *The Astrophysical Journal*, vol. 641, no. 1, pp. 447–457, 2006.

[246] D. A. Uzdensky and A. I. MacFadyen, "Magnetar-driven magnetic tower as a model for gamma-ray bursts and asymmetric supernovae," *Astrophysical Journal*, vol. 669, no. 1, pp. 546–560, 2007.

[247] S. E. Woosley, "Gamma-ray bursts from stellar mass accretion disks around black holes," *The Astrophysical Journal*, vol. 405, no. 1, pp. 273–277, 1993.

[248] P. Mösta, C. D. Ott, D. Radice, L. F. Roberts, E. Schnetter, and R. Haas, "A large-scale dynamo and magnetoturbulence in rapidly rotating core-collapse supernovae," *Nature*, vol. 528, no. 7582, pp. 376–379, 2015.

[249] O. Papish and N. Soker, "Exploding core collapse supernovae with jittering jets," *Monthly Notices of the Royal Astronomical Society*, vol. 416, no. 3, pp. 1697–1702, 2011.

[250] S. M. Couch and E. P. O'Connor, "High-resolution three-dimensional simulations of core-collapse supernovae in multiple progenitors," *The Astrophysical Journal*, vol. 785, no. 2, p. 123, 2014.

[251] S. M. Couch and C. D. Ott, "The role of turbulence in neutrino-driven core-collapse supernova explosions," *The Astrophysical Journal*, vol. 799, no. 1, article 5, 2015.

[252] L. Wang, D. Baade, P. Höflich, and J. C. Wheeler, "Spectropolarimetry of the type Ic supernova SN 2002ap in M74: more evidence for asymmetric core collapse," *Astrophysical Journal*, vol. 592, no. 1, pp. 457–466, 2003.

[253] A. G. Lyne and D. R. Lorimer, "High birth velocities of radio pulsars," *Nature*, vol. 369, no. 6476, pp. 127–129, 1994.

[254] B. W. Grefenstette, F. A. Harrison, S. E. Boggs et al., "Asymmetries in core-collapse supernovae from maps of radioactive ^{44}Ti in Cassiopeia A," *Nature*, vol. 506, no. 7488, pp. 339–342, 2014.

[255] J. M. Blondin, A. Mezzacappa, and C. Demarino, "Stability of standing accretion shocks, with an eye toward core-collapse supernovae," *Astrophysical Journal*, vol. 584, no. 2 I, pp. 971–980, 2003.

[256] S. Orlando, M. Miceli, M. L. Pumo, and F. Bocchino, "Modeling SNR cassiopeia a from the supernova explosion to its current age: the role of post-explosion anisotropies of Ejecta," *Astrophysical Journal*, vol. 822, no. 1, article 22, 2016.

[257] R. A. Fesen, "An optical survey of outlying ejecta in Cassiopeia A: evidence for a turbulent, asymmetric explosion," *The Astrophysical Journal, Supplement Series*, vol. 133, no. 1, pp. 161–186, 2001.

[258] J. C. Wheeler, J. R. Maund, and S. M. Couch, "The shape of Cas A," *The Astrophysical Journal*, vol. 677, no. 2, pp. 1091–1099, 2008.

[259] L. A. Lopez, E. Ramirez-Ruiz, D. Castro, and S. Pearson, "The galactic supernova remnant W49B likely originates from a jet-driven, core-collapse explosion," *The Astrophysical Journal*, vol. 764, no. 1, p. 50, 2013.

[260] R. A. Fesen and D. Milisavljevic, "An HST survey of the highest-velocity Ejecta in cassiopeia A," *Astrophysical Journal*, vol. 818, no. 1, article 17, 2016.

[261] J. C. Wheeler, V. Johnson, and A. Clocchiatti, "Analysis of late-time light curves of type IIb, Ib and Ic supernovae," *Monthly Notices of the Royal Astronomical Society*, vol. 450, no. 2, pp. 1295–1307, 2015.

[262] Z. Cano, K. Maeda, and S. Schulze, "Type Ib SN 1999dn as an example of the thoroughly mixed ejecta of Ib supernovae," *Monthly Notices of the Royal Astronomical Society*, vol. 438, no. 4, pp. 2924–2937, 2014.

[263] R. P. Harkness, J. C. Wheeler, B. Margon et al., "The early spectral phase of type Ib supernovae—evidence for helium," *The Astrophysical Journal*, vol. 317, pp. 355–367, 1987.

[264] L. B. Lucy, "Nonthermal excitation of helium in type Ib supernovae," *The Astrophysical Journal*, vol. 383, no. 1, pp. 308–313, 1991.

[265] H. Li and R. McCray, "The He I emission lines of SN 1987A," *The Astrophysical Journal*, vol. 441, pp. 821–829, 1995.

[266] C. Li, D. J. Hillier, and L. Dessart, "Non-thermal excitation and ionization in supernovae," *Monthly Notices of the Royal Astronomical Society*, vol. 426, no. 2, pp. 1671–1686, 2012.

[267] S. E. Woosley, N. Langer, and T. A. Weaver, "The presupernova evolution and explosion of helium stars that experience mass loss," *Astrophysical Journal*, vol. 448, no. 1, pp. 315–338, 1995.

[268] S. Hachinger, P. A. Mazzali, S. Taubenberger, W. Hillebrandt, K. Nomoto, and D. N. Sauer, "How much H and He is 'hidden' in SNe Ib/c? - I. Low-mass objects," *Monthly Notices of the Royal Astronomical Society*, vol. 422, no. 1, pp. 70–88, 2012.

[269] L. H. Frey, C. L. Fryer, and P. A. Young, "Can stellar mixing explain the lack of type ib supernovae in long-duration gamma-ray bursts?" *Astrophysical Journal Letters*, vol. 773, no. 1, article no. L7, 2013.

[270] L. Dessart, D. J. Hillier, C. Li, and S. Woosley, "On the nature of supernovae Ib and Ic," *Monthly Notices of the Royal Astronomical Society*, vol. 424, no. 3, pp. 2139–2159, 2012.

[271] Y. Liu, M. Modjaz, F. B. Bianco, and O. Graur, "Analyzing the largest spectroscopic data set of stripped supernovae to improve their identifications and constrain their progenitors," *The Astrophysical Journal*, vol. 827, no. 2, 2016.

[272] F. Taddia, J. Sollerman, G. Leloudas et al., "Early-time light curves of Type Ib/c supernovae from the SDSS-II Supernova Survey," *Astronomy & Astrophysics*, vol. 574, article A60, 2015.

[273] A. M. Soderberg, E. Nakar, E. Berger, and S. R. Kulkarni, "Late-time radio observations of 68 type Ibc supernovae: strong constraints on off-axis gamma-ray bursts," *Astrophysical Journal*, vol. 638, no. 2 I, pp. 930–937, 2006.

[274] L. E. Kay, J. P. Halpern, K. M. Leighly et al., "Spectropolarimetry of the peculiar Type IC supernovae 1998bw and 1997ef," *Bulletin of the American Astronomical Society*, vol. 30, article 1323, 1998.

[275] E. H. McKenzie and B. E. Schaefer, "The late-time light curve of SN 1998bw associated with GRB 980425," *Publications of the Astronomical Society of the Pacific*, vol. 111, no. 762, pp. 964–968, 1999.

[276] J. Gorosabel, V. Larionov, A. J. Castro-Tirado et al., "Detection of optical linear polarization in the SN 2006aj/XRF 060218 non-spherical expansion," *Astronomy & Astrophysics*, vol. 459, no. 3, pp. L33–L36, 2006.

[277] J. R. Maund, J. C. Wheeler, F. Patat, D. Baade, L. Wang, and P. Höflich, "Spectropolarimetry of SN 2006aj at 9.6 days," *Astronomy and Astrophysics*, vol. 475, no. 1, pp. L1–L4, 2007.

[278] M. Betoule, R. Kessler, J. Guy et al., "Improved cosmological constraints from a joint analysis of the SDSS-II and SNLS supernova samples," *Astronomy & Astrophysics*, vol. 568, article A22, 2014.

[279] M. M. Phillips, "The absolute magnitudes of type IA supernovae," *Astrophysical Journal*, vol. 413, no. 2, pp. L105–L108, 1993.

[280] C. T. Kowal, "Absolute magnitudes of supernovae," *The Astronomical Journal*, vol. 73, pp. 1021–1024, 1968.

[281] D. Branch and G. A. Tammann, "Type IA supernovae as standard candles," *Annual Review of Astronomy & Astrophysics*, vol. 30, pp. 359–389, 1992.

[282] A. Sandage and G. A. Tammann, "The Hubble diagram in V for supernovae of Type IA and the value of H(0) therefrom," *The Astrophysical Journal*, vol. 415, no. 1, pp. 1–9, 1993.

[283] M. Hamuy, M. M. Phillips, J. Maza, N. B. Suntzeff, R. A. Schommer, and R. Avilés, "A hubble diagram of distant Type Ia supernovae," *Astronomical Journal*, vol. 109, no. 1, pp. 1–13, 1995.

[284] M. Hamuy, M. M. Phillips, N. B. Suntzeff, R. A. Schommer, J. Maza, and R. Avilés, "The absolute luminosities of the calan/tololo type IA supernovae," *Astronomical Journal*, vol. 112, no. 6, pp. 2391–2397, 1996.

[285] M. Hamuy, M. M. Phillips, N. B. Suntzeff, R. A. Schommer, J. Maza, and R. Avilés, "The hubble diagram of the calan/tololo type IA supernovae and the value of HO," *Astronomical Journal*, vol. 112, no. 6, p. 2398, 1996.

[286] A. G. Riess, A. V. Filippenko, P. Challis et al., "Observational evidence from supernovae for an accelerating universe and a cosmological constant," *The Astronomical Journal*, vol. 116, no. 3, pp. 1009–1038, 1998.

[287] S. Perlmutter, G. Aldering, G. Goldhaber et al., "Measurements of Ω and Λ from 42 high-redshift supernovae," *The Astrophysical Journal*, vol. 517, no. 2, pp. 565–586, 1999.

[288] W. L. Freedman, B. F. Madore, B. K. Gibson et al., "Final results from the Hubble Space Telescope key project to measure the Hubble constant," *Astrophysical Journal Letters*, vol. 553, no. 1, pp. 47–72, 2001.

[289] X. Li, J. Hjorth, and R. Wojtak, "Cosmological parameters from supernovae associated with gamma-ray bursts," *The Astrophysical Journal*, vol. 796, no. 1, article L4, 2014.

[290] A. L. Piro and E. Nakar, "What can we learn from the rising light curves of radioactively powered supernovae?" *The Astrophysical Journal*, vol. 769, no. 1, article 67, 2013.

[291] S. E. Woosley and W. Zhang, "Models for GRBs and diverse transients," *Philosophical Transactions of the Royal Society A: Mathematical, Physical and Engineering Sciences*, vol. 365, no. 1854, pp. 1129–1139, 2007.

[292] D. Kasen and S. E. Woosley, "On the origin of the type Ia supernova width-luminosity relation," *The Astrophysical Journal*, vol. 656, no. 2, pp. 661–665, 2007.

[293] S. Savaglio, K. Glazebrook, and D. Le Borgne, "The galaxy population hosting gamma-ray bursts," *The Astrophysical Journal*, vol. 691, no. 1, pp. 182–211, 2009.

[294] P. Jakobsson, D. Malesani, J. Hjorth, J. P. U. Fynbo, and B. Milvang-Jensen, "Host galaxies of long gamma-ray bursts," in *Proceedings of the Gamma Ray Bursts 2010, GRB 2010*, pp. 265–270, usa, November 2010.

[295] E. M. Levesque, "The host galaxies of long-duration gamma-ray bursts," *Publications of the Astronomical Society of the Pacific*, vol. 126, no. 935, pp. 1–14, 2014.

[296] T. Krühler, D. Malesani, J. P. U. Fynbo et al., "GRB hosts through cosmic time: VLT/X-shooter emission-line spectroscopy of 96 γ-ray-burst-selected galaxies at $0.1 < z < 3.6$," *The Astronomy and Astrophysics*, vol. 581, article A125, 2015.

[297] S. Mao and H. J. Mo, "The nature of the host galaxies for gamma-ray bursts," *Astronomy and Astrophysics*, vol. 339, no. 1, pp. L1–L4, 1998.

[298] D. W. Hogg and A. S. Fruchter, "The faint-galaxy hosts of gamma-ray bursts," *Astrophysical Journal*, vol. 520, no. 1, pp. 54–58, 1999.

[299] S. G. Djorgovski, D. A. Frail, S. R. Kulkarni, J. S. Bloom, S. C. Odewahn, and A. Diercks, "The afterglow and the host galaxy of the dark burst GRB 970828," *The Astrophysical Journal*, vol. 562, no. 2, pp. 654–663, 2001.

[300] S. G. Djorgovski, J. S. Bloom, and S. R. Kulkarni, "The redshift and the host galaxy of GRB 980613: a gamma-ray burst from a merger-induced starburst?" *Astrophysical Journal*, vol. 591, no. 1, pp. L13–L16, 2003.

[301] E. Le Floc'h, P.-A. Duc, I. F. Mirabel et al., "Are the hosts of gamma-ray bursts sub-luminous and blue galaxies?" *Astronomy and Astrophysics*, vol. 400, no. 2, pp. 499–510, 2003.

[302] L. Christensen, J. Hjorth, and J. Gorosabel, "UV star-formation rates of GRB host galaxies," *Astronomy and Astrophysics*, vol. 425, no. 3, pp. 913–926, 2004.

[303] N. R. Tanvir, V. E. Barnard, A. W. Blain et al., "The submillimetre properties of gamma-ray burst host galaxies," *Monthly Notices of the Royal Astronomical Society*, vol. 352, no. 3, pp. 1073–1080, 2004.

[304] C. J. Conselice, J. A. Blackburne, and C. Papovich, "The luminosity, stellar mass, and number density evolution of field galaxies of known morphology from z = 0.5 to 3," *Astrophysical Journal*, vol. 620, no. 2, pp. 564–583, 2005.

[305] K. Z. Stanek, O. Y. Gnedin, J. F. Beacom et al., "Protecting life in the Milky Way: metals keep the GRBs away," *Acta Astronomica*, vol. 56, no. 4, pp. 333–345, 2006.

[306] J. M. C. Cerón, M. J. MichaŁowski, J. Hjorth, D. Watson, J. P. U. Fynbo, and J. Gorosabel, "Star formation rates and stellar masses in $z \sim 1$ gamma-ray burst hosts," *Astrophysical Journal*, vol. 653, no. 2, pp. L85–L88, 2006.

[307] C. Wainwright, E. Berger, and B. E. Penprase, "A morphological study of gamma-ray burst host galaxies," *The Astrophysical Journal*, vol. 657, no. 1 I, pp. 367–377, 2007.

[308] J. M. Castro Cerón, M. J. Michallowski, J. Hjorth et al., "On the distribution of stellar masses in gamma-ray burst host galaxies," *The Astrophysical Journal*, vol. 721, no. 2, pp. 1919–1927, 2010.

[309] A. S. Fruchter, A. J. Levan, L. Strolger et al., "Long γ-ray bursts and core-collapse supernovae have different environments," *Nature*, vol. 441, pp. 463–468, 2006.

[310] K. M. Svensson, A. J. Levan, N. R. Tanvir, A. S. Fruchter, and L.-G. Strolger, "The host galaxies of core-collapse supernovae and gamma-ray bursts," *Monthly Notices of the Royal Astronomical Society*, vol. 405, no. 1, pp. 57–76, 2010.

[311] J. S. Bloom, S. R. Kulkarni, and S. G. Djorgovski, "The observed offset distribution of gamma-ray bursts from their host galaxies: a robust clue to the nature of the progenitors," *Astronomical Journal*, vol. 123, no. 3, pp. 1111–1148, 2002.

[312] A. De Ugarte Postigo, J. P. U. Fynbo, C. C. Thöne et al., "The distribution of equivalent widths in long GRB afterglow spectra," *Astronomy and Astrophysics*, vol. 548, article no. A11, 2012.

[313] P. K. Blanchard, E. Berger, and W.-F. Fong, "The offset and host light distributions of long gamma-ray bursts: a new view from hst observations of swift bursts," *The Astrophysical Journal*, vol. 817, no. 2, article 144, 2016.

[314] J. X. Prochaska, J. S. Bloom, H.-W. Chen et al., "The host galaxy of GRB 031203: implications of its low metallicity, low redshift, and starburst nature," *The Astrophysical Journal*, vol. 611, no. 1, p. 200, 2004.

[315] J. Gorosabel, D. Pérez-Ramírez, J. Sollerman et al., "The GRB 030329 host: a blue low metallicity subluminous galaxy with intense star formation," *Astronomy and Astrophysics*, vol. 444, no. 3, pp. 711–721, 2005.

[316] J. Sollerman, G. Östlin, J. P. U. Fynbo, J. Hjorth, A. Fruchter, and K. Pedersen, "On the nature of nearby GRB/SN host galaxies," *New Astronomy*, vol. 11, no. 2, pp. 103–115, 2005.

[317] L. J. Kewley, W. R. Brown, M. J. Geller, S. J. Kenyon, and M. J. Kurtz, "SDSS 0809+1729: connections between extremely metal-poor galaxies and gamma-ray burst hosts," *Astronomical Journal*, vol. 133, no. 3, pp. 882–888, 2007.

[318] J. Lequeux, M. Peimbert, J. F. Rayo, A. Serrano, and S. Torres-Peimbert, "Chemical composition and evolution of irregular and blue compact galaxies," *Astronomy & Astrophysics*, vol. 80, pp. 155–166, 1979.

[319] E. D. Skillman, R. C. Kennicutt, and P. W. Hodge, "Oxygen abundances in nearby dwarf irregular galaxies," *The Astrophysical Journal*, vol. 347, pp. 875–882, 1989.

[320] D. Zaritsky, R. C. Kennicutt Jr., and J. P. Huchra, "H II regions and the abundance properties of spiral galaxies," *Astrophysical Journal*, vol. 420, no. 1, pp. 87–109, 1994.

[321] C. A. Tremonti, T. M. Heckman, G. Kauffmann et al., "The origin of the mass-metallicity relation: insights from 53,000 star-forming galaxies in the sloan digital sky survey," *The Astrophysical Journal*, vol. 613, no. 2, p. 898, 2004.

[322] C. Wolf and P. Podsiadlowski, "The metallicity dependence of the long-duration gamma-ray burst rate from host galaxy luminosities," *Monthly Notices of the Royal Astronomical Society*, vol. 375, no. 3, pp. 1049–1058, 2007.

[323] M. Modjaz, L. Kewley, R. P. Kirshner et al., "Measured metallicities at the sites of nearby broad-lined type Ic supernovae and implications for the supernovae gamma-ray burst connection," *Astronomical Journal*, vol. 135, no. 4, pp. 1136–1150, 2008.

[324] E. Le Floc'h, V. Charmandaris, W. J. Forrest, I. F. Mirabel, L. Armus, and D. Devost, "Probing cosmic star formation using long gamma-ray bursts: new constraints from the Spitzer Space Telescope," *Astrophysical Journal*, vol. 642, no. 2 I, pp. 636–652, 2006.

[325] M. J. Michałowski, J. Hjorth, J. M. Castro Cerón, and D. Watson, "The nature of GRB-selected submillimeter galaxies: hot and young," *Astrophysical Journal*, vol. 672, no. 2, pp. 817–824, 2008.

[326] J. Elliott, T. Krühler, J. Greiner et al., "The low-extinction afterglow in the solar-metallicity host galaxy of γ-ray burst 110918A," *Astronomy and Astrophysics*, vol. 556, article A23, 2013.

[327] P. Schady, T. Krühler, J. Greiner et al., "Super-solar metallicity at the position of the ultra-long GRB 130925A," *Astronomy and Astrophysics*, vol. 579, article no. A126, 2015.

[328] T. Krühler, J. Greiner, P. Schady et al., "The SEDs and host galaxies of the dustiest GRB afterglows," *Astronomy and Astrophysics*, vol. 534, article no. A108, 2011.

[329] J. Hjorth, D. Malesani, P. Jakobsson et al., "The optically unbiased gamma-ray burst host (tough) survey. I. Survey design and catalogs," *Astrophysical Journal*, vol. 756, no. 2, article no. 187, 2012.

[330] A. Rossi, S. Klose, P. Ferrero et al., "A deep search for the host galaxies of gamma-ray bursts with no detected optical

afterglow," *Astronomy and Astrophysics*, vol. 545, article no. A77, 2012.

[331] D. A. Perley, A. J. Levan, N. R. Tanvir et al., "A population of massive, luminous galaxies hosting heavily dust-obscured gamma-ray bursts: implications for the use of GRBs as tracers of cosmic star formation," *Astrophysical Journal*, vol. 778, no. 2, article no. 128, 2013.

[332] F. Mannucci, R. Salvaterra, and M. A. Campisi, "The metallicity of the long GRB hosts and the fundamental metallicity relation of low-mass galaxies," *Monthly Notices of the Royal Astronomical Society*, vol. 414, no. 2, pp. 1263–1268, 2011.

[333] D. Kocevski and A. A. West, "On the origin of the mass-metallicity relation for gamma-ray burst host galaxies," *The Astrophysical Journal Letters*, vol. 735, no. 1, article L8, 2011.

[334] E. Berger, D. B. Fox, S. R. Kulkarni, D. A. Frail, and S. G. Djorgovski, "The ERO host galaxy of GRB 020127: implications for the metallicity of GRB progenitors," *The Astrophysical Journal*, vol. 660, no. 1, pp. 504–508, 2007.

[335] D. A. Perley, R. A. Perley, J. Hjorth et al., "Connecting GRBs and ULIRGs: a sensitive, unbiased survey for radio emission from gamma-ray burst host galaxies at $0 < z < 2.5$," *Astrophysical Journal*, vol. 801, no. 2, article 102, 2015.

[336] D. Kocevski, A. A. West, and M. Modjaz, "Modeling the GRB host galaxy mass distribution: are GRBs unbiased tracers of star formation?" *Astrophysical Journal*, vol. 702, no. 1, pp. 377–385, 2009.

[337] J. F. Graham and A. S. Fruchter, "The metal aversion of long-duration gamma-ray bursts," *Astrophysical Journal*, vol. 774, no. 2, article no. 119, 2013.

[338] S. D. Vergani, R. Salvaterra, J. Japelj et al., "Are long gamma-ray bursts biased tracers of star formation? Clues from the host galaxies of the *Swift*/BAT6 complete sample of LGRBs: I. Stellar mass at $z < 1$," *The Astronomy and Astrophysics*, vol. 581, article A102, 2015.

[339] M. Trenti, R. Perna, and R. Jimenez, "The luminosity and stellar mass functions of GRB host galaxies: insight into the metallicity bias," *Astrophysical Journal*, vol. 802, no. 2, article no. 103, 2015.

[340] S. Schulze, R. Chapman, J. Hjorth et al., "The Optically Unbiased GRB Host (TOUGH) Survey. VII. The host galaxy luminosity function: probing the relationship between GRBs and star formation to redshift 6," *The Astrophysical Journal*, vol. 808, p. 73, 2015.

[341] J. F. Graham and A. S. Fruchter, "The relative rate of LGRB formation as a function of metallicity," https://arxiv.org/abs/1511.01079.

[342] P. L. Kelly, A. V. Filippenko, M. Modjaz, and D. Kocevski, "The host galaxies of fast-ejecta core-collapse supernovae," *Astrophysical Journal*, vol. 789, no. 1, article no. 23, 2014.

[343] G. Leloudas, S. Schulze, T. Krühler et al., "Spectroscopy of superluminous supernova host galaxies. A preference of hydrogen-poor events for extreme emission line galaxies," *Monthly Notices of the Royal Astronomical Society*, vol. 449, no. 1, pp. 917–932, 2015.

[344] C. Firmani, V. Avila-Reese, G. Ghisellini, and A. V. Tutukov, "Formation rate, evolving luminosity function, jet structure, and progenitors for long gamma-ray bursts," *Astrophysical Journal*, vol. 611, no. 2 I, pp. 1033–1040, 2004.

[345] P. A. Price and B. P. Schmidt, "Towards measuring the cosmic gamma-ray burst rate," in *Gamma-Ray Bursts: 30 Years of Discovery*, E. Fenimore and M. Galassi, Eds., vol. 727 of *American Institute of Physics Conference Series*, pp. 503–507, 2004.

[346] P. Natarajan, B. Albanna, J. Hjorth, E. Ramirez-Ruiz, N. Tanvir, and R. Wijers, "The redshift distribution of gamma-ray bursts revisited," *Monthly Notices of the Royal Astronomical Society*, vol. 364, no. 1, pp. L8–L12, 2005.

[347] R. Chary, E. Berger, and L. Cowie, "Spitzer observations of gamma-ray burst host galaxies: a unique window into high-redshift chemical evolution and star formation," *The Astrophysical Journal*, vol. 671, no. 1, pp. 272–277, 2007.

[348] J. Greiner, D. B. Fox, P. Schady et al., "Gamma-ray bursts trace uv metrics of star formation over $3 < z < 5$," *Astrophysical Journal*, vol. 809, no. 1, article 76, 2015.

[349] F. Y. Wang and Z. G. Dai, "Long GRBs are metallicity-biased tracers of star formation: evidence from host galaxies and redshift distribution," *The Astrophysical Journal*, vol. 213, no. 1, p. 15, 2014.

[350] J. Japelj, S. D. Vergani, R. Salvaterra et al., "Are LGRBs biased tracers of star formation? Clues from the host galaxies of the Swift/BAT6 complete sample of bright LGRBs. II: star formation rates and metallicities at $z < 1$," *Astronomy & Astrophysics*, vol. 590, article A129, 2016.

[351] L. Christensen, P. M. Vreeswijk, J. Sollerman, C. C. Thöne, E. Le Floc'H, and K. Wiersema, "IFU observations of the GRB 980425/SN 1998bw host galaxy: emission line ratios in GRB regions," *The Astronomy and Astrophysics*, vol. 490, no. 1, pp. 45–59, 2008.

[352] E. Le Floc'H, V. Charmandaris, K. Gordon et al., "The first infrared study of the close environment of a long gamma-ray burst," *The Astrophysical Journal*, vol. 746, no. 1, article 7, 2012.

[353] M. J. Michałowski, L. K. Hunt, E. Palazzi et al., "Spatially-resolved dust properties of the GRB 980425 host galaxy," *Astronomy & Astrophysics*, vol. 562, article A70, 2014.

[354] M. Arabsalmani, S. Roychowdhury, M. A. Zwaan, N. Kanekar, and M. J. Michałowski, "First measurement of H I 21 cm emission from a GRB host galaxy indicates a post-merger system," *Monthly Notices of the Royal Astronomical Society: Letters*, vol. 454, no. 1, pp. L51–L55, 2015.

[355] C. C. Thöne, J. P. U. Fynbo, G. Östlin et al., "Spatially resolved properties of the GRB 060505 host: implications for the nature of the progenitor," *Astrophysical Journal*, vol. 676, no. 2, pp. 1151–1161, 2008.

[356] E. M. Levesque, E. Berger, A. M. Soderberg, and R. Chornock, "Metallicity in the GRB 100316D/SN 2010bh host complex," *The Astrophysical Journal*, vol. 739, no. 1, p. 23, 2011.

[357] E. M. Levesque, R. Chornock, A. M. Soderberg, E. Berger, and R. Lunnan, "Host galaxy properties of the subluminous GRB 120422A/SN 2012bz," *The Astrophysical Journal*, vol. 758, no. 2, article 92, 2012.

[358] J. S. Bloom, S. G. Djorgovski, and S. R. Kulkarni, "The redshift and the ordinary host galaxy of GRB 970228," *Astrophysical Journal*, vol. 554, no. 2, pp. 678–683, 2001.

[359] J. X. Prochaska, H.-W. Chen, and J. S. Bloom, "Dissecting the circumstellar environment of γ-ray burst progenitors," *Astrophysical Journal*, vol. 648, no. 1, pp. 95–110, 2006.

[360] P. M. Vreeswijk, C. Ledoux, A. Smette et al., "Rapid-response mode VLT/UVES spectroscopy of GRB 060418. Conclusive evidence for UV pumping from the time evolution of Fe II and Ni II excited- and metastable-level populations," *Astronomy & Astrophysics*, vol. 468, no. 1, pp. 83–96, 2007.

[361] V. D'Elia, F. Fiore, E. J. A. Meurs et al., "UVES/VLT high resolution spectroscopy of GRB 050730 afterglow: probing the features of the GRB environment," *Astronomy and Astrophysics*, vol. 467, no. 2, pp. 629–639, 2007.

[362] J. P. U. Fynbo, T. Krühler, K. Leighly et al., "The mysterious optical afterglow spectrum of GRB140506A at $z = 0.889$," *Astronomy & Astrophysics*, vol. 572, article A12, 2014.

[363] R. A. Chevalier and Z.-Y. Li, "Wind interaction models for gamma-ray burst afterglows: the case for two types of progenitors," *The Astrophysical Journal*, vol. 536, no. 1, pp. 195–212, 2000.

[364] N. Mirabal, J. P. Halpern, R. Chornock et al., "GRB 021004: a possible shell nebula around a Wolf-Rayet star gamma-ray burst progenitor," *Astrophysical Journal*, vol. 595, no. 2, pp. 935–949, 2003.

[365] R. R. Treffers and Y.-H. Chu, "Galactic ring nebulae associated with Wolf-Rayet stars. V—the stellar wind-blown bubbles," *The Astrophysical Journal*, vol. 254, pp. 569–577, 1982.

[366] J. A. Toalá, M. A. Guerrero, G. Ramos-Larios, and V. Guzmán, "WISE morphological study of Wolf-Rayet nebulae," *The Astronomy and Astrophysics*, vol. 578, article A66, 2015.

[367] S. Schulze, S. Klose, G. Björnsson et al., "The circumburst density profile around GRB progenitors: a statistical study," *Astronomy & Astrophysics*, vol. 526, no. 3, article A23, 2011.

[368] A. J. Van Marie, N. Langer, and G. García-Segura, "Constraints on gamma-ray burst and supernova progenitors through circumstellar absorption lines," *Astronomy and Astrophysics*, vol. 444, no. 3, pp. 837–847, 2005.

[369] A. J. Van Marle, N. Langer, A. Achterberg, and G. García-Segura, "Forming a constant density medium close to long gamma-ray bursts," *Astronomy and Astrophysics*, vol. 460, no. 1, pp. 105–116, 2006.

[370] J. A. Baldwin, M. M. Phillips, and R. Terlevich, "Classification parameters for the emission-line spectra of extragalactic objects," *Publications of the Astronomical Society of the Pacific*, vol. 93, pp. 5–19, 1981.

[371] J. Brinchmann, M. Pettini, and S. Charlot, "New insights into the stellar content and physical conditions of star-forming galaxies at $z = 2$-3 from spectral modelling," *Monthly Notices of the Royal Astronomical Society*, vol. 385, no. 2, pp. 769–782, 2008.

[372] L. J. Kewley, C. Maier, K. Yabe et al., "The cosmic BPT diagram: confronting theory with observations," *Astrophysical Journal Letters*, vol. 774, no. 1, article no. L10, 2013.

[373] C. C. Steidel, G. C. Rudie, A. L. Strom et al., "Strong nebular line ratios in the spectra of $z \sim 2$-3 star forming galaxies: first results from KBSS-mosfire," *Astrophysical Journal*, vol. 795, no. 2, article 165, 2014.

[374] B. Paczyński, "Are gamma-ray bursts in star-forming regions?" *The Astrophysical Journal*, vol. 494, no. 1, pp. L45–L48, 1998.

[375] J. S. Bloom, S. Sigurdsson, and O. R. Pols, "The spatial distribution of coalescing neutron star binaries: implications for gamma-ray bursts," *Monthly Notices of the Royal Astronomical Society*, vol. 305, no. 4, pp. 763–769, 1999.

[376] C. L. Fryer, S. E. Woosley, and D. H. Hartmann, "Formation rates of black hole accretion disk gamma-ray bursts," *The Astrophysical Journal*, vol. 526, no. 1, pp. 152–177, 1999.

[377] K. Belczyński, T. Bulik, and W. Zbijewski, "Distribution of black hole binaries around galaxies," *Astronomy & Astrophysics*, vol. 355, no. 2, pp. 479–484, 2000.

[378] P. L. Kelly, R. P. Kirshner, and M. Pahre, "Long γ-ray bursts and Type Ic core-collapse supernovae have similar locations in hosts," *Astrophysical Journal*, vol. 687, no. 2, pp. 1201–1207, 2008.

[379] R.-P. Kudritzki and J. Puls, "Winds from hot stars," *Annual Review of Astronomy and Astrophysics*, vol. 38, no. 1, pp. 613–666, 2000.

[380] G. Meynet and A. Maeder, "Stellar evolution with rotation XI. Wolf-Rayet star populations at different metallicities," *Astronomy and Astrophysics*, vol. 429, no. 2, pp. 581–598, 2005.

[381] R. Hirschi, G. Meynet, and A. Maeder, "Yields of rotating stars at solar metallicity," *The Astronomy and Astrophysics*, vol. 433, no. 3, pp. 1013–1022, 2005.

[382] S.-C. Yoon and N. Langer, "Evolution of rapidly rotating metal-poor massive stars towards gamma-ray bursts," *Astronomy and Astrophysics*, vol. 443, no. 2, pp. 643–648, 2005.

[383] S. E. Woosley and A. Heger, "The progenitor stars of gamma-ray bursts," *The Astrophysical Journal*, vol. 637, no. 2, pp. 914–921, 2006.

[384] H. F. Song, G. Meynet, A. Maeder, S. Ekström, and P. Eggenberger, "Massive star evolution in close binaries: conditions for homogeneous chemical evolution," *Astronomy and Astrophysics*, vol. 585, article no. A120, 2016.

[385] I. Hunter, I. Brott, D. J. Lennon et al., "The VLT flames survey of massive stars: rotation and nitrogen enrichment as the key to understanding massive star evolution," *The Astrophysical Journal*, vol. 676, no. 1, p. L29, 2008.

[386] J.-C. Bouret, T. Lanz, D. J. Hillier et al., "Quantitative spectroscopy of O stars at low metallicity: O dwarfs in NGC 346," *The Astrophysical Journal*, vol. 595, no. 2, pp. 1182–1205, 2003.

[387] N. R. Walborn, N. I. Morrell, I. D. Howarth et al., "A CNO dichotomy among O2 giant spectra in the magellanic clouds," *Astrophysical Journal*, vol. 608, no. 2 I, pp. 1028–1038, 2004.

[388] F. Tramper, S. M. Straal, D. Sanyal et al., "Massive stars on the verge of exploding: the properties of oxygen sequence Wolf-Rayet stars," *Astronomy and Astrophysics*, vol. 581, article A110, 2015.

[389] P. Massey, "Massive stars in the local group: implications for stellar evolution and star formation," *Annual Review of Astronomy and Astrophysics*, vol. 41, pp. 15–56, 2003.

[390] B. A. Zauderer, E. Berger, R. Margutti et al., "Illuminating the darkest gamma-ray bursts with radio observations," *The Astrophysical Journal*, vol. 767, no. 2, article 161, 2013.

[391] S. Ekström, C. Georgy, P. Eggenberger et al., "Grids of stellar models with rotation I. Models from 0.8 to 120 $M_{\odot}$ at solar metallicity ($Z = 0.014$)," *Astronomy and Astrophysics*, vol. 537, article no. A146, 2012.

[392] C. Georgy, S. Ekström, G. Meynet et al., "Grids of stellar models with rotation: II. WR populations and supernovae/GRB progenitors at $Z = 0.014$," *The Astronomy and Astrophysics*, vol. 542, article A29, 2012.

[393] J. H. Groh, G. Meynet, C. Georgy, and S. Ekström, "Fundamental properties of core-collapse supernova and GRB progenitors: predicting the look of massive stars before death," *Astronomy and Astrophysics*, vol. 558, article no. A131, 2013.

[394] A. Maeder, "Stellar evolution with rotation IX. The effects of the production of asymmetric nebulae on the internal evolution," *Astronomy and Astrophysics*, vol. 392, no. 2, pp. 575–584, 2002.

[395] N. R. Tanvir, A. J. Levan, A. S. Fruchter et al., "A 'kilonova' associated with the short-duration γ-ray burst GRB 130603B," *Nature*, vol. 500, no. 7464, pp. 547–549, 2013.

[396] Z.-P. Jin, X. Li, Z. Cano, S. Covino, Y.-Z. Fan, and D.-M. Wei, "The light curve of the macronova associated with the long-short burst GRB 060614," *Astrophysical Journal Letters*, vol. 811, no. 2, article L22, 2015.

[397] M. Tanaka, K. Hotokezaka, K. Kyutoku et al., "Radioactively powered emission from black hole-neutron star mergers," *Astrophysical Journal*, vol. 780, no. 1, article no. 31, 2014.

[398] D. Kasen, N. R. Badnell, and J. Barnes, "Opacities and spectra of the *r*-process ejecta from neutron star mergers," *The Astrophysical Journal*, vol. 774, no. 1, article 25, 2013.

[399] E. P. Mazets, S. V. Golenetskii, V. N. Il'Inskii et al., "Catalog of cosmic gamma-ray bursts from the KONUS experiment data—parts I and II," *Astrophysics and Space Science*, vol. 80, no. 1, pp. 3–83, 1981.

[400] C. Kouveliotou, C. A. Meegan, G. J. Fishman et al., "Identification of two classes of gamma-ray bursts," *Astrophysical Journal*, vol. 413, no. 2, pp. L101–L104, 1993.

[401] E. Nakar, "Short-hard gamma-ray bursts," *Physics Reports*, vol. 442, no. 1-6, pp. 166–236, 2007.

[402] E. Berger, "Short-duration gamma-ray bursts," *Annual Review of Astronomy and Astrophysics*, vol. 52, pp. 43–105, 2014.

[403] J. Hjorth, J. Sollerman, J. Gorosabel et al., "GRB 050509B: constraints on short gamma-ray burst models," *Astrophysical Journal*, vol. 630, no. 2, pp. L117–L120, 2005.

[404] M. Della Valle, G. Chincarini, N. Panagia et al., "An enigmatic long-lasting γ-ray burst not accompanied by a bright supernova," *Nature*, vol. 444, no. 7122, pp. 1050–1052, 2006.

[405] J. P. Fynbo, D. Watson, C. C. Thöne et al., "No supernovae associated with two long-duration γ-ray bursts," *Nature*, vol. 444, no. 7122, pp. 1047–1049, 2006.

[406] N. Gehrels, J. P. Norris, S. D. Barthelmy et al., "A new γ-ray burst classification scheme from GRB 060614," *Nature*, vol. 444, no. 7122, pp. 1044–1046, 2006.

[407] W. Fong, E. Berger, and D. B. Fox, "Hubble space telescope observations of short gammaray burst host galaxies: morphologies, offsets, and local environments," *Astrophysical Journal*, vol. 708, pp. 9–25, 2010.

[408] W. Fong and E. Berger, "The locations of short gamma-ray bursts as evidence for compact object binary progenitors," *Astrophysical Journal*, vol. 776, no. 1, article 18, 2013.

[409] D. Eichler, M. Livio, T. Piran, and D. N. Schramm, "Nucleosynthesis, neutrino bursts and γ-rays from coalescing neutron stars," *Nature*, vol. 340, no. 6229, pp. 126–128, 1989.

[410] R. Narayan, B. Paczyński, and T. Piran, "Gamma-ray bursts as the death throes of massive binary stars," *The Astrophysical Journal*, vol. 395, no. 2, pp. L83–L86, 1992.

[411] W. H. Lee and E. Ramirez-Ruiz, "The progenitors of short gamma-ray bursts," *New Journal of Physics*, vol. 9, no. 1, p. 1, 2007.

[412] N. Gehrels, C. L. Sarazin, P. T. O'Brien et al., "A short big γ-ray burst apparently associated with an elliptical galaxy at redshift $z = 0.225$," *Nature*, vol. 437, pp. 851–854, 2005.

[413] L.-X. Li and B. Paczyński, "Transient events from neutron star mergers," *Astrophysical Journal*, vol. 507, no. 1, 1998.

[414] S. R. Kulkarni, "Modeling supernova-like explosions associated with gamma-ray bursts with short durations," https://arxiv.org/abs/astro-ph/0510256.

[415] S. Rosswog, "Mergers of neutron star-black hole binaries with small mass ratios: nucleosynthesis, gamma-ray bursts, and electromagnetic transients," *The Astrophysical Journal*, vol. 634, no. 2, pp. 1202–1213, 2005.

[416] B. D. Metzger, G. Martínez-Pinedo, S. Darbha et al., "Electromagnetic counterparts of compact object mergers powered by the radioactive decay of r-process nuclei," *Monthly Notices of the Royal Astronomical Society*, vol. 406, no. 4, pp. 2650–2662, 2010.

[417] B. D. Metzger and E. Berger, "What is the most promising electromagnetic counterpart of a neutron star binary merger?" *Astrophysical Journal*, vol. 746, no. 1, article no. 48, 2012.

[418] M. Tanaka, "Kilonova/macronova emission from compact binary mergers," *Advances in Astronomy*, vol. 2016, Article ID 6341974, 12 pages, 2016.

[419] J. M. Lattimer and D. N. Schramm, "Black-hole-neutron-star collisions," *The Astrophysical Journal*, vol. 192, pp. L145–L147, 1974.

[420] S. E. Woosley, J. R. Wilson, G. J. Mathews, R. D. Hoffman, and B. S. Meyer, "The r-process and neutrino-heated supernova ejecta," *The Astrophysical Journal*, vol. 433, no. 1, pp. 229–246, 1994.

[421] C. Freiburghaus, S. Rosswog, and F.-K. Thielemann, "R-process in neutron star mergers," *Astrophysical Journal*, vol. 525, no. 2, pp. L121–L124, 1999.

[422] J. Lippuner and L. F. Roberts, "R-process lanthanide production and heating rates in kilonovae," *The Astrophysical Journal*, vol. 815, article 82, 2015.

[423] B. D. Metzger and R. Fernández, "Red or blue? A potential kilonova imprint of the delay until black hole formation following a neutron star merger," *Monthly Notices of the Royal Astronomical Society*, vol. 441, no. 4, pp. 3444–3453, 2014.

[424] J. Hjorth, D. Watson, J. P. U. Fynbo et al., "The optical afterglow of the short γ-ray burst GRB 050709," *Nature*, vol. 437, no. 7060, pp. 859–861, 2005.

[425] Z.-P. Jin, K. Hotokezaka, X. Li et al., "The Macronova in GRB 050709 and the GRB-macronova connection," *Nature Communications*, vol. 7, Article ID 12898, 2016.

[426] A. M. Soderberg, E. Berger, M. Kasliwal et al., "The afterglow, energetics, and host galaxy of the short-hard gamma-ray burst 051221a," *Astrophysical Journal*, vol. 650, no. 1 I, pp. 261–271, 2006.

[427] B. Yang, Z.-P. Jin, X. Li et al., "A possible macronova in the late afterglow of the long-short burst GRB 060614," *Nature Communications*, vol. 6, article no. 7323, 2015.

[428] E. Berger, S. B. Cenko, D. B. Fox, and A. Cucchiara, "Discovery of the very red near-infrared and optical afterglow of the short-duration GRB 070724A," *Astrophysical Journal*, vol. 704, no. 1, pp. 877–882, 2009.

[429] D. Kocevski, C. C. Thöne, E. Ramirez-Ruiz et al., "Limits on radioactive powered emission associated with a short-hard GRB 070724A in a star-forming galaxy," *Monthly Notices of the Royal Astronomical Society*, vol. 404, no. 2, pp. 963–974, 2010.

[430] D. A. Perley, B. D. Metzger, J. Granot et al., "GRB 080503: implications of a naked short gamma-ray burst dominated by extended emission," *Astrophysical Journal*, vol. 696, no. 2, pp. 1871–1885, 2009.

[431] H. Gao, X. Ding, X.-F. Wu, Z.-G. Dai, and B. Zhang, "GRB 080503 late afterglow re-brightening: signature of a magnetar-powered merger-nova," *Astrophysical Journal*, vol. 807, no. 2, 2015.

[432] A. Rowlinson, K. Wiersema, A. J. Levan et al., "Discovery of the afterglow and host galaxy of the low-redshift short GRB 080905A," *Monthly Notices of the Royal Astronomical Society*, vol. 408, no. 1, pp. 383–391, 2010.

[433] E. Berger, W. Fong, and R. Chornock, "An r-process kilonova associated with the short-hard GRB 130603B," *Astrophysical Journal Letters*, vol. 774, no. 2, article no. L23, 2013.

[434] T. Piran, "The physics of gamma-ray bursts," *Reviews of Modern Physics*, vol. 76, no. 4, pp. 1143–1210, 2004.

[435] V. V. Uso, "Millisecond pulsars with extremely strong magnetic fields as a cosmological source of γ-ray bursts," *Nature*, vol. 357, no. 6378, pp. 472–474, 1992.

[436] N. Bucciantini, E. Quataert, J. Arons, B. D. Metzger, and T. A. Thompson, "Relativistic jets and long-duration gamma-ray bursts from the birth of magnetars," *Monthly Notices of the Royal Astronomical Society: Letters*, vol. 383, no. 1, pp. L25–L29, 2008.

[437] B. D. Metzger, D. Giannios, T. A. Thompson, N. Bucciantini, and E. Quataert, "The protomagnetar model for gamma-ray bursts," *Monthly Notices of the Royal Astronomical Society*, vol. 413, no. 3, pp. 2031–2056, 2011.

[438] M. J. Rees and P. Mészáros, "Unsteady outflow models for cosmological γ-ray bursts," *The Astrophysical Journal*, vol. 430, no. 2, pp. L93–L96, 1994.

[439] P. Mészáros, "Gamma-ray bursts: accumulating afterglow implications, progenitor clues, and prospects," *Science*, vol. 291, no. 5501, pp. 79–84, 2001.

[440] R. Sari, T. Piran, and R. Narayan, "Spectra and light curves of γ-ray burst afterglows," *The Astrophysical Journal*, vol. 497, no. 1, pp. L17–L20, 1998.

[441] S. Hümmer, P. Baerwald, and W. Winter, "Neutrino emission from gamma-ray burst fireballs, revised," *Physical Review Letters*, vol. 108, no. 23, Article ID 231101, 2012.

[442] Z. Li, "Note on the normalization of predicted gamma-ray burst neutrino flux," *Physical Review D*, vol. 85, no. 2, Article ID 027301, 2012.

[443] H.-N. He, R.-Y. Liu, X.-Y. Wang, S. Nagataki, K. Murase, and Z.-G. Dai, "Icecube nondetection of gamma-ray bursts: constraints on the fireball properties," *Astrophysical Journal*, vol. 752, article 29, 2012.

[444] E.-W. Liang, S.-X. Yi, J. Zhang, H.-J. Lü, B.-B. Zhang, and B. Zhang, "Constraining γ-ray burst initial lorentz factor with the afterglow onset feature and discovery of a tight Γ0-E γ, iso correlation," *The Astrophysical Journal*, vol. 725, no. 2, pp. 2209–2224, 2010.

[445] G. Ghirlanda, L. Nava, G. Ghisellini et al., "Gamma-ray bursts in the comoving frame," *Monthly Notices of the Royal Astronomical Society*, vol. 420, no. 1, pp. 483–494, 2012.

[446] J. Lü, Y.-C. Zou, W.-H. Lei et al., "Lorentz-factor-isotropic-luminosity/energy correlations of gamma-ray bursts and their interpretation," *Astrophysical Journal*, vol. 751, no. 1, article no. 49, 2012.

[447] A. Pe'er and F. Ryde, "A theory of multicolor blackbody emission from relativistically expanding plasmas," *The Astrophysical Journal*, vol. 732, no. 1, article 49, 2011.

[448] C. Lundman, A. Pe'er, and F. Ryde, "A theory of photospheric emission from relativistic, collimated outflows," *Monthly Notices of the Royal Astronomical Society*, vol. 428, no. 3, pp. 2430–2442, 2013.

[449] H. Ito, J. Matsumoto, S. Nagataki, D. C. Warren, and M. V. Barkov, "Photospheric emission from collapsar jets in 3D relativistic hydrodynamics," *Astrophysical Journal Letters*, vol. 814, no. 2, article no. L29, 2015.

[450] R. Santana, P. Crumley, R. A. Hernández, and P. Kumar, "Monte carlo simulations of the photospheric process," *Monthly Notices of the Royal Astronomical Society*, vol. 456, no. 1, pp. 1049–1065, 2016.

[451] A. Pe'er and F. Ryde, "Photospheric emission in gamma-ray bursts," https://arxiv.org/abs/1603.05058.

[452] B. Zhang and H. Yan, "The internal-collision-induced magnetic reconnection and turbulence (ICMART) model of gamma-ray bursts," *The Astrophysical Journal*, vol. 726, article 90, 2011.

[453] W. Deng, H. Zhang, B. Zhang, and H. Li, "Collision-induced magnetic reconnection and a unified interpretation of polarization properties of GRBs and blazars," *Astrophysical Journal Letters*, vol. 821, no. 1, article no. L12, 2016.

[454] B. Paczynski, "Gamma-ray bursters at cosmological distances," *The Astrophysical Journal*, vol. 308, pp. L43–L46, 1986.

[455] C. Thompson, "A model of gamma-ray bursts," *Monthly Notices of the Royal Astronomical Society*, vol. 270, no. 3, pp. 480–498, 1994.

[456] P. Mészáros and M. J. Rees, "Steep slopes and preferred breaks in γ-ray burst spectra: the role of photospheres and comptonization," *The Astrophysical Journal*, vol. 530, no. 1, pp. 292–298, 2000.

[457] S. A. Colgate, "Prompt gamma rays and X rays from supernovae," *Canadian Journal of Physics*, vol. 46, no. 10, pp. S476–S480, 1968.

[458] S. A. Colgate, "Early gamma rays from supernovae," *The Astrophysical Journal*, vol. 187, pp. 333–336, 1974.

[459] R. I. Klein and R. A. Chevalier, "X-ray bursts from Type II supernovae," *The Astrophysical Journal Letters*, vol. 223, pp. L109–L112, 1978.

[460] S. W. Falk, "Shock steepening and prompt thermal emission in supernovae," *The Astrophysical Journal Letters*, vol. 225, pp. L133–L136, 1978.

[461] L. Ensman and A. Burrows, "Shock breakout in SN 1987A," *Astrophysical Journal*, vol. 393, no. 2, pp. 742–755, 1992.

[462] X.-Y. Wang, Z. Li, E. Waxman, and P. Mészáros, "Nonthermal γ-ray/x-ray flashes from shock breakout in γ-ray burst-associated supernovae," *The Astrophysical Journal*, vol. 664, no. 2 I, pp. 1026–1032, 2007.

[463] C. D. Matzner, Y. Levin, and S. Ro, "Oblique shock breakout in supernovae and gamma-ray bursts. I. Dynamics and observational implications," *Astrophysical Journal*, vol. 779, no. 1, article no. 60, 2013.

[464] R. Barniol Duran, E. Nakar, T. Piran, and R. Sari, "The afterglow of a relativistic shock breakout and low-luminosity GRBs," *Monthly Notices of the Royal Astronomical Society*, vol. 448, pp. 417–428, 2015.

[465] K. Schawinski, S. Justham, C. Wolf et al., "Supernova shock breakout from a red supergiant," *Science*, vol. 321, no. 5886, pp. 223–226, 2008.

[466] S. Gezari, D. O. Jones, N. E. Sanders et al., "Galex detection of shock breakout in type IIP supernova PS1-13arp: implications for the progenitor star wind," *The Astrophysical Journal*, vol. 804, no. 1, p. 28, 2015.

[467] P. M. Garnavich, B. E. Tucker, A. Rest et al., "Shock breakout and early light curves of type II-P supernovae observed with *Kepler*," *Astrophysical Journal*, vol. 820, no. 1, 2016.

[468] A. Heger, C. L. Fryer, S. E. Woosley, N. Langer, and D. D. H. Hartmann, "How massive single stars end their life," *The Astrophysical Journal*, vol. 591, no. 1, pp. 288–300, 2003.

[469] R. A. Chevalier, "Common envelope evolution leading to supernovae with dense interaction," *Astrophysical Journal Letters*, vol. 752, no. 1, article L2, 2012.

[470] W. Zhang and C. L. Fryer, "The merger of a helium star and a black hole: gamma-ray bursts," *The Astrophysical Journal*, vol. 550, no. 1, pp. 357–367, 2001.

[471] C. L. Fryer and A. Heger, "Binary merger progenitors for gamma-ray bursts and hypernovae," *The Astrophysical Journal*, vol. 623, no. 1, pp. 302–313, 2005.

[472] M. M. Shara, S. M. Crawford, D. Vanbeveren, A. F. Moffat, D. Zurek, and L. Crause, "The spin rates of O stars in WR + O binaries - I. Motivation, methodology, and first results from SALT," *Monthly Notices of the Royal Astronomical Society*, vol. 464, no. 2, pp. 2066–2074, 2016.

[473] P. Podsiadlowski, N. Ivanova, S. Justham, and S. Rappaport, "Explosive common-envelope ejection: implications for γ-ray bursts and low-mass black-hole binaries," *Monthly Notices of the Royal Astronomical Society*, vol. 406, no. 2, pp. 840–847, 2010.

[474] H. Sana, S. E. de Mink, A. de Koter et al., "Binary interaction dominates the evolution of massive stars," *Science*, vol. 337, no. 6093, pp. 444–446, 2012.

[475] T. Linden, V. Kalogera, J. F. Sepinsky, A. Prestwich, A. Zezas, and J. S. Gallagher, "The effect of starburst metallicity on bright X-ray binary formation pathways," *The Astrophysical Journal*, vol. 725, no. 2, pp. 1984–1994, 2010.

[476] K. F. Neugent and P. Massey, "The close binary frequency of wolf-Rayet stars as a function of metallicity in M31 and M33," *Astrophysical Journal*, vol. 789, no. 1, article no. 10, 2014.

[477] N. Smith, W. Li, A. V. Filippenko, and R. Chornock, "Observed fractions of core-collapse supernova types and initial masses of their single and binary progenitor stars," *Monthly Notices of the Royal Astronomical Society*, vol. 412, no. 3, pp. 1522–1538, 2011.

[478] S. J. Smartt, "Progenitors of core-collapse supernovae," *Annual Review of Astronomy and Astrophysics*, vol. 47, pp. 63–106, 2009.

[479] J. J. Eldridge, M. Fraser, S. J. Smartt, J. R. Maund, and R. Mark Crockett, "The death of massive stars—II. Observational constraints on the progenitors of type Ibc supernovae," *Monthly Notices of the Royal Astronomical Society*, vol. 436, no. 1, pp. 774–795, 2013.

[480] Q. E. Goddard, N. Bastian, and R. C. Kennicutt, "On the fraction of star clusters surviving the embedded phase," *Monthly Notices of the Royal Astronomical Society*, vol. 405, no. 2, pp. 857–869, 2010.

[481] E. Silva-Villa, A. Adamo, and N. Bastian, "A variation of the fraction of stars that form in bound clusters within M83," *Monthly Notices of the Royal Astronomical Society*, vol. 436, pp. L69–L73, 2013.

[482] P. G. van Dokkum and C. Conroy, "A substantial population of low-mass stars in luminous elliptical galaxies," *Nature*, vol. 468, no. 7326, pp. 940–942, 2010.

[483] C. Conroy and P. G. Van Dokkum, "The stellar initial mass function in early-type galaxies from absorption line spectroscopy. II. Results," *Astrophysical Journal*, vol. 760, no. 1, article no. 71, 2012.

[484] F. Hammer, H. Flores, D. Schaerer, M. Dessauges-Zavadsky, E. Le Floc'h, and M. Puech, "Detection of Wolf-Rayet stars in host galaxies of gamma-ray bursts (GRBs): are GRBs produced by runaway massive stars ejected from high stellar density regions?" *Astronomy and Astrophysics*, vol. 454, no. 1, pp. 103–111, 2006.

[485] J. J. Eldridge, N. Langer, and C. A. Tout, "Runaway stars as progenitors of supernovae and gamma-ray bursts," *Monthly Notices of the Royal Astronomical Society*, vol. 414, no. 4, pp. 3501–3520, 2011.

[486] N. Yadav, D. Mukherjee, P. Sharma, and B. B. Nath, "Supernovae under microscope: how supernovae overlap to form superbubbles," https://arxiv.org/abs/1603.00815.

[487] P. A. Crowther, O. Schnurr, R. Hirschi et al., "The R136 star cluster hosts several stars whose individual masses greatly exceed the accepted 150 M. stellar mass limit," *Monthly Notices of the Royal Astronomical Society*, vol. 408, no. 2, pp. 731–751, 2010.

[488] P. A. Crowther, S. M. Caballero-Nieves, K. A. Bostroem et al., "The R136 star cluster dissected with Hubble Space Telescope/STIS. I. Far-ultraviolet spectroscopic census and the origin of He II λ1640 in young star clusters," *Monthly Notices of the Royal Astronomical Society*, vol. 458, no. 1, pp. 624–659, 2016.

[489] P. Podsiadlowski, P. C. Joss, and J. J. L. Hsu, "Presupernova evolution in massive interacting binaries," *Astrophysical Journal*, vol. 391, no. 1, pp. 246–264, 1992.

[490] C. J. Burrows, J. Krist, J. J. Hester et al., "Hubble space telescope observations of the SN 1987A triple ring nebula," *The Astrophysical Journal*, vol. 452, no. 2, p. 680, 1995.

[491] G. Sonneborn, C. Fransson, P. Lundqvist et al., "The evolution of ultraviolet emission lines from circumstellar material surrounding SN 1987A," *The Astrophysical Journal*, vol. 477, no. 2, pp. 848–864, 1997.

[492] P. Podsiadlowski, J. J. L. Hsu, P. C. Joss, and R. R. Ross, "The progenitor of supernova 1993J: a stripped supergiant in a binary system?" *Nature*, vol. 364, no. 6437, pp. 509–511, 1993.

[493] S. E. Woosley, R. G. Eastman, T. A. Weaver, and P. A. Pinto, "SN 1993J: a type IIb supernova," *Astrophysical Journal*, vol. 429, no. 1, pp. 300–318, 1994.

[494] J. R. Maund, S. J. Smartt, R. P. Kudritzki, P. Podsiadlowski, and G. F. Gilmore, "The massive binary companion star to the progenitor of supernova 1993J," *Nature*, vol. 427, no. 6970, pp. 129–131, 2004.

[495] O. D. Fox, K. Azalee Bostroem, S. D. Van Dyk et al., "Uncovering the putative B-star binary companion of the SN 1993J progenitor," *The Astrophysical Journal*, vol. 790, no. 1, p. 17, 2014.

[496] D. A. Kann, P. Schady, F. Olivares et al., "Highly Luminous Supernovae associated with Gamma-Ray Bursts I.: GRB 111209A/SN 2011kl in the Context of Stripped-Envelope and Superluminous Supernovae," https://arxiv.org/abs/1606.06791.

[497] I. Arcavi, W. M. Wolf, D. A. Howell et al., "Rapidly rising transients in the supernova—superluminous supernova gap," *Astrophysical Journal*, vol. 819, no. 1, article no. 35, 2016.

[498] N. E. Sanders, A. M. Soderberg, S. Valenti et al., "SN 2010ay is a luminous and broad-lined type ic supernova within a low-metallicity host galaxy," *The Astrophysical Journal*, vol. 756, no. 2, article 184, 2012.

[499] S. Q. Wang, L. J. Wang, Z. G. Dai, and X. F. Wu, "A unified energy-reservoir model containing contributions from 56Ni and neutron stars and its implication for luminous type Ic supernovae," *The Astrophysical Journal*, vol. 807, no. 2, p. 147, 2015.

[500] A. Lien, T. Sakamoto, S. D. Barthelmy et al., "The third swift burst alert telescope gamma-ray burst catalog," *The Astrophysical Journal*, vol. 829, no. 1, article 7, 2016.

[501] M. Demleitner, A. Accomazzi, G. Eichhorn et al., "ADS's dexter data extraction applet," in *Astronomical Data Analysis Software and Systems X*, F. R. Harnden Jr., F. A. Primini, and H. E. Payne, Eds., vol. 238 of *Astronomical Society of the Pacific Conference Series*, p. 321, 2001.

[502] O. Yaron and A. Gal-Yam, "WISeREP-an interactive supernova data repository," *Publications of the Astronomical Society of the Pacific*, vol. 124, no. 917, pp. 668–681, 2012.

10

Gamma-Ray Bursts: A Radio Perspective

Poonam Chandra

National Centre for Radio Astrophysics, Tata Institute of Fundamental Research, Pune University Campus, P.O. Box 3, Pune 411007, India

Correspondence should be addressed to Poonam Chandra; poonam@ncra.tifr.res.in

Academic Editor: WeiKang Zheng

Gamma-ray bursts (GRBs) are extremely energetic events at cosmological distances. They provide unique laboratory to investigate fundamental physical processes under extreme conditions. Due to extreme luminosities, GRBs are detectable at very high redshifts and potential tracers of cosmic star formation rate at early epoch. While the launch of *Swift* and *Fermi* has increased our understanding of GRBs tremendously, many new questions have opened up. Radio observations of GRBs uniquely probe the energetics and environments of the explosion. However, currently only 30% of the bursts are detected in radio bands. Radio observations with upcoming sensitive telescopes will potentially increase the sample size significantly and allow one to follow the individual bursts for a much longer duration and be able to answer some of the important issues related to true calorimetry, reverse shock emission, and environments around the massive stars exploding as GRBs in the early Universe.

1. Introduction

Gamma-ray bursts (GRBs) are nonrecurring bright flashes of γ-rays lasting from seconds to minutes. As we currently understand, in the standard GRB model a compact central engine is responsible for accelerating and collimating the ultra-relativistic jet-like outflows. The isotropic energy release in prompt γ-rays ranges from $\sim 10^{48}$ to $\sim 10^{54}$ ergs; see, for example, [1]. While the prompt emission spectrum is mostly nonthermal, presence of thermal or quasithermal components has been suggested for a handful of bursts [2]. Since the initial discovery of GRBs [3] till the discovery of GRB afterglows at X-ray, optical, and radio wavelengths three decades later [4–7], the origin of GRBs remained elusive. The afterglow emission confirmed that GRBs are cosmological in origin, ruling out multiple theories proposed favouring Galactic origin of GRBs; see, for example, [8].

In the *BATSE* burst population, the durations of GRBs followed bimodal distribution, short GRBs with duration less than 2 s and long GRBs lasting for more than 2 s [9]. Long GRBs are predominantly found in star forming regions of late type galaxies [10], whereas short bursts are seen in all kinds of galaxies [11]. Based on these evidences, the current understanding is that the majority of long GRBs originate in the gravitational collapse of massive stars [12], whereas at least a fraction of short GRBs form as a result of the merger of compact object binaries (see Berger [13] for a detailed review).

GRBs are detectable at very high redshifts. The highest redshift GRB is GRB 090429B with a photometric redshift of $z = 9.4$ [14]. However, the farthest known spectroscopically confirmed GRB is GRB 090423 at a redshift of $z = 8.23$ [15], indicating star formation must be taking place at such early epoch in the Universe [16]. At the same time, some GRBs at lower redshifts have revealed association with type Ib/c broad lined supernovae, for example, GRB 980425 associated with SN 1998bw [17].

Since the launch of the *Swift* satellite in November 2004 [18], the field of GRB has undergone a major revolution. Burst Alert Telescope (BAT) [19] on-board *Swift* has been localizing ~100 GRBs per year [20]. X-ray Telescope (XRT [21]) and Ultraviolet/Optical Telescope (UVOT [22]) on-board *Swift* slew towards the BAT localized position within minutes and provide uninterrupted detailed light curve at these bands. Before the launch of the *Swift*, due to the lack of dedicated instruments at X-ray and optical bands the afterglow coverage was sparse, which is no longer the case. *Swift*-XRT has revealed that central engine is capable of injecting energy into the forward shock at late times [23–25].

GRBs are collimated events. An achromatic jet break seen in all frequencies is an undisputed signature of it. However, the jet breaks are seen only in a few *Swift* bursts, for example, GRB 090426 [26], GRB 130603B [27], and GRB 140903A [28]. Many of the bursts have not shown jet breaks. It could be because *Swift* is largely detecting fainter bursts with an average redshift of >2, much larger than the detected by previous instruments [20]. The faintness of the bursts makes it difficult to see jet breaks. Some of the GRBs have also revealed chromatic jet breaks, for example, GRB 070125 [29].

An additional issue is the narrow coverage of the *Swift*-BAT in 15–150 keV range. Due to the narrow bandpass, the uncertainties associated in energetics are much larger since one needs to extrapolate to 1–10,000 keV bandpass to estimate the E_{iso}, which is a key parameter to evaluate the total released energy and other relations. Due to this constraint, it has been possible to catch only a fraction of traditional GRBs.

The *Swift* drawback was overcome by the launch of *Fermi* in 2008, providing observation over a broad energy range of over seven decades in energy coverage (8 keV–300 GeV). Large Area Telescope (LAT [30]) on-board *Fermi* is an imaging gamma-ray detector in 20 MeV–300 GeV range with a field of view of about 20% of the sky and Gamma-ray Burst Monitor (GBM) [31] on-board *Fermi* works in 150 keV–30 MeV and can detect GRBs across the whole of the sky. The highest energy photon detected from a GRB puts a stricter lower limit on the outflow Lorentz factor. *Fermi* has provided useful constraints on the initial Lorentz factor owing to its high energy coverage, for example, short GRB 090510 [32]. This is because to avoid pair production, the GRB jet must be moving towards the observer with ultra-relativistic speeds. Some of the key observations by *Fermi* had been (i) the delayed onset of high energy emission for both long and short GRBs [33–35], (ii) long lasting LAT emission [36], (iii) very high Lorentz factors (~1000) inferred for the detection of LAT high energy photons [33], (iv) significant detection of multiple emission components such as thermal component in several bright bursts [37–39], and (v) power-law [35] or spectral cut-off at high energies [40], in addition to the traditional band function [41].

While the GRB field has advanced a lot after nearly 5 decades of extensive research since the first discovery, there are many open questions about prompt emission, content of the outflow, afterglow emission, microphysics involved, detectability of the afterglow emission, and so forth. Resolving them would enable us to understand GRBs in more detail and also use them to probe the early Universe as they are detectable at very high redshifts. With the recent discoveries of gravitational waves (GWs) [42, 43], a new era of Gravitational Wave Astronomy has opened. GWs are ideal to probe short GRBs as they are the most likely candidates of GW sources with earth based interferometers.

In this paper, we aim to understand the GRBs with a radio perspective. Here we focus on limited problems which can be answered with more sensitive and extensive radio observations and modeling. By no means, this review is exhaustive in nature. In Section 2, we review the radio afterglow in general and out current understanding. In Section 3, we discuss some of the open issues in GRB radio afterglows. Section 4 lists the conclusion.

2. Afterglow Physics: A Radio Perspective and Some Milestones

In the standard afterglow emission model, the relativistic ejecta interacting with the circumburst medium gives rise to a forward shock moving into the ambient circumburst medium and a reverse shock going back into the ejecta. The jet interaction with the circumburst medium gives rise to mainly synchrotron emission in X-ray, optical, and radio bands. The peak of the spectrum moves from high to low observing frequencies over time due to the deceleration of the forward shock [44] (e.g., see Figure 1). Because of the relativistic nature of the ejecta, the spectral peak is typically below optical frequencies when the first observations commence, resulting in declining light curves at optical and X-ray frequencies. However, optically rising light curve has been seen in a handful of bursts after the launch of the *Swift* [45], for example, GRB 060418 [46].

The first radio afterglow was detected from GRB 970508 [7]. Since then the radio studies of GRB afterglows have increased our understanding of the afterglows significantly, for example, [47–49]. A major advantage of radio afterglow emission is that, due to slow evolution, it peaks in much later time and lasts longer, for months or even years (e.g., [50–52]). Thus unlike short-lived optical or X-ray afterglows, radio observations present the possibility of following the full evolution of the fireball emission from the very beginning till the nonrelativistic phase (see, e.g., [50–52]); also see GRB 030329 [53, 54]. Therefore, the radio regime plays an important role in understanding the full broadband spectrum. This constrains both the macrophysics of the jet, that is, the energetics and the circumburst medium density, as well as the microphysics, such as energy imparted in electrons and magnetic fields necessary for synchrotron emission [55]. Some of the phenomena routinely addressed through radio observations are interstellar scintillation, synchrotron self-absorption, forward shocks, reverse shocks, jet breaks, nonrelativistic transitions, and obscured star formation.

The inhomogeneities in the local interstellar medium manifest themselves in the form of interstellar scintillations and cause modulations in the radio flux density of a point source whose angular size is less than the characteristic angular size for scintillations [56]. GRBs are compact objects and one can see the signatures of interstellar scintillation at early time radio observations, when the angular size of the fireball is smaller than the characteristic angular scale for interstellar scintillation. This reflects influx modulations seen in the radio observations. Eventually due to relativistic expansion, the fireball size exceeds the characteristic angular scale for scintillations and the modulations quench. This can be utilised in determining the source size and the expansion speed of the blast wave [7]. In GRB 970508 and GRB 070125, the initial radio flux density fluctuations were interpreted as interstellar scintillations, which lead to an estimation of the upper limit on the fireball size [7, 29, 57]. In GRB 070125, the scintillation time scale and modulation intensity were consistent with those of diffractive scintillations, putting a tighter constraint on the fireball size [29].

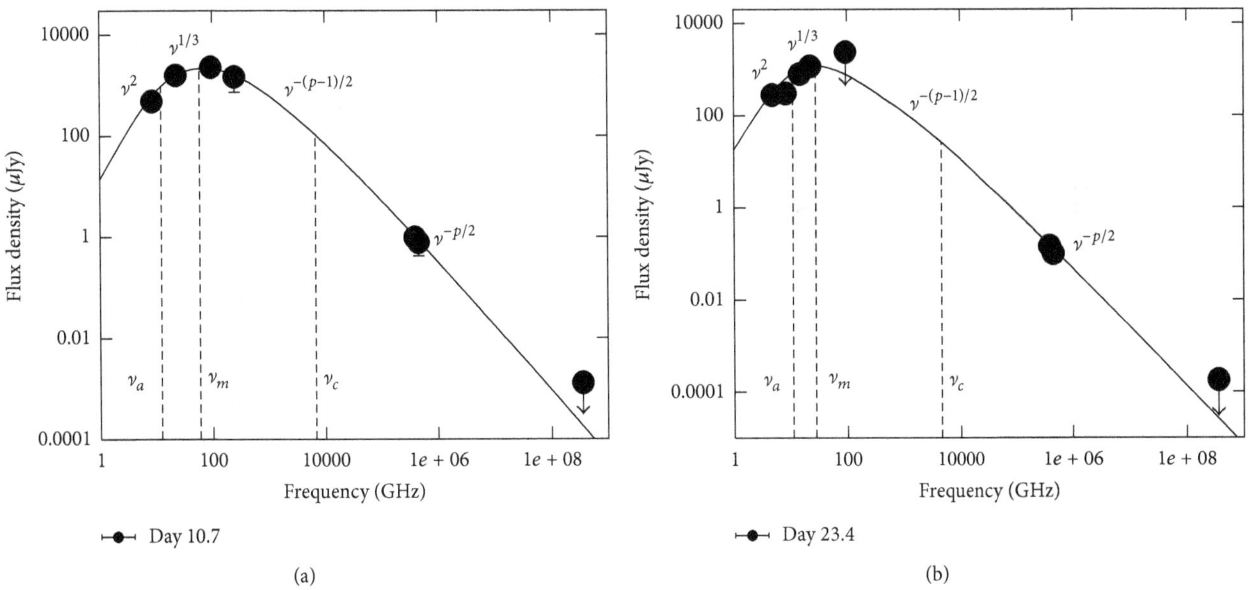

FIGURE 1: Multiwaveband spectra of GRB 070125 on day 10.7 and day 23.4. The spectra are in fast cooling regime. One can see that, between the spectra on day 10.7 and day 23.4, the peak has shifted to lower frequency. The figure is reproduced from Chandra et al. [29].

Very Long Baseline Interferometry (VLBI) radio observations also play a key role by providing evidence for the relativistic expansion of the jet using for bright GRBs. This provides microarcsecond resolution and directly constrains the source size and its evolution. So far this has been possible for a nearby ($z = 0.16$) GRB 030329 [58]. In this case, the source size measurements were combined with its long term light curves to better constrain the physical parameters [53, 54]. In addition, GRB 030329 also provided the first spectroscopic evidence for association of a GRB with a supernova. This confirmed massive stars origin of at least a class of GRBs.

Radio observations are routinely used in broadband modeling of afterglows and used to derive blast-wave parameters [1, 29, 59–61] (also see Figure 1). Early radio emission is synchrotron self-absorbed; radio observations uniquely constrain the density of the circumburst medium. Radio studies have also proven useful for inferring the opening angles of the GRB jets as their observational signature differs from those at higher wavelengths [50, 62–64]. Recently GRB 130427A, a nearby, high-luminosity event, was followed at all wavebands rigorously. It provided extremely good temporal (over 10 orders of magnitude) and spectral coverage (16 orders of magnitude in observing frequency [65, 66]). Radio observations started as early as 8 hours [67]. One witnessed reverse shock and its peak moving from high to low radio frequencies over time [67–70]. The burst is an ideal example to show how early to late-time radio observations can contribute significantly to our understanding of the physics of both the forward and reverse shocks.

Radio afterglows can be detected at high redshifts [16, 71] owing to the negative k-correction effect [72]. GRB 090423 at a redshift of 8.3 is the highest redshift (spectroscopically confirmed) known object in the Universe [15]. It was detected in radio bands for several tens of days [16]. The multiwaveband modeling indicated the $n\ 1\,\mathrm{cm}^{-3}$ density medium and the massive star origin of the GRB. This suggested that the star formation was taking place even at a redshift of 8.3.

The radio afterglow, due to its long-lived nature, is able to probe the time when the jet expansion has become subrelativistic and geometry has become quasispherical [50, 52, 73] and thus can constrain energetics independent of geometry. This is possible only in radio bands as it lasts for months or even years (e.g., [50–52]). GRB 970508 remained bright more than a year after the discovery, when the ejecta had reached subrelativistic speeds. This gave the most accurate estimate of the kinetic energy of the burst [50].

Reverse shock probes the ejecta and thus can potentially put constraints on the Lorentz factor and contents of the jet (e.g., [68, 69]). The shock moving into the ejecta will result in an optical flash in the first tens of seconds after the GRB under right conditions. The radio regime is also well suited to probe the reverse shock emission as well. Short-lived radio flares, most likely due to reverse shock, have also been detected from radio observations [16, 74–76] and seem more common in radio bands than in the optical bands. GRB 990123 was the first GRB in which the reverse shock was detected in optical [77] as well as in radio bands [74].

From the radio perspective, GRB 030329 holds a very important place. It was the first high-luminosity burst at low redshift with a spectroscopic confirmation of a supernova associated with it. So far this is the only GRB for which the source size has been measured with VLBI. The radio afterglow of GRB 030329 was bright and long lasting and has been detected for almost a decade at radio frequencies [52, 78]. This enabled one to perform broadband modeling in the different phases and has led to tighter constraints on the physical parameters [53, 54]. However, the absence of a counter

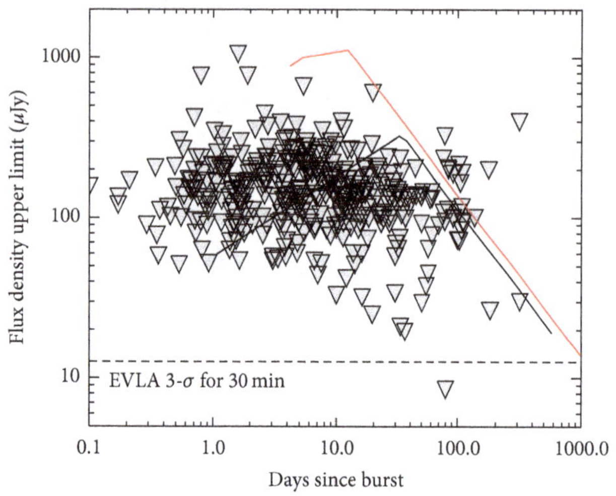

FIGURE 2: Plot of 3-σ upper limits at 8.5 GHz frequency band for all GRBs for which no afterglow was detected. The red line shows light curve of a rare, bright event GRB 980703 and the blue line shows the light curve of a more typical event GRB 980329. The detection fraction of radio afterglows in the first 10 days certainly appears to be mainly limited by the sensitivity. The black dashed line indicates 3-σ sensitivity of the JVLA in its full capacity for a 30-minute integration time. The figure is reproduced from [48].

jet poses serious question in our understanding of GRBs [79].

3. Open Problems in GRB Radio Afterglows

With various high sensitivity new and refurbished telescopes, for example, Atacama Large Millimetre Array (ALMA), Karl J. Jansky Very Large Array (JVLA), upgraded Giant Metrewave Radio Telescope (uGMRT), and upcoming telescopes, for example, Square Kilometre Array (SKA), the radio afterglow physics of GRBs is entering into new era, where we can begin to answer some of the open questions in the field, answers to which are long awaited. In this section, I discuss only some of those open problems in GRB science where radio measurements can play a crucial role.

This review is not expected to be exhaustive. We concentrate on only a few major issues.

3.1. Are GRBs Intrinsically Radio Weak? Since the launch of the *Swift*, the fractions of X-ray and optically detected afterglows have increased tremendously; that is, almost 93% of GRBs have a detected X-ray afterglow [80] and ~75% have detected optical afterglows [81, 82]. However, what is disconcerting is that the radio detection fraction has remained unchanged with only one-third of all GRBs being detected in radio bands [47, 48]. Chandra and Frail [48] attributed it to sensitivity limitation of the current telescopes (see Figure 2). This is because radio detected GRBs have flux densities typically ranging from a few tens of μJy to a few hundreds of μJy [48]. Even the largest radio telescopes have had the sensitivities close to a few tens of μJy, making the radio afterglow detection sensitivity limited. The newer generation radio telescopes should dramatically improve statistics of radio afterglows. For example, using numerical simulation of the forward shock, Burlon et al. [83] predict that the SKA-1 (SKA first phase) Mid band will be able to detect around 400–500 radio afterglows per sr^{-1} yr^{-1}.

The Five-hundred-meter Aperture Spherical radio Telescope (FAST) [84–86] is the largest worldwide single-dish radio telescope, being built in Guizhou province of China with an expected first light in Sep. 2016. FAST will continuously cover the radio frequencies between 70 MHz and 3 GHz. The radio afterglow of GRBs is one of the main focuses of FAST. Zhang et al. [84] have estimated the detectability with FAST of various GRBs like failed GRBs, low-luminosity GRBs, high-luminosity GRBs, and standard GRBs. They predict that FAST will be able to detect most of the GRBs other than subluminous ones up to a redshift of $z \leq 10$.

However, Hancock et al. [87] used stacking of radio visibility data of many GRBs and their analysis still resulted in nondetection. Based on this they proposed a class of GRBs which will produce intrinsically faint radio afterglow emission and have black holes as their central engine. GRBs with magnetars as central engine will produce radio bright afterglow emission. This is because the magnetar driven GRBs will have lower radiative efficiency and produce radio bright GRBs, whereas the black hole driven GRBs with their high radiative efficiency will use most of their energy budget in prompt emission and will be radio-faint. This is a very important aspect and may need to be addressed. And if true, it may reflect the nature of the central engine through radio measurements. JVLA at high radio frequencies and the uGMRT at low radio frequencies test this hypothesis. SKA will eventually be the ultimate instrument to distinguish between the sensitivity limitation and the intrinsic dimness of radio bursts [83].

3.2. Hyperenergetic GRBs. Accurate calorimetry is very important to understand the true nature of the GRBs. This includes prompt radiation energy in the form of γ-rays and kinetic energy in the form of shock powering the afterglow emission. Empirical constraints from models require that all long duration GRBs have the kinetic energies $\leq$ 10^{51} ergs. GRBs are collimated events; thus the jet opening angle is crucial to measure the true budget of the energies. While isotropic energies range of energies spread in four orders of magnitude (see Figure 3), the collimated nature of the jet makes the actual energies in much tighter range clustered around 10^{51} ergs [75, 88, 89]. However, it is becoming increasingly evident that the clustering may not be as tight as envisaged and the actual energy range may be much wider than anticipated earlier. A population of nearby GRBs have relativistic energy orders of magnitude smaller than a typical cosmological GRB; these are called subluminous GRBs, for example, GRB 980425 [25, 90]. *Fermi* has provided evidence for a class of hyperenergetic GRBs. These GRBs have total prompt and kinetic energy release, inferred via broadband modeling [61, 91], to be at least an order of magnitude above the canonical value of 10^{51} erg [1, 29, 48, 92]. The total energy budget of these hyperenergetic GRBs poses a significant challenge for some accepted progenitor models. The maximum energy release in magnetar models [93] is

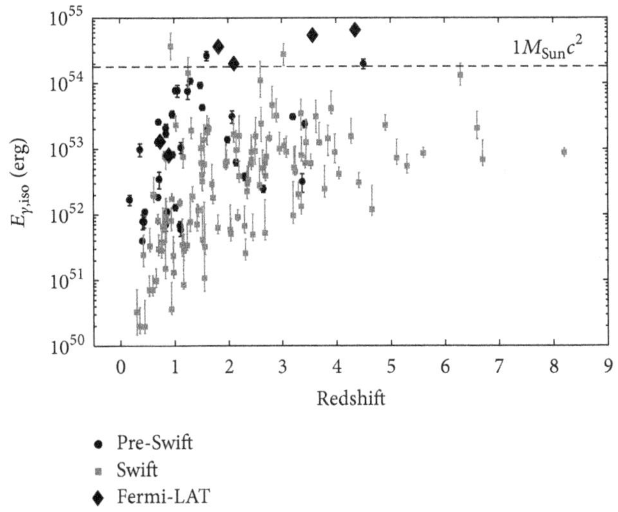

FIGURE 3: Isotropic prompt gamma-ray energy release ($E_{\gamma,\mathrm{iso}}$, in rest frame 1 keV–10 MeV bandpass) of GRBs with measured redshift. One can see a large range of $E_{\gamma,\mathrm{iso}}$. Reproduced from Cenko et al. [1].

3×10^{52} erg, set by the rotational energy of a maximally rotating stable neutron star [94, 95].

It has been very difficult to constrain the true prompt energy budget of the GRBs, mainly, for the following reasons. So far, *Swift* has been instrumental in detecting majority of the GRBs. However, peaks of the emission for various GRBs lie outside the narrow energy coverage of *Swift*-BAT (15–150 keV). In addition, extrapolation of 15–150 keV to 1–10,000 keV bandpass causes big uncertainties in the determination of prompt isotropic energies. With its huge energy coverage (8 keV–300 GeV), *Fermi* has overcome some of these limitations and provided unparalleled constraints on the spectral properties of the prompt emission. *Fermi* has been able to distinguish the true hyperenergetic bursts (such as GRB 090323, GRB 090902B, and GRB 090926A [1]; also see Figure 3). While *Swift* sample is biased towards faint bursts, *Fermi* sample is biased towards GRBs with very large isotropic energy releases (10^{54} erg), which even after collimation correction reach very high energies, for example, [1, 96], and provide some of the strongest constraints on possible progenitor models.

The uncertainty in jet structure in GRBs pose additional difficulty in constraining the energy budget of GRBs. Even after a jet break is seen, to convert it into opening angle, one needs density to convert it into the collimation angle. While some optical light curves can be used to constrain the circumburst density (e.g., Liang et al. [45]), radio SSA peak is easier to detect due to slow evolution in radio bands. With only one-third of sample being radio bright, this has been possible for only a handful of bursts. A larger radio sample at lower frequencies, at early times when synchrotron self-absorption (SSA) is still playing a major role, could be very useful. The uGMRT after upgrade will be able to probe this regime as SSA will be affecting the radio emission at longer wavelength for a longer time. However, the this works on the assumption that the entire relativistic outflow is collimated into a single uniform jet. While the proposed double-jet models for GRB 030329 [97, 98] and GRB 080319B [99] ease out the extreme efficiency requirements, it has caused additional concerns.

The ALMA also has an important role to play since GRB spectrum at early times peak at mm wavelengths, when it is the brightest. ALMA with its high sensitivity can detect such events at early times and give better estimation of the kinetic energy of the burst.

While X-ray and optical afterglows stay above detection limits only for weeks or months, radio afterglows of nearby bursts can be detected up to years [50, 100]. The longevity of radio afterglows also makes them interesting laboratories to study the dynamics and evolution of relativistic shocks. At late stages, the fireball would have expanded sideways so much that it would essentially make transition into nonrelativistic regime and become quasispherical and independent of the jet geometry; calorimetry can be employed to obtain the burst energetics [50, 52]. These estimates will be free of relativistic effects and collimation corrections. This regime is largely unexplored due to limited number of bursts staying above detection limit beyond subrelativistic regime. Several numerical calculations exist for the afterglow evolution starting from the relativistic phase and ending in the deep nonrelativistic phase [79, 101]. SKA with its μJy level sensitivity will be able to extend the current limits of afterglow longevity. This will provide us with an unprecedented opportunity to study the nonrelativistic regime of afterglow dynamics and thereby will be able to refine our understanding of relativistic to nonrelativistic transition of the blast-wave and changing shock microphysics and calorimetry in the GRBs. Burlon et al. [83] have computed that SKA1-MID will be able to observe 2% afterglows till the nonrelativistic (NR) transition but that the full SKA will routinely observe 15% of the whole GRB afterglow population at the NR transition.

3.3. Can Jet Breaks Be Chromatic? After the launch of *Swift*, one obtained a far better sampled optical and X-ray light curves, thus expected to witness achromatic jet breaks across the electromagnetic spectrum, a robust signature associated with a collimated outflow. Several groups conducted a comprehensive analysis of a large sample of light curves of *Swift* bursts in the X-rays [102–105] and optical [106] bands. Surprisingly fewer *Swift* bursts have shown this unambiguous signature of the jet collimation. Without these collimation angles, the true energy release from *Swift* events has remained highly uncertain. A natural explanation for absence of the jet breaks can be attributed to the high sensitivity of *Swift*. Due to its high sensitivity *Swift* is preferentially selecting GRBs with smaller isotropic gamma-ray energies and larger redshifts. This dictates that typical *Swift* events will have large opening angles, thus causing jet breaks to occur at much time than those of pre-*Swift* events. Since afterglow is already weak at later times, making jet break measurements is quite difficult [103, 107].

There have been some cases where chromatic jet breaks are also seen. For example, in GRB 070125, the X-ray jet break occurred around day 10, whereas the optical jet break

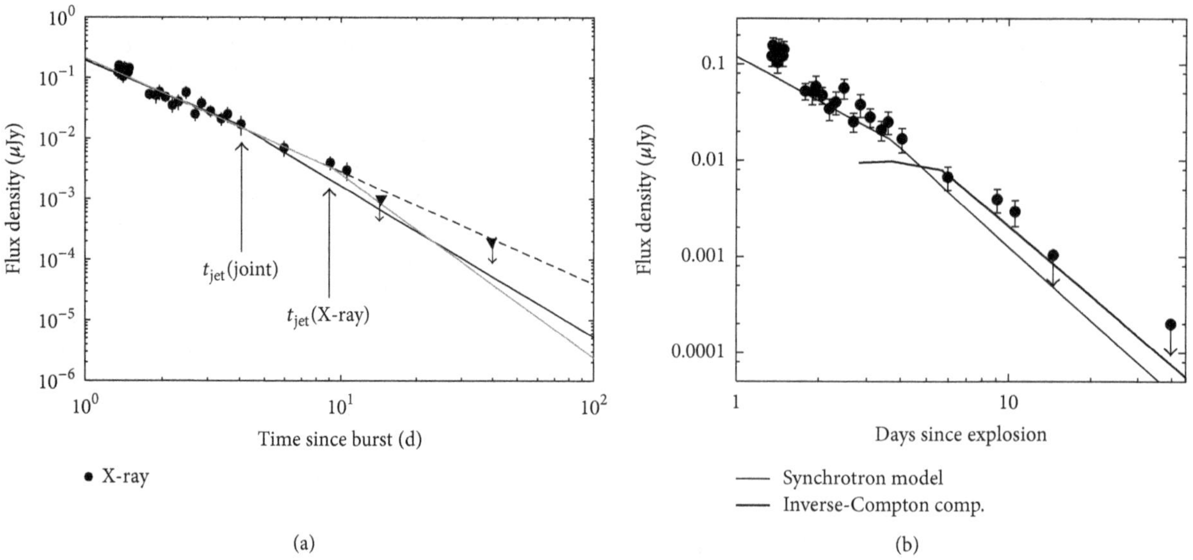

FIGURE 4: (a) X-ray light curve of GRB 070125. Best-fit single power-law models are shown with dashed lines, while the broken power-law models are shown in solid lines. The t_{jet}(joint) is the joint fit to optical and X-ray data and grey solid line t_{jet}(X-ray) is the independent fit. The independent fit shifts the jet break to ~9-10 days, which was found to be day 3 for optical bands. (b) Contribution of IC in the synchrotron model for the X-ray light curve of GRB 070125. The thin line represents the broadband model with the synchrotron component only. The thick line represents the IC light curve. One can see that IC effect can delay the jet breaks in X-ray bands [29].

occurred on day 3. Chandra et al. [29] attributed it to inverse Compton (IC) effect, which does not affect the photons at low energies but shifts the X-ray jet break at a later time (see Figure 4, [29]). As IC effects are dominant in high density medium, radio observations are an important indicator of the effectiveness of the IC effect. Chandra et al. [29] showed that, for a given density of GRB 070125, the estimated delay in X-ray jet break due to the IC effect is consistent with the observed delay. However, this area needs to be explored further for other GRBs. While high density bursts are likely to be brighter in radio bands, it may cause a burst to be a dark one in optical wavelength (Xin et al. [108] and references therein), which then make it difficult to detect the jet break simultaneously in several wavelengths. uGMRT and JVLA will be ideal instruments to probe IC effect and will potentially be able to explain the cause of chromaticity in some of the *Swift* bursts.

3.4. High-z GRBs and PoP III Stars. One of the major challenges of the observational cosmology is to understand the reionization of the Universe, when the first luminous sources were formed. So far quasar studies of the Gunn-Peterson absorption trough, the luminosity evolution of Lyman galaxies, and the polarization isotropy of the cosmic microwave background have been used as diagnostics. But they have revealed a complicated picture in which reionization took place over a range of redshifts.

The ultraviolet emission from young, massive stars (see Fan et al. [109] and references therein) appears to be the dominant source of reionization. However, none of these massive stars have been detected so far. Long GRBs, which are explosions of massive stars, are detectable out to large distances due to their extreme luminosities and thus are the potential signposts of the early massive stars. GRBs are predicted to occur at redshifts beyond those where quasars are expected; thus they could be used to study both the reionization history and the metal enrichment of the early Universe [110]. They could potentially reveal the stars that form from the first dark matter halos through the epoch of reionization [72, 111, 112]. The radio, infrared, and X-ray afterglow emission from GRBs are in principle observable out to $z = 30$ [72, 111–114]. Thus GRB afterglows make ideal sources to probe the intergalactic medium as well as the interstellar medium in their host galaxies at high z.

The fraction of detectable GRBs that lie at high redshift ($z > 6$) is, however, expected to be less than 10% [115, 116]. So far there are only 3 GRBs with confirmed measured redshifts higher than 6. These are GRB 050904 [117], GRB 080913 [118], and GRB 090423 [15]. Radio bands are ideal to probe GRB circumburst environments at high redshift because radio flux density show only a weak dependence on the redshift, due to the negative k-correction effect [72] (also see [47] and Figure 5). In k-correction effect, the afterglow flux density remains high because of the dual effects of spectral and temporal redshift, offsetting the dimming due to the increase in distance [111] (see Figure 5). GRB 050904 and GRB 090423 were detected in radio bands and radio observations of these bursts allowed us to put constraints on the density of the GRB environments at such high redshifts. While the density of GRB 090423 was $n \sim 1\,\mathrm{cm}^{-3}$ [16] (Figure 5), the density of GRB 050904 was $\sim 100\,\mathrm{cm}^{-3}$, indicating dense molecular cloud surrounding the GRB 050904 [119]. This revealed that these two high-z GRBs exploded in a very different environment.

ALMA will be a potential tool for selecting potential high-z bursts that would be suitable for intense follow-up across

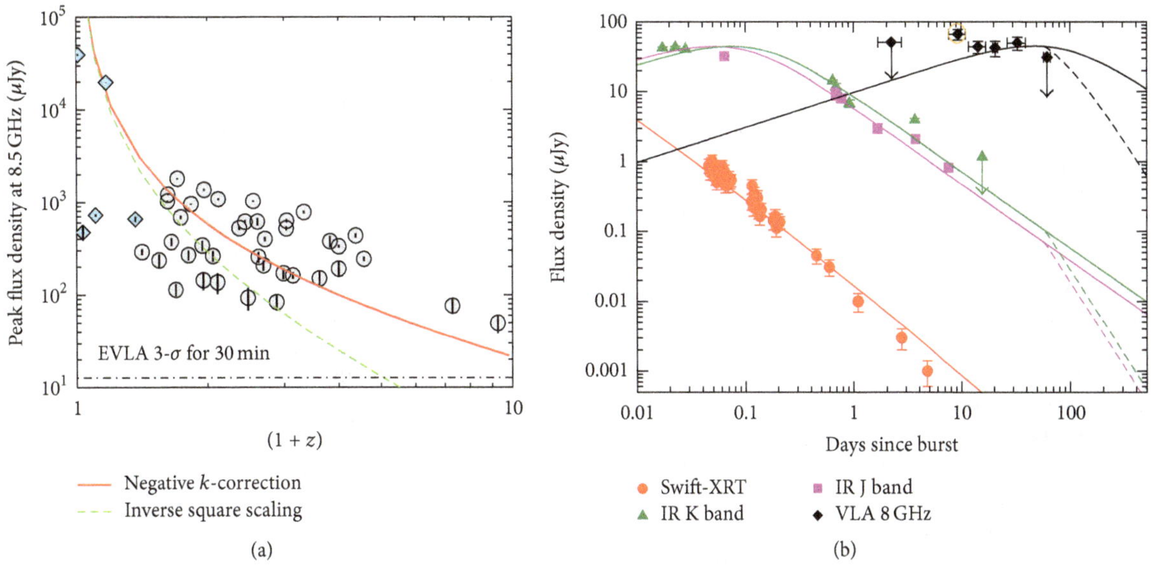

FIGURE 5: (a) The 8.5 GHz radio peak flux density versus $(1 + z)$ plot for radio afterglows with known redshifts. Blue diamonds are GRBs associated with supernovae, while the grey circles denote cosmological GRBs. The green dashed line indicates if the flux density scales as the inverse square of the luminosity distance. The red thick line is the flux density scaling in the canonical afterglow model which includes a negative k-correction effect, offsetting the diminution in distance (reproduced from [48]). (b) Multiwaveband afterglow modeling of highest redshift GRB 090423 at $z = 8.23$ (reproduced from [16]).

the electromagnetic spectrum. With an order of magnitude enhanced sensitivity the JLA will be able to study a high-z GRB for a longer timescale. For example, VLA can detect GRB 090423-like burst for almost 2 years. The uGMRT can also detect bright bursts up to a redshift of $z \sim 9$. These measurements will therefore obtain better density measurements and reveal the environments where massive stars were forming in the early Universe.

3.5. Reverse Shock. In a GRB explosion, there is a forward shock moving forward into the circumburst medium, as well as a reverse shock moving backwards into the ejecta [120]. The nearly self-similar behavior of a forward shock means that little information is preserved about the central engine properties that gave rise to the GRB. In contrast, the brightness of the short-lived reverse shock depends on the initial Lorentz factor and the magnetization of the ejecta. Thus, multifrequency observations of reverse shocks tell about the acceleration, the composition, and the strength and orientation of any magnetic fields in the relativistic outflows from GRBs [68, 69, 121–123]. In general, the reverse shock is expected to result in an optical flash in the first tens of seconds after the GRB [77], which makes it difficult to detect as robotic telescopes are required for fast triggers.

The discovery of a bright optical flash from GRB 990123 [77] leads to extensive searches for reverse shocks [124–127] in optical bands. One expected to see more evidences of reverse shocks in optical bands due to *Swift*-UVOOT; however, based on these efforts it seems that the incidence of optical reserve shocks is low. Since the peak of this emission moves to lower frequencies over time and can be probed at radio frequencies on a time scale of hours to days [74], the radio regime is well suited for studying early time reverse shock phenomena.

There have been several observational as well as theoretical studies of radio reverse shock emission in the literature after the first reverse shock detection in GRB 990123 [74]. Gao et al. [128], Kopač et al. [129], and Resmi and Zhang [130] have done comprehensive analytical and numerical calculations of radio reverse shock emissions and about their detectability. It has been shown [48, 67] that deep and fast monitoring campaigns of radio reverse shock emission could be achieved with the VLA for a number of bursts. JVLA radio frequencies are well suited as reverse shock emission is brighter in higher radio frequencies where self-absorption effects are relatively lesser. Radio afterglow monitoring campaigns in higher SKA bands (e.g., SKA1-Mid Band-4 and Band-5) will definitely be useful in exploring reverse shock characteristics [83].

Reverse shock is detectable in high redshift GRBs ($z \geq 6$) as well. Inoue et al. [131] have predicted that at mm bands the effects of time dilation almost compensate for frequency redshift, thus resulting in a near-constant observed peak frequency at a few hours after event and a flux density at this frequency that is almost independent of redshift. Thus ALMA mm band is ideal to look for reverse shock signatures at high redshifts. Burlon et al. [83] predict that SKA1-Mid will be able to detect a reverse shock from a GRB990123 like GRB at a redshift of ~10.

3.6. Connecting Prompt and Afterglow Physics. Swift is an ideal instrument for quick localization of GRBs and rapid follow-up and consequently redshift measurement [20, 132] and *Fermi* for the wideband spectral measurement during the prompt emission. However, good spectral and timing measurement covering early prompt to late afterglow phase is available for a few sources and rarely available for the short GRBs. Some of the key problems that can be addressed by the observation of the radio afterglows in connection with the prompt emission are (i) comparing the Lorentz factor estimation with both LAT detected GeV photons as well as from the reverse shock [133, 134]; (ii) comparison between nonthermal emission of both the prompt and afterglow emission, which would enable one to constrain the microphysics of the shocks accelerating electrons to ultra-relativistic energies eventually producing the observed radiation; (iii) detailed modeling of the afterglow observation of both long and short GRBs, which will enhance our knowledge about the circumburst medium surrounding the progenitors; (iv) current refurbished and upcoming radio telescopes with their finer sensitivity, which would play a key role in constraining the energetics of GRBs which is crucial in estimating the radiation efficiency of the prompt emission of GRBs. This would strengthen the understanding of the hardness-intensity correlation [135].

The recently launched *AstroSAT* satellite [136] carries several instruments enabling multiwavelength studies. The Cadmium Zinc Telluride Imager (CZTI) on-board *AstroSAT* can provide time resolved polarization measurements for bright GRBs and can act as a monitor above 80 keV [137, 138]. So far no other instrument has such capability to detect polarization. Hence, for a few selected bright GRBs, CZTI, in conjunction with ground based observatories like uGMRT and JVLA, and other space based facilities can provide a complete observational picture of a few bright GRBs from early prompt phase to late afterglow. This will provide us with a comprehensive picture of GRBs, thus enabling a good understanding of the emission mechanisms.

3.7. Some Other Unresolved Issues. So far I have discussed only that small fraction of on-axis GRBs, in which the jet is oriented along our line of sight. Due to large Lorentz factors, small opening angles of the collimated jets, we only detect a small fraction of GRBs [139]. Ghirlanda et al. [140] have estimated that, for every GRB detected, there must be 260 GRBs which one is not able to detect. However, their existence can be witnessed as "orphan afterglow" at late times when the GRB jet is decelerated and spread laterally to come into our line of sight. At such late times, the emission is expected to come only in radio bands. So far attempts to find such orphan radio afterglows have been unsuccessful [75, 141, 142]. Even if detected, disentangling the orphan afterglow emission from other classes will be very challenging. Soderberg et al. [141] carried out a survey towards the direction of 68 Type Ib/c supernovae looking for the orphan afterglows and put limit on GRB opening angles, $\theta_j > 0.8$ d. The detection of population of orphan afterglows with upcoming sensitive radio facilities is promising. This will give a very good handle on jet opening angles and on the total GRB rate whether beamed towards us or not.

The inspiral and merger of binary systems with black holes or neutron stars have been speculated as primary source of gravitational waves (GWs) for the ground based GW interferometers [143, 144]. The discovery of GWs from GW 150914 [42] and GW 151226 [43] with the Advanced LIGO detectors have provided the first observational evidence of the binary black hole systems inspiraling and merging. At least some of the compact binaries involving a neutron star are expected to give rise to radio afterglows of short GRBs. Electromagnetic counterparts of GW source, including emission in the radio bands, are highly awaited as they will, for the first time, confirm the hypothesis of binary merger scenario for GW waves. If localized at high energies, targeted radio observations can be carried out to study these events at late epochs.

Short GRBs arising from mergers of two neutron stars eject significant amount of mass in several components, including subrelativistic dynamical ejecta, mildly relativistic shock-breakout, and a relativistic jet [145]. Hotokezaka and Piran [145] have calculated the expected radio signals produced between the different components of the ejecta and the surrounding medium. The nature of radio emission years after GRB will provide invaluable information on the merger process [145] and the central products [146]. Fong et al. [146] have predicted that the formation of stable magnetar of energy 10^{53} erg during merger process will give rise to a radio transient a year later. They carried out search for radio emission from 9 short GRBs in rest frame times of 1–8 years and concluded that such a magnetar formation can be ruled out in at least half their sample.

In addition, radio observations can also probe the star formation and the metallicity of the GRB host galaxies when optical emissions are obscured by dust [147, 148].

4. Conclusions

In this article, I have reviewed the current status of the *Swift*/*Fermi* GRBs in context of their radio emission. With improved sensitivity of the refurbished radio telescopes, such as JVLA and uGMRT and upcoming telescopes like SKA, it will be possible to answer many open questions. The most crucial of them is the accurate calorimetry of the GRBs. Even after observing a jet break in the GRB afterglow light curves, which is an unambiguous signature of the jet collimation, one needs density estimation to convert the jet break epoch to collimation angle. The density information can be more effectively provided by the early radio measurements when the GRBs are still synchrotron self-absorbed. So far it has been possible for very limited cases because only one-third of the total GRBs have been detected in radio bands [48]. Sensitive radio measurements are needed to understand whether the low detection rate of radio afterglows is intrinsic to GRBs or the sensitivity limitations of the current telescopes are playing a major role. In the era of JVLA, uGMRT, ALMA, and upcoming SKA, this issue should be resolved. In addition, these sensitive radio telescopes will be crucial to detect radio afterglows at very high redshifts and provide unique constraints on the environments of the exploding massive stars in the early Universe. If GRBs are not intrinsically dim in radio bands and the sample is indeed sensitivity limited, then SKA is expected to detect almost 100% GRBs [83]. SKA will be able to study the individual bursts in great detail. This will also allow us to carry out various statistical analyses of the radio sample and drastically increase our overall understanding of the afterglow evolution from very early time to nonrelativistic regime. Detection of the orphan afterglow is due any time and will be novel in itself.

Competing Interests

The author declared that there are no competing interests.

Acknowledgments

The author thanks L. Resmi, Shabnam Iyyanni, A. R. Rao, Kuntal Misra, and D. Frail for many useful discussions in the past, which helped shape this article. The author acknowledges support from the Department of Science and Technology via SwarnaJayanti Fellowship Award (File no. DST/SJF/ PSA-01/2014-15). The author also acknowledges SKA Italy handbook (http://pos.sissa.it/cgi-bin/reader/conf.cgi?confid=215), where many of the SKA numbers on sensitivity, GRB detection rates, and so forth are taken.

References

[1] S. B. Cenko, D. A. Frail, F. A. Harrison et al., "Afterglow observations of Fermi large area telescope gamma-ray bursts and the emerging class of hyper-energetic events," *The Astrophysical Journal*, vol. 732, no. 1, p. 29, 2011.

[2] P. Kumar and B. Zhang, "The physics of gamma-ray bursts & relativistic jets," *Physics Reports*, vol. 561, pp. 1–109, 2015.

[3] R. W. Klebesadel, I. B. Strong, and R. A. Olson, "Observations of gamma-ray bursts of cosmic origin," *Astrophysical Journal*, vol. 182, article L85, 1973.

[4] E. Costa, F. Frontera, J. Heise et al., "Discovery of an X-ray afterglow associated with the γ-ray burst of 28 February 1997," *Nature*, vol. 387, no. 6635, pp. 783–785, 1997.

[5] J. van Paradijs, P. J. Groot, T. Galama et al., "Transient optical emission from the error box of the γ-ray burst of 28 February 1997," *Nature*, vol. 386, no. 6626, pp. 686–689, 1997.

[6] R. A. M. J. Wijers, M. J. Rees, and P. Meszaros, "Shocked by GRB 970228: the afterglow of a cosmological fireball," *Monthly Notices of the Royal Astronomical Society*, vol. 288, no. 4, pp. L51–L56, 1997.

[7] D. A. Frail, S. R. Kulkarnit, L. Nicastro, M. Feroci, and G. B. Taylor, "The radio afterglow from the γ-ray burst of 8 May 1997," *Nature*, vol. 389, no. 6648, pp. 261–263, 1997.

[8] G. J. Fishman and C. A. Meegan, "Gamma-ray bursts," *Annual Review of Astronomy & Astrophysics*, vol. 33, pp. 415–458, 1995.

[9] C. Kouveliotou, C. A. Meegan, G. J. Fishman et al., "Identification of two classes of gamma-ray bursts," *The Astrophysical Journal*, vol. 413, no. 2, pp. L101–L104, 1993.

[10] A. S. Fruchter, A. J. Levan, L. Strolger et al., "Long big γ-ray bursts and core-collapse supernovae have different environments," *Nature*, vol. 441, pp. 463–468, 2006.

[11] W. Fong, E. Berger, and D. B. Fox, "*Hubble Space Telescope* observations of short gamma-ray burst host galaxies: morphologies, offsets, and local environments," *The Astrophysical Journal*, vol. 708, no. 1, p. 9, 2010.

[12] S. E. Woosley and J. S. Bloom, "The supernova-γ-ray burst connection," *Annual Review of Astronomy and Astrophysics*, vol. 44, pp. 507–566, 2006.

[13] E. Berger, "Short-duration gamma-ray bursts," *Annual Review of Astronomy and Astrophysics*, vol. 52, no. 1, pp. 43–105, 2014.

[14] A. Cucchiara, A. J. Levan, D. B. Fox et al., "A photometric redshift of $z \sim 9.4$ for GRB 090429B," *The Astrophysical Journal*, vol. 736, no. 1, p. 7, 2011.

[15] N. R. Tanvir, D. B. Fox, A. J. Levan et al., "A big γ-ray burst at a redshift of $z \approx 8.2$," *Nature*, vol. 461, pp. 1254–1257, 2009.

[16] P. Chandra, D. A. Frail, D. Fox et al., "Discovery of radio afterglow from the most distant cosmic explosion," *The Astrophysical Journal Letters*, vol. 712, no. 1, pp. L31–L35, 2010.

[17] S. R. Kulkarni, D. A. Frail, M. H. Wieringa et al., "Radio emission from the unusual supernova 1998bw and its association with the γ-ray burst of 25 April 1998," *Nature*, vol. 395, no. 6703, pp. 663–669, 1998.

[18] N. Gehrels, G. Chincarini, P. Giommi et al., "The *Swift* γ-ray burst mission," *The Astrophysical Journal*, vol. 611, no. 2, p. 1005, 2004.

[19] S. D. Barthelmy, L. M. Barbier, J. R. Cummings et al., "The burst alert telescope (BAT) on the SWIFT midex mission," *Space Science Reviews*, vol. 120, no. 3-4, pp. 143–164, 2005.

[20] N. Gehrels, E. Ramirez-Ruiz, and D. B. Fox, "Gamma-ray bursts in the *Swift* era," *Annual Review of Astronomy and Astrophysics*, vol. 47, pp. 567–617, 2009.

[21] D. N. Burrows, J. E. Hill, J. A. Nousek et al., "The wift X-ray telescope," *Space Science Reviews*, vol. 120, no. 3-4, pp. 165–195, 2005.

[22] P. W. A. Roming, T. E. Kennedy, K. O. Mason et al., "The *Swift* ultra-violet/optical telescope," *Space Science Reviews*, vol. 120, no. 3, pp. 95–142, 2005.

[23] Z. G. Dai and T. Lu, "γ-ray burst afterglows and evolution of postburst fireballs with energy injection from strongly magnetic millisecond pulsars," *Astronomy and Astrophysics*, vol. 333, no. 3, pp. L87–L90, 1998.

[24] B. Zhang and P. Mészáros, "Gamma-ray bursts with continuous energy injection and their afterglow signature," *The Astrophysical Journal*, vol. 566, no. 2, pp. 712–722, 2002.

[25] E.-W. Liang, B.-B. Zhang, and B. Zhang, "A comprehensive analysis of *Swift* XRT data. II. Diverse physical origins of the shallow decay segment," *The Astrophysical Journal*, vol. 670, no. 1, pp. 565–583, 2007.

[26] A. N. Guelbenzu, S. Klose, A. Rossi et al., "GRB 090426: discovery of a jet break in a short burst afterglow," *Astronomy and Astrophysics*, vol. 531, article L6, 2011.

[27] W. Fong, E. Berger, B. D. Metzger et al., "short GRB 130603B: discovery of a jet break in the optical and radio afterglows, and a mysterious late-time x-ray excess," *The Astrophysical Journal*, vol. 780, no. 2, p. 118, 2014.

[28] E. Troja, T. Sakamoto, S. B. Cenko et al., "An achromatic break in the afterglow of the short GRB 140903A: evidence for a narrow jet," *The Astrophysical Journal*, vol. 827, no. 2, p. 102, 2016.

[29] P. Chandra, S. B. Cenko, D. A. Frail et al., "A comprehensive study of GRB 070125, a most energetic gamma-ray burst," *The Astrophysical Journal*, vol. 683, no. 2, p. 924, 2008.

[30] W. B. Atwood, A. A. Abdo, M. Ackermann et al., "The large area telescope on the *Fermi Gamma-Ray Space Telescope* mission," *The Astrophysical Journal*, vol. 697, no. 2, p. 1071, 2009.

[31] C. Meegan, G. Lichti, P. N. Bhat et al., "The fermi γ-ray burst monitor," *Astrophysical Journal*, vol. 702, no. 1, pp. 791–804, 2009.

[32] M. Ackermann, K. Asano, W. B. Atwood et al., "Fermi observations of GRB 090510: a short-hard γ-ray burst with an additional, hard power-law component from 10 Kev to GeV energies," *The Astrophysical Journal*, vol. 716, no. 2, p. 1178, 2010.

[33] A. A. Abdo, M. Ackermann, M. Ajello et al., "A limit on the variation of the speed of light arising from quantum gravity effects," *Nature*, vol. 462, pp. 331–334, 2009.

[34] A. A. Abdo, M. Ackermann, M. Arimoto et al., "Fermi observations of high-energy gamma-ray emission from GRB 080916C," *Science*, vol. 323, no. 5922, pp. 1688–1693, 2009.

[35] A. A. Abdo, M. Ackermann, M. Ajello et al., "*FERMI* observations of GRB 090902b: a distinct spectral component in the prompt and delayed emission," *The Astrophysical Journal Letters*, vol. 706, no. 1, p. L138, 2009.

[36] M. Ackermann, M. Ajello, K. Asano et al., "The first *Fermi*-LAT gamma-ray burst catalog," *The Astrophysical Journal Supplement Series*, vol. 209, no. 1, p. 11, 2013.

[37] S. Guiriec, V. Connaughton, M. S. Briggs et al., "Detection of a thermal spectral component in the prompt emission of GRB 100724B," *The Astrophysical Journal Letters*, vol. 727, no. 2, p. L33, 2011.

[38] M. Axelsson, L. Baldini, G. Barbiellini et al., "GRB110721A: an extreme peak energy and signatures of the photosphere," *The Astrophysical Journal Letters*, vol. 757, no. 2, p. L31, 2012.

[39] J. M. Burgess, R. D. Preece, V. Connaughton et al., "Time-resolved analysis of *FERMI* gamma-ray bursts with fast- and slow-cooled synchrotron photon models," *The Astrophysical Journal*, vol. 784, no. 1, p. 17, 2014.

[40] M. Ackermann, M. Ajello, K. Asano et al., "Detection of a spectral break in the extra hard component of GRB 090926A," *The Astrophysical Journal*, vol. 729, no. 2, p. 114, 2011.

[41] D. Band, J. Matteson, L. Ford et al., "BATSE observations of gamma-ray burst spectra. I—spectral diversity," *The Astrophysical Journal*, vol. 413, no. 1, pp. 281–292, 1993.

[42] B. P. Abbott, R. Abbott, T. D. Abbott et al., "GW150914: implications for the stochastic gravitational-wave background from binary black holes," *Physical Review Letters*, vol. 116, no. 13, Article ID 131102, 12 pages, 2016.

[43] B. P. Abbott, R. Abbott, T. D. Abbott et al., "Observation of gravitational waves from a binary black hole merger," *Physical Review Letters*, vol. 116, no. 6, Article ID 061102, 16 pages, 2016.

[44] R. Sari, T. Piran, and R. Narayan, "Spectra and light curves of gamma-ray burst afterglows," *The Astrophysical Journal*, vol. 497, no. 1, pp. L17–L20, 1998.

[45] E.-W. Liang, L. Li, H. Gao et al., "A comprehensive study of gamma-ray burst optical emission. II. Afterglow onset and late re-brightening components," *The Astrophysical Journal*, vol. 774, no. 1, p. 13, 2013.

[46] E. Molinari, S. D. Vergani, D. Malesani et al., "REM observations of GRB060418 and GRB 060607A: the onset of the afterglow and the initial fireball Lorentz factor determination," *Astronomy and Astrophysics*, vol. 469, no. 1, pp. L13–L16, 2007.

[47] P. Chandra and D. A. Frail, "Gamma ray bursts and their afterglow properties," *Bulletin of the Astronmical Society of India*, vol. 39, no. 3, pp. 451–470, 2011.

[48] P. Chandra and D. A. Frail, "A radio-selected sample of gamma-ray burst afterglows," *The Astrophysical Journal*, vol. 746, no. 2, p. 156, 2012.

[49] J. Granot and A. J. Van Der Horst, "Gamma-ray burst jets and their radio observations," *Publications of the Astronomical Society of Australia*, vol. 31, no. 1, article e008, 2014.

[50] D. A. Frail, E. Waxman, and S. R. Kulkarni, "A 450 day light curve of the radio afterglow of GRB 970508: fireball calorimetry," *The Astrophysical Journal*, vol. 537, no. 1, pp. 191–204, 2000.

[51] E. Berger, S. R. Kulkarni, and D. A. Frail, "The nonrelativistic evolution of GRBs 980703 and 970508: beaming-independent calorimetry," *Astrophysical Journal*, vol. 612, no. 2, pp. 966–973, 2004.

[52] A. J. van der Horst, A. Kamble, L. Resmi et al., "Detailed study of the GRB 030329 radio afterglow deep into the non-relativistic phase," *Astronomy & Astrophysics*, vol. 480, no. 1, pp. 35–43, 2008.

[53] J. Granot, E. Ramirez-Ruiz, and A. Loeb, "Implications of the measured image size for the radio afterglow of GRB 030329," *Astrophysical Journal*, vol. 618, no. 1, pp. 413–425, 2005.

[54] R. A. Mesler and Y. M. Pihlström, "Calorimetry of GRB 030329: simultaneous model fitting to the broadband radio afterglow and the observed image expansion rate," *The Astrophysical Journal*, vol. 774, no. 1, p. 77, 2013.

[55] R. A. M. J. Wijers and T. J. Galama, "Physical parameters of GRB 970508 and GRB 971214 from their afterglow synchrotron emission," *The Astrophysical Journal*, vol. 523, no. 1, p. 177, 1999.

[56] J. Goodman, "Radio scintillation of gamma-ray-burst afterglows," *New Astronomy*, vol. 2, no. 5, pp. 449–460, 1997.

[57] E. Waxman, S. R. Kulkarni, and D. A. Frail, "Implications of the radio afterglow from the gamma-ray burst of 1997 MAY 8," *Astrophysical Journal*, vol. 497, no. 1, pp. 288–293, 1998.

[58] G. Taylor, D. Frail, E. Berger, and S. Kulkarni, "High resolution observations of GRB 030329," *AIP Conference Proceedings*, vol. 727, pp. 324–327, 2004.

[59] F. A. Harrison, S. A. Yost, R. Sari et al., "Broadband observations of the afterglow of GRB 000926: observing the effect of inverse compton scattering," *The Astrophysical Journal*, vol. 559, no. 1, p. 123, 2001.

[60] A. Panaitescu and P. Kumar, "Fundamental physical parameters of collimated gamma-ray burst afterglows," *The Astrophysical Journal*, vol. 560, no. 1, pp. L49–L53, 2001.

[61] S. A. Yost, F. A. Harrison, R. Sari, and D. A. Frail, "A study of the afterglows of four gamma-ray bursts: constraining the explosion and fireball model," *The Astrophysical Journal*, vol. 597, no. 1, pp. 459–473, 2003.

[62] F. A. Harrison, J. S. Bloom, D. A. Frail et al., "Optical and radio observations of the afterglow from GRB 990510: evidence for a jet," *The Astrophysical Journal Letters*, vol. 523, no. 2, p. L121, 1999.

[63] E. Berger, R. Sari, D. A. Frail et al., "A jet model for the afterglow emission from GRB 000301C," *The Astrophysical Journal*, vol. 545, no. 1, p. 56, 2000.

[64] E. Berger, A. Diercks, D. A. Frail et al., "GRB 000418: a hidden jet revealed," *The Astrophysical Journal*, vol. 556, no. 2, p. 556, 2001.

[65] M. Ackermann, M. Ajello, K. Asano et al., "Fermi-LAT observations of the gamma-ray burst GRB 130427A," *Science*, vol. 343, no. 6166, pp. 42–47, 2014.

[66] A. Maselli, A. Melandri, L. Nava et al., "GRB 130427A: a nearby ordinary monster," *Science*, vol. 343, no. 6166, pp. 48–51, 2014.

[67] T. Laskar, E. Berger, B. A. Zauderer et al., "A reverse shock in GRB 130427A," *The Astrophysical Journal*, vol. 776, no. 2, p. 119, 2013.

[68] G. E. Anderson, A. J. van der horst, T. D. Staley et al., "Probing the bright radio flare and afterglow of GRB 130427A with the arcminute microkelvin imager," *Monthly Notices of the Royal Astronomical Society*, vol. 440, no. 3, pp. 2059–2065, 2014.

[69] D. A. Perley, S. B. Cenko, A. Corsi et al., "The afterglow of GRB 130427A from 1 to 10^{16} GHz," *The Astrophysical Journal*, vol. 781, no. 1, p. 37, 2014.

[70] A. J. van der Horst, Z. Paragi, A. G. de Bruyn et al., "A comprehensive radio view of the extremely bright gamma-ray burst 130427A," *Monthly Notices of the Royal Astronomical Society*, vol. 444, no. 4, pp. 3151–3163, 2014.

[71] D. A. Frail, P. B. Cameron, M. Kasliwal et al., "An energetic afterglow from a distant stellar explosion," *The Astrophysical Journal Letters*, vol. 646, no. 2, pp. L99–L102, 2006.

[72] B. Ciardi and A. Loeb, "Expected number and flux distribution of gamma-ray burst afterglows with high redshifts," *The Astrophysical Journal*, vol. 540, no. 2, pp. 687–696, 2000.

[73] D. A. Frail, A. M. Soderberg, S. R. Kulkarni et al., "Accurate calorimetry of GRB 030329," *Astrophysical Journal*, vol. 619, no. 2 I, pp. 994–998, 2005.

[74] S. R. Kulkarni, D. A. Frail, R. Sari et al., "Discovery of a radio flare from GRB 990123," *The Astrophysical Journal Letters*, vol. 522, no. 2, p. L97, 1999.

[75] E. Berger, S. R. Kulkarni, D. A. Frail, and A. M. Soderberg, "A radio survey of type Ib and Ic supernovae: searching for engine-driven supernovae," *The Astrophysical Journal*, vol. 599, no. 1, pp. 408–418, 2003.

[76] E. Nakar and T. Piran, "GRB 990123 revisited: further evidence of a reverse shock," *The Astrophysical Journal*, vol. 619, no. 2, pp. L147–L150, 2005.

[77] C. Akerlof, R. Balsano, S. Barthelmy et al., "Observation of contemporaneous optical radiation from a γ-ray burst," *Nature*, vol. 398, no. 6726, pp. 400–402, 1999.

[78] R. A. Mesler, Y. M. Pihlström, G. B. Taylor, and J. Granot, "VLBI and archival VLA and WSRT observations of the GRB 030329 radio afterglow," *The Astrophysical Journal*, vol. 759, no. 1, p. 4, 2012.

[79] F. De Colle, E. Ramirez-Ruiz, J. Granot, and D. Lopez-Camara, "Simulations of gamma-ray burst jets in a stratified external medium: dynamics, afterglow light curves, jet breaks, and radio calorimetry," *The Astrophysical Journal*, vol. 751, no. 1, article 57, 2012.

[80] P. A. Evans, A. P. Beardmore, K. L. Page et al., "Methods and results of an automatic analysis of a complete sample of *Swift*-XRT observations of GRBs," *Monthly Notices of the Royal Astronomical Society*, vol. 397, no. 3, pp. 1177–1201, 2009.

[81] D. A. Kann, S. Klose, B. Zhang et al., "The afterglows of *Swift*-era gamma-ray bursts. I. Comparing pre-*Swift* and *Swift*-era long/soft (type II) GRB optical afterglows," *The Astrophysical Journal*, vol. 720, no. 2, p. 1513, 2010.

[82] D. A. Kann, S. Klose, B. Zhang et al., "THE afterglows of *Swift*-era gamma-ray bursts. II. Type I GRB versus type II GRB optical afterglows," *The Astrophysical Journal*, vol. 734, no. 2, p. 96, 2011.

[83] D. Burlon, G. Ghirlanda, A. van der Horst et al., "The SKA view of gamma-ray bursts," https://arxiv.org/abs/1501.04629.

[84] Z.-B. Zhang, S.-W. Kong, Y.-F. Huang, D. Li, and L.-B. Li, "Detecting radio afterglows of gamma-ray bursts with FAST," *Research in Astronomy and Astrophysics*, vol. 15, no. 2, pp. 237–251, 2015.

[85] R. Nan, D. Li, C. Jin et al., "The five-hundred-meter aperture spherical radio telescope (FAST) project," *International Journal of Modern Physics D*, vol. 20, no. 6, pp. 989–1024, 2011.

[86] D. Li, R. Nan, and Z. Pan, "The five-hundred-meter aperture spherical radio telescope project and its early science opportunities," *Proceedings of the International Astronomical Union: Neutron Stars and Pulsars: Challenges and Opportunities after 80 Years*, vol. 8, no. 291, pp. 325–330, 2012.

[87] P. J. Hancock, B. M. Gaensler, and T. Murphy, "Two populations of gamma-ray burst radio afterglows," *Astrophysical Journal*, vol. 776, no. 2, article 106, 2013.

[88] D. A. Frail, S. R. Kulkarni, R. Sari et al., "Beaming in gamma-ray bursts: evidence for a standard energy reservoir," *The Astrophysical Journal Letters*, vol. 562, no. 1, p. L55, 2001.

[89] J. S. Bloom, D. A. Frail, and S. R. Kulkarni, "Gamma-ray burst energetics and the gamma-ray burst Hubble diagram: promises and limitations," *The Astrophysical Journal*, vol. 594, no. 2, pp. 674–683, 2003.

[90] A. M. Soderberg, S. R. Kulkarni, E. Berger et al., "The sub-energetic γ-ray burst GRB 031203 as a cosmic analogue to the nearby GRB 980425," *Nature*, vol. 430, no. 7000, pp. 648–650, 2004.

[91] A. Panaitescu and P. Kumar, "Properties of relativistic jets in gamma-ray burst afterglows," *The Astrophysical Journal*, vol. 571, no. 2 I, pp. 779–789, 2002.

[92] S. B. Cenko, D. A. Frail, F. A. Harrison et al., "The collimation and energetics of the brightest *Swift* gamma-ray bursts," *The Astrophysical Journal*, vol. 711, no. 2, p. 641, 2010.

[93] V. V. Usov, "Millisecond pulsars with extremely strong magnetic fields as a cosmological source of γ-ray bursts," *Nature*, vol. 357, no. 6378, pp. 472–474, 1992.

[94] T. A. Thompson, P. Chang, and E. Quataert, "Magnetar spin-down, hyperenergetic supernovae, and gamma-ray bursts," *Astrophysical Journal*, vol. 611, no. 1, pp. 380–393, 2004.

[95] B. D. Metzger, T. A. Thompson, and E. Quataert, "Proto-neutron star winds with magnetic fields and rotation," *The Astrophysical Journal*, vol. 659, no. 1, pp. 561–579, 2007.

[96] S. B. Cenko, M. Kasliwal, F. A. Harrison et al., "Multiwavelength observations of GRB 050820A: an exceptionally energetic event followed from start to finish," *The Astrophysical Journal*, vol. 652, no. 1, p. 490, 2006.

[97] E. Berger, S. R. Kulkarni, G. Pooley et al., "A common origin for cosmic explosions inferred from calorimetry of GRB030329," *Nature*, vol. 426, no. 6963, pp. 154–157, 2003.

[98] A. J. van der Horst, E. Rol, R. A. M. J. Wijers, R. Strom, L. Kaper, and C. Kouveliotou, "The radio afterglow of GRB 030329 at centimeter wavelengths: evidence for a structured jet or nonrelativistic expansion," *Astrophysical Journal*, vol. 634, no. 2 I, pp. 1166–1172, 2005.

[99] J. L. Racusin, S. V. Karpov, M. Sokolowski et al., "Broadband observations of the naked-eye big γ-ray burst GRB 080319B," *Nature*, vol. 455, pp. 183–188, 2008.

[100] L. Resmi, C. H. Ishwara-Chandra, A. J. Castro-Tirado et al., "Radio, millimeter and optical monitoring of GRB 030329 afterglow: constraining the double jet model," *Astronomy & Astrophysics*, vol. 440, no. 2, pp. 477–485, 2005.

[101] H. J. Van Eerten and A. I. MacFadyen, "Gamma-ray burst afterglow scaling relations for the full blast wave evolution," *Astrophysical Journal Letters*, vol. 747, article L30, 2012.

[102] A. Panaitescu, "Jet breaks in the X-ray light-curves of Swift gamma-ray burst afterglows," *Monthly Notices of the Royal Astronomical Society*, vol. 380, no. 1, pp. 374–380, 2007.

[103] D. Kocevski and N. Butler, "γ-ray burst energetics in the Swift ERA," *Astrophysical Journal*, vol. 680, no. 1, pp. 531–538, 2008.

[104] J. L. Racusin, E. W. Liang, D. N. Burrows et al., "Jet breaks and energetics of *Swift* gamma-ray burst X-ray afterglows," *The Astrophysical Journal*, vol. 698, no. 1, p. 43, 2009.

[105] N. Liang, W. K. Xiao, Y. Liu, and S. N. Zhang, "A cosmology-independent calibration of gamma-ray burst luminosity relations and the hubble diagram," *The Astrophysical Journal*, vol. 685, no. 1, pp. 354–360, 2008.

[106] X.-G. Wang, B. Zhang, E.-W. Liang et al., "How bad or good are the external forward shock afterglow models of gamma-ray bursts?" *The Astrophysical Journal Supplement Series*, vol. 219, no. 1, p. 9, 2015.

[107] R. Perna, R. Sari, and D. Frail, "Jets in γ-ray bursts: tests and predictions for the structured jet model," *The Astrophysical Journal*, vol. 594, no. 1 I, pp. 379–384, 2003.

[108] L. P. Xin, W. K. Zheng, J. Wang et al., "GRB 070518: a gamma-ray burst with optically dim luminosity," *Monthly Notices of the Royal Astronomical Society*, vol. 401, no. 3, pp. 2005–2011, 2010.

[109] X. Fan, C. L. Carilli, and B. Keating, "Observational constraints on cosmic reionization," *Annual Review of Astronomy and Astrophysics*, vol. 44, pp. 415–462, 2006.

[110] T. Totani, N. Kawai, G. Kosugi et al., "Implications for cosmic reionization from the optical afterglow spectrum of the gamma-ray burst 050904 at $z = 6.3$," *Publications of the Astronomical Society of Japan*, vol. 58, no. 3, pp. 485–498, 2006.

[111] D. Q. Lamb and D. E. Reichart, "Gamma-ray bursts as a probe of the very high redshift universe," *The Astrophysical Journal*, vol. 536, no. 1, pp. 1–18, 2000.

[112] L. J. Gou, P. Mészáros, T. Abel, and B. Zhang, "Detectability of long gamma-ray burst afterglows from very high redshifts," *The Astrophysical Journal*, vol. 604, no. 2, pp. 508–520, 2004.

[113] J. Miralda-Escudé, "Reionization of the intergalactic medium and the damping wing of the gunn-peterson trough," *The Astrophysical Journal*, vol. 501, no. 1, pp. 15–22, 1998.

[114] K. Ioka and P. Mészáros, "Radio afterglows of gamma-ray bursts and hypernovae at high redshift and their potential for 21 centimeter absorption studies," *Astrophysical Journal*, vol. 619, no. 2, pp. 684–696, 2005.

[115] D. A. Perley, S. B. Cenko, J. S. Bloom et al., "The host galaxies of Swift dark gamma-ray bursts: observational constraints on highly obscured and very high redshift GRBs," *Astronomical Journal*, vol. 138, no. 6, pp. 1690–1708, 2009.

[116] V. Bromm and A. Loeb, "High-redshift γ-ray bursts from population III progenitors," *Astrophysical Journal*, vol. 642, no. 1 I, pp. 382–388, 2006.

[117] N. Kawai, G. Kosugi, K. Aoki et al., "An optical spectrum of the afterglow of a γ-ray burst at a redshift of $z = 6.295$," *Nature*, vol. 440, no. 7081, pp. 184–186, 2006.

[118] J. Greiner, T. Krühler, J. P. U. Fynbo et al., "GRB 080913 at redshift 6.7," *The Astrophysical Journal*, vol. 693, no. 2, p. 1610, 2009.

[119] L.-J. Gou, D. B. Fox, and P. Mészáros, "Modeling GRB 050904: autopsy of a massive stellar explosion at $z = 6.29$," *The Astrophysical Journal*, vol. 668, no. 2, pp. 1083–1102, 2007.

[120] R. Sari and T. Piran, "GRB 990123: the optical flash and the fireball model," *Astrophysical Journal*, vol. 517, no. 2, pp. L109–L112, 1999.

[121] S. Kobayashi, "Light curves of gamma-ray burst optical flashes," *The Astrophysical Journal*, vol. 545, no. 2, pp. 807–812, 2000.

[122] B. Zhang, S. Kobayashi, and P. Mészáros, "γ-ray burst early optical afterglows: Implications for the initial lorentz factor and the central engine," *The Astrophysical Journal*, vol. 595, no. 2 I, pp. 950–954, 2003.

[123] E. Nakar and T. Piran, "Early afterglow emission from a reverse shock as a diagnostic tool for gamma-ray burst outflows," *Monthly Notices of the Royal Astronomical Society*, vol. 353, no. 2, pp. 647–653, 2004.

[124] C. Akerlof, R. Balsano, S. Barthelmy et al., "Prompt optical observations of gamma-ray bursts," *The Astrophysical Journal Letters*, vol. 532, no. 1, p. L25, 2000.

[125] P. W. A. Roming, P. Schady, D. B. Fox et al., "Very early optical afterglows of gamma-ray bursts: evidence for relative paucity of detection," *The Astrophysical Journal*, vol. 652, no. 2, p. 1416, 2006.

[126] E. S. Rykoff, F. Aharonian, C. W. Akerlof et al., "Looking into the fireball: Rotse-III AND *Swift* observations of early gamma-ray burst afterglows," *The Astrophysical Journal*, vol. 702, no. 1, p. 489, 2009.

[127] A. Gomboc, S. Kobayashi, C. G. Mundell et al., "Optical flashes, reverse shocks and magnetization," *AIP Conference Proceedings*, vol. 1133, pp. 145–150, 2009.

[128] H. Gao, W.-H. Lei, Y.-C. Zou, X.-F. Wu, and B. Zhang, "A complete reference of the analytical synchrotron external shock models of gamma-ray bursts," *New Astronomy Reviews*, vol. 57, no. 6, pp. 141–190, 2013.

[129] D. Kopač, C. G. Mundell, S. Kobayashi et al., "Radio flares from gamma-ray bursts," *The Astrophysical Journal*, vol. 806, no. 2, p. 179, 2015.

[130] L. Resmi and B. Zhang, "Gamma-ray burst reverse shock emission in early radio afterglows," *The Astrophysical Journal*, vol. 825, no. 1, p. 48, 2016.

[131] S. Inoue, K. Omukai, and B. Ciardi, "The radio to infrared emission of very high redshift gamma-ray bursts: probing early star formation through molecular and atomic absorption lines," *Monthly Notices of the Royal Astronomical Society*, vol. 380, no. 4, pp. 1715–1728, 2007.

[132] N. Gehrels and P. Mészáros, "γ-ray bursts," *Science*, vol. 337, no. 6097, pp. 932–936, 2012.

[133] L. Kidd and E. Troja, "The nature of the most extreme cosmic explosions: broadband studies of fermi LAT GRB afterglows," *American Astronomical Society Meeting Abstracts*, vol. 223, no. 223, 352.14, 2014.

[134] S. Iyyani, F. Ryde, M. Axelsson et al., "Variable jet properties in GRB 110721A: time resolved observations of the jet photosphere," *Monthly Notices of the Royal Astronomical Society*, vol. 433, no. 4, pp. 2739–2748, 2013.

[135] L. Amati, F. Frontera, M. Tavani et al., "Intrinsic spectra and energetics of BeppoSAX γ-ray bursts with known redshifts," *Astronomy and Astrophysics*, vol. 390, no. 1, pp. 81–89, 2002.

[136] K. P. Singh, S. N. Tandon, P. C. Agrawal et al., "ASTROSAT mission," in *Space Telescopes and Instrumentation: Ultraviolet to Gamma Ray*, vol. 9144 of *Proceedings of SPIE*, Montréal, Canada, June 2014.

[137] A. R. Rao, "Hard X-ray spectro-polarimetry of Black Hole sources?" in *Proceedings of the Recent Trends in the Study of Compact Objects (RETCO-II): Theory and Observation*, I. Chattopadhyay, A. Nandi, S. Das, and S. Mandal, Eds., vol. 12 of *ASI Conference Series*, 2015.

[138] V. Bhalerao, D. Bhattacharya, A. R. Rao, and S. Vadawale, "GRB 151006A: astrosat CZTI detection," *GRB Coordinates Network, Circular Service*, no. 18422, p. 1, 2015.

[139] J. E. Rhoads, "How to tell a jet from a balloon: a proposed test for beaming in gamma-ray bursts," *The Astrophysical Journal Letters*, vol. 487, no. 1, p. L1, 1997.

[140] G. Ghirlanda, D. Burlon, G. Ghisellini et al., "GRB orphan afterglows in present and future radio transient surveys," *Publications of the Astronomical Society of Australia*, vol. 31, article e022, 2014.

[141] A. M. Soderberg, E. Nakar, E. Berger, and S. R. Kulkarni, "Late-time radio observations of 68 type Ibc supernovae: strong constraints on off-axis gamma-ray bursts," *Astrophysical Journal*, vol. 638, no. 2, pp. 930–937, 2006.

[142] M. F. Bietenholz, F. De Colle, J. Granot, N. Bartel, and A. M. Soderberg, "Radio limits on off-axis grb afterglows and vlbi observations of sn 2003 gk," *Monthly Notices of the Royal Astronomical Society*, vol. 440, no. 1, pp. 821–832, 2014.

[143] K. S. Thorne, "Gravitational radiation," in *Three Hundred Years of Gravitation*, S. W. Hawking and W. Israel, Eds., pp. 330–458, Cambridge University Press, Cambridge, UK, 1987.

[144] B. F. Schutz, "Sources of gravitational radiation: coalescing binaries," *Advances in Space Research*, vol. 9, no. 9, pp. 97–101, 1989.

[145] K. Hotokezaka and T. Piran, "Mass ejection from neutron star mergers: different components and expected radio signals," *Monthly Notices of the Royal Astronomical Society*, vol. 450, no. 2, pp. 1430–1440, 2015.

[146] W.-F. Fong, B. D. Metzger, E. Berger, and F. Ozel, "Radio constraints on long-lived magnetar remnants in short gamma-ray bursts," https://arxiv.org/abs/1607.00416.

[147] J. F. Graham, A. S. Fruchter, E. M. Levesque et al., "High metallicity LGRB hosts," https://arxiv.org/abs/1511.00667.

[148] J. Greiner, M. J. Michałowski, S. Klose et al., "Probing dust-obscured star formation in the most massive gamma-ray burst host galaxies," *Astronomy & Astrophysics*, vol. 593, article A17, 12 pages, 2016.

11

Distribution Inference for Physical and Orbital Properties of Jupiter's Moons

F. B. Gao, [1] **X. H. Zhu,** [2] **X. Liu,** [1] **and R. F. Wang** [1]

[1] *School of Mathematical Science, Yangzhou University, Yangzhou 225002, China*
[2] *Department of Mathematics, Shanghai University, Shanghai 200444, China*

Correspondence should be addressed to F. B. Gao; gaofabao@sina.com

Academic Editor: Geza Kovacs

According to the physical and orbital characteristics in Carme group, Ananke group, and Pasiphae group of Jupiter's moons, the distributions of physical and orbital properties in these three groups are investigated by using one-sample Kolmogorov–Smirnov nonparametric test. Eight key characteristics of the moons are found to mainly obey the Birnbaum–Saunders distribution, logistic distribution, Weibull distribution, and *t* location-scale distribution. Furthermore, for the moons' physical and orbital properties, the probability density curves of data distributions are generated; the differences of three groups are also demonstrated. Based on the inferred results, one can predict some physical or orbital features of moons with missing data or even new possible moons within a reasonable range. In order to better explain the feasibility of the theory, a specific example is illustrated. Therefore, it is helpful to predict some of the properties of Jupiter's moons that have not yet been discovered with the obtained theoretical distribution inference.

1. Introduction

There are 69 (the number has been refreshed to 79 by a team from Carnegie Institution for Science in July 2018. https://sites.google.com/carnegiescience.edu/sheppard/moons/jupitermoons) confirmed moons of Jupiter, around 65 of which have been well investigated [1, 2]. Considering the formation of Jupiter's moons is influenced by diverse factors, which results in their physical characteristics differing greatly [3], Jupiter's moons are divided into two basic categories: regular and irregular. The regular satellites are so named because they have prograde and near-circular orbits of low inclination, and they are in turn split into two groups: Inner satellites and Galilean [4]. The irregular satellites are actually the objects whose orbits are far more distant and eccentric. They form families that share similar orbits (semi-major axis, inclination, and eccentricity) and composition. These families, which are considered to be part of collisions, arise when the larger parent bodies were shattered by impacts from asteroids captured by Jupiter's gravitational field. That is to say, at the early time of moons' formation of the Jupiter, mass of the original moon's ring was still sufficient to absorb the asteroid's power and put it into orbit. So, part of the irregular moons might be created by the captured asteroids and then collided with other moons [5, 6], thus forming the various groups we see today. The identification of satellite families is tentative (please see [7, 8] for more details), and these families bear the names of their largest members. The most detailed modelling of the collisional origin of the families was reported in [9, 10].

According to this identification scheme [11], 60 moons were classified into 8 different groups, including Small Inner Regulars and Rings, Galileans, Themisto group, Carpo group, Himalia group, Carme group, and Ananke group as well as Pasiphae group, in addition to 9 satellites that do not belong to any of previous groups. The detailed information about all the groups of Jupiter's moons can be found in Appendix A.

In recent years, many scientists have paid considerable attention to astronomical observation, physical research, and deep space exploration of small bodies, including asteroids, comets, and satellites, and so on. For example, planetary scientist Carry collected mass and volume estimates of 17 near-Earth asteroids, 230 main-belt and Trojan asteroids, 12 comets, and 28 trans-Neptunian objects from the known literature [12]. The accuracy and biases affecting the methods used to estimate these quantities were discussed and best-estimates were strictly selected. For the asteroids in retrograde orbit, there are at least 50 known moons of Jupiter's that are retrograde, some of which are thought to be asteroids or comets that originally formed near the gas giant and were captured when they got too close. Kankiewicz and Włodarczyk selected the 25 asteroids with the best-determined orbital elements and then estimated their dynamical lifetimes by using the latest observational data, including astrometry and physical properties [13]. However, few researchers have tried to extrapolate the distribution of Jupiter's satellites through statistical methods as yet.

In this paper, distributions of physical and orbital properties for the moons of Jupiter will be conducted by using one-sample Kolmogorov-Smirnov (K-S) test and maximum likelihood estimation [14–16]. Based on the analysis of satellites' data, it is found surprisingly that the physical and orbital characteristics obey some distribution, such as Birnbaum-Saunders distribution [17], logistic distribution, Weibull distribution, and t location-scale distribution. Furthermore, the probability density curves of the data distribution are generated, and the differences of physical and orbital characteristics in the three groups are presented. In addition, the results of theoretical inference results are then proved to be feasible through one concrete example. Therefore, the results may be helpful to astronomers to discover new moons of Jupiter in the future.

2. Method of Distribution Inference

In statistics, the K-S test, one type of nonparametric test, is used to determine whether a sample comes from a population with a specific distribution. The null hypothesis of one-sample K-S test is that the Cumulative Distribution Function (CDF) of the data follows the adopted CDF. For one-sample case, null distribution of statistic can be obtained from the null hypothesis that the sample is extracted from a reference distribution. The two-sided test for "unequal" CDF tests the null hypothesis against the alternative that the CDF of the data is different from the adopted CDF. The test statistic is the maximum absolute difference between the empirical CDF calculated by x and the hypothetical CDF:

$$D_n = \sup_x \left| F_n(x) - F(x) \right|, \quad (1)$$

where $F(x)$ is a given CDF and

$$F_n(x) = \frac{1}{n}\sum_{i=1}^{n} I_{(-\infty,x]}(X_i) \quad (2)$$

is the empirical distribution function of the observations X_i. Here $I_{(-\infty,x]}(X_i)$ is the indicator function with the following form

$$I_{(-\infty,x]}(X_i) = \begin{cases} 1, & X_i \le x, \\ 0, & otherwise. \end{cases} \quad (3)$$

According to Glivenko–Cantelli theorem [18], if the sample comes from distribution $F(x)$, then D_n will almost surely converge to zero when $n \longrightarrow \infty$. Therefore, we only focus on three satellite groups that have more than ten satellites in their groups, respectively. Of all these groups, Themisto group and Carpo group only contain one satellite, respectively, and only 4 moons were found separately in Small Inner Regulars and Rings, as well as in Galileans group. Moreover, there are 5 moons in Himalia group. For these 5 groups, there is no sufficient data for distribution inference, so we will focus on the Carme group, the Ananke group, and the Pasiphae group.

In the following sections, in order to use the one-sample K-S test, three sets of observed data from Jupiter's moons will be tested against some commonly used distributions in statistics. The list of these distributions is shown in Table 1.

For the 9 continuous distributions in Table 1, the one-sample K-S test will be used to select the distribution with the highest confidence level. In order to characterize these distributions with well-defined parameter values, maximum likelihood estimation is also used. In addition, the parameter values of these distributions can be calculated from the observed data. However, when confidence level (typically set to 0.05) decreases, the rejection domain of the test becomes smaller, so the observed values that initially fall into the rejection domain may eventually fall into the acceptance domain. This situation will bring some trouble in practical application. To this end, we adopt p value, which represents the obtained confidence level by using the one-sample K-S test. In addition, the use of p value not only avoids determining the level of significance in advance, but also makes it easy to draw conclusions about the test by comparing the p value and significance level of the test. If the p value is greater than 0.05, we declare that the null hypothesis can be accepted. Furthermore, if the p values of several distributions are all greater than 0.05, the distribution with the largest p value should be selected, and the corresponding distribution will be the most appropriate one to fit the observed data. Our results can be found in the tables in Appendix B.

3. Distribution Inference of Satellite Groups

In this section, distributions of several diverse physical and orbital properties for the Carme group, Ananke group, and Pasiphae group are inferred sequentially.

3.1. Carme Group. There are 15 moons in the Carme group (please see Appendix A.). Due to the lack of enough data from S/2010 J1, the number of adopted moons is 14, which means the length of each data set is 14 from the mathematical perspective. In addition, considering all the mean densities have been calculated at being 2.60 g/cm^3, the surface gravity of Carme and other moons in this group is 0.017 m/s^2 and 0.001 m/s^2, respectively [19]. Therefore, these data obviously do not obey the distributions.

Based on the previous method of statistical distribution inference and MATLAB 2016a (Intel Core i5-3230 M, CPU 2.60 GHz), distribution inferences in the Carme group can

TABLE 1: A list of common distributions.

Name of Selected Distributions	Parameters	Meaning
Beta	a	first shape parameter
	b	second shape parameter
Birnbaum-Saunders	β	scale parameter
	γ	shape parameter
Gamma	a	shape parameter
	b	scale parameter
Logistic	μ	mean
	σ	scale parameter
Nakagami	μ	shape parameter
	ω	scale parameter
Normal	μ	mean
	σ	standard deviation
Rice	s	noncentrality parameter
	σ	scale parameter
t location-scale	μ	location parameter
	σ	scale parameter
	υ	shape parameter
Weibull	A	scale parameter
	B	shape parameter

TABLE 2: The distribution inference in each physical and orbital characteristic (Carme group).

Characteristics	Distribution inference	Parameter estimates	*p*-value
Semi-major axis (10^7 km)	logistic distribution	$\mu = 2.3326$, $\sigma = 0.00651346$	0.9988
Mean orbit velocity (10^3 km/h)	logistic distribution	$\mu = 8.21989$, $\sigma = 0.0112805$	0.6607
Orbit eccentricity (10^{-1})	Birnbaum-Saunders distribution	$\beta = 2.54254$, $\gamma = 0.0330888$	0.6245
Inclination of orbit (10^2∘)	*t* location-scale distribution	$\mu = 1.6511$, $\sigma = 0.0017$, $\upsilon = 0.8751$	0.6662
Equatorial radius (km)	*t* location-scale distribution	$\mu = 1.65708$, $\sigma = 0.440683$, $\upsilon = 1.14501$	0.7119
Escape velocity (km/h)	*t* location-scale distribution	$\mu = 7.32997$, $\sigma = 1.58419$, $\upsilon = 1.06821$	0.6619

be found in Table 2, and the last column represents the p value. The smaller the p value, the greater the significance because it tells us that hypothesis under consideration may not be sufficient to explain the observations. The hypothesis will be rejected if any of these probabilities is less than or equal to a small, fixed but arbitrarily predefined threshold value. The null hypothesis here refers to data obeying a particular distribution, and the alternative hypothesis assumes that the data does not obey the distribution.

From Table 2, semi-major axis and mean orbit velocity obey the logistic distribution, of which parameter μ denotes the average orbital eccentricity and parameter σ plays a key role in representing the variance of the data set, through the variance formula $\sigma^2\pi^2/3$, so the variances of these logistic distributions are 1.3957E-04 and 4.1863E-04, respectively. As the parameter μ increases, it indicates that the average semi-major axis and mean orbit velocity of the Carme group increase. As the parameter σ increases, the data in these two characteristics gradually disperse, and the discrepancy between the data and the average value increases. The meaning of decreasing the parameter μ and σ shares the same principle as that of increasing the parameters. The orbit eccentricity follows the Birnbaum–Saunders distribution, which is unimodal with a median of β. The mean value and the variance of the distribution can be calculated by the following relationships:

$$\mu = \beta\left(1 + \frac{\gamma^2}{2}\right),$$

Table 3: The distribution inference in each orbital or physical characteristic (Ananke group).

Characteristics	Distribution inference	Parameter estimates	p-value
Semi-major axis (10^7 km)	t location-scale distribution	$\mu = 2.11665$, $\sigma = 0.00915557$, $\upsilon = 1.52402$	0.9304
Mean orbit velocity (10^3 km/h)	Weibull distribution	$A = 2.29923$, $B = 21.9834$	0.9788
Orbit eccentricity (10^{-1})	t location-scale distribution	$\mu = 8.72481$, $\sigma = 0.0203158$, $\upsilon = 1.36328$	0.6017
Inclination of orbit (10^2∘)	t location-scale distribution	$\mu = 1.4879$, $\sigma = 0.0127$, $\upsilon = 1.4538$	0.9075
Equatorial radius (km)	t location-scale distribution	$\mu = 1.80724$, $\sigma = 0.660297$, $\upsilon = 1.19615$	0.7231
Mass (10^{13} kg)	Weibull distribution	$A = 38.3573$, $B = 0.368827$	0.4304
Surface gravity ($(\mathrm{m/s^2})/10^3$)	t location-scale distribution	$\mu = 0.167044$, $\sigma = 0.0636087$, $\upsilon = 1.40815$	0.4210
Escape velocity (km/h)	t location-scale distribution	$\mu = 8.24903$, $\sigma = 2.61085$, $\upsilon = 1.15837$	0.7262

$$\sigma^2 = (\beta\gamma)^2\left(1 + \frac{5\gamma^2}{4}\right). \tag{4}$$

Therefore, the mean value of orbital eccentricity can be calculated to be 0.2543 and the variance is 7.08E-9.

The inclination of orbit, equatorial radius, and escape velocity are subject to the t location-scale distribution, which contains the scale parameter σ, the location parameter μ, and the shape parameter υ. Without loss of generality, we assume that the data vector x obeys the t location-scale distribution, and then we have $(x - \mu)/\sigma \sim t(\upsilon)$, which obeys Student's t-distribution; here υ represents the degrees of freedom. As can also be seen in Table 2, the inclination of orbit obeys the t location-scale distribution with parameters (1.6511, 0.0017, 0.8751) and the mean inclination is 165.11°. The equatorial radius follows the t location-scale distribution with parameters (1.65708, 0.440683, 1.14501). Therefore, the moons in Carme group have an average equatorial radius of 1.65708 km. In addition, the escape velocity characteristic obeys the t location-scale distribution with parameters (7.32997, 1.58419, 1.06821) and the average escape velocity is 7.32997 km/h. As the parameter μ changes, the average equatorial radius and escape velocity of the group also change accordingly. In the t location-scale distribution, the variance is $\sigma^2\upsilon/(\upsilon - 2)$, and when the shape parameter υ is greater than two, the variance of the distribution is defined. Therefore, the variances of these two specific t location-scale distributions cannot be defined.

3.2. Ananke Group. There are 10 moons in the Ananke group, of which we do not have enough data about S/2010 J2. Thus, the length of each characteristic of the remaining moons in this group will be 9.

From our discussion on the t location-scale distribution in Carme group, it is easy to understand the distribution inference of semi-major axis, mean orbit velocity, equatorial radius, surface gravity, and escape velocity in the Ananke group (see Table 3). However, compared with Table 2, it is noted that the orbit eccentricity and mass properties in the Ananke group are subject to the Weibull distribution. Parameters A and B represent the scale and shape parameters of the distribution [20], respectively, which together determine the mean and variance of the distribution.

3.3. Pasiphae Group. There are 19 moons in the Pasiphae group; due to the lack of data about three moons, S/2011 J2, S/2017 J1, and S/2016 J1, the remaining 16 will be studied in this subsection.

The distributions in Table 4 are inferred to be similar to the distribution in Tables 2 and 3. Therefore, we can easily understand the parameters of these distributions.

4. Comparison of Data Properties

Table 5 is given according to the previous distribution inference.

Based on the previous distribution inference, the properties of the moons' data can be compared more specifically and conveniently. According to the distribution of the specific parameters, we get the following probability density function (PDF) diagram.

As shown in Figure 1, the semi-major axis in the Ananke group is the smallest, followed by the Carme group and the Pasiphae group. In addition, the PDF of the Pasiphae group is relatively flat, indicating a large dispersion in semi-major axis around Jupiter, while the data in the Carme and Ananke groups differ slightly. As can be seen from Figure 2, the

TABLE 4: The distribution inference in each orbital or physical characteristic (Pasiphae group).

Characteristics	Distribution inference	Parameter estimates	p-value
Semi-major axis (10^7 km)	t location-scale distribution	$\mu = 2.38241$, $\sigma = 0.0273627$, $v = 0.730397$	0.3730
Mean orbit velocity (10^3 km/h)	logistic distribution	$\mu = 8.22803$, $\sigma = 0.226046$	0.4015
Orbit eccentricity (10^{-1})	logistic distribution	$\mu = 2.95251$, $\sigma = 0.599373$	0.9550
Inclination of orbit ($10^2\circ$)	logistic distribution	$\mu = 1.5136$, $\sigma = 0.0339$	0.8987
Equatorial radius (km)	Weibull distribution	$A = 3.69781$, $B = 0.781648$	0.0859
Mass (10^{13} kg)	Weibull distribution	$A = 72.5121$, $B = 0.266585$	0.1096
Surface gravity ((m/s^2)/10^3)	Weibull distribution	$A = 3.09876$, $B = 0.843794$	0.0594
Escape velocity (km/h)	Weibull distribution	$A = 17.0302$, $B = 0.807236$	0.0722

TABLE 5: Distribution inference summary.

Characteristics	Carme Group	Ananke Group	Pasiphae Group
Semi-major axis	logistic distribution	t location-scale distribution	t location-scale distribution
Mean orbit velocity	logistic distribution	t location-scale distribution	logistic distribution
Orbit eccentricity	Birnbaum–Saunders distribution	Weibull distribution	logistic distribution
Inclination of orbit	t location-scale distribution	t location-scale distribution	logistic distribution
Equatorial radius	t location-scale distribution	t location-scale distribution	Weibull distribution
Mass	None	Weibull distribution	Weibull distribution
Surface gravity	None	t location-scale distribution	Weibull distribution
Escape velocity	t location-scale distribution	t location-scale distribution	Weibull distribution

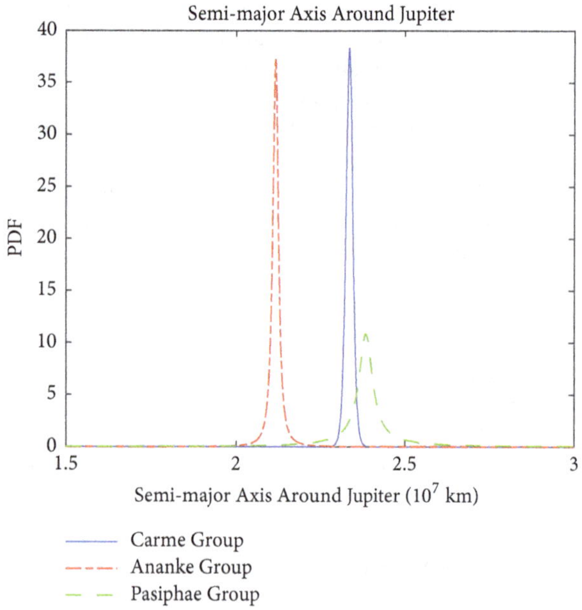

FIGURE 1: The PDF curves of semi-major axis around Jupiter.

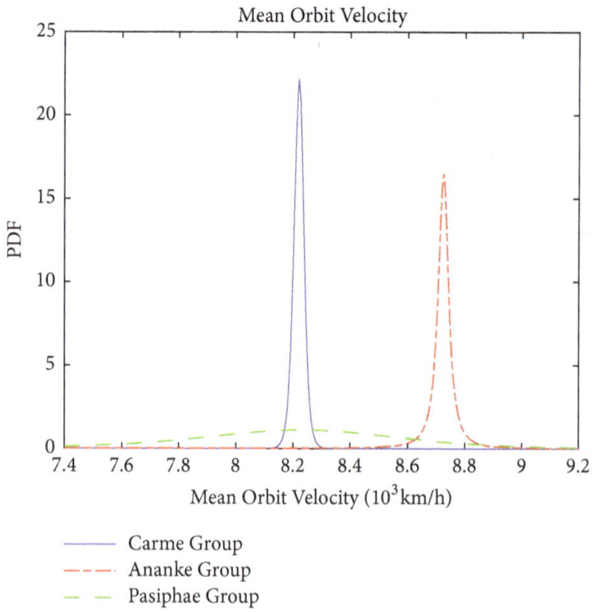

FIGURE 2: The PDF curves of mean orbit velocity.

trend of mean orbital velocity distribution in Carme group is the closest to each other, with the mean value being the smallest among the three groups, followed with the Pasiphae and Ananke groups. Obviously, the curve of the Pasiphae group is the flattest, which shows the mean orbit velocity in the Pasiphae group differing greatly. Figure 3 shows that the mean value in the Ananke group is the smallest, followed by the Carme and Pasiphae groups. It is obvious that the PDF curve of the Pasiphae group is flatter compared to the others, which means the orbital eccentricity of the Pasiphae group has a relatively large dispersion. In Figure 4, although the inclinations of orbit in the Ananke group and Pasiphae group are relatively close in value compared with the Carme group, and the data looks more dispersed than those in Carme group, the inclination of orbit in the Ananke group has the same distribution as the Carme group.

Figure 5 shows the PDF curves of equatorial radii of these three groups. Data attributes are similar; most of the moons'

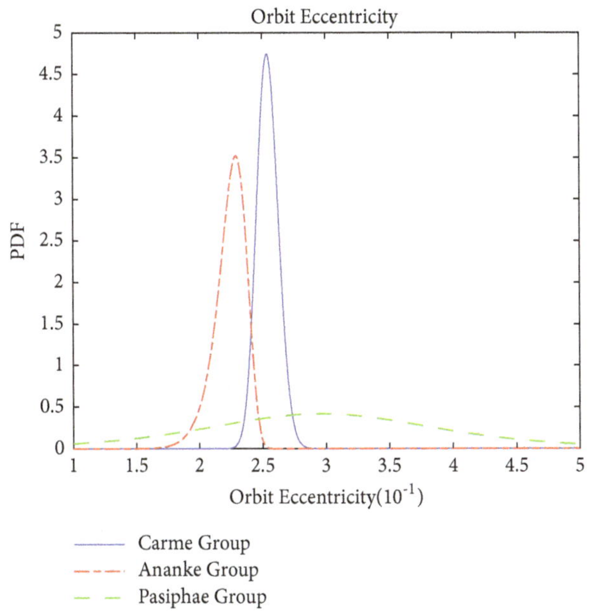

FIGURE 3: The PDF curves of orbit eccentricity.

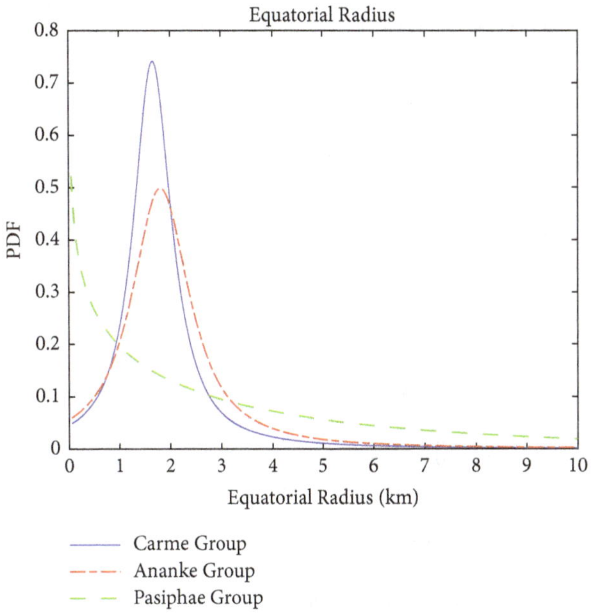

FIGURE 5: The PDF curves of equatorial radius.

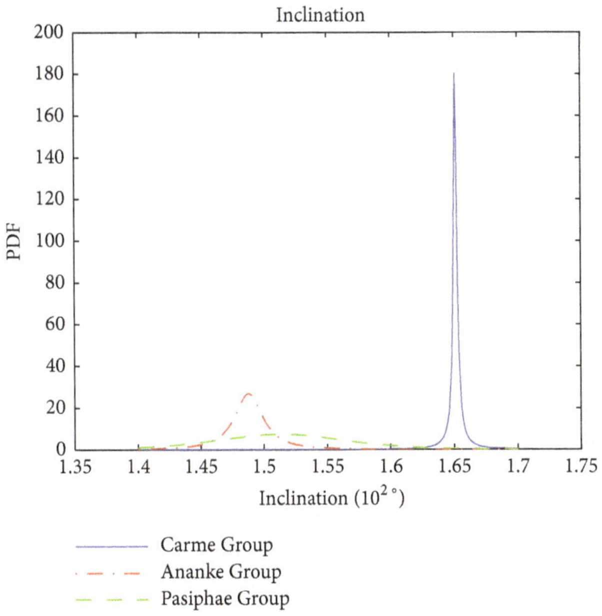

FIGURE 4: The PDF curves of inclination.

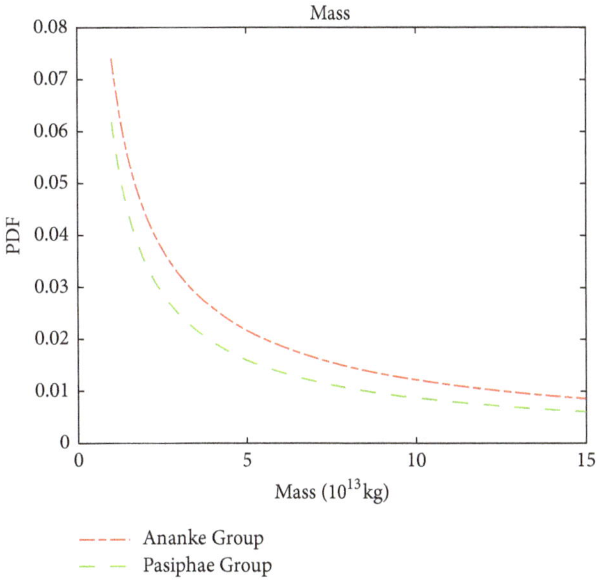

FIGURE 6: The PDF curves of mass.

radiuses are less than 4 km. Figure 6 indicates similarities between the Ananke and Pasiphae groups. As can be seen from these curves, most of the moons in these two groups are of relatively small mass.

Figure 7 illustrates the surface gravity PDF curves in the Ananke and Pasiphae groups. The difference in density between the two groups indicates that the surface gravity of the Ananke moons is higher than that of the Pasiphae group. The PDF plots corresponding to the escape velocity are shown in Figure 8, where it is clear that the Pasiphae group is flatter than the other two groups. The Carme group also has similar escape velocity density to the Ananke group, but the former average escape velocity is smaller.

5. Verification of Rationality of Theoretical Results

In this section, we take the semi-major axis and mean orbit velocity of the moons in the Carme group as concrete examples to illustrate the rationality of the statistical inferences in the preceding sections. As can be seen in Table 2, semi-major axis (*sma*) and mean orbit velocity (*mov*) obey the logistic distribution with parameters (2.3326, 0.00651346)

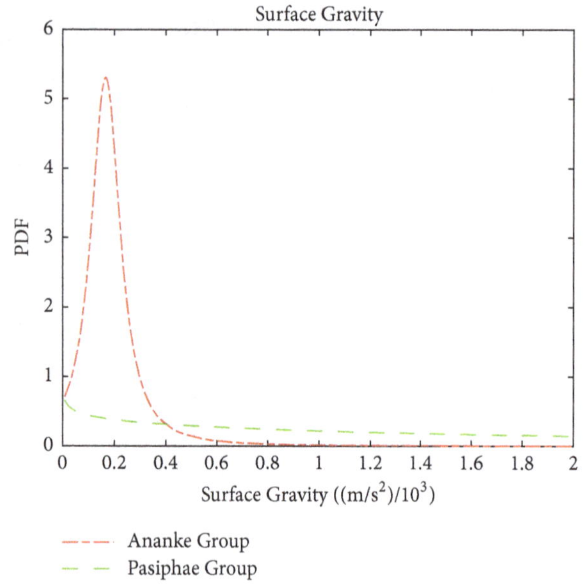

FIGURE 7: The PDF curves of surface gravity.

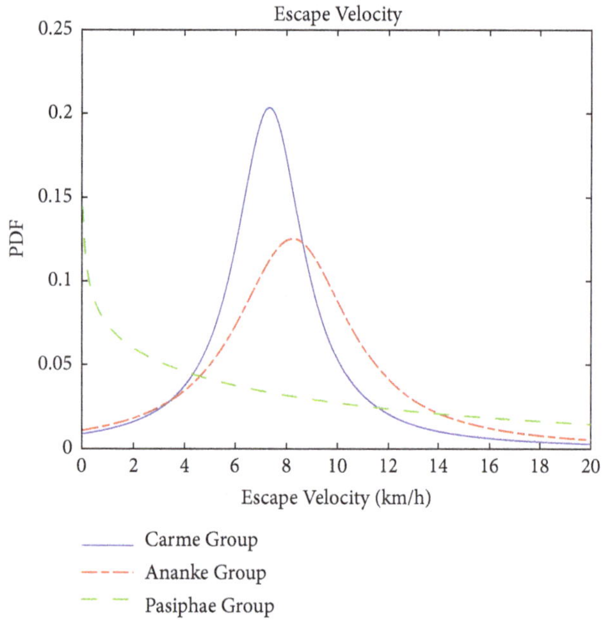

FIGURE 8: The PDF curves of escape velocity.

and (8.21989, 0.0112805), respectively. So the corresponding predicted PDFs can be written as

$$f_{pre,sma}(v;\mu,\sigma) = \frac{e^{-(v-2.3326)/0.00651346}}{0.00651346\left(1+e^{-(v-2.3326)/0.00651346}\right)^2} \tag{5}$$

and

$$f_{pre,mov}(v;\mu,\sigma) = \frac{e^{-(v-8.21989)/0.00112805}}{0.00112805\left(1+e^{-(v-8.21989)/0.00112805}\right)^2}. \tag{6}$$

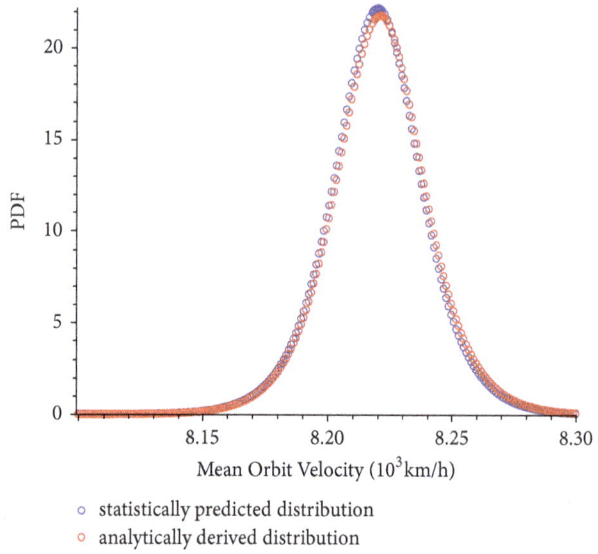

FIGURE 9: Comparison of statistical prediction distribution and analytical derivation distribution.

Note that the orbital and physical properties are not independent of each other. For instance, the semi-major axis r is related to the mean orbit velocity v through the relation $v = \sqrt{GM/r}$, where GM is the mass parameter. Then the PDF of mean orbit velocity can also be derived analytically as follows

$$\begin{aligned} f_{ana,mov}(v;\mu,\sigma) &= 2\frac{GM}{v^3} f_{pre,sma}\left(\frac{GM}{v^2}; 2.3326, 0.00651346\right) \\ &= 2\frac{GM}{v^3}\frac{e^{-(GM-2.3326v^2)/0.00651346v^2}}{0.00651346\left(1+e^{-(GM-2.3326v^2)/0.00651346v^2}\right)^2} \end{aligned} \tag{7}$$

Although the PDF of mean orbit velocity obtained by different methods has different mathematical representations, from Figure 9, we can find that $f_{ana,mov}(v;\mu,\sigma)$ (PDF curve represented by red circles) obtained by the analytical method is in good agreement with $f_{pre,mov}(v;\mu,\sigma)$ (PDF curve represented by blue circles) obtained by statistical inference.

However, what we need to pay attention to here is that some physical features and orbital elements are mixed and can be linked by some mathematical formulas similar to the above. Theoretically, the distribution of another variable in the formula can be solved by a known defined distribution; although this may be a complex process because the probability density function may contain some transcendental functions and gamma functions, it can be achieved. Yet the known distribution becomes uncertain now; that is, the distribution exists with a certain probability. So, there will be a certain risk when we calculate the distribution of the linked variable based on this uncertain distribution, especially when the possibility of the inferred distribution is not very high.

In addition, in order to further show that the results of the KS test agree well with the actual observed results,

TABLE 6: Distribution inference of Carme group without Erinome (S/2000 J4).

	p-value	h	Parameter Values
Beta	4.02E-05	1	$a = 0.202265$ $b = 0.173517$
Birnbaum-Saunders	0.0307	1	$\beta = 2.37288$ $\gamma = 0.943409$
Gamma	0.0308	1	$a = 1.05985$ $b = 3.11366$
Logistic	0.0463	1	$\mu = 1.95282$ $\sigma = 1.72324$
Nakagami	0.0039	1	$\mu = 0.282272$ $\omega = 43.5177$
Normal	0.0037	1	$\mu = 3.3$ $\sigma = 5.94531$
Rice	5.73E-08	1	$s = 0.138623$ $\sigma = 4.66385$
t location-scale	0.6806	0	$\mu = 1.67257$ $\sigma = 0.487941$ $v = 1.15623$
Weibull	0.0698	0	$A = 3.06061$ $B = 0.897697$

Note: $h = 0$ and $h = 1$ indicate acceptance of the null hypothesis and rejection of the null hypothesis, respectively.

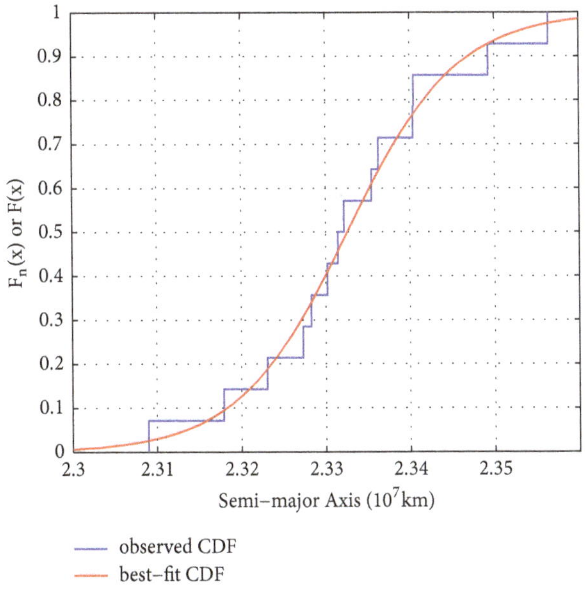

FIGURE 10: Comparison of the observed CDF and the best-fit CDF of semi-major axis.

we also compared the best-fit CDFs (the CDFs of inferred statistically) and the observed CDFs. From Figures 10 and 11, it can be seen that the best-fit CDF and the observed CDF of semi-major axis agree better than the case of mean orbital velocity. This should be due to the fact that the p value of the former is 0.9988, which is obviously larger than that of the latter 0.6607.

6. Application of the Distribution Inference

As mentioned previously, identification of irregular satellite families is tentative. However, after the distribution was inferred, it is found that these features of moons in the three different groups obey some selected distributions. Furthermore, the obtained population distribution can also be used to predict the characteristic data. Considering the rationality of proving this prediction method, suppose some characteristic data of a moon is unknown in a given irregular moon group. One specific feature of the moon can be predicted by using other moons' data and one-sample K-S method. Here is an example:

Assume that some characteristic data of the moon Erinome (S/2000 J4) in Carme group is poorly known. Now, we try to predict the equatorial radius of the Erinome. First, we use one-sample K-S method to find the most appropriate continuous distribution for these data. The inference results are as shown in Table 6.

From the p values displayed in Table 6, it becomes clear that the corresponding p value 0.6806 is the largest and $h = 0$, so the best distribution for the remaining characteristic data in the Carme group is t location-scale distribution with the PDF

$$f(x \mid \mu, \sigma, v) = \frac{\Gamma((v+1)/2)}{\sigma\sqrt{\pi v}\,\Gamma(v/2)} \left[\frac{v + ((x-\mu)/\sigma)^2}{v}\right]^{-(v+1)/2}, \quad -\infty < x < +\infty, \tag{8}$$

where $\Gamma(\cdot)$ is the Gamma function.

TABLE 7: Information of Jupiter's moons (source: the Jupiter satellite and moon page. Carnegie Institution. http://home.dtm.ciw.edu/users/sheppard/satellites/jupsatdata.html).

Number	Name	Designation	The Year of Discovery
		Small Inner Regulars and Rings	
XVI	Metis		1979
XV	Adrastea		1979
V	Amalthea		1892
XIV	Thebe		1979
		Galileans	
I	Io		1610
II	Europa		1610
III	Ganymede		1610
IV	Callisto		1610
		Themisto Prograde Irregular Group	
XVIII	Themisto	S/2000 J1	2000
		Himalia Prograde Irregular Group	
XIII	Leda		1974
VI	Himalia		1904
X	Lysithea		1938
VII	Elara		1905
LIII	Dia	S/2000 J11	2000
		Carpo Prograde Irregular Group	
XLVI	Carpo	S/2003 J20	2003
		Retrograde Irregular Groups	
		—	
		Ananke Retrograde Irregular Group	
LII		S/2010 J2	2010
XXXIII	Euanthe	S/2001 J7	2001
XXIX	Thyone	S/2001 J2	2001
XL	Mneme	S/2003 J21	2003
XXII	Harpalyke	S/2000 J5	2000
XXX	Hermippe	S/2001 J3	2001
XXVII	Praxidike	S/2000 J7	2000
XLII	Thelxinoe	S/2003 J22	2003
LX		S/2003 J3	2003
XXIV	Iocaste	S/2000 J3	2000
XII	Ananke		1951
		Carme Retrograde Irregular Group	
XLIII	Arche	S/2002 J1	2002
XXXVIII	Pasithee	S/2001 J6	2001
L	Herse	S/2003 J17	2003
XXI	Chaldene	S/2000 J10	2000
XXXVII	Kale	S/2001 J8	2001
XXVI	Isonoe	S/2000 J6	2000
XXXI	Aitne	S/2001 J11	2001
XXV	Erinome	S/2000 J4	2000
LI		S/2010 J1	2010
XX	Taygete	S/2000 J9	2000
XI	Carme		1938
XXIII	Kalyke	S/2000 J2	2000
XLVII	Eukelade	S/2003 J1	2003
LVII		S/2003 J5	2003
XLIV	Kallichore	S/2003 J11	2003

Table 7: Continued.

Number	Name	Designation	The Year of Discovery
		Pasiphae or No Strong Clustering Retrograde Irregular Group	
XXXIV	Euporie	S/2001 J10	2001
LV		S/2003 J18	2003
LIV		S/2016 J1	2016
XXXV	Orthosie	S/2001 J9	2001
XLV	Helike	S/2003 J6	2003
LVIII		S/2003 J15	2003
XXXII	Eurydome	S/2001 J4	2001
XXVIII	Autonoe	S/2001 J1	2001
LVI		S/2011 J2	2011
XXXVI	Sponde	S/2001 J5	2001
LIX		S/2017 J1	2017
VIII	Pasiphae		1908
XIX	Megaclite	S/2000 J8	2000
IX	Sinope		1914
XXXIX	Hegemone	S/2003 J8	2003
XLI	Aoede	S/2003 J7	2003
XVII	Callirrhoe	S/1999 J1	1999
XLVIII	Cyllene	S/2003 J13	2003
XLIX	Kore	S/2003 J14	2003
		The new Jupiter satellites discovered but yet to be numbered or named. All numbered or named satellites are in the table above	
		S/2003 J2	2003
		S/2003 J4	2003
		S/2003 J9	2003
		S/2003 J10	2003
		S/2003 J12	2003
		S/2003 J16	2003
		S/2003 J19	2003
		S/2003 J23	2003
		S/2011 J1	2011

TABLE 8: Carme group.

		Beta	Birnbaum-Saunders	Gamma	Logistic	Nakagami	Normal	Rice	t location-scale	Weibull
Semi-major axis (10^7 km)	p-value	3.55E-04	0.9898	null	0.9987	0.0136	0.9891	0.9882	null	0.7911
	h	1	0	null	0	1	0	0	null	0
	parameters	$a = 27.2522$ $b = 0.272087$	$\beta = 2.33268$ $\gamma = 0.00500346$	Matrix is close to singular.	$\mu = 2.3326$ $\sigma = 0.00651346$	$\mu = 9985.19$ $\omega = 5.44166$	$\mu = 2.33271$ $\sigma = 0.0121133$	$s = 2.33268$ $\sigma = 0.0116728$	The calculated Hessian is not positive definite.	$A = 2.33856$ $B = 201.843$
Mean orbit velocity (10^3 km/h)	p-value	8.80E-03	0.5488	null	0.6607	7.2972E-5	0.5598	0.5461	The calculated Hessian is not positive definite.	0.2983
	h	1	0	null	0	0	0	0	null	0
	parameters	$a = 68.2281$ $b = 0.245874$	$\beta = 8.22086$ $\gamma = 0.00232688$	Matrix is close to singular.	$\mu = 8.21989$ $\sigma = 0.0112805$	$\mu = 46161.3$ $\omega = 67.5832$	$\mu = 8.22088$ $\sigma = 0.0198551$	$s = 8.22086$ $\sigma = 0.0191329$	MLE (Maximum Likelihood Estimation) did not converge.	$A = 8.23058$ $B = 452.354$
Orbit eccentricity (10^{-1})	p-value	4.80E-03	0.6245	4.0418E-7	0.5695	0.6128	0.6118	0.6002	null	0.3726
	h	1	0	0	0	0	0	0	null	0
	parameters	$a = 4.79853$ $b = 0.247005$	$\beta = 2.54254$ $\gamma = 0.0330888$	$a = 909.093$ $b = 0.00279831$	$\mu = 2.53584$ $\sigma = 0.0506215$	$\mu = 226.315$ $\omega = 6.47876$	$\mu = 2.54393$ $\sigma = 0.0880083$	$s = 2.54251$ $\sigma = 0.0848305$	The calculated Hessian is not positive definite.	$A = 2.58664$ $B = 30.5827$
Inclination of orbit (10^2∘)	p-value	0.0994	0.0984	0.0991	0.2697	0.0998	0.1005	0.1005	0.6662	0.2271
	h	0	0	0	0	0	0	0	0	0
	parameters	$a = 32931.5$ $b = 166729$	$\beta = 1.6494$ $\gamma = 0.005040$	$a = 39419.4189$ $b = 0.0000418$	$\mu = 1.6502$ $\sigma = 0.004132$	$\mu = 9866.0175$ $\omega = 2.7205$	$\mu = 1.6494$ $\sigma = 0.008298$	$s = 1.6494$ $\sigma = 0.008298$	$\mu = 1.6511$ $\sigma = 0.0017$ $\upsilon = 0.8751$	$A = 1.6533$ $B = 212.8357$
Equatorial radius (km)	p-value	1.78E-05	0.0254	0.0249	0.0374	0.0024	0.0024	2.01E-08	0.7119	0.0574
	h	1	1	1	1	1	1	1	0	0
	parameters	$a = 0.206976$ $b = 0.185628$	$\beta = 2.32292$ $\gamma = 0.913758$	$a = 1.10481$ $b = 2.87704$	$\mu = 1.90719$ $\sigma = 1.606$	$\mu = 0.286619$ $\omega = 40.5921$	$\mu = 3.17857$ $\sigma = 5.73011$	$s = 0.116687$ $\sigma = 4.50454$	$\mu = 1.65708$ $\sigma = 0.440683$ $\upsilon = 1.14501$	$A = 2.98203$ $B = 0.911426$
Mass (10^{13} kg)	p-value	6.83E-05	8.58E-05	9.31E-04	0.0028	6.02E-04	3.10E-04	6.68E-12	null	0.0478
	h	1	1	1	1	1	1	1	null	1
	parameters	$a = 0.0841809$ $b = 0.128804$	$\beta = 153.808$ $\gamma = 6.88668$	$a = 0.160966$ $b = 5891.52$	$\mu = 151.943$ $\sigma = 941.547$	$\mu = 0.0699297$ $\omega = 1.44943e+7$	$\mu = 948.336$ $\sigma = 3522.9$	$s = 113.882$ $\sigma = 2491.35$	MLE did not converge.	$A = 31.522$ $B = 0.309664$
Surface gravity ($(m/s^2)/10^2$)	p-value	1.64E-05	0.0171	0.0232	0.0162	0.0023	0.0024	3.21E-08	null	0.0346
	h	1	1	1	1	1	1	1	null	1
	parameters	$a = 0.211325$ $b = 0.186944$	$\beta = 0.180961$ $\gamma = 0.87834$	$a = 1.16534$ $b = 0.208399$	$\mu = 0.149513$ $\sigma = 0.120199$	$\mu = 0.298776$ $\omega = 0.224286$	$\mu = 0.242857$ $\sigma = 0.421927$	$s = 0.001165$ $\sigma = 0.334886$	The calculated Hessian is not positive definite.	$A = 0.232103$ $B = 0.933741$
Escape velocity (km/h)	p-value	1.40E-05	0.0126	0.0141	0.0287	0.0015	0.0018	1.37E-08	0.6619	0.0358
	h	1	1	1	1	1	1	1	0	1
	parameters	$a = 0.210438$ $b = 0.186748$	$\beta = 10.5296$ $\gamma = 0.871682$	$a = 1.16971$ $b = 12.0298$	$\mu = 8.93572$ $\sigma = 7.99281$	$\mu = 0.296761$ $\omega = 769.929$	$\mu = 14.0714$ $\sigma = 24.8177$	$s = 0.703387$ $\sigma = 19.6157$	$\mu = 7.32997$ $\sigma = 1.58419$ $\upsilon = 1.06821$	$A = 13.4252$ $B = 0.932058$

Table 9: Ananke group.

		Beta	Birnbaum-Saunders	Gamma	Logistic	Nakagami	Normal	Rice	tlocation-scale	Weibull
Semi-major axis (10^7 km)	p-value	0.0328	0.1347	0.1371	0.6785	0.1395	0.1364	0.1421	0.9304	0.684
	h	1	0	0	0	0	0	0	0	0
	parameters	$a = 17.098$ $b = 0.178068$	$\beta = 2.1074$ $\gamma = 0.0142595$	Matrix is close to singular.	$\mu = 2.1139$ $\sigma = 0.0123376$	$\mu = 2.10761$ $\omega = 0.0311751$	$\mu = 2.10761$ $\sigma = 0.0311751$	$s = 2.1074$ $\sigma = 0.0295767$	$\mu = 2.11665$ $\sigma = 0.00915557$ $\upsilon = 1.52402$	$A = 2.11821$ $B = 141.939$
Mean orbit velocity (10^3 km/h)	p-value	4.30E-3	0.91	0.9211	0.9652	0.9312	0.9485	0.9408	0.7032	0.9788
	h	1	0	0	0	0	0	0	0	0
	parameters	$a = 2.619$ $b = 0.185662$	$\beta = 2.24183$ $\gamma = 0.052298$	$a = 368.072$ $b =$ 0.006.09908	$\mu = 2.24867$ $\sigma = 0.0683413$	$\mu = 92.6977$ $\omega = 5.05313$	$\mu = 2.2449$ $\sigma = 0.1227135$	$s = 2.24187$ $\sigma = 0.116491$	MLE did not converge.	$A = 2.29923$ $B = 21.9834$
Orbit eccentricity (10^{-1})	p-value	8.99E-05	0.1634	0.1616	0.6557	0.1598	0.151	0.1581	0.6017	0.0821
	h	1	0	0	0	0	0	0	0	0
	parameters	$a = 8.26353$ $b = 0.198125$	$\beta = 8.75391$ $\gamma = 0.0087315$	Matrix is close to singular.	$\mu = 8.73706$ $\sigma = 0.0313777$	$\mu = 3235.86$ $\omega = 76.6427$	$\mu = 8.75424$ $\sigma = 0.0813835$	$s = 8.7539$ $\sigma = 0.0772086$	$\mu = 8.72481$ $\sigma = 0.0203158$ $\upsilon = 1.36328$	$A = 8.79917$ $B = 82.6918$
Inclination of orbit ($10^2 \circ$)	p-value	0.5027	0.4904	0.4991	0.7114	0.5077	0.5167	0.5167	0.9075	0.6297
	h	0	0	0	0	0	0	0	0	0
	parameters	$a = 1951.25$ $b = 11193.4$	$\beta = 1.4841$ $\gamma = 0.0210$	$a = 2286.47$ $b = 0.000649$	$\mu = 1.4863$ $\sigma = 0.0154$	$\mu = 575.3654$ $\omega = 2.2045$	$\mu = 1.4844$ $\sigma = 0.0309$	$s = 1.4841$ $\sigma = 0.0308$	$\mu = 1.4879$ $\sigma = 0.0127$ $\upsilon = 1.4538$	$A = 1.4986$ $B = 52.7307$
Equatorial radius (km)	p-value	7.43E-04	0.5045	0.4118	0.2689	0.158	0.1036	2.40E-03	0.7231	0.4889
	h	1	0	0	0	0	0	1	0	0
	parameters	$a = 0.225054$ $b = 0.144156$	$\beta = 2.32358$ $\gamma = 0.827948$	$a = 1.45368$ $b = 2.11188$	$\mu = 2.18108$ $\sigma = 1.39766$	$\mu = 0.395632$ $\omega = 23.241$	$\mu = 3.07$ $\sigma = 3.91806$	$s = 0.126333$ $\sigma = 3.40798$	$\mu = 1.80724$ $\sigma = 0.660297$ $\upsilon = 1.19615$	$A = 3.18574$ $B = 1.07908$
Mass (10^{13} kg)	p-value	0.0025	0.0341	0.0668	0.03	0.037	6.00E-03	2.74E-08	null	0.4304
	h	1	1	0	1	1	1	1	null	0
	parameters	$a = 0.0964716$ $b = 0.106068$	$\beta = 60.0159$ $\gamma = 4.33548$	$a = 0.219599$ $b = 1411.32$	$\mu = 78.0193$ $\sigma = 298.807$	$\mu = 0.0927262$ $\omega = 898663$	$\mu = 309.925$ $\sigma = 944.346$	$s = 22.4209$ $\sigma = 670.178$	Maximum likelihood estimation did not converge.	$A = 38.3573$ $B = 0.368827$
Surface gravity ((m/s2)/10^2)	p-value	7.52E-04	0.2802	0.2227	0.3042	0.0982	0.0933	3.50E-03	0.421	0.2732
	h	1	0	0	0	0	0	1	0	0
	parameters	$a = 0.246544$ $b = 0.14788$	$\beta = 0.198988$ $\gamma = 0.732471$	$a = 1.82168$ $b = 0.137236$	$\mu = 0.190652$ $\sigma = 0.101607$	$\mu = 0.482908$ $\omega = 0.129$	$\mu = 0.25$ $\sigma = 0.271825$	$s = 0.0041522$ $\sigma = 0.253955$	$\mu = 0.167044$ $\sigma = 0.0636087$ $\upsilon = 1.40815$	$A = 0.270122$ $B = 1.21203$
Escape velocity (km/h)	p-value	6.67E-04	0.3295	0.2788	0.2429	0.1138	0.0989	1.70E-03	0.7262	0.3567
	h	1	0	0	0	0	0	1	0	0
	parameters	$a = 0.230441$ $b = 0.145225$	$\beta = 10.6647$ $\gamma = 0.778473$	$a = 1.58042$ $b = 8.66859$	$\mu = 9.82501$ $\sigma = 5.97056$	$\mu = 0.418467$ $\omega = 445.3$	$\mu = 13.7$ $\sigma = 16.9184$	$s = 0.531109$ $\sigma = 14.9178$	$\mu = 8.24903$ $\sigma = 2.61085$ $\upsilon = 1.15837$	$A = 14.419$ $B = 1.11631$

Table 10: Pasiphae group.

		Beta	Birnbaum-Saunders	Gamma	Logistic	Nakagami	Normal	Rice	t location-scale	Weibull
Semi-major axis (10^7 km)	p-value	0.0012	0.1039	0.1069	0.2952	0.1092	0.1238	0.1125	0.373	0.1848
	h	1	0	0	0	0	0	0	0	0
	parameters	$a = 4.22446$ $b = 0.262587$	$\beta = 2.28804$ $\gamma = 0.068061$	$a = 222.433$ $b = 0.0103103$	$\mu = 2.32$ $\sigma = 0.0823946$	$\mu = 57.3364$ $\omega = 5.28181$	$\mu = 2.29334$ $\sigma = 0.154515$	$s = 2.28844$ $\sigma = 0.14977$	$\mu = 2.38241$ $\sigma = 0.0273627$ $\upsilon = 0.730397$	$A = 2.35522$ $B = 23.6908$
Mean orbit velocity (10^3 km/h)	p-value	9.38E-04	0.2085	0.2025	0.4015	0.1961	0.2048	0.1906	null	0.1794
	h	1	0	0	0	0	0	0	null	0
	parameters	$a = 3.19406$ $b = 0.286862$	$\beta = 8.27432$ $\gamma = 0.0465993$	$a = 455.023$ $b = 0.0182041$	$\mu = 8.22803$ $\sigma = 0.226046$	$\mu = 112.484$ $\omega = 68.7679$	$\mu = 8.28331$ $\sigma = 0.406311$	$s = 8.27394$ $\sigma = 0.393632$	The calculated Hessian is not positive definite.	$A = 8.48321$ $B = 19.8563$
Orbit eccentricity (10^{-1})	p-value	0.0132	0.3639	0.5771	0.955	0.7272	0.9138	0.8914	null	0.9072
	h	1	0	0	0	0	0	0	null	0
	parameters	$a = 0.766602$ $b = 0.237408$	$\beta = 2.66229$ $\gamma = 0.426688$	$a = 6.69584$ $b = 0.434085$	$\mu = 2.95251$ $\sigma = 0.599373$	$\mu = 2.02049$ $\omega = 9.47173$	$\mu = 2.90656$ $\sigma = 1.04493$	$s = 2.68061$ $\sigma = 1.06913$	MLE did not converge.	$A = 3.24837$ $B = 3.30468$
Inclination of orbit (10^2∘)	p-value	0.5345	0.5633	0.5423	0.8987	0.5206	0.5002	0.5002	0.5004	0.2133
	h	0	0	0	0	0	0	0	0	0
	parameters	$a = 554.177$ $b = 3091.06$	$\beta = 1.5191$ $\gamma = 0.0389$	$a = 655.7110$ $b = 0.0023$	$\mu = 1.5136$ $\sigma = 0.0339$	$\mu = 162.6092$ $\omega = 2.3148$	$\mu = 1.5203$ $\sigma = 0.0599$	$s = 1.5191$ $\sigma = 0.0599$	$\mu = 1.5203$ $\sigma = 0.0599$ $\upsilon = 5657176.32$	$A = 1.5508$ $B = 23.9875$
Equatorial radius (km)	p-value	1.85E-04	0.0224	0.0377	0.0153	0.0139	0.0079	2.09E-08	null	0.0859
	h	1	1	1	1	1	1	1	null	0
	parameters	$a = 0.203821$ $b = 0.201972$	$\beta = 2.80607$ $\gamma = 1.20376$	$a = 0.770973$ $b = 5.83678$	$\mu = 2.50052$ $\sigma = 2.9666$	$\mu = 0.242429$ $\omega = 81.8925$	$\mu = 4.5$ $\sigma = 8.10876$	$s = 0.192009$ $\sigma = 6.39782$	The calculated Hessian is not positive definite.	$A = 3.69781$ $B = 0.781648$
Mass (10^{13} kg)	p-value	0.0012	0.0051	0.0111	0.0047	0.0091	8.16E-04	4.36E-11	null	0.1096
	h	1	1	1	1	1	1	1	null	0
	parameters	$a =$ 0.0820299 $b = 0.133182$	$\beta = 185.784$ $\gamma = 8.45355$	$a = 0.147015$ $b = 17059.2$	$\mu = 722.941$ $\sigma = 2437.85$	$\mu = 0.0664969$ $\omega = 6.36374E+7$	$\mu = 2507.97$ $\sigma = 7838.61$	$s = 215.387$ $\sigma = 5639.23$	The calculated Hessian is not positive definite.	$A = 72.5121$ $B = 0.266585$
Surface gravity ($(m/s^2)/10^2$)	p-value	9.34E-5	0.0186	0.0238	0.0139	0.0062	0.0055	2.15E-08	null	0.0594
	h	1	1	1	1	1	1	1	null	0
	parameters	$a = 0.216072$ $b = 0.206344$	$\beta = 0.232932$ $\gamma = 1.08145$	$a = 0.896179$ $b = 0.390547$	$\mu = 0.205059$ $\sigma = 0.214293$	$\mu = 0.270001$ $\omega = 0.44625$	$\mu = 0.35$ $\sigma = 0.587651$	$s = 0.0105316$ $\sigma = 0.472316$	The calculated Hessian is not positive definite.	$A = 0.309876$ $B = 0.843794$
Escape velocity (km/h)	p-value	1.55E-04	0.0157	0.0296	0.0142	0.011	0.0074	2.39E-08	null	0.0722
	h	1	1	1	1	1	1	1	null	0
	parameters	$a = 0.208766$ $b = 0.203804$	$\beta = 12.9337$ $\gamma = 1.14416$	$a = 0.822159$ $b = 24.4022$	$\mu = 11.3591$ $\sigma = 12.8644$	$\mu = 0.253336$ $\omega = 1566.94$	$\mu = 20.0625$ $\sigma = 35.2429$	$s = 0.946989$ $\sigma = 27.9843$	The calculated Hessian is not positive definite.	$A = 17.0302$ $B = 0.807236$

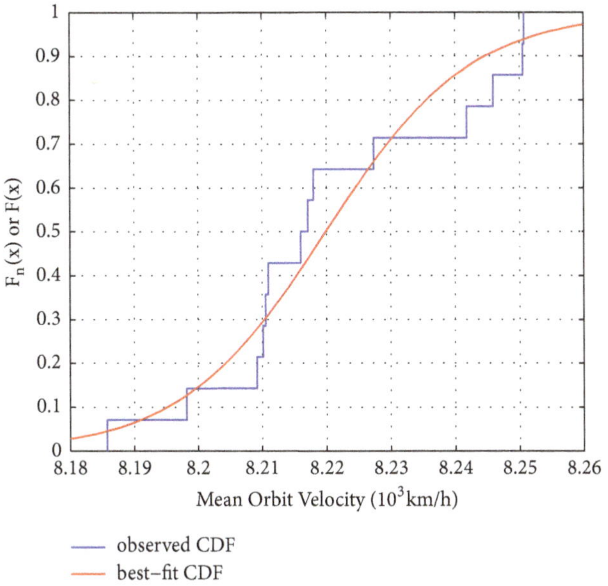

FIGURE 11: Comparison of the observed CDF and best-fit CDF of mean orbit velocity.

The distribution implies that average equatorial radius of the moons in Carme group is 1.67257 km with confidence interval $[1.23204, 2.1131]$ (we believe that the confidence interval will be smaller and shorter with continuous progress of observation technology) under the level of significance 0.05. For the true value of Erinome's equatorial radius being 1.6 km, it is easy to find that we can use the estimated value to predict the equatorial radius or as a reference to study other relevant physical and orbital characteristics.

7. Conclusions

By using the one-sample K-S nonparametric test method of statistical inference, the distribution laws of the physical and orbital properties of Jupiter's moons are investigated statistically in this paper. The physical and orbital characteristics of Jupiter's moons are found to obey the Birnbaum–Saunders distribution, the logistic distribution, the Weibull distribution, and the t location-scale distribution.

In addition, the probability density curves of the data distributions are also generated, and the differences in the physical and orbital characteristics of the three groups are explained in more detail.

Through a specific example, we find that some moons' missing data can be inferred by using the aforementioned distributional model and probability density function. More importantly, with the help of the distribution, it can be even helpful to predict the physical or orbital features of the undiscovered moon.

If future observations will allow for the expansion of the number of Jupiter's moons, we believe that the distribution laws will be slightly modified as potential newly discovered distribution functions fit the increased sample better, and the distributions will probably tend to be more uniform; i.e., some of the different properties follow the same distribution obviously, but these will not change dramatically over a long period of time.

Appendix

A.

See Table 7.

B. Distribution Inference Results

See Tables 8, 9, and 10

Acknowledgments

The authors acknowledge the support of National Natural Science Foundation of China (NSFC) through grant Nos. 11672259, 11302187, and 11571301; the Ministry of Land and Resources Research of China in the Public Interest through grant No. 201411007; and Top-notch Academic Programs Project of Jiangsu Higher Education Institutions (TAPP) through grant No. PPZY2015B109.

References

[1] S. S. Sheppard, *The Jupiter satellite and moon page*, Carnegie Institution, 2017.

[2] D. B. Jupiter, "moons," *NASA Planetary Science Division*, 2017, https://solarsystem.nasa.gov/planets/jupiter/moons.

[3] D. Jewitt and N. Haghighipour, "Irregular satellites of the planets: Products of capture in the early solar system," *Annual Review of Astronomy and Astrophysics*, vol. 45, pp. 261–295, 2007.

[4] S. S. Sheppard, "Outer irregular satellites of the planets and their relationship with asteroids, comets and Kuiper Belt objects," *Proceedings of the International Astronomical Union*, vol. 1, no. 229, pp. 319–334, 2005.

[5] B. W. Carroll and D. A. Ostlie, *An introduction to modern astrophysics*, Pearson, 2nd edition, 2006.

[6] Elkins-Tanton LT, "Jupiter and Saturn," *Facts on File*, 2010.

[7] D. Jewitt, S. Sheppard, and C. Porco, "Jupiter's outer satellites and Trojans," in *Jupiter. The planet, satellites and magnetosphere*, F. Bagenal, T. E. Dowling, and W. B. McKinnon, Eds., vol. 1, pp. 263–280, Cambridge planetary science, Cambridge, UK, 2004.

[8] Wikipedians, "Moons of Jupiter," in *Jupiter*, pp. 78–94, PediaPress GmbH, 2018.

[9] D. Nesvorný, J. L. A. Alvarellos, L. Dones, and H. F. Levison, "Orbital and collisional evolution of the irregular satellites," *The Astronomical Journal*, vol. 126, no. 1, pp. 398–429, 2003.

[10] D. Nesvorný, C. Beaugé, and L. Dones, "Collisional origin of families of irregular satellites," *The Astronomical Journal*, vol. 127, no. 3, pp. 1768–1783, 2004.

[11] S. S. Sheppard and D. C. Jewitt, "An abundant population of small irregular satellites around Jupiter," *Nature*, vol. 423, no. 6937, pp. 261–263, 2003.

[12] B. Carry, "Density of asteroids," *Planetary and Space Science*, vol. 73, no. 1, pp. 98–118, 2012.

[13] P. Kankiewicz and I. Włodarczyk, "Dynamical lifetimes of asteroids in retrograde orbits," *Monthly Notices of the Royal Astronomical Society*, vol. 468, no. 4, pp. 4143–4150, 2017.

[14] R. V. Hogg and A. T. Craig, *Introduction to mathematical statistics*, Pearson Education Asia Limited and Higher Education Press, 2004.

[15] H. Xie, X. Cui, B. Wan, and J. Zhang, "Statistical analysis of radio interference of 1000 kV UHV AC double-circuit transmission lines in foul weather," *CSEE Journal of Power and Energy Systems*, vol. 2, no. 2, pp. 47–55, 2016.

[16] R. R. Wilcox, "Some practical reasons for reconsidering the Kolmogorov-Smirnov test," *British Journal of Mathematical and Statistical Psychology*, vol. 50, no. 1, pp. 9–20, 1997.

[17] Z. W. Birnbaum and S. C. Saunders, "A new family of life distributions," *Journal of Applied Probability*, vol. 6, no. 2, pp. 319–327, 1969.

[18] K. B. Athreya and S. N. Lahiri, *Measure Theory and Probability Theory*, Springer Texts in Statistics, Springer, New York, NY, USA, 2006.

[19] List of moons in orbit around Jupiter, 2018, https://www.universeguide.com/planet/jupiter#themoons.

[20] R. Jiang and D. N. P. Murthy, "A study of Weibull shape parameter: Properties and significance," *Reliability Engineering & System Safety*, vol. 96, no. 12, pp. 1619–1626, 2011.

12

High-Order Analytic Expansion of Disturbing Function for Doubly Averaged Circular Restricted Three-Body Problem

Takashi Ito

National Astronomical Observatory of Japan, Osawa 2-21-1, Mitaka, Tokyo 181-8588, Japan

Correspondence should be addressed to Takashi Ito; tito@cfca.nao.ac.jp

Academic Editor: Elbaz I. Abouelmagd

Terms in the analytic expansion of the doubly averaged disturbing function for the circular restricted three-body problem using the Legendre polynomial are explicitly calculated up to the fourteenth order of semimajor axis ratio (α) between perturbed and perturbing bodies in the inner case ($\alpha < 1$), and up to the fifteenth order in the outer case ($\alpha > 1$). The expansion outcome is compared with results from numerical quadrature on an equipotential surface. Comparison with direct numerical integration of equations of motion is also presented. Overall, the high-order analytic expansion of the doubly averaged disturbing function yields a result that agrees well with the numerical quadrature and with the numerical integration. Local extremums of the doubly averaged disturbing function are quantitatively reproduced by the high-order analytic expansion even when α is large. Although the analytic expansion is not applicable in some circumstances such as when orbits of perturbed and perturbing bodies cross or when strong mean motion resonance is at work, our expansion result will be useful for analytically understanding the long-term dynamical behavior of perturbed bodies in circular restricted three-body systems.

1. Introduction

In the long tradition of celestial mechanics, the restricted three-body problem has occupied a fundamental role. In this problem, the mass of one of the three bodies is assumed to be small enough so that it does not affect the motion of the other two bodies. The restricted three-body problem is often considered on a rotating coordinate where central body and perturbing body are always located on the x-axis. See Szebehely [1] for more general characteristics of the restricted three-body problem.

Among many variants of the restricted three-body problem, its circular version called the circular restricted three-body problem (hereafter referred to as CR3BP) has been studied particularly well, and it makes a basis for understanding solar system dynamics and many other fields in celestial mechanics. In this system, the perturbing body lies on a circular orbit around central body. As is well known, the degree of freedom of CR3BP becomes unity and the system turns into integrable once we average the disturbing function of the system by mean anomalies of perturbed and perturbing bodies. Theories of the so-called classical Lidov–Kozai cycle have been developed based on the integrable characteristics of the doubly averaged CR3BP [2–4], where stationary points of argument of pericenter g around $\pm\pi/2$ appear when the vertical component of the angular momentum of perturbed body is smaller than a certain value. The present paper deals with the doubly averaged CR3BP.

In the classical theory of the Lidov–Kozai cycle, the doubly averaged disturbing function is expanded using the Legendre polynomials of even orders. Putting $\alpha = a/a'$, where a is the semimajor axis of perturbed body and a' is that of perturbing body, only the lowest-order terms up to $O(\alpha^2)$ are considered in many of the studies along this line (e.g., [5–9]); it is the quadruple-order approximation. Recent studies of the so-called eccentric Lidov–Kozai mechanism (e.g., [10–13]) that deal with the eccentric restricted three-body problem (ER3BP), where the orbit of perturbing body has a finite eccentricity, are based on the octupole-order approximation of disturbing function up to $O(\alpha^3)$. It is now getting better known that the inclusion of octupole terms in the disturbing function substantially changes the dynamical behavior of ER3BP. Even in CR3BP, the quadruple-order approximation is not accurate enough when α is large. To eliminate this shortcoming, in the early days, Kozai [3] expanded the doubly averaged disturbing function of the inner CR3BP up to $O(\alpha^8)$. More recently, Laskar and Boué [14] calculated analytic expansions of the general three-body disturbing function together with a practical method to compute the Hansen coefficients. Laskar and Boué [14] explicitly showed expressions of secular disturbing function up to $O(\alpha^{14})$ for planar problems and up to $O(\alpha^5)$ for spatial problems.

In the present paper we will show specific expressions of the analytic expansion of the doubly averaged spatial disturbing function up to $O(\alpha^{14})$ for the inner CR3BP ($\alpha < 1$) and up to $O(\alpha'^{15})$ for the outer CR3BP ($\alpha' = \alpha^{-1} < 1$) using the Legendre polynomials. As most readers are aware, very wide varieties of studies have been already done on the analytic expansion of the doubly averaged disturbing function of CR3BP. Compared with previous literature, the present paper intentionally aims to be rather expository. The major purpose of this paper is to explicitly show expressions of the high-order analytic expansion of the doubly averaged disturbing function for CR3BP so that readers with interest can consult the high-order terms without going through algebraic manipulation by themselves. We also aim at analytically reproducing local extremums that secular disturbing function of CR3BP intrinsically has, particularly when α or α' is large. Thus, even on a very basic subject like this, we have felt it advisable to give more details than would otherwise be necessary.

In Section 2 we give a brief description of the disturbing function of the three-body problem that we consider in the present paper. Section 3 goes to double averaging and analytic expansion of the disturbing function for CR3BP: general procedure (Section 3.1) and specific forms (from Sections 3.2 to 3.8). In Section 4 we show a comparison between the results obtained by the analytic expansion and by numerical quadrature. We also carried out direct numerical integration of equations of motion for comparison and show its result in Section 5. Section 6 is devoted to summary and discussion.

For readers' convenience, before getting into the main sections let us quickly write down the basic equations of motion of the system that we deal with in the present paper. The differential equation that we will consider is the simple classical Newtonian equation of motion

$$\frac{d^2\mathbf{r}}{dt^2} + \mu\frac{\mathbf{r}}{r^3} = \nabla R, \tag{1}$$

where $\mathbf{r}$ indicates the position vector of the perturbed body and μ relates to the central mass. The disturbing function that plays a central role in this paper is denoted as R. As for a literal definition of the doubly averaged disturbing function, particularly its direct part, we use the following one:

$$\langle\langle R\rangle_{l'}\rangle_l = \frac{\mu'}{4\pi^2}\iint_0^{2\pi}\frac{dl\,dl'}{\Delta}, \tag{2}$$

where μ' is related to the mass of perturbing body, l and l' denote mean anomaly of perturbed and perturbing bodies, respectively, and Δ is the osculating distance of the two bodies in space. Consult later sections for detailed definitions of the variables in the above equations. Needless to say, the considered system contains only three bodies.

2. Disturbing Function

In the present paper we categorize CR3BP in two cases: (i) the inner case where the orbit of the perturbed body is located inside that of a perturbing body ($\alpha < 1$), such as the Sun-asteroid-Jupiter system, and (ii) the outer case where the orbit of the perturbed body is located outside that of a perturbing body ($\alpha' = \alpha^{-1} < 1$) such as the Sun-Neptune-TNO system (TNO = Trans-Neptunian Object). The coorbital case ($\alpha = 1$) is out of the scope of the present paper.

Following the long-term convention of celestial mechanics, in the present paper we express the disturbing function R of CR3BP in relative coordinates where the origin is located on the primary body (Figure 1(a)). In this coordinate system, the disturbing function that describes the perturbation on the motion of an object with mass m due to the motion of another mass m' has the following general form (e.g., [15, p. 228]):

$$R = \frac{\mu'}{\Delta} - \mu'\frac{\mathbf{r}\cdot\mathbf{r}'}{r'^3}, \tag{3}$$

where $\mu' = \mathcal{G}m'$ with the gravitational constant $\mathcal{G}$, $\mathbf{r}$ is the position vector of the mass m with respect to the central body, $\mathbf{r}'$ is the position vector of the mass m' with respect to the central body, and $\Delta = |\mathbf{r}' - \mathbf{r}|$. In what follows we will consider only the first term of the right-hand side of (3) which is often referred to as the direct part. The second term is called the indirect part. As is well known, the indirect part makes no contribution to long-term dynamics of the system because it vanishes after the double averaging procedure, unless nonnegligible mean motion resonances are at work and we cannot simply employ the double averaging procedure.

When designating S as the angle between the vectors $\mathbf{r}$ and $\mathbf{r}'$, it is also well known that $1/\Delta$ on the right-hand side of (3) can be expanded using the Legendre polynomials P_j as

$$\frac{1}{\Delta} = \frac{1}{r'}\sum_{j=0}^{\infty}\left(\frac{r}{r'}\right)^j P_j(\cos S), \tag{4}$$

when $r < r'$ (i.e., the inner case), and

$$\frac{1}{\Delta} = \frac{1}{r}\sum_{j=0}^{\infty}\left(\frac{r'}{r}\right)^j P_j(\cos S), \tag{5}$$

when $r > r'$ (i.e., the outer case). Once again, it is well known that the terms of $j = 0$ and $j = 1$ in (4) and (5) do not contribute to secular motion of the bodies, as they will vanish or become constant after the double averaging procedure. Hence in the remaining part of this paper we will just consider terms with $j \geq 2$ in (4) and (5).

Readers find the expressions of the disturbing functions (3), (4), and (5) and their derivations in many textbooks such as Brouwer and Clemence [16], Danby [17], or Murray

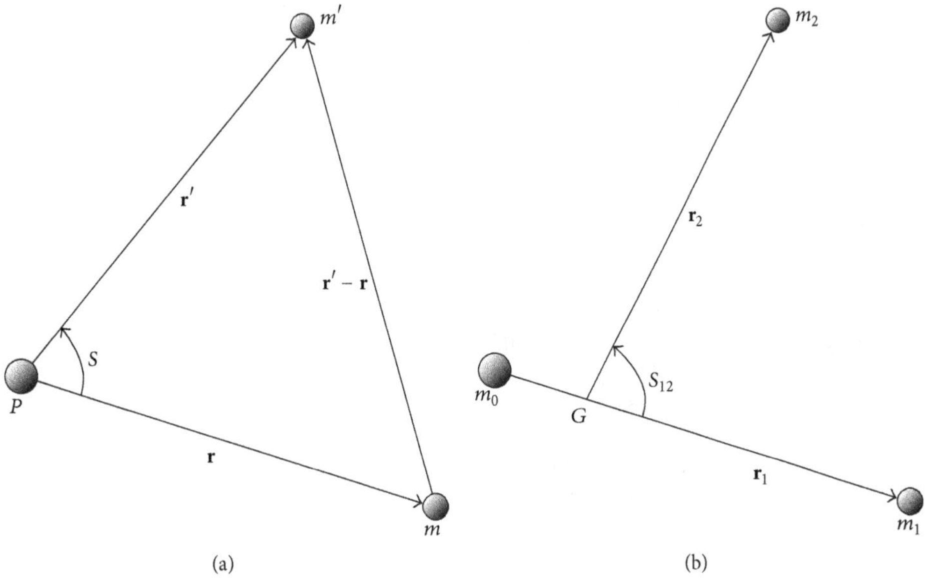

FIGURE 1: A schematic illustration of the three-body configuration considered here in two kinds of coordinate. (a) Relative coordinate. Both the vectors $\mathbf{r}$ and $\mathbf{r}'$ originates from the central mass denoted as P. (b) Jacobi coordinate. The vector $\mathbf{r}_1$ is originated from the primary mass (m_0), and the vector $\mathbf{r}_2$ is originated from the barycenter of the primary mass and the secondary mass (m_1) denoted as G.

and Dermott [15]. In the inner case, the direct part of the disturbing function μ'/Δ can be derived also in a more general way. Consider a general three-body system with three masses: primary m_0, secondary m_1, and tertiary m_2. Now we use the Jacobi coordinate (e.g., [18]): measuring m_1's position vector $\mathbf{r}_1$ from m_0, measuring m_2's position vector $\mathbf{r}_2$ from the barycenter of m_0 and m_1, and S_{12} is the angle between the vectors $\mathbf{r}_1$ and $\mathbf{r}_2$. Naturally $\mathbf{r}_1$ and $\mathbf{r}_2$ have different origins, and the angle S_{12} is different from S in general (see Figure 1(b)). We assume $r_1 < r_2$. In this coordinate system, the equations of motion of m_1 and m_2 become (see [16, 19], for detailed derivation)

$$\widetilde{m}_1 \frac{d^2\mathbf{r}_1}{dt^2} = \frac{\partial F}{\partial \mathbf{r}_1}, \qquad \widetilde{m}_2 \frac{d^2\mathbf{r}_2}{dt^2} = \frac{\partial F}{\partial \mathbf{r}_2}, \tag{6}$$

where

$$\widetilde{m}_1 = \frac{m_0 m_1}{m_0 + m_1}, \qquad \widetilde{m}_2 = \frac{(m_0 + m_1) m_2}{m_0 + m_1 + m_2} \tag{7}$$

are reduced masses used in the Jacobi coordinate system (e.g., [20, 21]), and F is the common force function

$$F = \mathcal{G}\left[\frac{m_0 m_1}{r_1} + \frac{(m_0 + m_1) m_2}{r_2} + \frac{1}{r_2}\sum_{j=2}^{\infty} M_j \left(\frac{r_1}{r_2}\right)^j P_j(\cos S_{12})\right], \tag{8}$$

where

$$M_j = \frac{m_0 m_1 m_2 \left(m_0^{j-1} - (-m_1)^{j-1}\right)}{(m_0 + m_1)^j} \tag{9}$$

is the mass factor. Using the force function F, this system can be written in a canonical form governed by a Hamiltonian. By expressing a_1 and a_2 as the semimajor axis of the orbits of the secondary and the tertiary (e.g., [22–25]), the Hamiltonian $\mathcal{H}$ becomes

$$\mathcal{H} = \frac{\mathcal{G} m_0 m_1}{2a_1} + \frac{\mathcal{G}(m_0 + m_1) m_2}{2a_2} + \frac{\mathcal{G}}{r_2}\sum_{j=2}^{\infty} M_j \left(\frac{r_1}{r_2}\right)^j P_j(\cos S_{12}). \tag{10}$$

The first and the second terms of $\mathcal{H}$ in (10) drive the Keplerian motion of the secondary and tertiary mass, respectively. Note that the third term which represents the mutual interaction of the secondary and the tertiary does not include terms of $j = 0$ or $j = 1$. This is a typical consequence of the use of the Jacobi coordinate that separates the motions of three bodies into two separate binaries and their interactions by a single infinite series.

Now, let us think about the limit where m_1 is infinitesimally small; this would correspond to the inner R3BP. In this case, we divide the force function F in (8) by $\widetilde{m}_1$ in (7) for normalization by mass before taking the limit. Then the third term of F becomes

$$\frac{F_{3\mathrm{rd}}}{\widetilde{m}_1} = \frac{\mathcal{G} m_2}{r_2}\sum_{j=2}^{\infty} \frac{m_0^{j-1} - (-m_1)^{j-1}}{(m_0 + m_1)^{j-1}} \left(\frac{r_1}{r_2}\right)^j P_j(\cos S_{12}). \tag{11}$$

Now we take the limit of $m_1 \to 0$, and the position of G in Figure 1(b) gets overlapped with the position of m_0. Thus we can replace $\mathbf{r}_1$ for $\mathbf{r}$, $\mathbf{r}_2$ for $\mathbf{r}'$, S_{12} for S, and m_2 for m' and will end up with an expression equivalent to the direct part of the disturbing function R of the inner case written in the relative coordinate (4).

On the other hand, deriving the form of the disturbing function of the outer case written in the relative coordinate (5) by taking the mass-less limit ($m_2 \to 0$) of Hamiltonian (10) is difficult, if not impossible. In the outer case, the origin of the position vector of the tertiary ($\mathbf{r}_2$) is the barycenter of the primary and secondary (G in Figure 1(b)). But G would not be overlapped with the position of the primary regardless of the value of the tertiary's mass m_2, unless $m_1 \to 0$. Therefore, in the present paper, we will not mention the conversion of the general three-body Hamiltonian into the outer disturbing function written in the relative coordinate. In modern celestial mechanics, more and more methods for expanding disturbing function without using the relative coordinate are becoming available (e.g., [14, 26, 27]).

3. Doubly Averaged Disturbing Function for CR3BP

3.1. General Form. From (3) and (4), the direct part of the disturbing function for the inner CR3BP where $r' = a'$ becomes as follows:

$$R = \frac{\mu'}{a'} \sum_{n=1}^{\infty} \left(\frac{r}{a'}\right)^{2n} P_{2n}(\cos S), \tag{12}$$

where we ignore the term of $n = 0$. In (12) we also ignore all the terms including the odd Legendre polynomials ($P_1, P_3, P_5, P_7, P_9, \ldots$) because they all vanish after the averaging procedure using mean anomaly of the perturbing body. Note that in the remaining part of this paper we will not consider the indirect part of the disturbing function either, as they do not have any secular dynamical contributions in nonresonant systems. Therefore we just use the variable R for representing the entire part of the disturbing function.

Assuming there is no major resonant relationship between the mean motions of perturbed and perturbing bodies, we now try to get the double average of R (12) over mean anomalies of both the bodies. Nonexistence of a resonant relationship means that the mean anomalies of perturbed and perturbing bodies (referred to as l and l' in what follows) are independent from each other. The procedure to carry out double averaging of R is straightforward as follows: Let us pick up the nth term of R and name it as R_{2n}. We have

$$R_{2n} = \frac{\mu'}{a'} \left(\frac{r}{a'}\right)^{2n} P_{2n}(\cos S). \tag{13}$$

First we average R_{2n} by mean anomaly of the perturbing body l', as

$$\langle R_{2n} \rangle_{l'} = \frac{\mu'}{a'} \left(\frac{r}{a'}\right)^{2n} \langle P_{2n} \rangle_{l'}, \tag{14}$$

where

$$\langle P_{2n} \rangle_{l'} = \frac{1}{2\pi} \int_0^{2\pi} P_{2n}(\cos S)\, dl'. \tag{15}$$

The angle S is expressed by orbital angles through a relationship (e.g., [3, Eq. (7) in p. 592])

$$\begin{aligned} \cos S &= \cos(f+g)\cos(f'+g') \\ &\quad + \cos i \sin(f+g)\sin(f'+g'), \end{aligned} \tag{16}$$

where f, f' are true anomalies of the perturbed and perturbing bodies, g, g' are arguments of pericenter of the perturbed and perturbing bodies, and i is their mutual inclination measured at the node of the two orbits. We choose the orbital plane of the perturbing body as a reference plane for the entire system, and then g and g' can be measured from the mutual node. Note that g' is not actually defined in CR3BP. Therefore, in (16) we regard $f' + g'$ as a single, fast-moving variable when we carry out averaging of (15). Practically, we can simply replace $\int dl'$ for $\int df'$ in the discussion here.

To obtain $\langle P_{2n} \rangle_{l'}$ of (15), we calculate the average of $\cos^{2n} S$ by l' as

$$\left\langle \cos^{2n} S \right\rangle_{l'} = \frac{1}{2\pi} \int_0^{2\pi} \cos^{2n} S\, dl'. \tag{17}$$

Then we average $\langle R_{2n} \rangle_{l'}$ of (14) by mean anomaly of the perturbed body l, as

$$\langle\langle R_{2n} \rangle_{l'} \rangle_l = \frac{\mu'}{a'} \left(\frac{a}{a'}\right)^{2n} \frac{1}{2\pi} \int_0^{2\pi} \left(\frac{r}{a}\right)^{2n} \langle P_{2n} \rangle_{l'}\, dl. \tag{18}$$

If we switch the integration variable from l to eccentric anomaly u, (18) becomes

$$\begin{aligned} &\langle\langle R_{2n} \rangle_{l'} \rangle_l \\ &\quad = \frac{\mu'}{a'} \left(\frac{a}{a'}\right)^{2n} \frac{1}{2\pi} \int_0^{2\pi} (1 - e\cos u)^{2n+1} \langle P_{2n} \rangle_{l'}\, du. \end{aligned} \tag{19}$$

We can obtain the doubly averaged disturbing function for the outer CR3BP in the same way as above. From (3) and (5), the direct part of the disturbing function R' for the outer CR3BP becomes as follows:

$$R' = \frac{\mu'}{r} \sum_{n=1}^{\infty} \left(\frac{a'}{r}\right)^{2n} P_{2n}(\cos S). \tag{20}$$

Note that our definition of $1/\Delta$ for the outer case (5), hence also in (20), may be different from what is seen in conventional textbooks (e.g., [15, Eq. (6.22) in p. 229]): Roles of the dashed quantities (r', μ') may be the opposite. This difference comes from the fact that conventional textbooks always assume $r/r' < 1$, while we assume $r/r' > 1$ for the outer problem. This is because we make it a rule to always use dash ($'$) for the quantities of perturbing body, whether it is located inside or outside the perturbed body.

Similar to the procedures that we went through for the inner CR3BP, we again assume that there is no major resonant relationship between mean motions of perturbed and perturbing bodies in the outer CR3BP. We then try to get the double average of R' over mean anomalies of both the bodies. Let us pick the nth term of R' in (20) and name it R'_{2n}. We have

$$R'_{2n} = \frac{\mu'}{r}\left(\frac{a'}{r}\right)^{2n} P_{2n}(\cos S). \quad (21)$$

First we average R'_{2n} by mean anomaly of the perturbing body l', as

$$\langle R'_{2n}\rangle_{l'} = \frac{\mu'}{r}\left(\frac{a'}{r}\right)^{2n} \langle P_{2n}\rangle_{l'}, \quad (22)$$

where $\langle P_{2n}\rangle_{l'}$ is already defined in (15).

Then we average $\langle R'_{2n}\rangle_{l'}$ in (22) by mean anomaly of the perturbed body l, as

$$\langle\langle R'_{2n}\rangle_{l'}\rangle_l = \frac{\mu'}{a'}\left(\frac{a'}{a}\right)^{2n+1} \frac{1}{2\pi}\int_0^{2\pi}\left(\frac{a}{r}\right)^{2n+1} \langle P_{2n}\rangle_{l'}\, dl. \quad (23)$$

If we switch the integration variable from l to true anomaly f, (23) becomes

$$\langle\langle R'_{2n}\rangle_{l'}\rangle_l = \frac{\mu'}{a'}\left(\frac{a'}{a}\right)^{2n+1} \frac{\left(1-e^2\right)^{-2n+1/2}}{2\pi} \cdot \int_0^{2\pi}\left(1+e\cos f\right)^{2n-1} \langle P_{2n}\rangle_{l'}\, df. \quad (24)$$

Note that $\langle\langle R'_{2n}\rangle_{l'}\rangle_l$ has the order of $O(\alpha'^{2n+1})$ as in (23) and (24), not $O(\alpha'^{2n})$.

In Sections 3.2 to 3.8 we show the expressions of $\langle \cos^{2n} S\rangle_{l'}$ in (17), $\langle P_{2n}\rangle_{l'}$ in (15), the Legendre polynomial $P_{2n}(x)$ in its original form, $\langle\langle R_{2n}\rangle_{l'}\rangle_l$ in (19), and $\langle\langle R'_{2n}\rangle_{l'}\rangle_l$ in (24) for $2n = 2, 4, 6, 8, 10, 12, 14$. We used Maple™ for algebraic manipulation to obtain the series of expressions. Note that in what follows we use $\sin i$ instead of $\cos i$ because it generally makes the formulas simpler. For this reason, some of the expressions look apparently different from what was presented in the previous literature in spite of their equivalence.

3.2. $2n = 2$. At this order, the corresponding component of the disturbing function for the inner problem R_2 is of $O(\alpha^2)$, and that for the outer problem R'_2 is of $O(\alpha'^3)$.

We just describe the resulting expressions of the expansion as follows: Let us emphasize again that the dashed quantities such as μ' and a' are those of the perturbing body, whether its orbit is located inside or outside of the orbit of the perturbed body.

$$\left\langle \cos^2 S\right\rangle_{l'} = \frac{1}{4}\sin^2 i\cos 2(f+g) - \frac{1}{4}\sin^2 i + \frac{1}{2}, \quad (25)$$

$$P_2(x) = \frac{3}{2}x^2 - \frac{1}{2}, \quad (26)$$

$$\langle P_2\rangle_{l'} = \frac{3}{8}\sin^2 i\cos 2(f+g) - \frac{3}{8}\sin^2 i + \frac{1}{4}, \quad (27)$$

$$\langle\langle R_2\rangle_{l'}\rangle_l = \frac{\mu'}{a'}\left(\frac{a}{a'}\right)^2 \cdot \left[\frac{15}{16}e^2\sin^2 i\cos 2g - \frac{1}{16}\left(3e^2+2\right)\left(3\sin^2 i - 2\right)\right], \quad (28)$$

$$\langle\langle R'_2\rangle_{l'}\rangle_l = \frac{\mu'}{a'}\left(\frac{a'}{a}\right)^3 \left(1-e^2\right)^{-3/2}\left[-\frac{3}{8}\sin^2 i + \frac{1}{4}\right]. \quad (29)$$

The expression in (28) shows the leading term of the doubly averaged disturbing function that causes the classical (circular) Lidov–Kozai cycle in the inner CR3BP that we have often seen in the previous literature. Meanwhile, the expression in (29) shows the leading term of the doubly averaged disturbing function for the outer CR3BP, but somehow we do not see it often. We should note that the leading term of the doubly averaged disturbing function for the outer problem (29) does not contain dependence on g. The g-dependence in the doubly averaged outer CR3BP first shows up in the next order: $2n = 4$. This is why the Lidov–Kozai mechanism for the outer CR3BP is more subtle than the inner one, particularly when α' is small, and perhaps this is why we rarely see the expression in the literature.

3.3. $2n = 4$. At this order, the corresponding component of the disturbing function for the inner problem R_4 is of $O(\alpha^4)$, and that for the outer problem R'_4 is of $O(\alpha'^5)$.

$$\left\langle \cos^4 S\right\rangle_{l'} = \frac{3}{64}\sin^4 i\cos 4(f+g) - \frac{3}{16}\sin^2 i\left(\sin^2 i - 2\right)\cos 2(f+g) + \frac{9}{64}\sin^4 i - \frac{3}{8}\sin^2 i + \frac{3}{8}, \quad (30)$$

$$P_4(x) = \frac{35}{8}x^4 - \frac{15}{4}x^2 + \frac{3}{8}, \quad (31)$$

$$\langle P_4\rangle_{l'} = \frac{105}{512}\sin^4 i\cos 4(f+g) - \frac{15}{128}\sin^2 i\left(7\sin^2 i - 6\right)\cos 2(f+g) + \frac{315}{512}\sin^4 i - \frac{45}{64}\sin^2 i + \frac{9}{64}, \quad (32)$$

$$\langle\langle R_4\rangle_{l'}\rangle_l = \frac{\mu'}{a'}\left(\frac{a}{a'}\right)^4\left[\frac{6615}{4096}e^4\sin^4 i\cos 4g - \frac{315}{1024}e^2 \cdot \sin^2 i\left(e^2+2\right)\left(7\sin^2 i - 6\right)\cos 2g + \frac{9}{4096}\left(15e^4+40e^2+8\right) \cdot \left(35\sin^4 i + 40\sin^2 i + 8\right)\right], \quad (33)$$

$$\langle\langle R_4'\rangle_{l'}\rangle_l = \frac{\mu'}{a'}\left(\frac{a'}{a}\right)^5 \left(1-e^2\right)^{-7/2}\left[-\frac{45}{512}e^2\sin^2 i \cdot \left(7\sin^2 i - 6\right)\cos 2g + \frac{9}{1024}\left(3e^2+2\right) \cdot \left(35\sin^4 i - 40\sin^2 i + 8\right)\right]. \quad (34)$$

Note that we now see the g-dependence of the doubly averaged disturbing function for the outer CR3BP in the expression of (34).

3.4. $2n = 6$. At this order, the corresponding component of the disturbing function for the inner problem R_6 is of $O(\alpha^6)$, and that for the outer problem R_6' is of $O(\alpha'^7)$.

$$\left\langle \cos^6 S\right\rangle_{l'} = \frac{5}{512}\sin^6 i\cos 6(f+g) - \frac{15}{256}\sin^4 i\left(\sin^2 i - 2\right)\cos 4(f+g) + \frac{15}{512}\sin^2 i\left(5\sin^4 i - 16\sin^2 i + 16\right)\cos 2(f+g) - \frac{5}{256}\left(\sin^2 i - 2\right)\left(5\sin^4 i - 8 \cdot \sin^2 i + 8\right), \quad (35)$$

$$P_6(x) = \frac{231}{16}x^6 - \frac{315}{16}x^4 + \frac{105}{16}x^2 - \frac{5}{16}, \quad (36)$$

$$\langle P_6\rangle_{l'} = \frac{1155}{8192}\sin^6 i\cos 6(f+g) - \frac{315}{4096}\sin^4 i\left(11 \cdot \sin^2 i - 10\right)\cos 4(f+g) + \frac{525}{8192}\sin^2 i\left(33\sin^4 i - 48\sin^2 i + 16\right)\cos 2(f+g) - \frac{5775}{4096}\sin^6 i + \frac{4725}{2048} \cdot \sin^4 i - \frac{525}{512}\sin^2 i + \frac{25}{256}, \quad (37)$$

$$\langle\langle R_6\rangle_{l'}\rangle_l = \frac{\mu'}{a'}\left(\frac{a}{a'}\right)^6 \left[\frac{495495}{131072}e^6\sin^6 i\cos 6g - \frac{10395}{65536}e^4\sin^4 i\left(3e^2+10\right)\left(11\sin^2 i - 10\right)\cos 4g + \frac{1575}{131072}e^2\sin^2 i\left(15e^4+80e^2+48\right) \cdot \left(33\sin^4 i - 48\sin^2 i + 16\right)\cos 2g - \frac{25}{65536}\left(35e^6+210e^4+168e^2+16\right) \cdot \left(231\sin^6 i - 378\sin^4 i + 168\sin^2 i - 16\right)\right], \quad (38)$$

$$\langle\langle R_6'\rangle_{l'}\rangle_l = \frac{\mu'}{a'}\left(\frac{a'}{a}\right)^7 \left(1-e^2\right)^{-11/2}\left[-\frac{1575}{65536}e^4\sin^4 i \cdot \left(11\sin^2 i - 10\right)\cos 4g + \frac{2625}{32768}e^2\sin^2 i\left(e^2+2\right) \cdot \left(33\sin^4 i - 48\sin^2 i + 16\right)\cos 2g - \frac{25}{32768}\left(15e^4+40e^2+8\right) \cdot \left(231\sin^6 i - 378\sin^4 i + 168\sin^2 i - 16\right)\right]. \quad (39)$$

3.5. $2n = 8$. At this order, the corresponding component of the disturbing function for the inner problem R_8 is of $O(\alpha^8)$, and that for the outer problem R_8' is of $O(\alpha'^9)$.

$$\left\langle \cos^8 S\right\rangle_{l'} = \frac{35}{16384}\sin^8 i\cos 8(f+g) - \frac{35}{2048}\sin^6 i \cdot \left(\sin^2 i - 2\right)\cos 6(f+g) + \frac{35}{4096}\sin^4 i\left(7\sin^4 i - 24\sin^2 i + 24\right)\cos 4(f+g) - \frac{35}{2048}\sin^2 i\left(\sin^2 i - 2\right)\left(7\sin^4 i - 16\sin^2 i + 16\right)\cos 2(f+g) + \frac{1225}{16384}\sin^8 i - \frac{175}{512}\sin^6 i + \frac{315}{512}\sin^4 i - \frac{35}{64}\sin^2 i + \frac{35}{128}, \quad (40)$$

$$P_8(x) = \frac{6435}{128}x^8 - \frac{3003}{32}x^6 + \frac{3465}{64}x^4 - \frac{315}{32}x^2 + \frac{35}{128}, \quad (41)$$

$$\langle P_8\rangle_{l'} = \frac{225225}{2097152}\sin^8 i\cos 8(f+g) - \frac{15015}{262144}\sin^6 i \cdot \left(15\sin^2 i - 14\right)\cos 6(f+g) + \frac{24255}{524288}\sin^4 i\left(65 \cdot \sin^4 i - 104\sin^2 i + 40\right)\cos 4(f+g) - \frac{11025}{262144} \cdot \sin^2 i\left(143\sin^6 i - 286\sin^4 i + 176\sin^2 i - 32\right) \cdot \cos 2(f+g) + \frac{7882875}{2097152}\sin^8 i - \frac{525525}{65536}\sin^6 i + \frac{363825}{65536}\sin^4 i - \frac{11025}{8192}\sin^2 i + \frac{1225}{16384}, \quad (42)$$

$$\langle\langle R_8\rangle_{l'}\rangle_l = \frac{\mu'}{a'}\left(\frac{a}{a'}\right)^8 \left[\frac{2737609875}{268435456}e^8\sin^8 i\cos 8g - \frac{10735725}{33554432}e^6\sin^6 i\left(3e^2+14\right)\left(15\sin^2 i - 14\right) \cdot \cos 6g + \frac{17342325}{67108864}e^4\sin^4 i\left(e^4+8e^2+8\right)\left(65\sin^4 i - 104\sin^2 i + 40\right)\cos 4g - \frac{606375}{33554432}e^2\sin^2 i\left(7e^6 + 70e^4 + 112e^2 + 32\right)\left(143\sin^6 i - 286\sin^4 i\right.\right.$$

$$
\begin{aligned}
&+ 176\sin^2 i - 32\big)\cos 2g + \frac{1225}{268435456}\big(315e^8 \\
&+ 3360e^6 + 6048e^4 + 2304e^2 + 128\big)\big(6435\sin^8 i \\
&- 13728\sin^6 i + 9504\sin^4 i - 2304\sin^2 i + 128\big)\Big],
\end{aligned}
\tag{43}
$$

$$
\begin{aligned}
\langle\langle R'_8\rangle_{l'}\rangle_l &= \frac{\mu'}{a'}\left(\frac{a'}{a}\right)^9\left(1-e^2\right)^{-15/2}\Big[-\frac{105105}{16777216}e^6 \\
&\cdot \sin^6 i\big(15\sin^2 i - 14\big)\cos 6g + \frac{169785}{16777216}e^4\sin^4 i \\
&\cdot \big(3e^2 + 10\big)\big(65\sin^4 i - 104\sin^2 i + 40\big)\cos 4g \\
&- \frac{77175}{16777216}e^2\sin^2 i\big(15e^4 + 80e^2 + 48\big)\big(143\sin^6 i \\
&- 286\sin^4 i + 176\sin^2 i - 32\big)\cos 2g \\
&+ \frac{1225}{33554432}\big(35e^6 + 210e^4 + 168e^2 + 16\big) \\
&\cdot \big(6435\sin^8 i - 13728\sin^6 i + 9504\sin^4 i \\
&- 2304\sin^2 i + 128\big)\Big].
\end{aligned}
\tag{44}
$$

3.6. $2n = 10$. At this order, the corresponding component of the disturbing function for the inner problem R_{10} is of $O(\alpha^{10})$, and that for the outer problem R'_{10} is of $O(\alpha'^{11})$.

$$
\begin{aligned}
\left\langle \cos^{10} S\right\rangle_{l'} &= \frac{63}{131072}\sin^{10} i\cos 10(f+g) - \frac{315}{65536} \\
&\cdot \sin^8 i\big(\sin^2 i - 2\big)\cos 8(f+g) + \frac{315}{131072}\sin^6 i\big(9 \\
&\cdot \sin^4 i - 32\sin^2 i + 32\big)\cos 6(f+g) - \frac{315}{16384}\sin^4 i \\
&\cdot \big(\sin^2 i - 2\big)\big(3\sin^4 i - 8\sin^2 i + 8\big)\cos 4(f+g) \\
&+ \frac{315}{65536}\sin^2 i\big(21\sin^8 i - 112\sin^6 i + 240\sin^4 i \\
&- 256\sin^2 i + 128\big)\cos 2(f+g) - \frac{63}{65536}\big(\sin^2 i \\
&- 2\big)\big(63\sin^8 i - 224\sin^6 i + 352\sin^4 i - 256\sin^2 i \\
&+ 128\big),
\end{aligned}
\tag{45}
$$

$$
\begin{aligned}
P_{10}(x) &= \frac{46189}{256}x^{10} - \frac{109395}{256}x^8 + \frac{45045}{128}x^6 \\
&- \frac{15015}{128}x^4 + \frac{3465}{256}x^2 - \frac{63}{256},
\end{aligned}
\tag{46}
$$

$$
\begin{aligned}
\langle P_{10}\rangle_{l'} &= \frac{2909907}{33554432}\sin^{10} i\cos 10(f+g) \\
&- \frac{765765}{16777216}\sin^8 i\big(19\sin^2 i - 18\big)\cos 8(f+g) \\
&+ \frac{405405}{33554432}\sin^6 i\big(323\sin^4 i - 544\sin^2 i + 224\big) \\
&\cdot \cos 6(f+g) - \frac{135135}{4194304}\sin^4 i\big(323\sin^6 i - 714 \\
&\cdot \sin^4 i + 504\sin^2 i - 112\big)\cos 4(f+g) \\
&+ \frac{72765}{16777216}\sin^2 i\big(4199\sin^8 i - 10608\sin^6 i + 9360 \\
&\cdot \sin^4 i - 3328\sin^2 i + 384\big)\cos 2(f+g) \\
&- \frac{183324141}{16777216}\sin^{10} i + \frac{241215975}{8388608}\sin^8 i \\
&- \frac{14189175}{524288}\sin^6 i + \frac{2837835}{262144}\sin^4 i - \frac{218295}{131072}\sin^2 i \\
&+ \frac{3969}{65536},
\end{aligned}
\tag{47}
$$

$$
\begin{aligned}
\langle\langle R_{10}\rangle_{l'}\rangle_l &= \frac{\mu'}{a'}\left(\frac{a}{a'}\right)^{10}\Big[\frac{256592689353}{8589934592}e^{10}\sin^{10} i \\
&\cdot \cos 10g - \frac{9646341705}{4294967296}e^8\sin^8 i\big(e^2 + 6\big)\big(19\sin^2 i \\
&- 18\big)\cos 8g + \frac{89594505}{8589934592}e^6\sin^6 i\big(15e^4 + 160e^2 \\
&+ 224\big)\big(323\sin^4 i - 544\sin^2 i + 224\big)\cos 6g \\
&- \frac{36891855}{1073741824}e^4\sin^4 i\big(5e^6 + 70e^4 + 168e^2 + 80\big) \\
&\cdot \big(323\sin^6 i - 714\sin^4 i + 504\sin^2 i - 112\big)\cos 4g \\
&+ \frac{2837835}{4294967296}e^2\sin^2 i\big(21e^8 + 336e^6 + 1008e^4 \\
&+ 768e^2 + 128\big)\big(4199\sin^8 i - 10608\sin^6 i \\
&+ 9360\sin^4 i - 3328\sin^2 i + 384\big)\cos 2g \\
&- \frac{3969}{4294967296}\big(693e^{10} + 11550e^8 + 36960e^6 \\
&+ 31680e^4 + 7040e^2 + 256\big)\big(46189\sin^{10} i \\
&- 121550\sin^8 i + 114400\sin^6 i - 45760\sin^4 i \\
&+ 7040\sin^2 i - 256\big)\Big],
\end{aligned}
\tag{48}
$$

$$
\begin{aligned}
\langle\langle R'_{10}\rangle_{l'}\rangle_l &= \frac{\mu'}{a'}\left(\frac{a'}{a}\right)^{11}\left(1-e^2\right)^{-19/2} \\
&\cdot \Big[-\frac{6891885}{4294967296}e^8\sin^8 i\big(19\sin^2 i - 18\big)\cos 8g \\
&+ \frac{1216215}{1073741824}e^6\sin^6 i\big(3e^2 + 14\big)\big(323\sin^4 i
\end{aligned}
$$

$$- 544\sin^2 i + 224\Big)\cos 6g - \frac{8513505}{268435456}e^4\sin^4 i\Big(e^4 + 8e^2 + 8\Big)\Big(323\sin^6 i - 714\sin^4 i + 504\sin^2 i - 112\Big)\cos 4g + \frac{654885}{536870912}e^2\sin^2 i\Big(7e^6 + 70e^4 + 112e^2 + 32\Big)\Big(4199\sin^8 i - 10608\sin^6 i + 9360\sin^4 i - 3328\sin^2 i + 384\Big)\cos 2g - \frac{3969}{2147483648}\Big(315e^8 + 3360e^6 + 6048e^4 + 2304e^2 + 128\Big)\Big(46189\sin^{10} i - 121550\sin^8 i + 114400\sin^6 i - 45760\sin^4 i + 7040\sin^2 i - 256\Big)\Big]. \quad (49)$$

3.7. $2n = 12$. At this order, the corresponding component of the disturbing function for the inner problem R_{12} is of $O(\alpha^{12})$, and that for the outer problem R'_{12} is of $O(\alpha'^{13})$.

$$\Big\langle \cos^{12} S\Big\rangle_{l'} = \frac{231}{2097152}\sin^{12} i\cos 12(f+g) - \frac{693}{524288}\cdot\sin^{10} i\Big(\sin^2 i - 2\Big)\cos 10(f+g) + \frac{693}{1048576}\sin^8 i\cdot\Big(11\sin^4 i - 40\sin^2 i + 40\Big)\cos 8(f+g) - \frac{1155}{524288}\cdot\sin^6 i\Big(\sin^2 i - 2\Big)\Big(11\sin^4 i - 32\sin^2 i + 32\Big)\cos 6(f + g) + \frac{3465}{2097152}\sin^4 i\Big(33\sin^8 i - 192\sin^6 i + 448\cdot\sin^4 i - 512\sin^2 i + 256\Big)\cos 4(f+g) - \frac{693}{262144}\cdot\sin^2 i\Big(\sin^2 i - 2\Big)\Big(33\sin^8 i - 144\sin^6 i + 272\sin^4 i - 256\sin^2 i + 128\Big)\cos 2(f+g) + \frac{53361}{1048576}\sin^{12} i - \frac{43659}{131072}\sin^{10} i + \frac{121275}{131072}\sin^8 i - \frac{5775}{4096}\sin^6 i + \frac{10395}{8192}\sin^4 i - \frac{693}{1024}\sin^2 i + \frac{231}{1024},$$

$$P_{12}(x) = \frac{676039}{1024}x^{12} - \frac{969969}{512}x^{10} + \frac{2078505}{1024}x^8 - \frac{255255}{256}x^6 + \frac{225225}{1024}x^4 - \frac{9009}{512}x^2 + \frac{231}{1024},$$

$$\langle P_{12}\rangle_{l'} = \frac{156165009}{2147483648}\sin^{12} i\cos 12(f+g) - \frac{20369349}{536870912}\sin^{10} i\Big(23\sin^2 i - 22\Big)\cos 10(f+g) + \frac{32008977}{1073741824}\sin^8 i\Big(161\sin^4 i - 280\sin^2 i + 120\Big)\cdot\cos 8(f+g) - \frac{2807805}{536870912}\sin^6 i\Big(3059\sin^6 i - 7182\sin^4 i + 5472\sin^2 i - 1344\Big)\cos 6(f+g) + \frac{10405395}{2147483648}\sin^4 i\Big(7429\sin^8 i - 20672\sin^6 i + 20672\sin^4 i - 8704\sin^2 i + 1280\Big)\cos 4(f+g) - \frac{2081079}{268435456}\sin^2 i\Big(7429\sin^{10} i - 22610\sin^8 i + 25840\sin^6 i - 13600\sin^4 i + 3200\sin^2 i - 256\Big)\cdot\cos 2(f+g) + \frac{36074117079}{1073741824}\sin^{12} i - \frac{14115958857}{134217728}\sin^{10} i + \frac{16804712925}{134217728}\sin^8 i - \frac{294819525}{4194304}\sin^6 i + \frac{156080925}{8388608}\sin^4 i - \frac{2081079}{1048576}\cdot\sin^2 i + \frac{53361}{1048576},$$

$$\langle\langle R_{12}\rangle_{l'}\rangle_l = \frac{\mu'}{a'}\left(\frac{a}{a'}\right)^{12}\Big[\frac{203026224075675}{2199023255552}e^{12}\sin^{12} i\cdot\cos 12g - \frac{1059267256047}{549755813888}e^{10}\sin^{10} i\Big(3e^2 + 22\Big)\cdot\Big(23\sin^2 i - 22\Big)\cos 10g + \frac{361861484985}{1099511627776}e^8\sin^8 i\cdot\Big(3e^4 + 40e^2 + 72\Big)\Big(161\sin^4 i - 280\sin^2 i + 120\Big)\cdot\cos 8g - \frac{6348447105}{549755813888}e^6\sin^6 i\Big(5e^6 + 90e^4 + 288e^2 + 192\Big)\Big(3059\sin^6 i - 7182\sin^4 i + 5472\sin^2 i - 1344\Big)\cos 6g + \frac{1238242005}{2199023255552}e^4\cdot\sin^4 i\Big(45e^8 + 960e^6 + 4032e^4 + 4608e^2 + 1280\Big)\cdot\Big(7429\sin^8 i - 20672\sin^6 i + 20672\sin^4 i - 8704\sin^2 i + 1280\Big)\cos 4g - \frac{72837765}{274877906944}e^2\cdot\sin^2 i\Big(99e^{10} + 2310e^8 + 11088e^6 + 15840e^4 + 7040e^2 + 768\Big)\Big(7429\sin^{10} i - 22610\sin^8 i + 25840\sin^6 i - 13600\sin^4 i + 3200\sin^2 i - 256\Big)\cdot\cos 2g + \frac{53361}{1099511627776}\Big(3003e^{12} + 72072e^{10} + 360360e^8 + 549120e^6 + 274560e^4 + 39936e^2$$

$$+ 1024)\left(676039\sin^{12}i - 2116296\sin^{10}i + 2519400\sin^{8}i - 1414400\sin^{6}i + 374400\sin^{4}i - 39936\sin^{2}i + 1024\right)\Big],$$

$$\langle\langle R'_{12}\rangle_{l'}\rangle_{l} = \frac{\mu'}{a'}\left(\frac{a'}{a}\right)^{13}\left(1-e^{2}\right)^{-23/2} \cdot \Big[-\frac{224062839}{549755813888}e^{10}\sin^{10}i\left(23\sin^{2}i - 22\right)\cos 10g + \frac{1760493735}{549755813888}e^{8}\sin^{8}i\left(e^{2}+6\right)\left(161\sin^{4}i - 280\sin^{2}i + 120\right)\cos 8g - \frac{92657565}{549755813888}e^{6}\sin^{6}i \cdot \left(15e^{4} + 160e^{2} + 224\right)\left(3059\sin^{6}i - 7182\sin^{4}i + 5472\sin^{2}i - 1344\right)\cos 6g + \frac{343378035}{274877906944}e^{4} \cdot \sin^{4}i\left(5e^{6} + 70e^{4} + 168e^{2} + 80\right)\left(7429\sin^{8}i - 20672\sin^{6}i + 20672\sin^{4}i - 8704\sin^{2}i + 1280\right) \cdot \cos 4g - \frac{114459345}{137438953472}e^{2}\sin^{2}i\left(21e^{8} + 336e^{6} + 1008e^{4} + 768e^{2} + 128\right)\left(7429\sin^{10}i - 22610\sin^{8}i + 25840\sin^{6}i - 13600\sin^{4}i + 3200\sin^{2}i - 256\right)\cos 2g + \frac{53361}{274877906944}\left(693e^{10} + 11550e^{8} + 36960e^{6} + 31680e^{4} + 7040e^{2} + 256\right)\left(676039\sin^{12}i - 2116296\sin^{10}i + 2519400\sin^{8}i - 1414400\sin^{6}i + 374400\sin^{4}i - 39936\sin^{2}i + 1024\right)\Big]. \tag{50}$$

3.8. $2n = 14$. At this order, the corresponding component of the disturbing function for the inner problem R_{14} is of $O(\alpha^{14})$, and that for the outer problem R'_{14} is of $O(\alpha'^{15})$.

$$\left\langle\cos^{14}S\right\rangle_{l'} = \frac{429}{16777216}\sin^{14}i\cos 14(f+g) - \frac{3003}{8388608}\sin^{12}i\left(\sin^{2}i - 2\right)\cos 12(f+g) + \frac{3003}{16777216}\sin^{10}i\left(13\sin^{4}i - 48\sin^{2}i + 48\right) \cdot \cos 10(f+g) - \frac{3003}{4194304}\sin^{8}i\left(\sin^{2}i - 2\right)\left(13\sin^{4}i - 40\sin^{2}i + 40\right)\cos 8(f+g) + \frac{3003}{16777216} \cdot \sin^{6}i\left(143\sin^{8}i - 880\sin^{6}i + 2160\sin^{4}i - 2560\sin^{2}i + 1280\right)\cos 6(f+g) - \frac{3003}{8388608}\sin^{4}i\left(\sin^{2}i - 2\right)\left(143\sin^{8}i - 704\sin^{6}i + 1472\sin^{4}i - 1536\sin^{2}i + 768\right)\cos 4(f+g) + \frac{3003}{16777216}\sin^{2}i\left(429\sin^{12}i - 3168\sin^{10}i + 10080\sin^{8}i - 17920\sin^{6}i + 19200\sin^{4}i - 12288\sin^{2}i + 4096\right)\cos 2(f+g) - \frac{429}{4194304}\left(\sin^{2}i - 2\right)\left(429\sin^{12}i - 2376\sin^{10}i + 5832\sin^{8}i - 7936\sin^{6}i + 6528\sin^{4}i - 3072\sin^{2}i + 1024\right),$$

$$P_{14}(x) = \frac{5014575}{2048}x^{14} - \frac{16900975}{2048}x^{12} + \frac{22309287}{2048} \cdot x^{10} - \frac{14549535}{2048}x^{8} + \frac{4849845}{2048}x^{6} - \frac{765765}{2048}x^{4} + \frac{45045}{2048}x^{2} - \frac{429}{2048},$$

$$\langle P_{14}\rangle_{l'} = \frac{2151252675}{34359738368}\sin^{14}i\cos 14(f+g) - \frac{557732175}{17179869184}\sin^{12}i\left(27\sin^{2}i - 26\right)\cos 12(f+g) + \frac{870062193}{34359738368}\sin^{10}i\left(45\sin^{2}i - 44\right)\left(5\sin^{2}i - 4\right) \cdot \cos 10(f+g) - \frac{189143955}{8589934592}\sin^{8}i\left(1035\sin^{6}i - 2530\sin^{4}i + 2024\sin^{2}i - 528\right)\cos 8(f+g) + \frac{693527835}{34359738368}\sin^{6}i\left(3105\sin^{8}i - 9200\sin^{6}i + 9936\sin^{4}i - 4608\sin^{2}i + 768\right)\cos 6(f+g) - \frac{328513185}{17179869184}\sin^{4}i\left(6555\sin^{10}i - 21850\sin^{8}i + 27968\sin^{6}i - 17024\sin^{4}i + 4864\sin^{2}i - 512\right) \cdot \cos 4(f+g) + \frac{19324305}{34359738368}\sin^{2}i\left(334305\sin^{12}i - 1188640\sin^{10}i + 1664096\sin^{8}i - 1157632\sin^{6}i + 413440\sin^{4}i - 69632\sin^{2}i + 4096\right)\cos 2(f+g) - \frac{922887397575}{8589934592}\sin^{14}i + \frac{1674869721525}{4294967296}\sin^{12}i$$

$$- \frac{602953099749}{1073741824}\sin^{10}i + \frac{218461268025}{536870912}\sin^8 i - \frac{10402917525}{67108864}\sin^6 i + \frac{985539555}{33554432}\sin^4 i - \frac{19324305}{8388608}\sin^2 i + \frac{184041}{4194304},$$

$$\langle\langle R_{14}\rangle_{l'}\rangle_l = \frac{\mu'}{a'}\left(\frac{a}{a'}\right)^{14}\left[\frac{20856061239960375}{70368744177664}e^{14} \cdot \sin^{14}i\cos 14g - \frac{186452654763375}{35184372088832}e^{12}\sin^{12}i\left(3e^2 + 26\right)\left(27\sin^2 i - 26\right)\cos 12g + \frac{32318460158985}{70368744177664} \cdot e^{10}\sin^{10}i\left(5e^4 + 80e^2 + 176\right)\left(45\sin^2 i - 44\right) \cdot \left(5\sin^2 i - 4\right)\cos 10g - \frac{1405150441695}{17592186044416}e^8\sin^8 i \cdot \left(7e^6 + 154e^4 + 616e^2 + 528\right)\left(1035\sin^6 i - 2530\sin^4 i + 2024\sin^2 i - 528\right)\cos 8g + \frac{672028472115}{70368744177664}e^6\sin^6 i\left(21e^8 + 560e^6 + 3024e^4 + 4608e^2 + 1792\right)\left(3105\sin^8 i - 9200\sin^6 i + 9936\sin^4 i - 4608\sin^2 i + 768\right)\cos 6g - \frac{318329276265}{35184372088832}e^4\sin^4 i\left(11e^{10} + 330e^8 + 2112e^6 + 4224e^4 + 2816e^2 + 512\right)\left(6555\sin^{10}i - 21850\sin^8 i + 27968\sin^6 i - 17024\sin^4 i + 4864\sin^2 i - 512\right)\cos 4g + \frac{328513185}{70368744177664}e^2 \cdot \sin^2 i\left(429e^{12} + 13728e^{10} + 96096e^8 + 219648e^6 + 183040e^4 + 53248e^2 + 4096\right)\left(334305\sin^{12}i - 1188640\sin^{10}i + 1664096\sin^8 i - 1157632\sin^6 i + 413440\sin^4 i - 69632\sin^2 i + 4096\right)\cos 2g - \frac{184041}{17592186044416}\left(6435e^{14} + 210210e^{12} + 1513512e^{10} + 3603600e^8 + 3203200e^6 + 1048320e^4 + 107520e^2 + 2048\right)\left(5014575\sin^{14}i - 18201050\sin^{12}i + 26209512\sin^{10}i - 18992400\sin^8 i + 7235200\sin^6 i - 1370880\sin^4 i + 107520\sin^2 i - 2048\right)\right],$$

$$\langle\langle R'_{14}\rangle_{l'}\rangle_l = \frac{\mu'}{a'}\left(\frac{a'}{a}\right)^{15}\left(1 - e^2\right)^{-27/2} \cdot \left[-\frac{7250518275}{70368744177664}e^{12}\sin^{12}i\left(27\sin^2 i - 26\right) \cdot \cos 12g + \frac{11310808509}{35184372088832}e^{10}\sin^{10}i\left(3e^2 + 22\right) \cdot \left(45\sin^2 i - 44\right)\left(5\sin^2 i - 4\right)\cos 10g - \frac{27047585565}{17592186044416}e^8\sin^8 i\left(3e^4 + 40e^2 + 72\right) \cdot \left(1035\sin^6 i - 2530\sin^4 i + 2024\sin^2 i - 528\right) \cdot \cos 8g + \frac{99174480405}{35184372088832}e^6\sin^6 i\left(5e^6 + 90e^4 + 288e^2 + 192\right)\left(3105\sin^8 i - 9200\sin^6 i + 9936\sin^4 i - 4608\sin^2 i + 768\right)\cos 6g - \frac{46977385455}{70368744177664}e^4\sin^4 i\left(45e^8 + 960e^6 + 4032e^4 + 4608e^2 + 1280\right)\left(6555\sin^{10}i - 21850\sin^8 i + 27968\sin^6 i - 17024\sin^4 i + 4864\sin^2 i - 512\right) \cdot \cos 4g + \frac{251215965}{17592186044416}e^2\sin^2 i\left(99e^{10} + 2310e^8 + 11088e^6 + 15840e^4 + 7040e^2 + 768\right) \cdot \left(334305\sin^{12}i - 1188640\sin^{10}i + 1664096\sin^8 i - 1157632\sin^6 i + 413440\sin^4 i - 69632\sin^2 i + 4096\right)\cos 2g - \frac{184041}{8796093022208}\left(3003e^{12} + 72072e^{10} + 360360e^8 + 549120e^6 + 274560e^4 + 39936e^2 + 1024\right)\left(5014575\sin^{14}i - 18201050\sin^{12}i + 26209512\sin^{10}i - 18992400\sin^8 i + 7235200\sin^6 i - 1370880\sin^4 i + 107520\sin^2 i - 2048\right)\right]. \tag{51}$$

4. Comparison with Numerical Quadrature

To graphically show the validity of the high-order analytic expansion of the doubly averaged disturbing function that we have presented, let us carry out numerical quadrature for comparison. The literal definition of the doubly averaged disturbing function, particularly its direct part, becomes from (3)

$$\langle\langle R\rangle_{l'}\rangle_l = \frac{\mu'}{4\pi^2}\iint_0^{2\pi}\frac{dl\,dl'}{\Delta}. \tag{52}$$

Since the orbit of the perturbing body is circular in CR3BP, we can turn the double integral of (52) into a single integral using the complete elliptic integral of the first kind (e.g., [28–30]). To compare the numerical quadrature result with that from the analytic expansion of disturbing function, we draw equipotential curves on the $(x, y) = (e\cos g, e\sin g)$ plane. Since g is an angle that intrinsically rotates, use of this polar-type coordinate is reasonable; therefore, it has often been used in the previous literature (e.g., [31–33]). All the numerical quadratures including the calculation of elliptic integral presented in this section are achieved using the functions implemented in GNU Scientific Library (GSL; http://www.gnu.org/software/gsl/).

As mentioned, the doubly averaged CR3BP makes an integrable system with just one degree of freedom. Then the state of the system is determined by just two constant parameters: α (or α') and (square of) normalized vertical component of the perturbed body's angular momentum k^2 defined as (e.g., [2, 34–37])

$$k^2 = \left(1 - e^2\right)\cos^2 i. \tag{53}$$

We selected several combinations of (α, k^2) and drew equipotential curves using the analytically expanded doubly averaged disturbing function, as well as using the numerical quadrature (Figure 2). Note that from the definition of (53), the theoretical largest value of e is achieved when $\cos i = 1$ as $e_{\max} = \sqrt{1-k^2}$. Similarly, the theoretical smallest value of $\cos i$ is achieved when $e = 0$ as $|\cos i|_{\min} = k$.

Figures 2(a) and 2(b) are for the inner case, and Figures 2(c)–2(f) are for the outer case. In each panel of Figure 2, the partial circle with red represents a line where the orbits of perturbed and perturbing bodies cross at the ascending node of the perturbed body. The partial circle with blue represents a line where the orbits of perturbed and perturbing bodies cross at the descending node of the perturbed body. These two circles are functions just of a and a' (i.e., functions of α), and the relationship between a, e, g, and a' is expressed as (e.g., [37–40])

$$a' = \frac{a\left(1 - e^2\right)}{1 \pm e\cos g}. \tag{54}$$

In the denominator of the right-hand term of (54), the positive sign takes place when the orbit intersect happens at the ascending node of perturbed body (when $f = -g$, where f denotes perturbed body's true anomaly). The negative sign takes place when the encounter happens at the descending node of perturbed body (when $f = \pi - g$). Using the variables $x = e\cos g$ and $y = e\sin g$, (54) can be rewritten as

$$\left(x \pm \frac{1}{2\alpha}\right)^2 + y^2 = \left(1 - \frac{1}{2\alpha}\right)^2, \tag{55}$$

and it is more obvious that (55) represents a pair of circles on the (x, y) plane.

Note that the orbit-crossing lines introduce singularities in N-body Hamiltonian (e.g., [41, 42]), and R (or R') is continuous but not regular across the orbit-crossing lines. Therefore the analytic expansion of the doubly averaged disturbing function is not applicable to the region very close to the orbit-crossing lines (e.g., [43]). Note also that each panel in Figure 2 has an individual contour interval, which represents a fixed interval of disturbing potential in each of the systems. Therefore, regions with dense contours imply that the potential gradient is steep there.

Since the high-order analytic expansion of disturbing function must have its significance when α (or α') is large, we chose relatively large α (or α') so that we cannot ignore their high-order powers. For example, when $2n = 14$ and $\alpha = 0.50$, $\alpha^{14} = 6.103515625 \times 10^{-5}$ which we may ignore. But when $2n = 14$ and $\alpha = 0.90$, $\alpha^{14} = 0.22876792454961$, which is not ignorable. In Figure 2(a) where $(\alpha, k^2) = (0.90, 0.900)$, effect of the high-order analytic expansion is well exhibited. As is widely known, there is no stationary point for argument of pericenter g while $k^2 > 0.6$ in the doubly averaged inner CR3BP at the quadruple-order approximation ($2n = 2$). However when α becomes large, g acquires stationary points at $g = \pm\pi/2$ even when $k^2 > 0.6$. This is what we see in the quadrature result in the leftmost panel in Figure 2(a). Kozai [3] derived expressions of high-order analytic expansion of the doubly averaged disturbing function for the inner CR3BP up to $2n = 8$. But the stationary points of g seen in the rightmost panel of Figure 2(a) (for quadrature) are not reproduced even by using the $2n = 8$ analytic terms. We confirmed that the stationary points at $g = \pm\pi/2$ in this parameter set $(\alpha, k^2) = (0.90, 0.900)$ first appear in the analytic approximation up to $2n = 12$, and the value of e at the stationary points is better reproduced using the analytic expansion of $2n = 14$, as seen in Figure 2(a). Note that other sets of stationary points at $g = 0, \pi$ seen in the quadrature result in the leftmost panel are not reproduced by the analytic expansion of disturbing function, since they are out of the orbit-crossing lines.

Note that we did not draw all the contours near the outer boundary ($e = e_{\max}$) indicated by the dashed circles, particularly in the panels in the left three columns. This is mainly because the data size of the figure would be too large if we drew all the contours with a fixed interval of disturbing potential toward the outer boundary, and also because the contour intervals would become too narrow. Moreover, we believe that we can grab major topological patterns of disturbing potential in each panel even if we draw contours only partially, at least inside the orbit-crossing lines.

Figure 2(b) is for $(\alpha, k^2) = (0.90, 0.400)$. They exemplify some limitations of the high-order analytic expansion of disturbing function that we have carried out. Although a saddle point seen in the quadrature result at the origin $(x, y) = (0, 0)$ is reproduced in the results of the analytic expansion of disturbing function, we see spurious local extremums along the y-axis near the boundary of $e = e_{\max}$ in the $2n = 14$ case. The spurious local extremums are not so remarkable in the $2n = 8$ case (although they surely exist), but there seem to exist spurious saddle points around

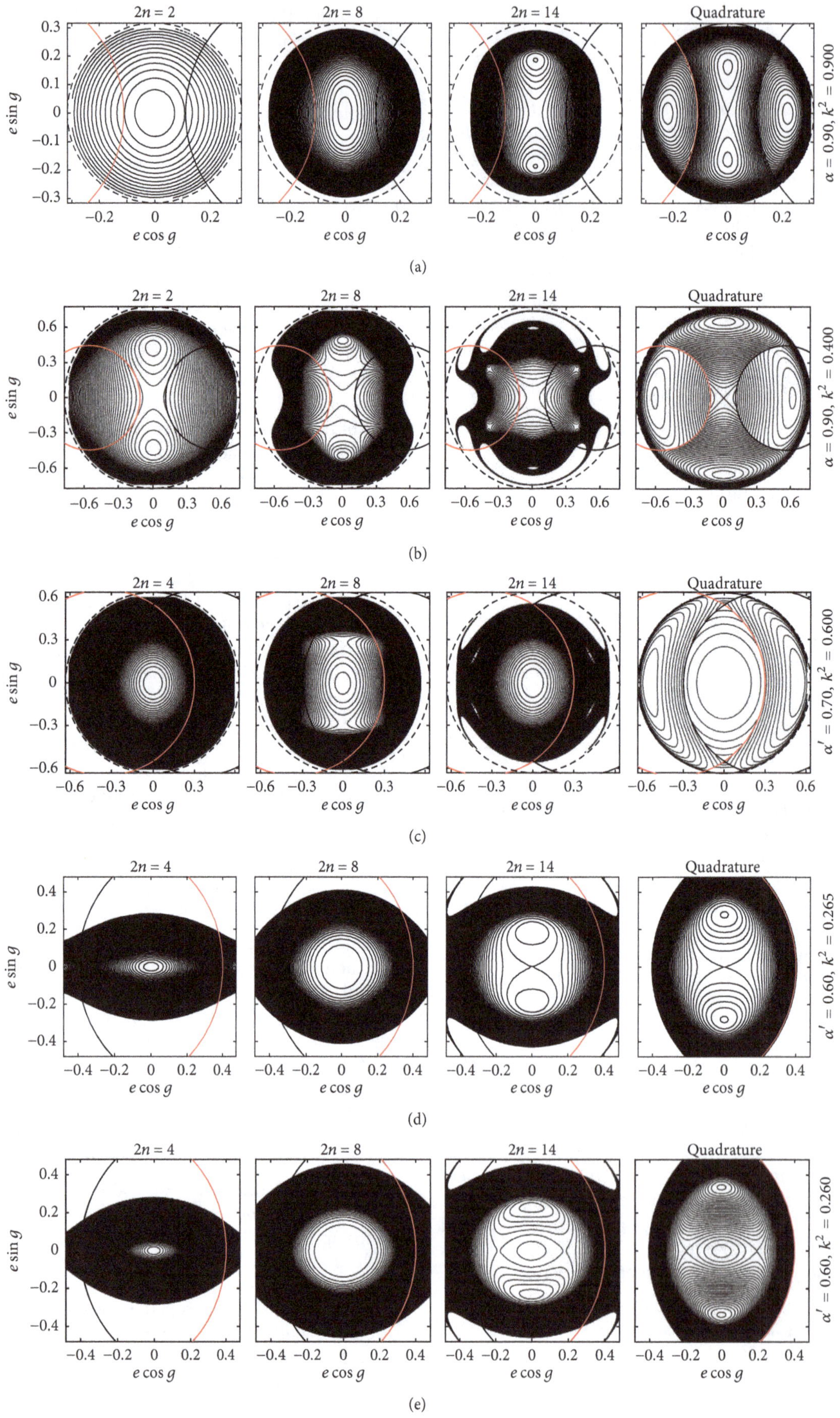

FIGURE 2: Continued.

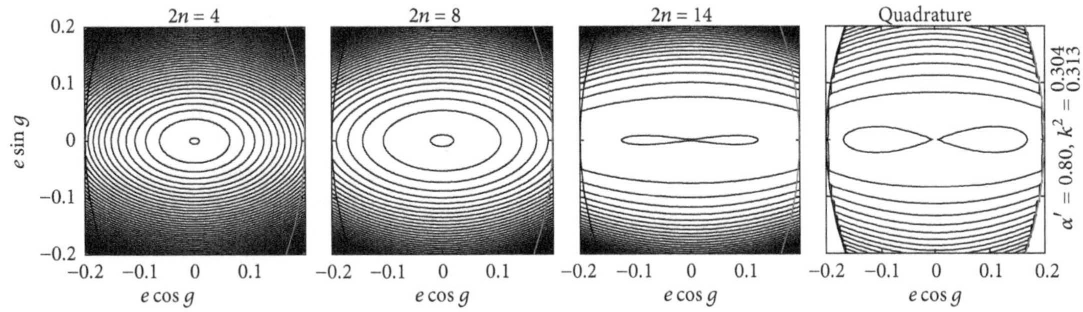

(f)

FIGURE 2: Equipotential curves computed through the doubly averaged disturbing function and plotted on the plane $(x, y) = (e\cos g, e\sin g)$. Columns are, from the left, $2n = 2$ (for (a), (b)) or $2n = 4$ (for (c), (d), (e), and (f)), $2n = 8$, $2n = 14$, and the numerical quadrature defined by (52). Rows are, from the top, (a) $(\alpha, k^2) = (0.90, 0.900)$, (b) $(\alpha, k^2) = (0.90, 0.400)$, (c) $(\alpha', k^2) = (0.70, 0.600)$, (d) $(\alpha', k^2) = (0.60, 0.265)$, (e) $(\alpha', k^2) = (0.60, 0.260)$, and (f) $(\alpha', k^2) = (0.80, 0.304)$ for the analytic expansion and $(\alpha', k^2) = (0.80, 0.313)$ for the numerical quadrature. The black dashed circles represent the theoretical maximum eccentricity ($e_{\max}$) of each of the systems. The red and the blue partial circles represent the conditions where the orbits of perturbed and perturbing bodies cross at the ascending node (red) and at the descending node (blue) of the perturbed body.

$y = e\sin g \sim \pm 0.3$ that do not appear in the quadrature result. Thus we say that the analytic expansion of the doubly averaged disturbing function for the inner CR3BP, at least up to these orders, is applicable to systems with moderate eccentricity when α is large. Note that it apparently seems that the leftmost panel for the analytic expansion of the order of $2n = 2$ looks the closest to the quadrature result. However, we think this is just a coincidence, because the $2n = 2$ expansion includes only the terms of $\cos 2g$ as in (28), and $\langle\langle R_2 \rangle_{l'}\rangle_l$ would not be influenced by terms of $\cos 2pg$ with $p \geq 2$.

Figure 2(c) is for the doubly averaged outer CR3BP when $(\alpha', k^2) = (0.70, 0.600)$. In this case, the high-order analytic expansion of the doubly averaged disturbing function ($2n = 14$) reproduces the quadrature result inside the orbit-crossing lines. However, the medium-order expansion ($2n = 8$) yields artificial saddle points along the y-axis. We confirmed that these spurious saddle points show up in the expansion of $2n = 10$ too (panels are not shown here). We thus say that the highest-order analytic expansion of $2n = 14$ pushes these saddle points out of the orbit-crossing lines, and the equipotential curves become similar to the quadrature result, at least inside the orbit-crossing lines where e is moderate.

In Figure 2(d) when $(\alpha', k^2) = (0.60, 0.265)$, we see three local extremums in the quadrature result: a pair of local minima along the y-axis together with a saddle point at the origin $(0, 0)$. These local extremums cannot be reproduced by the low-order analytic expansion such as $2n = 4$ or $2n = 8$; the local extremums show up only in the expansion of $2n = 12$ or higher, although the locations of the local minima along the y-axis are slightly different from the quadrature result. When k^2 becomes even smaller, the difference in the topology of the equipotential curves increases, as seen in Figure 2(e) where $(\alpha', k^2) = (0.60, 0.260)$. A pair of local minima along the y-axis persists, but the origin $(0, 0)$ becomes a local maximum and a pair of saddle points takes place along the x-axis. We confirmed that this topology is only reproduced when using the analytic expansion of disturbing function at our highest-order ($2n = 14$), although the locations of the local extremums are slightly different from the quadrature result.

When α' gets even larger and k^2 is in a certain range of moderate values, a pair of local minima show up along the x-axis inside the orbit-crossing lines together with a saddle point at the origin $(0, 0)$. This is the case of Figure 2(f) when $(\alpha', k^2) = (0.80, 0.304)$. The topology of the local extremums is reproduced by the analytic expansion of disturbing function at our highest-order ($2n = 14$). However, we should note a point here. In the quadrature result of the rightmost panel in Figure 2(f), the pair of local minima along the x-axis appear when the value of k^2 is slightly different ($k^2 = 0.313$) from the analytic result ($k^2 = 0.304$). We have not yet figured out the cause of the discrepancy of the k^2 values that are required for the local minima to appear in the analytic expansion result and in the numerical quadrature result. It is possible that even higher-order analytic expansions of disturbing function ($2n = 16, 18, 20, \ldots$) may diminish the discrepancy.

As a summary of this section, we say that the high-order analytic expansion of the doubly averaged disturbing function of CR3BP that has been presented in the present paper overall seems valid inside the orbit-crossing lines even when α (or α') is large, as long as eccentricity of perturbed body is moderate. Under these conditions, local extremums of the doubly averaged disturbing function for CR3BP are quantitatively reproduced by the high-order analytic expansion.

5. Comparison with Numerical Integration

In Section 4 we inspected the validity of the high-order analytic expansion of the doubly averaged disturbing function by comparing its outcome with that from numerical quadrature. The numerical quadrature defined by the double integral (52) yields a numerical proxy of an integrable CR3BP system with one degree of freedom, and it is safe for us to directly compare the quadrature result with that from the analytic expansion of disturbing function. However, equations of motion of

Table 1: Configuration of the numerical integration presented in Section 5. Δt is nominal stepsize (days), $t_{\rm int}$ is interval time of output (years), and T is total integration period (million years). The parameter combinations of α (or α') and k^2 from (a) to (f) are common to both in Figures 2, 3, 6, and 7.

	α or α'	k^2	m'/m_0	Δt	$t_{\rm int}$	T
(a)	$\alpha = 0.90$	0.900	3.0404326×10^{-6}	2	500	10
(b)	$\alpha = 0.90$	0.400	3.0404326×10^{-6}	2	500	10
(c)	$\alpha' = 0.70$	0.600	3.0404326×10^{-6}	2	500	10
(d)	$\alpha' = 0.60$	0.265	4.7739597×10^{-5}	20	1000	200
(e)	$\alpha' = 0.60$	0.260	4.7739597×10^{-5}	20	1000	200
(f)	$\alpha' = 0.80$	0.313	9.5479194×10^{-6}	4	500	30

three-body systems are generally not integrable, having short-term oscillations and various mean motion resonances intact without being averaged out. To demonstrate the practical usefulness of our analytic expansion of the doubly averaged disturbing function, we carried out direct numerical integration of equations of motion using the same parameter combinations α (or α') and k^2 as in the previous section. Based on the comparison, we will again give considerations on the validity of the high-order analytic expansion of the doubly averaged disturbing function for CR3BP.

5.1. Model and Method. We consider motion of small solar system bodies with infinitesimally small mass (asteroid, comet, etc.) undergoing gravitational perturbation from a major planet on a circular orbit. Using the same six sets of parameters in the framework of CR3BP as in Figure 2, we calculate orbit propagation of the small bodies by numerical integration for ten to two hundred million years. The parameter sets are tabulated in Table 1. The main result of the numerical integration is presented and discussed in Section 5.2.

In Table 1, mass ratios between perturbing body and central body (m'/m_0) for (a), (b), and (c) are close to the ratio of the Earth-Moon total mass and the solar mass (~1/314 of the mass ratio between Jupiter and the Sun). The m'/m_0 value for (f) is nearly 1/100 of the mass ratio of Jupiter and the Sun. The reason for these small m'/m_0 values is to avoid occurrence of orbital instability of the small body due to close encounters with the perturbing planet near the orbit-crossing lines. This is necessary because α and α' are large in these four cases. We also chose small values for the stepsize of numerical integration Δt for these cases by the same reason. As for (d) and (e), the m'/m_0 value is close to 1/20 of the mass ratio of Jupiter and the Sun. Since α' is relatively small in these two cases, the probability of close encounters between perturbed and perturbing bodies is lower, and we can choose a larger mass ratio m'/m_0 as well as larger stepsize Δt. Here, readers should recall the fact that the influence of the mass ratio m'/m_0 is limited to the timescale of orbital evolution of perturbed body, and it does not affect the topology of relative orbit, particularly in the doubly averaged CR3BP. It is obvious from the function form of the doubly averaged disturbing function (19) or (24) where the perturber's mass serves just as a constant coefficient through μ'. Therefore, the trajectory shape of the perturbed body remains the same even if we change the mass of the perturber from m' to pm' (where p is a positive real number); only the timescale of orbital evolution would change from t to t/p. This is why we can arbitrarily change the mass ratio m'/m_0 in numerical integration as long as what we need to know is the secular topology of trajectories of the perturbed body, not its time evolution sequence.

In the numerical integration, the semimajor axis of the perturbing body $a' = 5.2025217$ AU is common to all the cases, which is a substitute for Jupiter's semimajor axis. The perturber's mean anomaly (practically equivalent to mean longitude) l' is set to zero at time $t = 0$. By the definition of CR3BP, the perturber's eccentricity e' and inclination i' are both set to zero. As for the perturbed body's initial orbital elements, semimajor axis a is automatically determined from the value of α or α'. We set the initial values of eccentricity e and argument of pericenter g of perturbed body along the x- and y-axis on the $(e\cos g, e\sin g)$ plane: we selected e of perturbed body from $e = 0$ to $e_{\rm max}$ with the interval of 0.01. As for g, we used four values of $g = 0, \pi/2, \pi$, and $3\pi/2$. The values of the perturbed body's i are automatically determined by relation (53). Initial values of the perturbed body's longitude of ascending node (h) are set to 0, and those for its initial mean anomaly are set to π.

The differential equation to be solved is the simple classical Newtonian equation of motion of the perturbed body. For the inner case, it is

$$\frac{d^2\mathbf{r}}{dt^2} + \mu\frac{\mathbf{r}}{r^3} = \nabla R, \tag{56}$$

with $\mu = \mathcal{G}m_0$. For the outer case we just replace R for R'. As for the numerical integration scheme, we use the regularized mixed-variable symplectic method [44] based on the so-called Wisdom–Holman map [20, 45]. In this scheme the stepsize of numerical integration is automatically reduced when a close encounter between two bodies happens, trying to keep the calculation accuracy high. The Δt values listed in Table 1 are the nominal stepsizes used when no close encounter takes place between bodies. We have implemented this scheme based on the code SWIFT [44] and used it in our previous studies where close encounters between asteroids and terrestrial planets take place very often [46, 47], as well as where almost no close encounter happens [48, 49]. We have confirmed that our implementation of this scheme does not have a practical problem in accuracy or efficiency when applied to these dynamical systems. We adopt the Gaussian units where the Gaussian gravitational constant $\sqrt{\mathcal{G}} = 0.01720209895$, the unit of time is an ephemeris day, and the unit of length is the astronomical unit (AU), so that the mass of the primary body becomes unity (e.g., [16, 50]).

5.2. Integration Result. We show the main results of our numerical integration in Figure 3 whose parameters are described in Table 1. The perturbed body's $(e\cos g, e\sin g)$ values are plotted on the panels by black and green dots (see below for the reason for this distinction) with the fixed output interval $t_{\rm int}$ designated in Table 1. We chose $t_{\rm int}$ empirically so that the output data amount does not get too

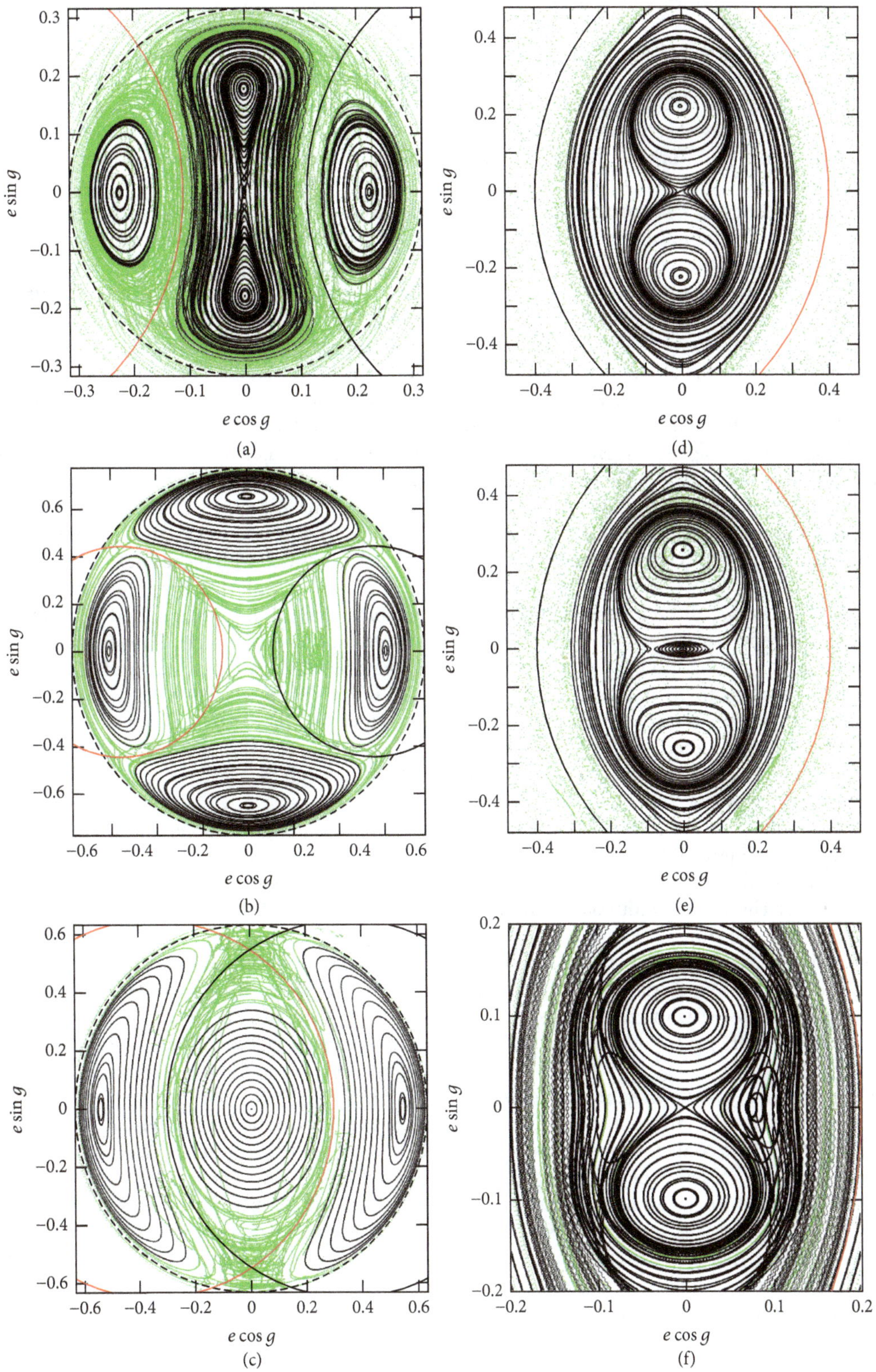

FIGURE 3: Plots of the perturbed body's ($e \cos g, e \sin g$) values obtained from the direct numerical integration whose parameters are listed in Table 1. As for the meaning of the black dashed circles and of the red and blue partial circles, see the caption of Figure 2. The green dots represent the trajectories of bodies that experienced orbital instability during the integration period. Stable trajectories are drawn by black dots.

large, while avoiding any kind of aliasing in the plots. The following settings are all common to both Figures 2 and 3: the coordinate system $(x, y) = (e \cos g, e \sin g)$, the ranges of x- and y-axis, meaning of the black dashed circles, and that of the red and blue partial circles.

When inspecting Figure 3, we would like readers to notice two points: One is that we drew trajectories of perturbed bodies in black only when their orbital motion remains stable from the beginning to the end of the integration period T. This is to facilitate the comparison between the quadrature result and the integration result. Here we define the "stable" orbital motion if the semimajor axis a does not exhibit any major changes throughout the integration period T. Otherwise we categorize the orbital motion as "unstable." This categorization is done by the naked eye, plotting the time sequence of a for all the perturbed bodies used in the integration (320 bodies per panel, 1920 bodies in total). Browsing through all the plots of a, we subjectively distinguish "stable" and "unstable" orbits. Hence there can be some ambiguity in the categorization. However, let us say that the difference between stable and unstable orbits is rather clear in Figure 3, and the subjective distinction between the black and the green trajectories does not affect the discussion here.

The second point is that we did not draw some of the trajectories that are largely overlapped with others. This is to prevent the panels from becoming too busy with too many similar trajectories, as well as to keep the figure file size reasonably small. As we mentioned before, starting points of the numerical integration are aligned on the x- and y-axis as $g = 0, \pi/2, \pi, 3\pi/2$ with a fixed interval of eccentricity. Due to the intrinsic symmetry that the disturbing function has, different initial condition can generate almost identical trajectories. For example, in Figure 3(c) the initial conditions $(e, g) = (0.03, 0)$ and $(e, g) = (0.03, \pi)$ generate almost identical trajectories. For eliminating redundancy in figure and reducing the file size, we did not draw the trajectory from the initial condition $(e, g) = (0.03, \pi)$ and just drew that from $(e, g) = (0.03, 0)$. We carried out this kind of selection of overlapped trajectories with great care so that our subjective selection never alters general topological patterns in the panels.

Comparing Figures 3 and 2, we say that their agreement is overall good. In Figure 3(a) for the inner CR3BP, existence of stationary points at $g = \pm\pi/2$ is confirmed by the numerical integration. The eccentricity value of the stationary points is close to what the analytic expansion and the numerical quadrature had yielded in Figure 2(a). Existence of stationary points at $g = 0, \pi$ outside the orbit-crossing lines is also confirmed, although most of the trajectories near the orbit-crossing lines become unstable due to close encounters with the perturbing body.

Similarly in Figure 3(b), existence of stationary points at $g = \pm\pi/2$ as well as those at $g = 0, \pi$ is confirmed by the numerical integration. There must be a saddle point at the origin $(x, y) = (0, 0)$, but most of the trajectories near the origin are unstable in Figure 3(b) due to the proximity to the orbit-crossing lines, and it is not easy to visually confirm it.

In Figure 3(c) for the outer CR3BP, existence of a stationary point at the origin $(x, y) = (0, 0)$ as well as those at $g = 0, \pi$ is confirmed by the numerical integration. It is clearly seen that trajectories can easily become unstable around the two intersections of the red and blue orbit-crossing lines along the y-axis.

In Figures 3(d) and 3(e), the existence of stationary points is again confirmed by the numerical integration. We see a pair of stationary points at $g = \pm\pi/2$ in both panels. The origin $(x, y) = (0, 0)$ in Figure 3(d) makes a saddle point, as suggested by the quadrature (Figure 2(d)). The origin $(0, 0)$ in Figure 3(e) makes a local maximum, which is also consistent with the quadrature result (Figure 2(e)). A pair of saddle points seen along the x-axis in Figure 3(e) has also been suggested by the quadrature (Figure 2(e)). Although the location of the saddle points seen in Figure 3(e) is slightly closer to the origin $(0, 0)$ compared with that in Figure 2(e), we would say they are consistent. In both the panels of Figures 3(d) and 3(e), trajectories get unstable and the green dots are scattered as they approach the orbit-crossing lines.

As we have seen in Figures 3(a)–3(e), the result from the direct numerical integration of the equations of motion overall agrees well with the result from the numerical quadrature, as well as with that from the analytic expansion of the doubly averaged disturbing function. We thus say that this comparison largely justifies the validity of use of the analytic expansion of the doubly averaged disturbing function for CR3BP presented in Section 3.

However, in Figure 3(f) we encounter with a typical system where the use of doubly averaged disturbing function, either through analytic expansion or quadrature, is not appropriate. As is clearly seen, the topological pattern of trajectories in Figure 3(f) is very different from the equipotential contours shown in Figure 2(f). A pair of stationary points shows up at $g = \pm\pi/2$ in Figure 3(f) that we do not see in Figure 2(f). Also, although we see stationary points along the x-axis in Figure 3(f), the shape of the trajectories is fairly different from what is observed in Figure 2(f). We then plotted the whole phase space for this system ($e \le e_{\rm max}$) in Figure 4. As we see, the trajectories outside the orbit-crossing lines are very similar between Figure 4(a) (quadrature) and Figure 4(b) (integration). But the complicated trajectory pattern inside the orbit-crossing lines takes place only in the numerical integration result (Figure 4(b)).

So far we have not completely figured out the cause of the difference in trajectory patterns between Figures 2(f) and 3(f). We have confirmed that there is no problem in accuracy of the numerical integration that is produced in Figure 3. As far as we have been able to investigate, the apparently peculiar trajectories seen in Figure 3(f) are possibly related to the existence of a mean motion resonance. We selected some of the periodic orbits observed inside the orbit-crossing lines in Figure 4(b) and plotted them in separate panels in Figure 5 (the panels in its left column). For each of the periodic orbits, we calculated an argument $5\lambda' - 7\lambda + 2g$, where λ' is mean longitude of the perturbing body and λ is that of the perturbed body. As a result, we found that in the top two trajectories (Figures 5(a) and 5(b)) the argument $5\lambda' - 7\lambda + 2g$ librates around π, although the libration

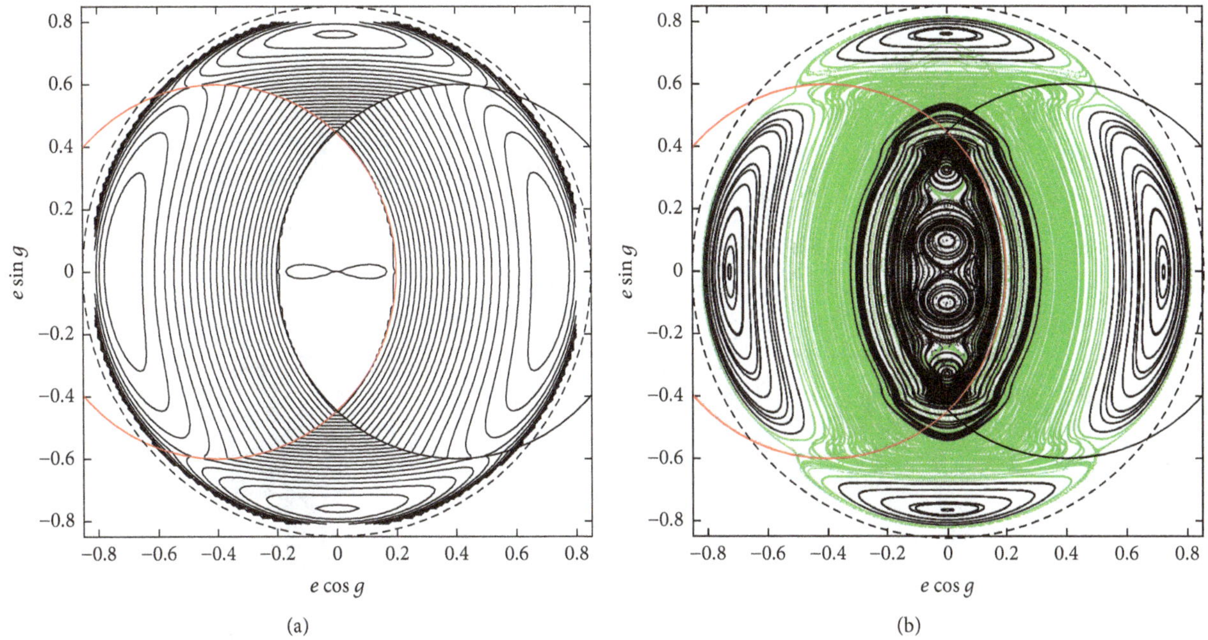

FIGURE 4: (a) Extended plot of Figure 2(f) up to $e = e_{max}$ (numerical quadrature). (b) Extended plot of Figure 3(f) (numerical integration). Note that we set the contour interval in (a) ten times larger than in Figure 2(f) for a better visibility of plots. The interval of initial eccentricity in (b) is twice larger than in Figure 3(f) for the same reason.

amplitude is relatively large (see the panels in the middle column of Figure 5). This indicates that the systems in Figures 5(a) and 5(b) are trapped in the 5 : 7 mean motion resonance. System (c) seems close to this resonance too. It is well known that the existence of strong mean motion resonance can substantially change the shape of a disturbing potential, together with the location of local extremums, compared with nonresonant cases (e.g., [51–53]). To deal with resonant systems like Figure 3(f) rigorously, we would need to employ different methods that are particular to each of the resonant systems (e.g., [54, 55]).

Although all the five perturbed bodies whose trajectories are shown in Figure 5 started with the same initial semimajor axis (a_0 = 6.503152 AU), we see a systematic difference in the average value of the semimajor axis a between 5(a–c) and 5(d, e) (see the panels in the right column of Figure 5). The short-term oscillation of a with relatively large amplitude seen in Figures 5(a), 5(b), and 5(c) is probably related to the 5 : 7 mean motion resonance. It is possible that the oscillation of a seen in Figures 5(d) and 5(e) is also related to this resonance or others that we have not yet found.

We will pursue the above problem in our forthcoming publications. Several possibilities might account for the peculiar trajectories seen in Figure 4(b), such as the effect of high-order mean motion resonance (e.g., [56]) and its combination with the Lidov–Kozai libration (e.g., [57]). A kind of secondary resonance may be embedded in this system (e.g., [58]). We should be aware that more and more mean motion resonances show up as α or α' approaches 1 (e.g., [59]), and treatment of such systems would need greater care.

For readers' interest, we also plotted trajectories $(\sin i \cos h, \sin i \cos h)$ of the perturbed body where i is its inclination and h is its longitude of ascending node. The six panels from (a) to (f) in Figure 6 correspond to each in Figure 3 with the same color pattern (stable orbits are drawn in black, and unstable orbits are drawn in green). Recalling the fact that the theoretical smallest value of $\cos i$ takes place when $e = 0$ as $|\cos i|_{min} = k$ (see Section 4), it is clear that the theoretical largest value of $\sin i$ is $|\sin i|_{max} = \sqrt{1-k^2}$ when $e = 0$ (note that this value happens to be the same as e_{max}). We drew the outer boundary $|\sin i|_{max} = \sqrt{1-k^2}$ by the black dashed circles in Figure 6.

In the doubly averaged CR3BP system that we now consider, eccentricity e and inclination i are correlated with each other through the definition of k^2 in (53). However, since longitude of ascending node h is not included in the disturbing function R, h does not have a particular correlation with eccentricity e, inclination i, or argument of pericenter g. The panels of Figures 6(b) and 6(c) typically exemplify this circumstance where we see the trajectories of perturbed bodies which are not closed on the $(\sin i \cos h, \sin i \cos h)$ plane; i and h are not correlated, and these trajectories do not compose equipotential contours.

For more demonstrations of the time variation of i and h, we picked one body per each of the panels from Figure 6(a) to Figure 6(f) and showed the time variation of their e, i, g, and h in Figure 7. Initial values of (e, g) of each of the bodies are written in the caption of Figure 7. Our choice of the particular body among the 320 bodies in each of the panels is almost arbitrary: we basically chose the bodies whose argument of pericenter librates around $g = \pi/2$ or 0 (a, b, c, d, and f), but we chose a body whose argument of pericenter circulates from 0 to 2π (e).

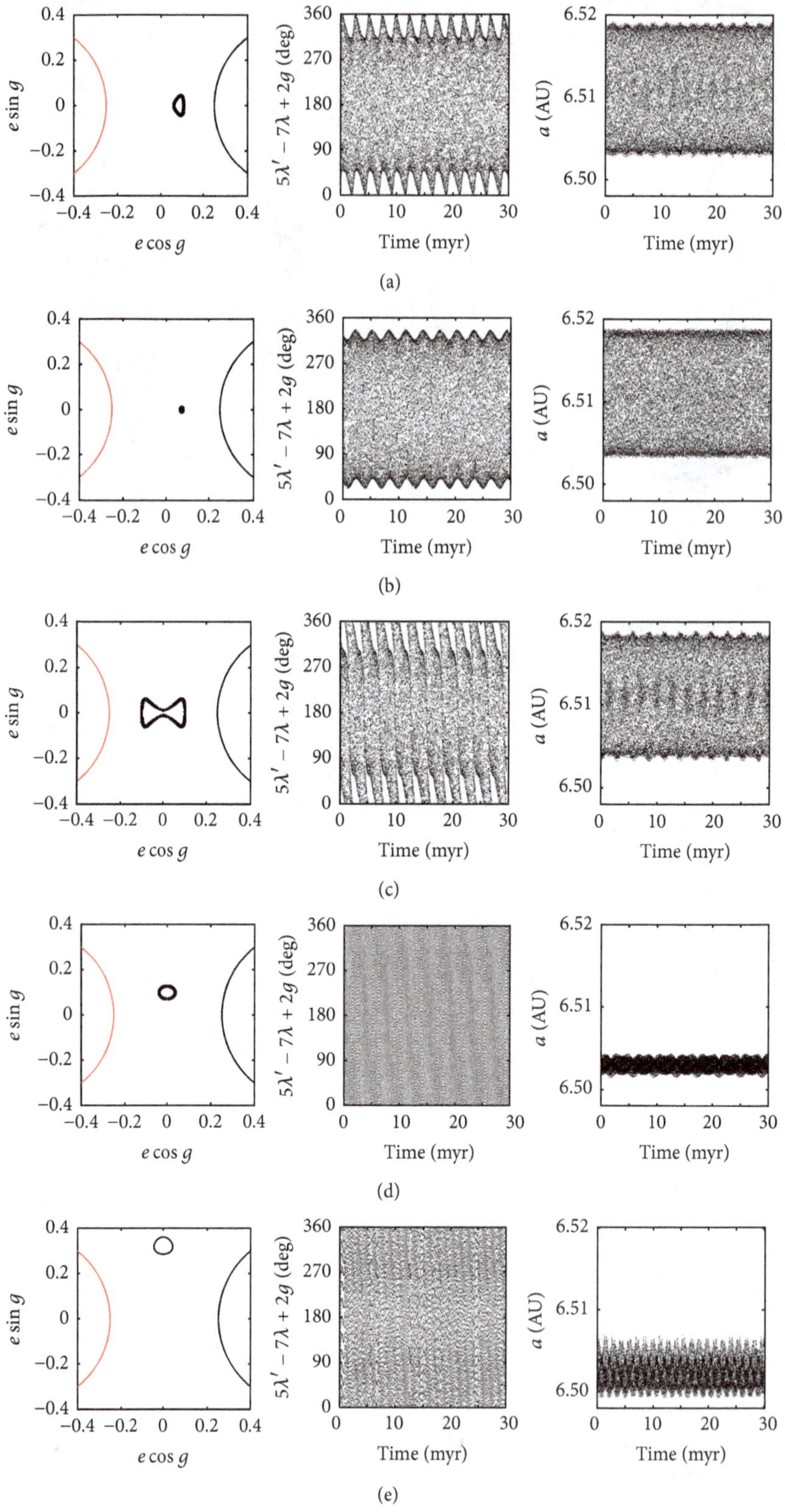

FIGURE 5: Some of the peculiar periodic orbits seen in Figure 4(b). The initial e and g are as follows: (a) $(e, g) = (0.06, 0)$, (b) $(e, g) = (0.08, 0)$, (c) $(e, g) = (0.10, 0)$, (d) $(e, g) = (0.07, \pi/2)$, and (e) $(e, g) = (0.36, \pi/2)$. The left column: the usual $(e \cos g, e \sin g)$ plots. As for the meaning of the red and blue partial circles, see the caption of Figure 2. The middle column: the argument $5\lambda' - 7\lambda + 2g$ plotted with a time interval of 1500 years. The right column: osculating semimajor axis a plotted with an interval of 1500 years. All the orbital elements are calculated on the heliocentric coordinate centered on the central mass.

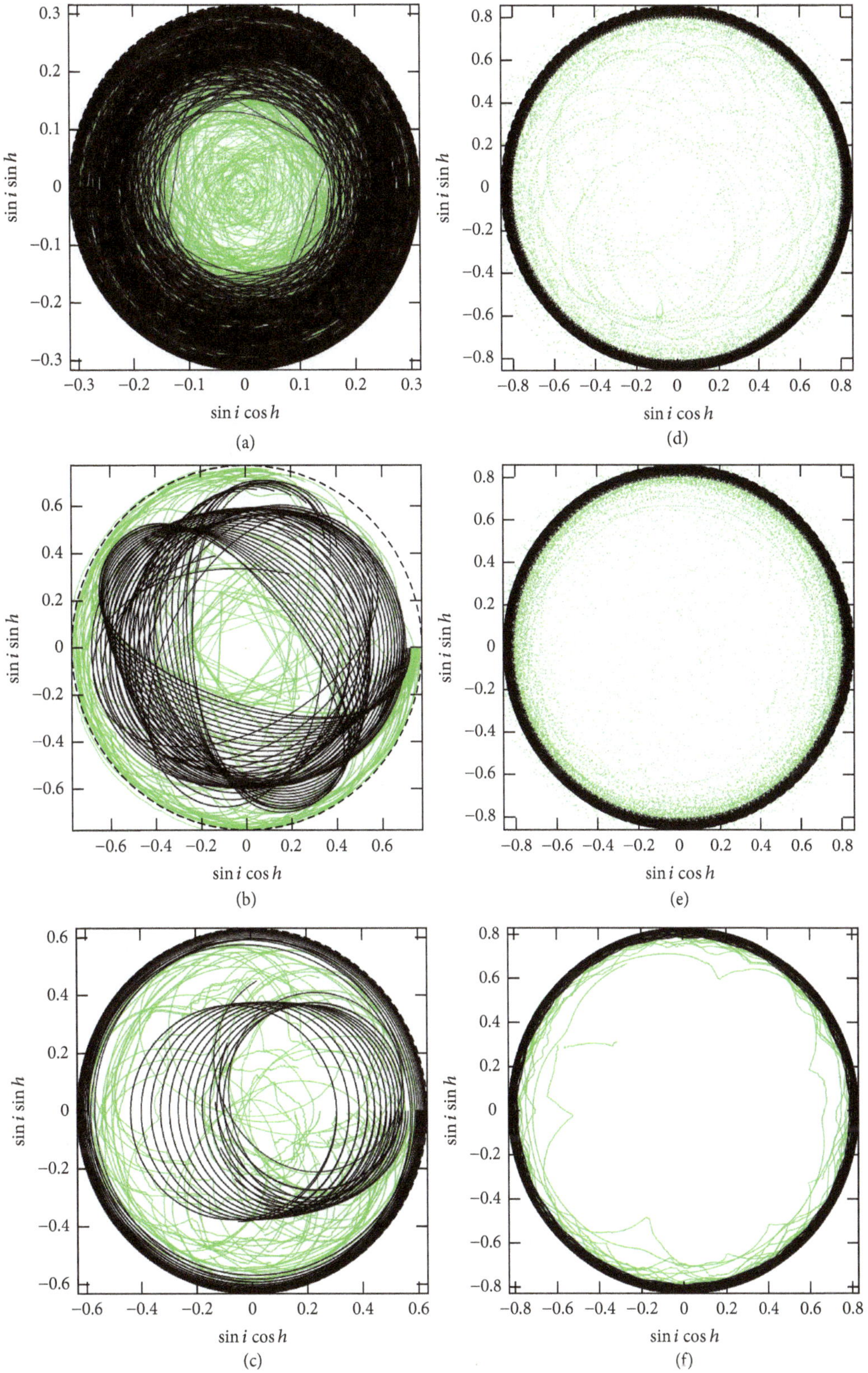

FIGURE 6: Plots of the perturbed body's ($\sin i \cos h, \sin i \sin h$) values obtained from the direct numerical integration whose parameters are listed in Table 1. The black dashed circles represent the theoretical maximum $\sin i$ value (which is the same as e_{max}). Meanings of the black dots and the green dots remain the same as in Figure 3.

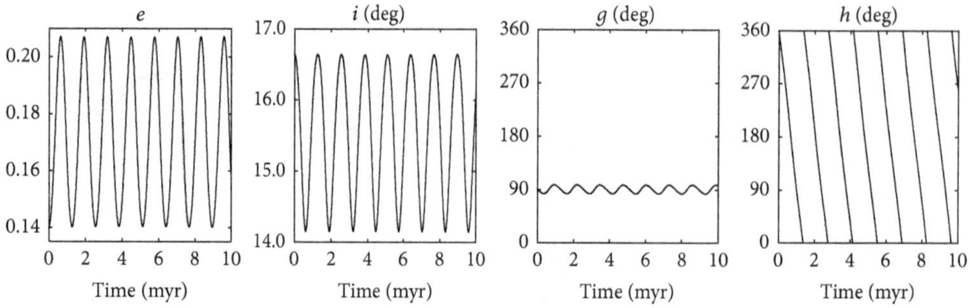

(a)

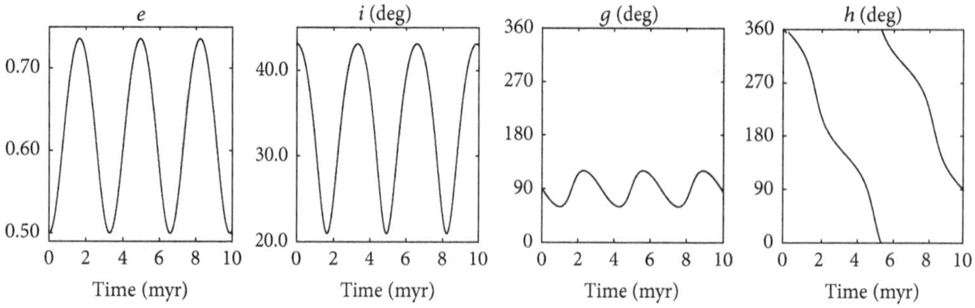

(b)

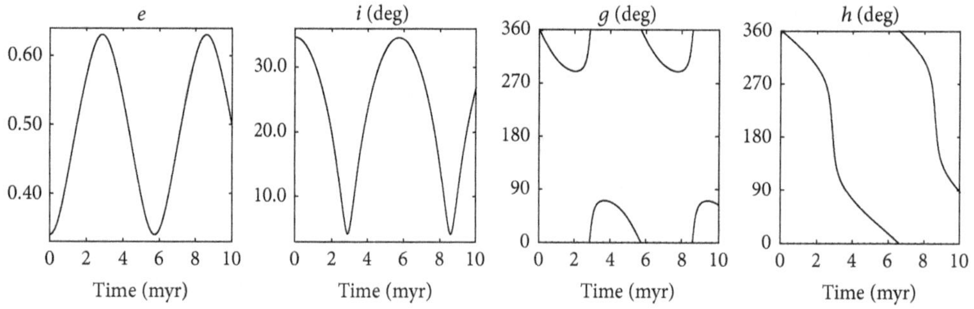

(c)

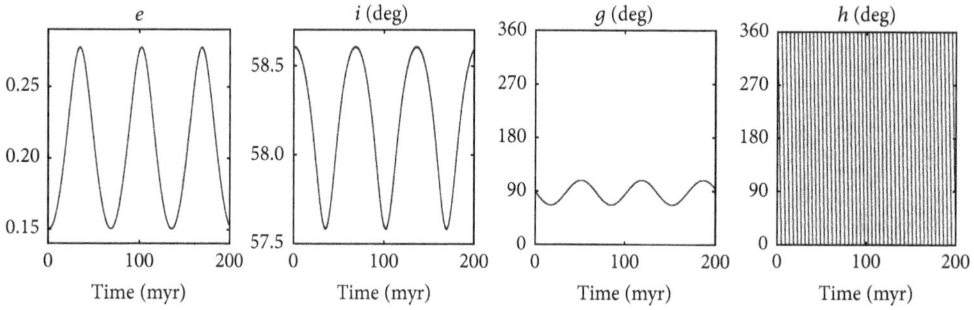

(d)

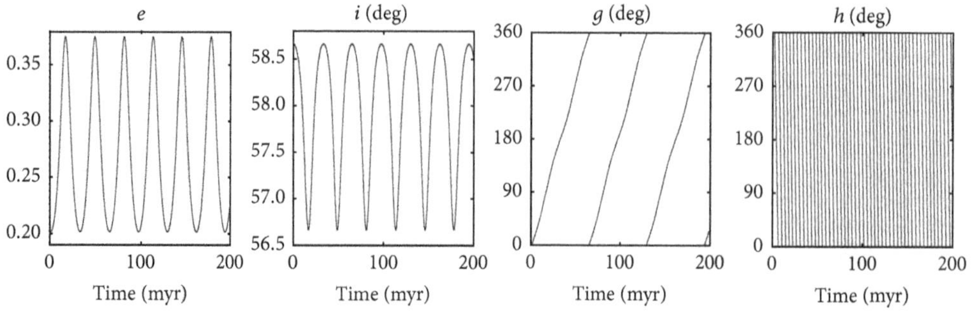

(e)

FIGURE 7: Continued.

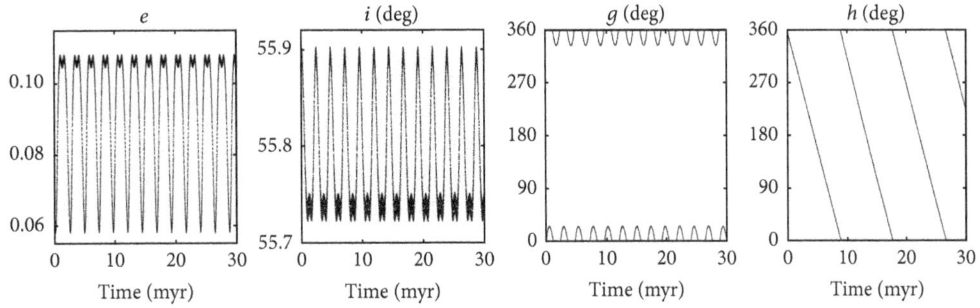

(f)

FIGURE 7: Variation of e, i, g, and h of representative bodies in Figures 6(a)–6(f). Initial values of e and g for each body are as follows: (a) $(e, g) = (0.14, \pi/2)$, (b) $(e, g) = (0.50, \pi/2)$, (c) $(e, g) = (0.34, 0)$, (d) $(e, g) = (0.15, \pi/2)$, (e) $(e, g) = (0.20, 0)$, and (f) $(e, g) = (0.06, 0)$. Note that the unit of g and h is degree, not radian, for consistency with the unit of i. All the elements are plotted with a time interval of 1500 years and are calculated on the heliocentric coordinate centered on the central mass.

The panels in Figure 7 immediately tell us the correlation between eccentricity e and inclination i, which is an obvious outcome of the preservation of k^2 in (53). The argument of pericenter g is correlated to them, as we have seen in the study of the doubly averaged disturbing function in Section 3. The rightmost panels show the fact that only longitude of ascending node h has no correlation with other elements; it shows a monotonic decrease in all the panels.

We do not show the analytic counterpart of Figure 6 or Figure 7 because they are out of the scope of the present paper. Readers with particular interest may consult classic literature such as Moiseev [34, 60] or Lidov [2] for how to formally obtain analytic solutions of time evolution of h. For example, in Lidov [2], the author uses the mutual dependence between e, i, and g and constructs a definite integral for quadrature that formally depicts the function form of h depending on time. Incidentally, if we limit ourselves to the doubly averaged inner CR3BP at the quadruple-order analytic approximation ($2n = 2$; see Section 3.2), analytic solution of h and that of g (hence those of e and i) using elliptic functions have already been derived in much of the literature (e.g., [3, 5–7, 61, 62]).

6. Summary and Discussion

Through the comparison of the results obtained from analytic expansion of the doubly averaged disturbing function, numerical quadrature, and direct numerical integration of the equations of motion, overall we validated the use of our high-order analytic expansion of the doubly averaged disturbing function both for the inner and outer CR3BP, even when the semimajor axis ratio α (or α') is large. Our result tells us that the high-order analytic expansion is not applicable in several circumstances: For example, when a perturbed body's eccentricity approaches its theoretical largest value, when orbits of perturbed and perturbing bodies cross, and when strong mean motion resonance is at work. Particularly, occurrence of strong mean motion resonance can significantly change global topology of disturbing potential, and the doubly averaged method itself goes out of use. Nevertheless, we think our original objective to analytically exemplify dynamical characteristics of doubly averaged disturbing function for CR3BP in this paper is largely fulfilled. We thus conclude that the high-order analytic expansion of the doubly averaged disturbing function is suitable for expository purposes to account for the dynamical characteristics of secular CR3BP even when α (or α') is large, as long as the system is free from serious mean motion resonance.

Let us just mention a point regarding one limitation of the high-order analytic expansion of the doubly averaged disturbing function. As seen in Sections 4 and 5, sometimes the locations of local extremums of the analytically expanded doubly averaged disturbing function differ from those obtained by the numerical quadrature and numerical integration, particularly in the outer problem such as in Figure 2(d) or 2(e). In the outer case, the factor $(1-e^2)^{-2n+1/2}$ in (24) substantially enhances the magnitude of high-order terms in the analytically expanded disturbing function when eccentricity e is large. For example, when $e = 0.5$ the factor becomes ~8.65 ($2n = 8$), ~27.3 ($2n = 12$), and ~48.6 ($2n = 14$). When $e = 0.8$, the factor reaches ~2127 ($2n = 8$), ~126626 ($2n = 12$), and ~977049 ($2n = 14$), possibly causing artificial oscillation of the higher-order analytic terms in the large-e region.

Incidentally, we would like readers to be aware of a recent study that revealed that solutions of the three-body problem obtained from the double averaging procedure can substantially deviate from true solution due to accumulation of the effect of short-term oscillation [63]. This can happen particularly when the mass of the perturbing body is not negligibly small compared to the primary mass.

A Maple script that yields our analytic expansion in the present paper (see (25)–(51)) is available from the author upon request.

Competing Interests

The author declares no competing financial interests associated with this article.

Acknowledgments

The author has benefited from stimulating enlightenment through discussions with Katsuhito Ohtsuka, Arika Higuchi,

Akihiko Fujii, Ruslan Salyamov, Fumi Yoshida, and Renu Malhotra. Detailed and constructive review by Yolande McLean has considerably improved the English presentation of this paper. Algebraic manipulations with Maple, numerical quadrature operations, and orbit propagation by numerical integration carried out in the present paper were all performed at Center for Computational Astrophysics (CfCA), National Astronomical Observatory of Japan (NAOJ). This study is supported by a pair of JSPS Kakenhi Grants, JP25400458/2013–2016 and JP16K05546/2016–2018.

References

[1] V. Szebehely, *Theory of Orbits*, Academic Press, New York, NY, USA, 1967.

[2] M. L. Lidov, "The evolution of orbits of artificial satellites of planets under the action of gravitational perturbations of external bodies," *Planetary and Space Science*, vol. 9, no. 10, pp. 719–759, 1962.

[3] Y. Kozai, "Secular perturbations of asteroids with high inclination and eccentricity," *The Astronomical Journal*, vol. 67, pp. 591–598, 1962.

[4] I. Shevchenko, *The Lidov-Kozai Effect: Applications in Exoplanet Research and Dynamical Astronomy*, vol. 441 of *Astrophysics and Space Science Library*, Springer, Dordrecht, The Netherlands, 2017.

[5] H. Kinoshita and H. Nakai, "Secular perturbations of fictitious satellites of Uranus," *Celestial Mechanics & Dynamical Astronomy*, vol. 52, no. 3, pp. 293–303, 1991.

[6] H. Kinoshita and H. Nakai, "Analytical solution of the Kozai resonance and its application," *Celestial Mechanics & Dynamical Astronomy*, vol. 75, no. 2, pp. 125–147, 1999.

[7] H. Kinoshita and H. Nakai, "General solution of the Kozai mechanism," *Celestial Mechanics & Dynamical Astronomy*, vol. 98, no. 1, pp. 67–74, 2007.

[8] V. I. Prokhorenko, "A geometric study of solutions to restricted circular double-averaged three-body problem," *Cosmic Research*, vol. 39, no. 6, pp. 583–593, 2001.

[9] M. Nagasawa, S. Ida, and T. Bessho, "Formation of hot planets by a combination of planet scattering, tidal circularization, and the Kozai mechanism," *The Astrophysical Journal*, vol. 678, no. 1, pp. 498–508, 2008.

[10] E. B. Ford, B. Kozinsky, and F. A. Rasio, "Secular evolution of hierarchical triple star systems," *The Astrophysical Journal*, vol. 535, no. 1, pp. 385–401, 2000.

[11] B. Katz, S. Dong, and R. Malhotra, "Long-term cycling of Kozai-Lidov cycles: extreme eccentricities and inclinations excited by a distant eccentric perturber," *Physical Review Letters*, vol. 107, no. 18, Article ID 181101, 2011.

[12] S. Naoz, W. M. Farr, Y. Lithwick, F. A. Rasio, and J. Teyssandier, "Secular dynamics in hierarchical three-body systems," *Monthly Notices of the Royal Astronomical Society*, vol. 431, no. 3, pp. 2155–2171, 2013.

[13] S. Naoz, "The eccentric Kozai-Lidov effect and its applications," *Annual Review of Astronomy and Astrophysics*, vol. 54, no. 1, pp. 441–489, 2016.

[14] J. Laskar and G. Boué, "Explicit expansion of the three-body disturbing function for arbitrary eccentricities and inclinations," *Astronomy & Astrophysics*, vol. 522, no. 5, article A60, 2010.

[15] C. D. Murray and S. F. Dermott, *Solar System Dynamics*, Cambridge University Press, Cambridge, UK, 1999.

[16] D. Brouwer and G. M. Clemence, *Methods of Celestial Mechanics*, Academic Press, New York, NY, USA, 1961.

[17] J. M. Danby, *Fundamentals of Celestial Mechanics*, Willmann-Bell, Richmond, Va, USA, 2nd edition, 1992.

[18] H. C. Plummer, *An Introductory Treatise on Dynamical Astronomy*, Dover, New York, NY, USA, 1960.

[19] W. M. Smart, *Celestial Mechanics*, Longmans, London, UK, 1954.

[20] J. Wisdom and M. Holman, "Symplectic maps for the *N*-body problem," *The Astronomical Journal*, vol. 102, no. 4, pp. 1528–1538, 1991.

[21] P. Saha and S. Tremaine, "Long-term planetary integration with individual time steps," *The Astronomical Journal*, vol. 108, no. 5, pp. 1962–1969, 1994.

[22] R. S. Harrington, "Dynamical evolution of triple stars," *The Astronomical Journal*, vol. 73, pp. 190–194, 1968.

[23] Y. Krymolowski and T. Mazeh, "Studies of multiple stellar systems—II. Second-order averaged Hamiltonian to follow long-term orbital modulations of hierarchical triple systems," *Monthly Notices of the Royal Astronomical Society*, vol. 304, no. 4, pp. 720–732, 1999.

[24] H. Beust and A. Dutrey, "Dynamics of the young multiple system GG Tauri. II. Relation between the stellar system and the circumbinary disk," *Astronomy & Astrophysics*, vol. 446, pp. 137–154, 2006.

[25] J. P. S. Carvalho, R. V. de Moraes, A. F. B. A. Prado, and O. C. Winter, "Analysis of the secular problem for triple star systems," *Journal of Physics: Conference Series*, vol. 465, no. 1, Article ID 012010, 2013.

[26] R. A. Broucke, "Expansion of the third-body disturbing function," *Journal of Guidance, Control, and Dynamics*, vol. 4, no. 3, pp. 346–348, 1981.

[27] R. A. Mardling, "New developments for modern celestial mechanics—I. General coplanar three-body systems. Application to exoplanets," *Monthly Notices of the Royal Astronomical Society*, vol. 435, no. 3, pp. 2187–2226, 2013.

[28] M. E. Bailey, "Is there a dense primordial cloud of comets just beyond Pluto?" in *Proceedings of the Asteroids, Comets, Meteors*, pp. 383–386, Uppsala Universitet, Uppsala, Sweden, June 1983.

[29] T. Quinn, S. Tremaine, and M. Duncan, "Planetary perturbations and the origin of short-period comets," *The Astrophysical Journal*, vol. 355, no. 2, pp. 667–679, 1990.

[30] M. E. Bailey, J. E. Chambers, and G. Hahn, "Origin of sungrazers: a frequent cometary end-state," *Astronomy & Astrophysics*, vol. 257, pp. 315–322, 1992.

[31] P. Michel and F. Thomas, "The Kozai resonance for near-Earth asteroids with semimajor axis smaller than 2 AU," *Astronomy & Astrophysics*, vol. 307, pp. 310–318, 1996.

[32] F. Thomas and A. Morbidelli, "The Kozai resonance in the outer solar system and the dynamics of long-period comets," *Celestial Mechanics & Dynamical Astronomy*, vol. 64, no. 3, pp. 209–229, 1996.

[33] X.-S. Wan and T.-Y. Huang, "An exploration of the Kozai resonance in the Kuiper Belt," *Monthly Notices of the Royal Astronomical Society*, vol. 377, no. 1, pp. 133–141, 2007.

[34] N. D. Moiseev, "On some basic simplified schemes of celestial mechanics, obtained by means of averaging of restricted circular problem of three bodies 2. On averaged variants of restricted circular three bodies of the spatial bodies," *Publication of the State Institute of Astronomy Named after P. K. Sternberg*, vol. 15, no. 1, pp. 100–117, Moscow State University named after

Lomonosov M. V., Scientific Memoirs, Release 96, Originally in Russian, 1945.

[35] K. Ohtsuka, T. Sekiguchi, D. Kinoshita et al., "Apollo asteroid 2005 UD: split nucleus of (3200) Phaethon?" *Astronomy & Astrophysics*, vol. 450, no. 3, pp. L25–L28, 2006.

[36] K. Ohtsuka, H. Arakida, T. Ito et al., "Apollo asteroids 1566 Icarus and 2007 MK6: Icarus family members?" *The Astrophysical Journal*, vol. 668, no. 1, pp. L71–L74, 2007.

[37] P. B. Babadzhanov, G. I. Kokhirova, and Yu. V. Obrubov, "The potentially hazardous asteroid 2007CA19 as the parent of the η-Virginids meteoroid stream," *Astronomy & Astrophysics*, vol. 579, article A119, 2015.

[38] P. B. Babadzhanov and Yu. V. Obrubov, "Evolution of short period meteoroid streams," *Celestial Mechanics & Dynamical Astronomy*, vol. 54, no. 1, pp. 111–127, 1992.

[39] G. F. Gronchi and A. Milani, "The stable Kozai state for asteroids and comets: with arbitrary semimajor axis and inclination," *Astronomy & Astrophysics*, vol. 341, no. 3, pp. 928–935, 1999.

[40] P. Farinella, L. Foschini, Ch. Froeschlé et al., "Probable asteroidal origin of the Tunguska Cosmic Body," *Astronomy & Astrophysics*, vol. 377, no. 3, pp. 1081–1097, 2001.

[41] R. McGehee, "von Zeipel's theorem on singularities in celestial mechanics," *Expositiones Mathematicae*, vol. 4, no. 4, pp. 335–345, 1986.

[42] G. F. Gronchi and A. Milani, "Averaging on Earth-crossing orbits," *Celestial Mechanics & Dynamical Astronomy*, vol. 71, no. 2, pp. 109–136, 1998.

[43] H. Beust, "Orbital clustering of distant Kuiper belt objects by hypothetical Planet 9. Secular or resonant?" *Astronomy & Astrophysics*, vol. 590, article L2, 2016.

[44] H. F. Levison and M. J. Duncan, "The long-term dynamical behavior of short-period comets," *Icarus*, vol. 108, no. 1, pp. 18–36, 1994.

[45] J. Wisdom and M. Holman, "Symplectic maps for the n-body problem: stability analysis," *The Astronomical Journal*, vol. 104, no. 5, pp. 2022–2029, 1992.

[46] R. G. Strom, R. Malhotra, T. Ito, F. Yoshida, and D. A. Kring, "The origin of planetary impactors in the inner solar system," *Science*, vol. 309, no. 5742, pp. 1847–1850, 2005.

[47] T. Ito and R. Malhotra, "Asymmetric impacts of near-Earth asteroids on the Moon," *Astronomy & Astrophysics*, vol. 519, article A63, 2010.

[48] T. Ito and S. M. Miyama, "An estimation of upper limit masses of v Andromedae planets," *The Astrophysical Journal*, vol. 552, no. 1, pp. 372–379, 2001.

[49] T. Ito and K. Tanikawa, "Long-term integrations and stability of planetary orbits in our solar system," *Monthly Notices of the Royal Astronomical Society*, vol. 336, no. 2, pp. 483–500, 2002.

[50] G. M. Clemence, "The system of astronomical constants," *Annual Review of Astronomy and Astrophysics*, vol. 3, no. 1, pp. 93–112, 1965.

[51] Y. Kozai, "Secular perturbations of resonant asteroids," *Celestial Mechanics & Dynamical Astronomy*, vol. 36, no. 1, pp. 47–69, 1985.

[52] M. Yoshikawa, "A survey of the motions of asteroids in the commensurabilities with Jupiter," *Astronomy & Astrophysics*, vol. 213, pp. 436–458, 1989.

[53] M. Moons, "Review of the dynamics in the Kirkwood gaps," *Celestial Mechanics & Dynamical Astronomy*, vol. 65, no. 1-2, pp. 175–204, 1996.

[54] G. E. O. Giacaglia, "Secular motion of resonant asteroids," SAO Special Report 278, Smithsonian Institution, Astrophysical Observatory, Cambridge, Mass, USA, 1968.

[55] G. E. O. Giacaglia, "Resonance in the restricted problem of three bodies," *The Astronomical Journal*, vol. 74, pp. 1254–1261, 1969.

[56] M. Saillenfest, M. Fouchard, G. Tommei, and G. B. Valsecchi, "Long-term dynamics beyond Neptune: secular models to study the regular motions," *Celestial Mechanics & Dynamical Astronomy*, vol. 126, no. 4, pp. 369–403, 2016.

[57] T. Gallardo, "The occurrence of high-order resonances and Kozai mechanism in the scattered disk," *Icarus*, vol. 181, no. 1, pp. 205–217, 2006.

[58] S. J. Kortenkamp, R. Malhotra, and T. Michtchenko, "Survival of Trojan-type companions of Neptune during primordial planet migration," *Icarus*, vol. 167, no. 2, pp. 347–359, 2004.

[59] T. Gallardo, "Atlas of the mean motion resonances in the Solar System," *Icarus*, vol. 184, no. 1, pp. 29–38, 2006.

[60] N. D. Moiseev, "On some basic simplified schemes of celestial mechanics, obtained by means of averaging of restricted circular problem of three bodies 1. On averaged variants of restricted circular three bodies of the plane problem," *Publication of the State Institute of Astronomy named after P. K. Sternberg*, vol. 15, no. 1, pp. 75–99, Moscow State University named after Lomonosov M. V., Scientific Memoirs, Release 96. Originally in Russian, 1945.

[61] Y. F. Gordeeva, "Time-dependence of orbital elements in long-period oscillations in the three-body boundary-value problem," *Cosmic Research*, vol. 6, pp. 450–453, 1968.

[62] M. A. Vashkov'yak, "Evolution of the orbits of distant satellites of Uranus," *Astronomy Letters*, vol. 25, no. 7, pp. 476–481, 1999.

[63] L. Luo, B. Katz, and S. Dong, "Double-averaging can fail to characterize the long-term evolution of Lidov–Kozai Cycles and derivation of an analytical correction," *Monthly Notices of the Royal Astronomical Society*, vol. 458, no. 3, pp. 3060–3074, 2016.

13

A Novel Scanning Method Applied to New-Style Solar Telescope based on Autoguiding System

Zhi-ming Song [1,2] and Zhong-quan Qu [2]

[1] *University of Chinese Academy of Sciences, Beijing 100049, China*
[2] *Yunnan Observatories, Chinese Academy of Science, Kunming 650011, Yunnan, China*

Correspondence should be addressed to Zhi-ming Song; 339255245@qq.com

Academic Editor: Michael Kueppers

To expand field of view (FOV) of telescope, the method of special scanning often is used, but, for some telescopes with special structure in optics and machine, the conventional scanning methods are unsuitable. This paper proposes a novel scanning method based on autoguiding system so as to expand the FOV of fiber array solar optical telescope (FASOT) in possession of the special structure in optics and machine. Meanwhile, corresponding experiments are conducted in the FASOT prototype, FASOT-1B, in order to demonstrate that, for both FASOT and FASOT-1B, the proposed scanning method is feasible. First of all, on the basis of the software and hardware characteristics of FASOT and FASOT-1B, the three key technologies related to the proposed scanning method are described: quickly locating and pointing the first scanning step, the closed-loop controlling of multistep scanning, and the disturbance suppression of every scanning step based on Kalman filter. Afterwards, experiments are conducted and corresponding results show that the proposed scanning method is robust for the random disturbances forced on every scanning step and able to meet the scanning requirement of both FASOT and FASOT-1B .

1. Introduction

Fiber array solar optical telescope (FASOT) is a pioneer of Chinese giant solar telescope (CGST) [1], and it is a telescope which will be capable of conducting real-time, high-efficiency, high-precision spectropolarimetry of multiple magnetosensitive lines over a two-dimensional field of view, i.e., giving real-time 3D stokes measurements of multiple lines. Therefore, FASOT will act as a very efficient 3D spectropolarimeter and be capable of observing and inversing physical quantities in multiple heights of the solar atmosphere, especially the physical quantities associated with magnetic field [2, 3].

The optical path of FASOT is depicted in Figure 1. Obviously, the optical path of FASOT is different from those of general-purpose telescopes. First of all, the light from solar is collected by the guiding optics of main telescope labeled as 1 in Figure 1 and then inputted to a field stop located at the Cassegrain focus plane of main telescope that splits the FOV into two parts. One smaller part (0.5'×0.5') passes the light directly into the polarimetric system labeled as 2 in Figure 1 for the polarimetric measurement, and the remnant FOV is reflected vertically and used by the monitoring system labeled as 5 in Figure 1. Afterwards, the light modulated by the polarimeter is split into two beams with opposite polarization states (ordinary and extraordinary beams) and transmitted into integral field spectrographs and their detectors using integral field unit (IFU) with optical fibers so as to conduct two beams' spectroimaging polarimetry. Due to the compact and symmetrical optics configuration of FASOT up to the polarization modulator, the additional polarization from instrument is minimized. On the other hand, a novel polarization demodulation technique named reduced optical switching demodulation [4] is adopted by FASOT to improve the polarimetric sensitivity and reduce the integration time. These technologies mentioned above will make FASOT obtain a polarimetric noise level on the order of $8.0 \times 10^{-4} Ic$.

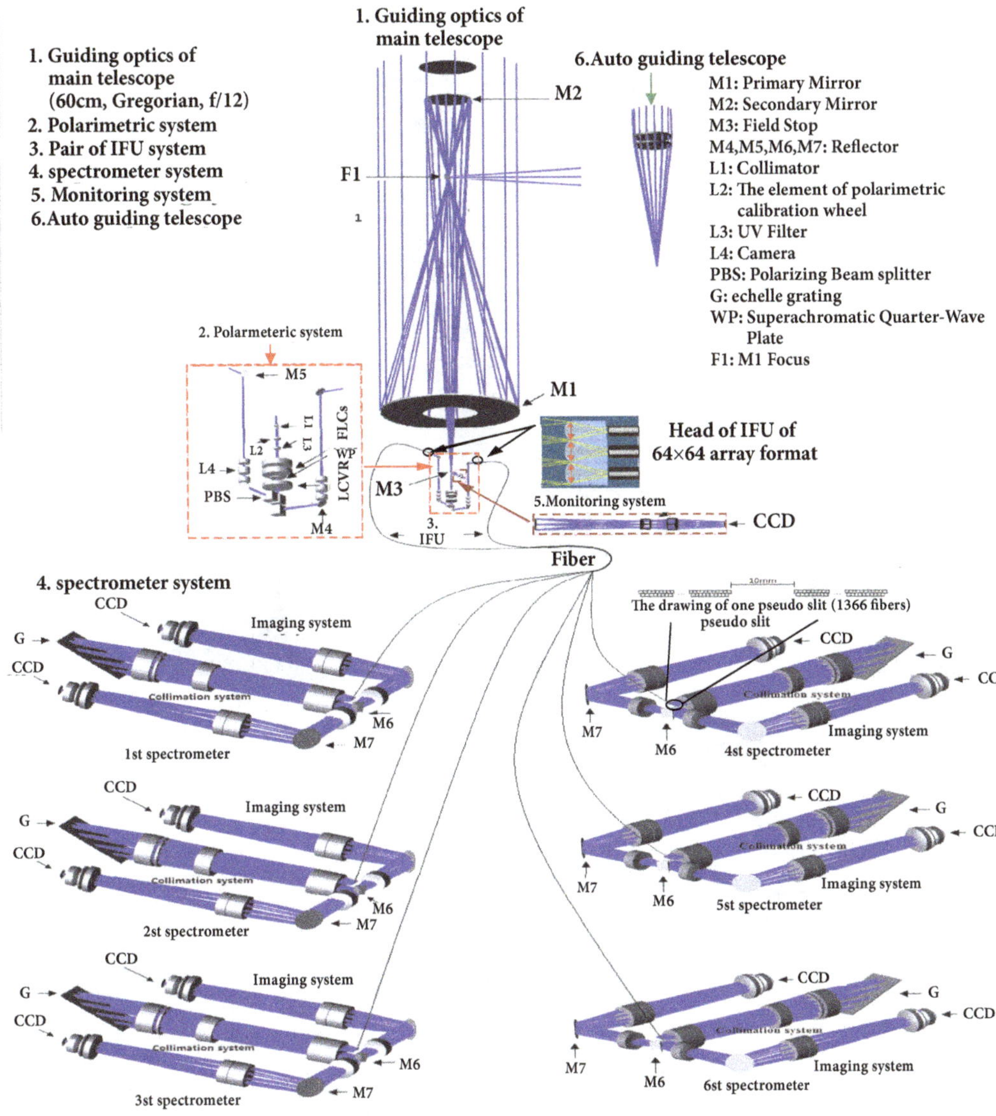

Figure 1: The optical path of FASOT.

Currently, for FASOTs, the polarimetric system and IFU labeled as 2 and 3 in Figure 1, a small FOV, 0.5'×0.5', is obtained. However, based on designing requirement, FASOT should be in possession of the ability to observe the local region of solar about 3'×3' and inverse 2D polarization spectrum image of the region. In other words, FASOT should have the capacity to scan a 2D space. However, because of the compact and symmetrical optics configuration of FASOT dedicated to improve polarimetric sensitivity, the conventional scanning methods are unsuitable for FASOT, such as the slit scanning method usually used by general-purpose telescope [5–7], the scanning method of using tip-tilt secondary mirror [8], and the method adding a rotary dual-wedge prism [9] between the secondary mirror and the field stop. Therefore, for FASOT, a special scanning method distinguishing from those mentioned above should be adopted.

Recently, a special scanning method of rotating entire telescope has been adopted by the visible infrared imaging radiometer suite (VIIRS) [10], and the method can efficiently suppresses stray light and improves polarization sensitivity. So, in consideration of the traits of FASOT, the scanning method similar to VIIRS will be adopted. But, the distinction between the two scanning methods of FASOT and VIIRS is that the method of FASOT will simultaneously rotate guiding optics, polarimetric system and autoguiding system. Furthermore, the rotation of FASOT mainly relies on the closed-loop controlling of its autoguiding system.

Normally, the autoguiding systems [11–13] used by general-purpose solar telescope just have some conventional functions such as monitoring full-disk solar image and closed-loop tracking. In contrast to those functions, FASOT will integrate the function of scanning 2D space into its conventional autoguiding system and construct a novel scanning system based on autoguiding system. It should be emphasized that the proposed scanning method has not been reported in the published literatures. What is more, some

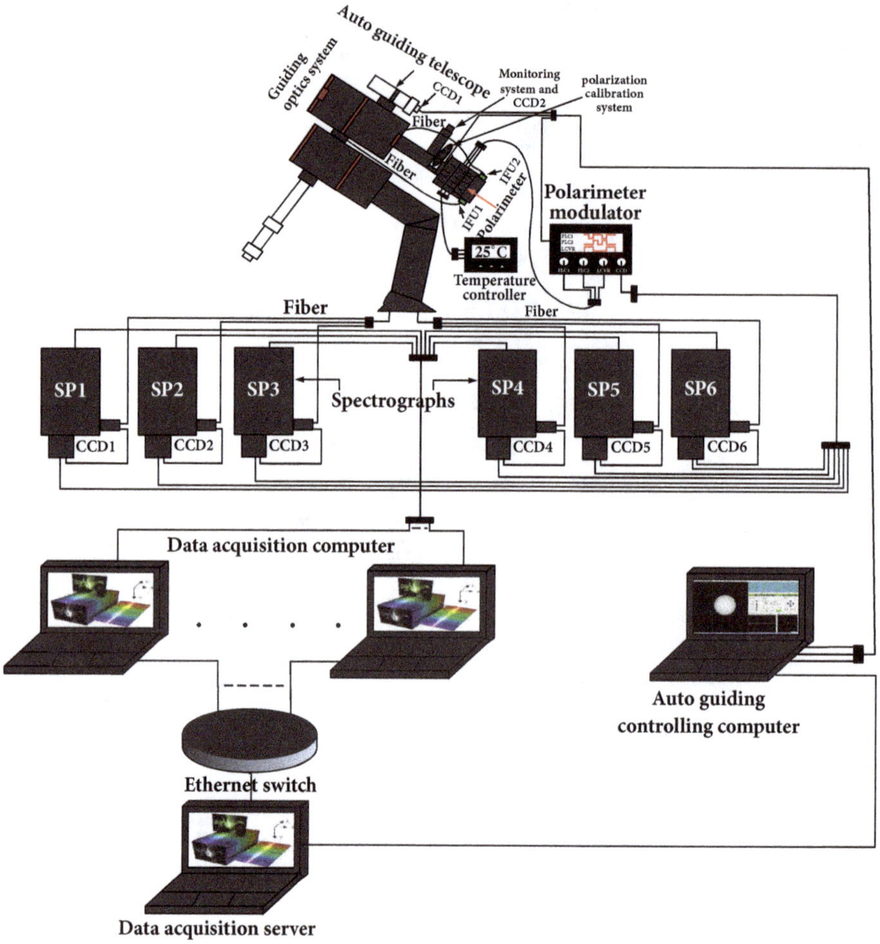

FIGURE 2: The structure diagram of controlling system of FASOT.

techniques used by the proposed scanning method also are novel and are not found in the published literatures. For example, in the process of multistep scanning, Kalman filter is used to suppress the random disturbances and strengthen the scanning performance. Therefore, this paper will be beneficial for the people anxious to expand the FOV of their telescope with a similar structure to FASOT.

This paper is organized as follows: Section 2 briefly introduces the structures of controlling system of FASOT and its prototype FASOT-1B and elucidates the reason why the proposed scanning system first is implemented in FASOT-1B. Section 3 describes the principle of our proposed scanning system. Three key technologies related to our proposed scanning system are introduced in Section 4 to Section 6, respectively. Experiments are conducted, and corresponding results are obtained in Section 7. The conclusions are given in Section 8.

2. Brief Introduction of FASOT and Its Prototype FASOT-1B

The structure diagram of controlling system of FASOT is shown in Figure 2. As shown in Figure 2, the system consists of the guiding optics system, autoguiding system (including autoguiding telescope, corresponding detector, and controlling computer), polarimetric system (including the polarimeter, its modulator, its temperature controller, and polarization calibration system), IFU, optical fibers, data acquisition system (including spectrographs, data acquisition computers, and server), and the mount of telescope.

In this system, autoguiding system will be taken as the central part of entire system, and besides the functions of monitoring full-disk solar image, closed-loop tracking, controlling the modulator of polarimeter, and sending the instruction of data acquisition, it also is of the capacity to scan a 2D space of local region of full-disk solar image.

It needs to be emphasized that to implement the scanning method based on autoguiding system, four conditions must be met, and they are as follows.

First, the total weight of the target object used to scan should be light, namely, a small rotary inertia. For example, For FASOT, the total weight of its guiding optics, polarimetric system, and autoguiding system is just about 120KG much less than the loading capacity of its mount, 300 Kg.

Second, the mount of telescope should have a perfect performance. For example, For FASOT, its mount is excellent in performance as a result of the fact that its tracking accuracy

TABLE 1: The distinctions between FASOT and its prototype FASOT-1B.

System composition	FASOT	FASOT-1B
Guiding optics system	600mm Focal length: 7200mm Field of view: 6'×6' (Gregory system)	400mm Focal length:3200mm Field of view: 2.1'×2.1' (RC system)
Aoto guiding telescope	The aperture: 80mm Focal ratio: 1:6	The aperture: 80mm Focal ratio: 1:6
Aoto guiding image sensor	SCOMS chip Resolution: 2048∗2048 Pixel size: 5.5um	SCOMS chip Resolution: 2048∗2048 Pixel size: 5.5um
The number of micro lens of IFU	2×64×64	2×11×11
The number of optical fibers	2×64×64	2×11×11
The number of spectrograph	6	1
The number of detector of the spectrographs	12	1
Data acquisition computer	6	1
Data acquisition server	1	0
The type of the mount	DDM160 of Astrosysteme Austria German equatorial mount	GM3000HSP of Micron 10 German equatorial mount
Software interface of the mount	SDK based on c/c++ and ASCOM	ASCOM
The tracking accuracy of the mount	⩽ 0.25" RMS	⩽1" RMS
The pointing accuracy of the mount	<8" RMS	<20" RMS
Polarimetric system	FASOT-1B is similar to FASOT	FASOT-1B is similar to FASOT
Loading capacity of the mount (only instrument)	300 Kg	100 Kg
The total weight of the target object used to scan	120Kg	70Kg

≤ 0.25" RMS/ 5 minutes, peak track error ≤ 0.68", and driving mode is direct-driven.

Third, the accuracy of autoguiding system should be viable. For example, for FASOT, the aperture of its autoguiding telescope is 80mm, focal ratio is 1:6, the resolution of autoguiding image sensor is 2048×2048, and pixel size is 5.5um.

Fourth, the mount of telescope should support secondary development, namely, supplying corresponding software interfaces. For example, for FASOT, its mount supports the software interfaces, such as the SDK based on C\C++ and the Astronomy Component Object Model (ASCOM).

So, on the basis of the information mentioned above, the scanning method based on autoguiding system is suitable for FASOT.

However, until now, some parts of FASOT still are in shaping. But, to ensure that the development of FASOT is smooth, a prototype of FASOT, named FASOT-1B shown in Figure 7, has been developed. The prototype is almost similar to FASOT in the optical path and the structure of control system. But some distinctions still exist between them and are displayed in Table 1. From Table 1, we can see that although the performance of the mount of FASOT-1B is not as good as the one of FASOT, it still meets the four conditions of implementing the scanning based on autoguiding system. Therefore, the proposed scanning method will first be developed and then applied to FASOT-1B. In other word, if the proposed scanning method is suitable for FASOT-1B, it must be able to be directly transported to FASOT and even has a more excellent performance as a result of that the performance of the mount of FASOT is superior to the one of FASOT-1B. Actually, This is also a good way to speed up the development of FASOT.

3. The Principle of the Proposed Scanning Method

As shown in Table 1, the software interface of the mount of FASOT-1B just supports ASCOM. Therefore, the designing of the proposed scanning method considers not only the

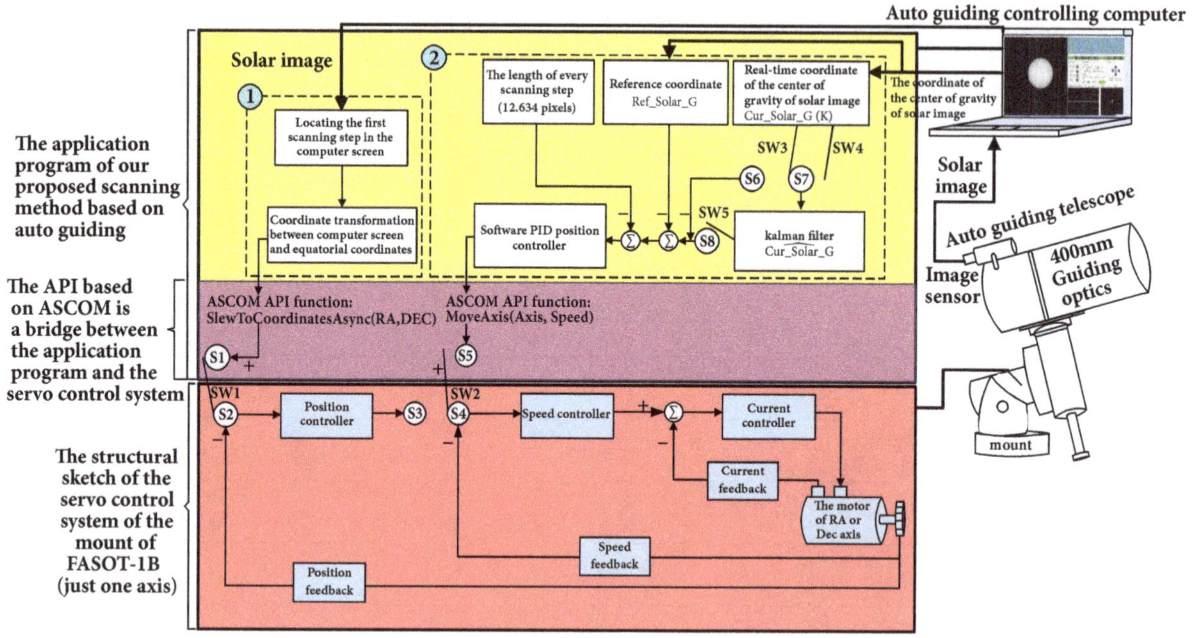

FIGURE 3: The software structure diagram of the proposed scanning method.

structure of the servo control system integrated into the mount of FASOT-1B, but also the relationship between the servo control system and the software interface, ASCOM. The structure diagram of combining the servo control system of the mount of FASOT-1B with its software interface to implement the proposed scanning method is shown in Figure 3. From Figure 3, we can see that the red part is the structural sketch of the servo control system of the mount of FASOT-1B, and the servo control system is a common three-ring control system, including current, speed, and position rings. The yellow parts are the software structure of our proposed scanning method integrated into the software of autoguiding system, and the purple part, ASCOM, is a bridge between the yellow part (scanning function algorithm) and the red part (servo control system).

The principle of the proposed scanning method based on autoguiding system is shown in Figure 4. The white disk is full-disk solar image acquired by autoguiding image sensor. The center point A' of yellow cross is the center of autoguiding image sensor, and it is also the optical axis of autoguiding telescope, guiding optics of FASOT-1B, polarimetric system, and IFU. Before beginning multistep scanning, telescope should be controlled and pointed to the first scanning step (in Figure 4, the first scanning step is the point B'), namely, implementing the function which quickly locates and points the first scanning step (a detailed introduction of this function will be described in Section 4. On the other hand, its software module diagram is shown in the yellow part of Figure 3 and labeled as ①). Afterwards, an entire procedure of multistep scanning will be carried out.

In Section 1, we have had the knowledge about the FOV of IFU being just 0.5'×0.5'. Therefore, to expand the FOV to 3'×3', the number of the scanning step should be equal to 35. As shown in Figure 4, the small red rectangles surrounded

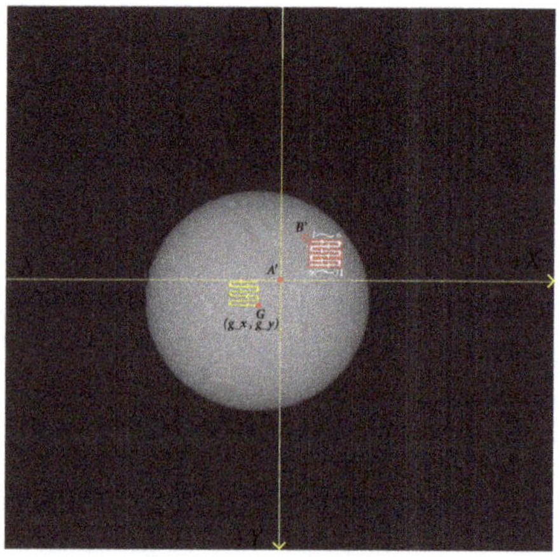

FIGURE 4: The principle of the proposed scanning method based on autoguiding system.

by the big one represent the corresponding positions of every scanning step, and the white arrow represents the direction of scanning. The point G(g_x,g_y) is the coordinate of the center of gravity of full-disk solar image, and the yellow arrow next to the point G(g_x,g_y) shows that the movement direction of the point G is contrary to the direction of its scanning.

Based on FASOTs, the requirements of the observation and data process, polarization modulation, and image acquisitions should be conducted after every scanning step, and then the acquired images will be reconfigured into an entire image which reflects the whole scanning areas. These requirements result in that scanning should be of the following specifications:

(1) The scanning step length should be 0.5'; namely, Ls=0.5'

(2) The accuracy between adjacent scanning steps should be less than or equal to 1"; namely, Ac≤1"

(3) The time of every scanning step should be as fast as possible

To implement the specifications mentioned above, the following strategies are adopted (a detailed software module diagram is shown in the yellow part of Figure 3 and labeled as ②).

First, as shown in Figure 3, S6 is connected by SW3, and S4 is connected to S5 by SW2.

Second, the reference coordinate of the center of gravity of full-disk solar image is calculated, **Ref_Solar_G(Ref_Solar_G_x,Ref_Solar_G_y)**, before beginning every scanning step.

Third, the coordinate of the center of gravity of full-disk solar image is calculated in real-time, $Cur\_Solar\_G(k)(Cur\_Solar\_G\_x(k), Cur\_Solar\_G\_y(k))$, after beginning every scanning step.

Fourth, because every scanning step should meet the two specifications, Ls=0.5' and Ac≤1", when the scanning direction is horizontal X, the two errors which determine whether a horizontal scanning step was finished are calculated in real-time and expressed as

$$H\_Error\_x(k) = \mathbf{12.6932} - |Cur\_Solar\_G\_x(k) - \mathbf{Ref\_Solar\_G\_x}| \tag{1}$$

$$H\_Error\_y(k) = |Cur\_Solar\_G\_y(k) - \mathbf{Ref\_Solar\_G\_y}| \tag{2}$$

where k is the sequence number of measurement of $Cur\_Solar\_G(i)$, $i = 1 \cdots k \cdots N$, and 12.6932 is the number of pixels of a scanning step length, which corresponds to 0.5'.

Fifth, the two errors, $H\_Error\_x(k)$ and $H\_Error\_y(k)$, are sent to corresponding software PID position controllers of RA and DEC axes, respectively. Although Figure 3 just depicts one of the two axes, RA and DEC, the two axes are operated in the process of scanning simultaneously. Furthermore, $H\_Error\_x(k)$ is sent to DEC axis, and $H\_Error\_y(k)$ is sent to RA axis. On the other hand, the communication process between the two errors and the two axes depends on the Application Programming Interface (API) based on ASCOM, MoveAxis(RA or DEC,Speed).

Sixth, when both $H\_Error\_x(k)$ and $H\_Error\_y(k)$ are less than or equal to 0.4 pixels, which corresponds to 1", a scanning step of horizontal X is over. In other words, a scanning step is over as long as the terminating condition is met. Afterwards, polarization modulation and data acquisition are conducted.

Finally, for the remnant scanning steps of horizontal X, the strategies mentioned above are repeated. But, when next new scanning step starts, the reference coordinate, Ref_Solar_G, is updated again, and the coordinate, $Cur\_Solar\_G(k)$, is recalculated in real-time.

It is clear that our proposed scanning method is to combine autoguiding system with the speed ring of the servo control system (RA axis and DEC axis) in order to construct a more flexible and software-controlled close-loop controlling system.

In contrast to the direction of horizontal X, when the scanning direction is vertical Y, the two errors which determine whether a vertical scanning step was finished are calculated in real-time and expressed as follows:

$$V\_Error\_x(k) = |Cur\_Solar\_G\_x(k) - \mathbf{Ref\_Solar\_G\_x}| \tag{3}$$

$$V\_Error\_y(k) = \mathbf{12.6932} - |Cur\_Solar\_G\_y(k) - \mathbf{Ref\_Solar\_G\_y}| \tag{4}$$

Afterwards, remnant procedures are similar to those of horizontal X mentioned above.

Obviously, both $Cur\_Solar\_G\_x(k)$ and $Cur\_Solar\_G\_y(k)$ play an important role in the process of every scanning step. But they often are disturbed by wind and other factors forced on telescope (wind is dominant), resulting in that it is difficult to smoothly and steadily finish the specifications of every scanning step.

With respect to the disturbance problem, Kalman filter will be used so as to suppress the random disturbances and strengthen the robustness of every scanning step. The detailed introduction of Kalman filter will be described in Section 5.

In the end, the program flow chart of our proposed scanning method is shown in Figure 5. It needs to be emphasized that our program is based on Microsoft Foundation Classes (MFC), and multithreading and parallel processing are used for every scanning step.

4. To Quickly Locate and Point the First Scanning Step

With respect to FASOT, the destination of expanding its FOV is to observe more the solar local areas of interest. In other words, the optical axis of FASOT and its autoguiding telescope should first be pointed to a certain point, which is located within the solar local areas of interest. As shown in Figure 4, the optical axis is the point A', and the certain point located within the area of interest is B', namely, the first scanning step. Therefore, for the proposed scanning method, the first step is to move A' to B'.

Taking into consideration the fact that B' is likely to locate on any position of full-disk solar image, a Cartesian coordinate system A'XY is constructed (as shown in Figure 4, A' will be taken as the origin of the coordinate system), and then the distance between A' and B' can be obtained and expressed as the coordinate, (x,y). Afterwards, a coordinate transformation between Cartesian and equatorial coordinates will be conducted in order to obtain the equatorial coordinate of the point B' (RA,DEC), and then the API based on ASCOM, SlewToCoordinatesAsync (RA,DEC), is used to drive the FASOT's mount to move A' to B'. On the other hand, to implement the function mentioned above, as shown in

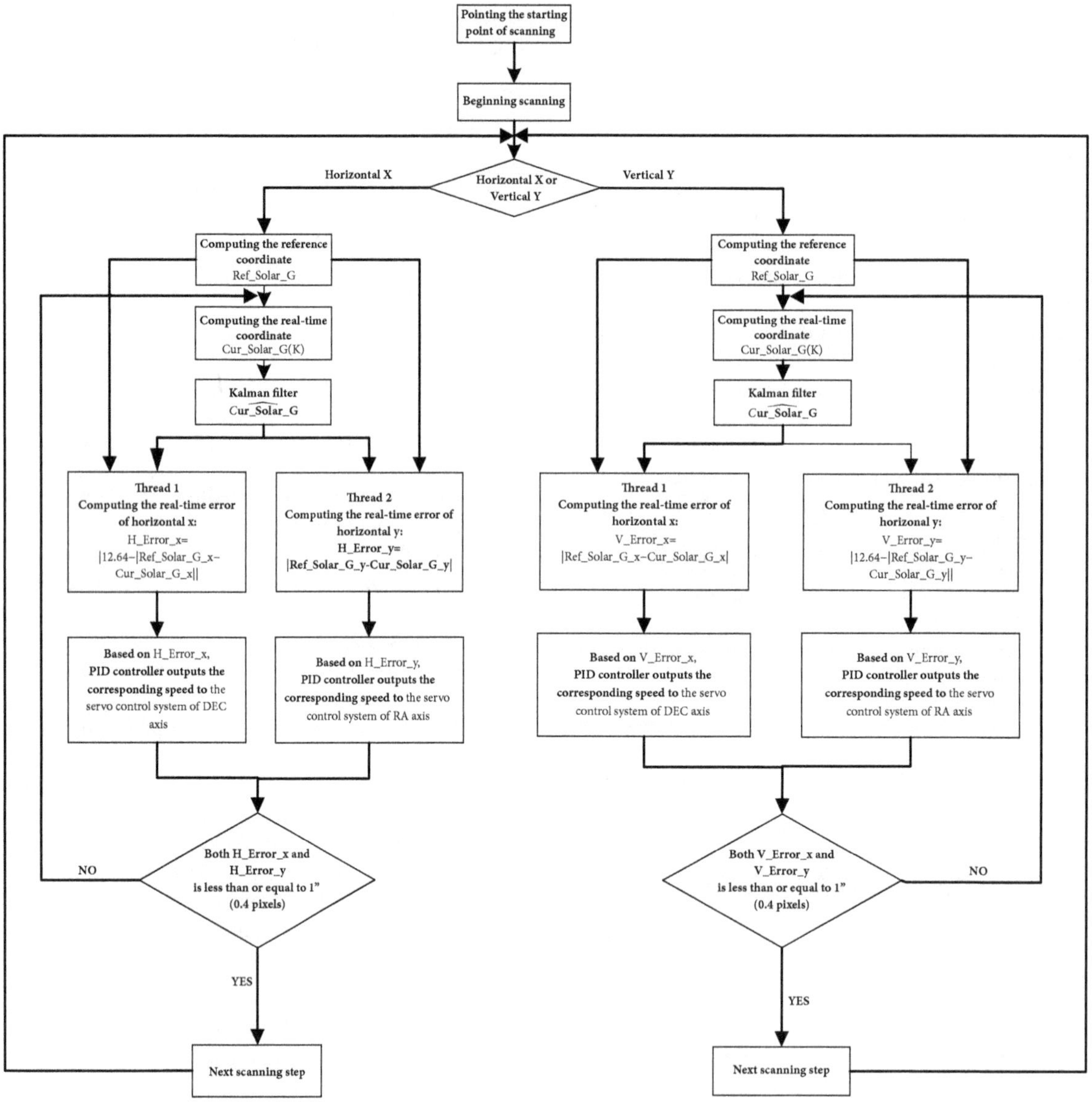

FIGURE 5: The program flow chart of our proposed scanning method.

Figure 3, the scanning software should first implement the following operations: connecting S1 to S2 and S3 to S4.

In the following, the coordinate transformation between Cartesian and equatorial coordinates will be described.

As shown in Figure 6, the big disk represents celestial sphere. Eq-Eq' is celestial equator. P is celestial pole. Z is zenith. O is observation site. Taking into consideration the fact that the distance between earth and solar is much larger than the diameter of earth, earth will be taken as the center of the celestial sphere, namely, the point o. 0A' is the focal length of autoguiding telescope, and its value is equal to f. A' is the center of autoguiding image sensor, and it is also the projection point of optic axis of autoguiding telescope (as shown in Figure 4, in the Cartesian coordinate system, A'XY, A' is taken as the origin of the system). A is the point located in actual solar surface currently being pointed by the optic axis of autoguiding telescope, and its equatorial coordinate is (A,D). B is another point located in actual solar surface which the optic axis of autoguiding telescope is anxious to point, and its equatorial coordinate is (A',D'). AT is the tangent line of AP, and, on the basis of AT, a Cartesian coordinates system TAT' will be constructed. It is clear that B' is the projection point of B and B", and it is also the first scanning step (as shown in Figure 4). Therefore, based on the theorems and formulas of spherical and projection triangle [14], the two transformation formulas between Cartesian and equatorial coordinates can be obtained:

$$\tan\left(A - A'\right) = \frac{y \sec(D)}{f - x \tan(D)} \tag{5}$$

$$\tan\left(D'\right) = \frac{x + f \tan(D)}{f - x \tan(D)} \cos\left(A - A'\right) \tag{6}$$

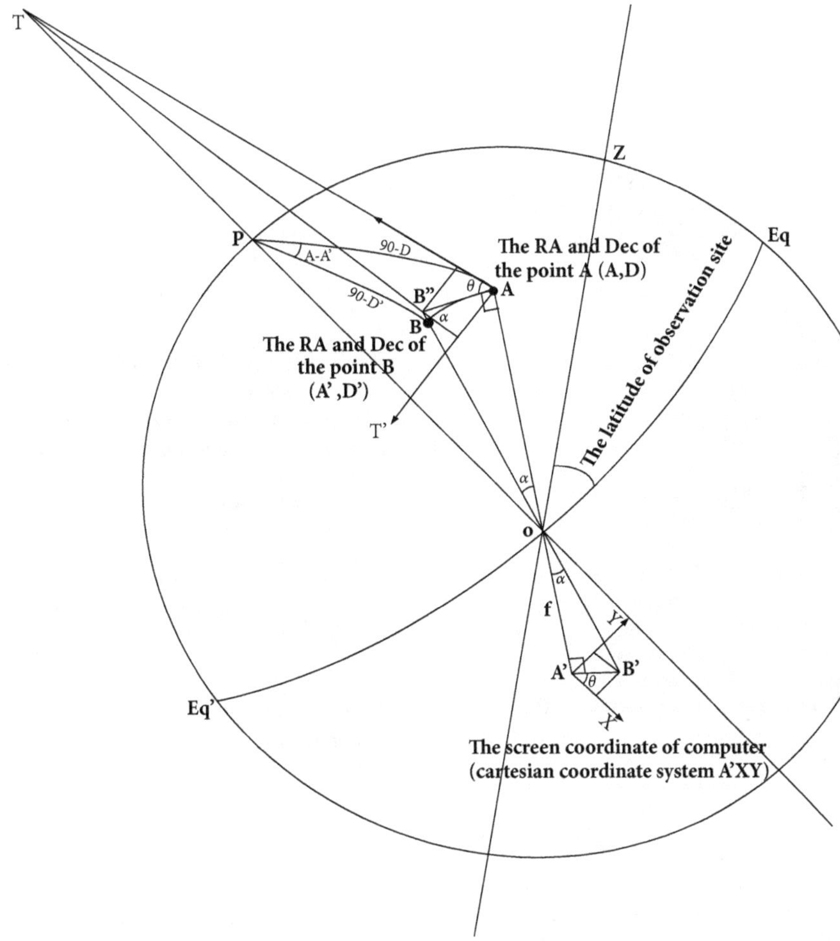

FIGURE 6: The coordinate transformation between Cartesian and equatorial coordinates.

where (A,D) is the equatorial coordinate of the point A currently being pointed by the optic axis of autoguiding telescope, and it is a known quantity as a result of the fact that it can be obtained through the API based on ASCOM. (x,y) is the Cartesian coordinate of the point B', and it is a known quantity as a result of the fact that it can be obtained through controlling computer screen of autoguiding system (as shown in Figure 8). f is the focal length of the autoguiding telescope, and it is also a known quantity. Therefore, based on the known quantities, the equatorial coordinate of the point B, (A',D'), can be obtained. Because B' is the projection point of B, its equatorial coordinate is also (A',D').

Obviously, this is an efficient method to quickly locate and point the first scanning step. However, this method is unsuitable for the multistep scanning described in Section 3. The reasons are as follows.

First, this method depends on the pointing accuracy of the used mount, and the accuracy often is not satisfactory. For example, as shown in Table 1, the pointing accuracy of FASOT-1B can not meet the specification of every scanning step, Ac≤1". However, with respect to the first scanning step, the pointing accuracy of FASOT-1B is sufficient as a result of the fact that it is ok as long as the first scanning step is pointed to the point which is located within the solar local area of interest.

Second, the speed regulation of this method is terrible, resulting in that after pointing every scanning step, the corresponding vibrations exists in telescope, and polarization modulation and data acquisition can not be conducted immediately. Obviously, the time of every scanning step also will be elongated as a result of waiting for the end of the vibrations.

Third, this method is not software-controlled, and its pointing procedure depends on the servo control system integrated into the mount, resulting in the fact that the movement velocity of every scanning step can not be flexibly controlled, and the capacity to resist disturbance is terrible.

5. The Closed-Loop Controlling of Every Scanning Step

As mentioned in Section 3, to meet the specifications of every scanning step, a new closed-loop controlling system combining autoguiding system with servo control system of the mount is constructed and its structure diagram is shown in Figure 3. In Section 3, the principle of this closed-loop

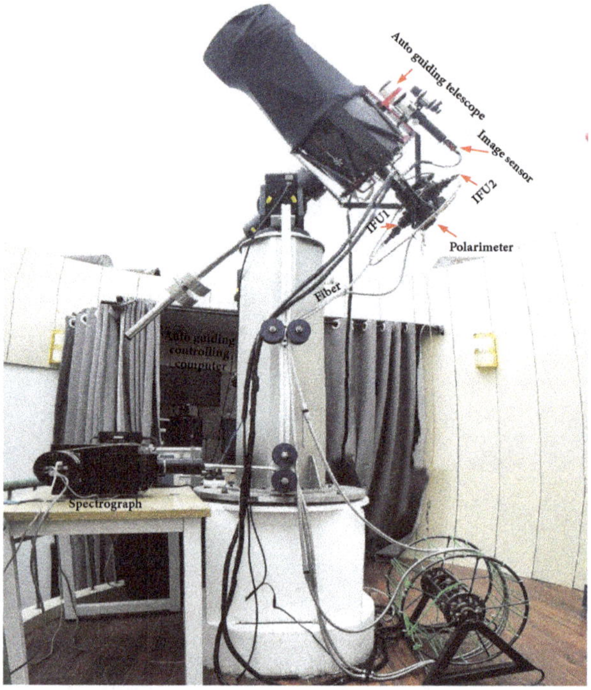

FIGURE 7: The physical map of FASOT-1B.

controlling system and its flow chart of program also have been described. Its core ideal is to construct a more flexible and software-controlled closed-loop controlling system in order to flexibly regulate the speed of every scanning step and fulfill the scanning specifications. On the other hands, it is convenient for the software-controlled closed-loop controlling system to integrate some efficient algorithms into its interior to strengthen its robustness. For example, in our scanning method, Kalman filter is used to strengthen the robustness of every scanning step.

6. Kalman Filter of Every Scanning Step

In every scanning step, due to the random disturbances from wind and other factors forced on telescope (wind is dominant), the corresponding vibrations exist in the guiding optic of main telescope, resulting in that autoguiding telescope also will be disturbed. There is no doubt that the disturbances will have a terrible influence on the performance of our multistep scanning. Therefore, corresponding measures must be taken. Kalman filter is chosen by us to improve the performance of our multistep scanning after investigating a lot of published literatures. Although Kalman filter has been applied to some telescopes and systems to strengthen their robustness [15–18], its application in telescope scanning has not been found. In the following, a detailed description about the application of Kalman filter in our scanning will be given.

Kalman filter is a statistical estimation method, which is autoregressive, linear, and unbiased. Through iteration, the optimal estimation of observation data can obtained. In other words, the disturbances forced on the observation data will be suppressed.

For our proposed multistep scanning, the application of Kalman filter is as follows.

First, for every scanning step, Kalman filter is used to obtain the optimal estimation coordinate from the real-time coordinate of the center of gravity of full-disk solar image, namely, estimating $\widehat{Cur\_Solar\_G}(k/k)$, $k = \mathbf{1,2....N}$ from $Cur\_Solar\_G(k)$, $k = \mathbf{1,2....N}$.

Second, $Cur\_Solar\_G(k)$ shown in (1)-(4) will be replaced by $\widehat{Cur\_Solar\_G}(k/k)$.

Finally, the remnant the procedures are similar to those described in Section 3.

The equations of traditional Kalman filter are as follows.

State and observation equations are

$$\mathbf{X}(\boldsymbol{k}) = \mathbf{AX}(\boldsymbol{k}-1) + \mathbf{B}\boldsymbol{U}(\boldsymbol{k}-1) + \boldsymbol{V}(\boldsymbol{k}) \quad (7)$$

$$\mathbf{Y}(\boldsymbol{k}) = \mathbf{HX}(\boldsymbol{k}) + \boldsymbol{W}(\boldsymbol{k}) \quad (8)$$

where **A** is state-transition matrix. **B** is control matrix. **V** is state noise which obeys a Gaussian distribution $N(0, {\sigma_W}^2)$. **H** is observation matrix. **W** is observation noise which obeys a Gaussian distribution $N(0, {\sigma_V}^2)$.

Time renewal equation is

$$\widehat{\mathbf{X}_1(k)} = \mathbf{AX}\left(\hat{k}-1\right) + \mathbf{B} \quad (9)$$

$$\mathbf{P}_1(k) = \mathbf{AP}(k-1)\mathbf{A}^T + \mathbf{Q} \quad (10)$$

where **A** is state-transition matrix. **B** is control matrix. **Q** is covariance matrix of **W**.

Observation renewal equation is

$$\mathbf{K}(k) = \mathbf{P}_1(k)\mathbf{H}^T\left[\mathbf{HP}_1(k)\mathbf{H}^T + \mathbf{R}\right]^{-1} \quad (11)$$

$$\widehat{\mathbf{X}(k)} = \widehat{\mathbf{X}_1(k)} + \mathbf{K}(\mathrm{k})\left[\mathbf{Y}(k) - \widehat{\mathbf{X}_1(k)}\right] \quad (12)$$

$$\mathbf{P}(k) = [\mathbf{I} - \mathbf{K}(k)\mathbf{H}]\mathbf{P}_1(k) \quad (13)$$

where **H** is observation matrix. **I** unit matrix. **K** is Kalman gain. **R** is covariance matrix of **V**. $\widehat{\mathbf{X}(k)}$ is the estimation value of Kalman filter.

Obviously, on the basis of the equations mentioned above, the scanning state and observation equations (7)-(8) are first constructed. Afterwards, the matrixes mentioned above, (**A**, **B**, **H**, **W**, **Q**, **V**, and **R**), are obtained. Next, the estimation data of Kalman filter, $\widehat{\mathbf{Cur\_Solar\_G}}(k/k)$, is calculated by iteration equations (9)-(13). It needs to be emphasized that in our proposed multistep scanning, the scanning direction contains horizontal X and vertical Y. Therefore, different scanning direction has its own state and observation equations.

Taking the scanning of horizontal X for example, the scanning procedure is that the RA axis and DEC axis are simultaneously operated based on the two errors, H_Error_x and H_Error_y. Furthermore, this operation results in the change of the real-time coordinate of the center of gravity of full-disk solar image, $Cur\_Solar\_G(k)$, at the same movement size and speed as the two axes. Therefore, the scanning movement equation of *Cur_Solar_G*(k), namely, its state equation, can be expressed as follows:

$$\begin{bmatrix} X_S(\boldsymbol{k}) \\ X_V(\boldsymbol{k}) \end{bmatrix} = \begin{bmatrix} 1 & T \\ 0 & T \end{bmatrix} \begin{bmatrix} X_S(\boldsymbol{k}-1) \\ X_V(\boldsymbol{k}-1) \end{bmatrix} + \begin{bmatrix} \frac{1}{2T^2} \\ T \end{bmatrix} \boldsymbol{U}(\boldsymbol{k}-1) + \boldsymbol{V}(\boldsymbol{k}) \tag{14}$$

where k is the sequence number of measurement of *Cur_Solar_G*(i), $i = 1 \cdots k \cdots N$, and T is its measurement cycle. $X_S(k)$ is the movement size of *Cur_Solar_G*(k). $X_V(k)$ is the movement speed of *Cur_Solar_G*(k). $U(k)$ is the movement acceleration of *Cur_Solar_G*(k). $\boldsymbol{V}(\boldsymbol{k})$ is the random disturbances enforced on telescope, and suppose that it obeys the Gaussian distribution $N(0,\sigma_V^2)$.

On the other hand, the scanning observation equation can be expressed as follows:

$$\begin{bmatrix} Y_S(\boldsymbol{k}) \\ Y_V(\boldsymbol{k}) \end{bmatrix} = \begin{bmatrix} 1 & 0 \end{bmatrix} \begin{bmatrix} X_S(\boldsymbol{k}) \\ X_V(\boldsymbol{k}) \end{bmatrix} + \boldsymbol{W}(\boldsymbol{k}) \tag{15}$$

where Y_S(k) and Y_V(k) are the observation data of autoguiding image sensor for $X_S(k)$ and $X_V(k)$. $\boldsymbol{W}(\mathbf{k})$ is observation noise of autoguiding image sensor, and suppose that it obeys a Gaussian distribution $N(0,\sigma_W^2)$.

Due to the fact that *Cur_Solar_G*(k) consists of *Cur_Solar_G_x*(k) and *Cur_Solar_G_y*(k), therefore, (14) and (15) can be expanded as follows:

$$\begin{bmatrix} \mathrm{X_{S\_X}}(\boldsymbol{k}) \\ \mathrm{X_{V\_X}}(\boldsymbol{k}) \\ \mathrm{X_{S\_Y}}(\boldsymbol{k}) \\ \mathrm{X_{V\_Y}}(\boldsymbol{k}) \end{bmatrix} = \begin{bmatrix} \mathbf{1} & \mathbf{T} & 0 & 0 \\ 0 & 1 & 0 & 0 \\ 0 & 0 & 1 & \mathrm{T} \\ \mathbf{0} & \mathbf{0} & \mathbf{0} & \mathbf{1} \end{bmatrix} \begin{bmatrix} \mathrm{X_{S\_X}}(\boldsymbol{k}-1) \\ \mathrm{X_{V\_X}}(\boldsymbol{k}-1) \\ \mathrm{X_{S\_Y}}(\boldsymbol{k}-1) \\ \mathrm{X_{V\_Y}}(\boldsymbol{k}-1) \end{bmatrix} + \begin{bmatrix} \frac{1}{2\mathrm{T}^2} \\ \mathrm{T} \\ \frac{1}{2\mathrm{T}^2} \\ \mathrm{T} \end{bmatrix} \boldsymbol{U}(\boldsymbol{k}-1) + \boldsymbol{V}(\boldsymbol{k}) \tag{16}$$

$$\begin{bmatrix} \mathrm{Y_S}(\boldsymbol{k}) \\ \mathrm{Y_V}(\boldsymbol{k}) \end{bmatrix} = \begin{bmatrix} 1 & 0 & 0 & 0 \\ 0 & 0 & 1 & 0 \end{bmatrix} \begin{bmatrix} \mathrm{X_{S\_X}}(\boldsymbol{k}) \\ \mathrm{X_{V\_X}}(\boldsymbol{k}) \\ \mathrm{X_{S\_Y}}(\boldsymbol{k}) \\ \mathrm{X_{V\_Y}}(\boldsymbol{k}) \end{bmatrix} + \boldsymbol{W}(\boldsymbol{k}) \tag{17}$$

Obviously, from the two equations, (16) and (17), we can obtain the state-transition matrix **A**, the control matrix **B**, and the observation matrix **H**. Generally, $\boldsymbol{V}$ is associated with the disturbances forced on telescope in the procedure of scanning movement, and $\boldsymbol{W}$ is related to the noise of autoguiding image sensor. However, in our scanning movement, autoguiding telescope and its image sensor are located on the top of the guiding optics of main telescope (as shown in the Figure 7). Therefore, the disturbances forced on telescope will have a similar influence on autoguiding telescope and its image sensor. In other words, the disturbances can be observed by autoguiding system and seen as the noise of autoguiding image sensor. On the basis of the above analysis, in our Kalman filter equation (16), $\boldsymbol{V}$ can be seen as a very small quantity or zero, and in (17), $\boldsymbol{W}$ can be seen as the accumulation of the disturbances forced on telescope and autoguiding image sensor's own noise.

On the other hand, we have had the knowledge that, for **W**, the influence of wind is dominant. What is more, different levels of wind result in different **W**. Furthermore, our scanning will be stopped when the level of wind is stronger than level three wind. Therefore, in different levels of wind (from level one to level three), three groups of the real-time coordinates of the center of gravity of full-disk solar image are obtained, namely, *Cur_Solar_G_1*, *Cur_Solar_G_2*, and *Cur_Solar_G_3*. In the meantime, based on the three groups of coordinates, three groups of **W** will be obtained, namely, **W**1, **W**2, and **W**3. Finally, when the Kalman filter method mentioned above is applied to our every scanning step of horizontal X, the different **W** will be selected based on corresponding the level of wind.

With respect to the scanning of vertical Y, the strategies about using Kalman filter are similar to the one of horizontal X mentioned above.

In the end, the things which we need to point to are as follows.

First, as shown in Figure 3, when Kalman filter is used, S4 is connected to S5 by SW2, S7 is connected by SW3, and S8 is connected by SW5.

Second, the level of wind is determined by the speed of wind, and the relationship between them is shown in Table 2. Furthermore, a digital anemometer is used by us to measure the speed of wind and select different **W** in the procedure of scanning.

7. Experiments and Results of Scanning

To demonstrate whether the proposed scanning method can meet its specifications ((1) the scanning step length should be 0.5'; (2) the accuracy between adjacent scanning steps should be less than or equal to 1"; (3) the time of every scanning step should be as fast as possible), experiments have been conducted in FASOT prototype FASOT-1B. Because the performance of the mount of FASOT is superior to the one of FASOT-1B, if the scanning specifications can be met by FASOT-1B, then they also can be did by FASOT. Figure 7 displays the picture of real product of FASOT-1B, and Figure 8 displays the graphical user interface of controlling system of our proposed scanning method based on autoguiding system on the computer screen.

Because our proposed scanning method is based on the real-time error of the center of gravity of full-disk solar image, the trajectory of the center of gravity of full-disk solar image can be used to verify whether every scanning step is smooth and accurate. In the meantime, the trajectory also can be

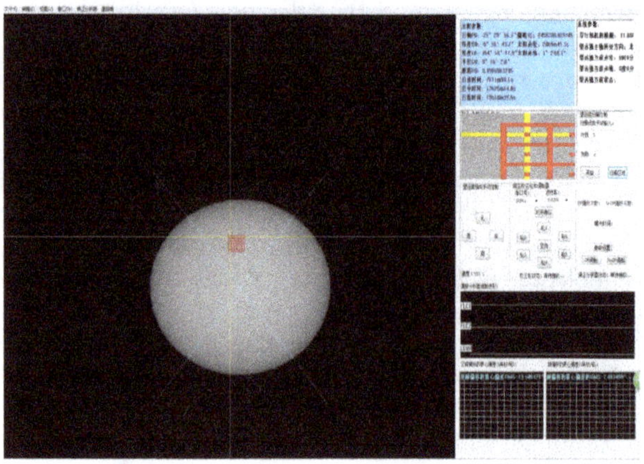

Figure 8: The graphical user interface of controlling system of our proposed scanning method based on autoguiding system on the computer screen.

Table 2: The relationship between the level of wind and speed of wind.

The level of wind	The speed of wind (m/s)
1	0.3-1.6
2	1.6-3.4
3	3.4-5.5

used to verify whether the scanning specifications can be met.

First of all, we need to verify whether the Kalman filter can strengthen the robustness of every scanning step or not. Therefore, the structure of Figure 3 should be that S5 is connected to S4 by SW2, SW3 is connected to S6, SW4 is connected to S7, and SW5 is not connected to S8. In other words, the estimated data of Kalman filter, $\mathrm{Cur\_}\widehat{\mathrm{Solar}}\mathrm{\_G}(k/k)$, just is taken as a comparison variable with $\mathrm{Cur\_Solar\_G}(k)$, instead of a feedback variable. Therefore, in distinct levels of wind (from level one to level three), some experiments have been conducted, and corresponding data also has been obtained. To better show the performance of Kalman filter, the group of data with remarkable random disturbances, which is obtained in level 3 wind, is used and displayed in Figure 9. From Figure 9(b), it is clear that the center of gravity of full-disk solar image without Kalman filter (the black line shown in Figure 9(b)) will be off its trajectory when some strong random disturbances exist in the process of scanning. In contrast to the result without Kalman filter, the center of gravity of full-disk solar image with Kalman filter (the red line shown in Figure 9(b)) almost keeps its trajectory. What is more, compared with the trajectory without Kalman filter, the one with Kalman filter is more smooth and steady. Therefore, we can conclude that Kalman filter can correctly estimate the position of the center of gravity of full-disk solar image, and if the output of Kalman filter, $\mathrm{Cur\_}\widehat{\mathrm{Solar}}\mathrm{\_G}(k/k)$, can be applied to the closed-loop control system of our proposed scanning method, our proposed scanning method will be of more strong performance.

Second, the Kalman filter is applied to the closed-loop controlling system of our proposed scanning method (the structure of Figure 3 should be that S5 is connected to S4 by SW2, SW3 is connected to S7, and SW5 is connected to S8). Afterwards, some experiments similar to those mentioned above also have been conducted, and then the group of data with remarkable random disturbances, which is obtained in level 3 wind, is used and displayed in Figure 10.

Because, currently, the Kalman filter has been used, the scanning procedure should be smooth and steady. Therefore, the movement accuracy and time between adjacent scanning steps are what we want to obtain. With respect to the accuracy, if the number of pixels between adjacent scanning steps is within the range $[12.2932, 12.6932]$, the accuracy of every scanning step can be met. Therefore, in Figure 10(b), the trajectory of the center of gravity of full-disk solar image just contains points, which belong to the beginning and end of every scanning step. Based on Figure 10(b), accuracy of every scanning step can be obtained, and it is clear that their accuracy are satisfactory. On the other hand, to obtain the movement time of every scanning step, the timer based software is used, and corresponding results show that all their time is less than or equal to 4s.

Finally, the method of quickly locating and pointing the first scanning step described in Section 4 is also tested, and results show that the first scanning step always is pointed to the point, which is located within the solar local area of interest.

Obviously, from the analyses mentioned above, we may come to the conclusion that, for FASOT-1B, the proposed scanning method can meet the specifications of every scanning step. In other words, this method can be directly transplanted to FASOT after it is built.

8. Conclusion

On the basis of the special structure of FASOT in optics and machine, this paper proposes a novel scanning method based on autoguiding system so as to expand the field of view of observation of FASOT. In the meantime, the proposed scanning method also is turned out to be feasible in the FASOT prototype FASOT-1B. In other words, this method can be directly transplanted to FASOT after it is built. On the other hand, in addition to giving a detailed introduction to the principle of our proposed scanning method, this paper also provides an efficient method based on Kalman filter to suppress the random disturbances forced on every scanning step so as to strengthen the scanning performance.

As a closing remark, because FASOT has a special structure which is distinct from the existing solar telescopes and our proposed scanning method based on autoguiding system has not been reported in the published literatures, the key technologies described in our proposed scanning method are able to be taken as an efficient reference when the solar telescopes with a similar structure to FASOT are constructed, especially in the situation where the mount of the solar telescopes supports the software protocols based on ASCOM.

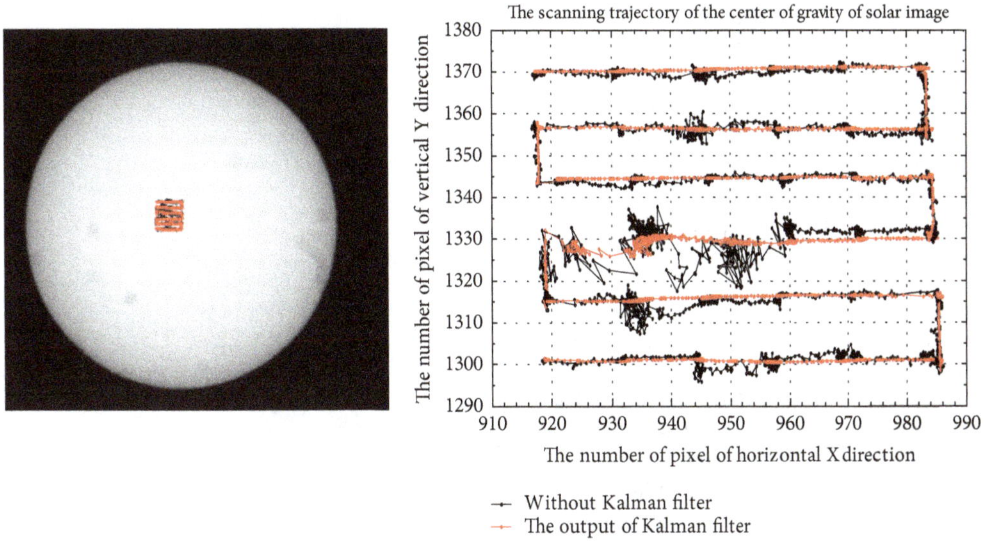

(a) The track of the actual picture (b) The track of multistep scanning with and without Kalman filters

FIGURE 9: The comparison of the track of the center of gravity of full-disk solar image.

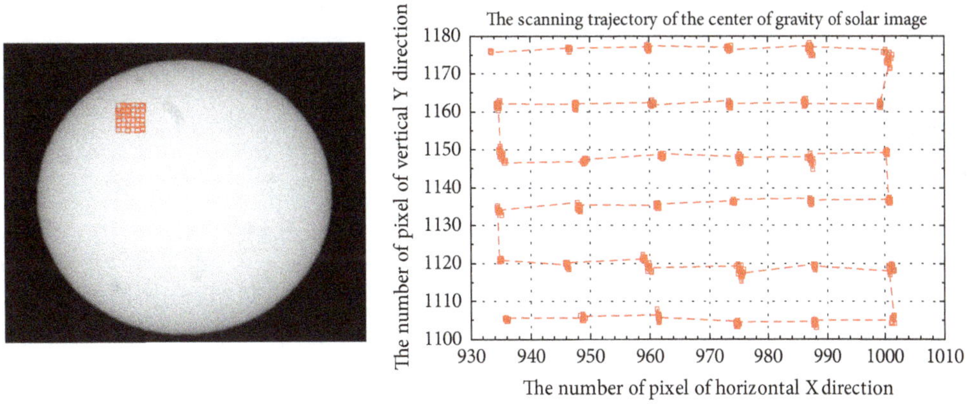

(a) The actual scanning position (b) The track of multistep scanning using Kalman filter

FIGURE 10: The track of the center of gravity of full-disk solar image using Kalman filter.

Acknowledgments

This work is supported by the National Natural Science Foundation of China (Grant nos. 11527804 and 11703087)

References

[1] Y. Deng, Z. Liu, and C. Group, "The Chinese Giant Solar Telescope (CGST)," in *Proceedings of the Second ATST-EAST Meeting: Magnetic Fields from the Photosphere to the Corona*, 2012.

[2] G. T. Dun and Z. Q. Qu, "Design of the polarimeter for the fibre arrayed solar optical telescope," *Acta Astronomica Sinica*, vol. 53, no. 1, pp. 342–352, 2012.

[3] Z. Q. Qu, "A fiber arrayed solar optical telescope," in *Proceedings of the Conference Series of Astronomical Society of the Pacific*, vol. 437, pp. 423–431, 2011.

[4] Z. Q. Qu, G. T. Dun, L. Chang et al., "Spectro-imaging polarimetry of the local corona during solar eclipse," *Solar Physics*, vol. 292, no. 2, p. 37, 2017.

[5] Y. Changchun, L. Zhenggang, C. Yuchao, and X. Jun, "The design of a spectrum scanning observation system for the new vacuum solar telescope," *Astronomical Research and Technology*, vol. 13, no. 2, pp. 257–265, 2016.

[6] F. F. Yue, X. U. Jun, and R. L. Zhang, "Analysis of the slit-scanning control system of the 1 meter solar telescope of YNAO," *Astronomical Research & Technology*, vol. 6, no. 4, pp. 319–326, 2009.

[7] K. Ichimoto, B. Lites, D. Elmore et al., *Polarization Calibration of the Solar Optical Telescope Onboard Hinode*, The Hinode Mission, Springer, NY, USA, 2008.

[8] C. Pernechele, F. Bortoletto, and K. Reif, "Position control for active secondary mirror of a two-mirror telescope," *Telescope Control Systems II. International Society for Optics and Photonics*, vol. 3112, pp. 172–181, 1997.

[9] Z. Jingguo, L. Feng, and H. Qitai, "Design and realization of an airborne LiDAR dual-wedge scanning system," *Infrared and Laser Engineering*, vol. 45, no. 5, 2016.

[10] R. E. Wolfe, G. Lin, M. Nishihama et al., "Suomi NPP VIIRS prelaunch and on-orbit geometric calibration and characterization," *Journal of Geophysical Research: Atmospheres*, vol. 118, no. 20, 2013.

[11] C. H. Wu and Q. S. Zhu, "The tracking and guiding method for full solar disk vector magnetograph telescope," *Astronomical Research & Technology*, vol. 4, no. 2, p. 147, 2007.

[12] W. Thompson and M. Carter, "EUV full-sun imaging and pointing calibration of the SOHO/CDS," *Solar Physics*, vol. 178, no. 1, pp. 71–83, 1998.

[13] X. Jiang, H. U. Ke-Liang, and J. B. Lin, "Tracking and guiding for full solar disk image using large CCD-array," *Optics & Precision Engineering*, vol. 16, no. 9, pp. 1589–1594, 2008.

[14] L. G. Taff, *Computational Spherical Astronomy*, Wiley, 1981.

[15] Y. Yang, N. Rees, and T. Chuter, "Reduction of encoder measurement errors in UKIRT telescope control system using a Kalman filter," *IEEE Transactions on Control Systems Technology*, vol. 10, no. 1, pp. 149–157, 2002.

[16] T. Erm and S. Sandrock, "Adaptive periodic error correction for the VLT," in *Proceedings of the Large Ground-based Telescopes (International Society for Optics and Photonics)*, pp. 900–909, 2003.

[17] T. Erm and S. Sandrock, "Adaptive correction of periodic errors improves telescope performance," in *Proceedings of the 2005 IEEE American Control Conference*, vol. 6, pp. 3776-3777, 2005.

[18] N. Tian, J. Sun, and Z. Liu, "Real-time light-spot positioning for target observation and aiming based on monocular vision," *Infrared & Laser Engineering*, 2014.

14

Capability of the HAWC Gamma-Ray Observatory for the Indirect Detection of Ultrahigh-Energy Neutrinos

Hermes León Vargas, Andrés Sandoval, Ernesto Belmont, and Rubén Alfaro

Instituto de Física, Universidad Nacional Autónoma de México, Apartado Postal 20-364, 01000 Ciudad de México, Mexico

Correspondence should be addressed to Hermes León Vargas; hermes.leon.vargas@cern.ch

Academic Editor: Dieter Horns

The detection of ultrahigh-energy neutrinos, with energies in the PeV range or above, is a topic of great interest in modern astroparticle physics. The importance comes from the fact that these neutrinos point back to the most energetic particle accelerators in the Universe and provide information about their underlying acceleration mechanisms. Atmospheric neutrinos are a background for these challenging measurements, but their rate is expected to be negligible above ≈1 PeV. In this work we describe the feasibility to study ultrahigh-energy neutrinos based on the Earth-skimming technique, by detecting the charged leptons produced in neutrino-nucleon interactions in a high mass target. We propose to detect the charged leptons, or their decay products, with the High Altitude Water Cherenkov (HAWC) observatory and use as a large-mass target for the neutrino interactions the Pico de Orizaba volcano, the highest mountain in Mexico. In this work we develop an estimate of the detection rate using a geometrical model to calculate the effective area of the observatory. Our results show that it may be feasible to perform measurements of the ultrahigh-energy neutrino flux from cosmic origin during the expected lifetime of the HAWC observatory.

1. Introduction

The first evidence of ultrahigh-energy neutrinos (in the PeV energy range) from extraterrestrial origin was recently reported [1]. This opened a new field in astroparticle physics that will allow the identification and characterization of the most powerful particle accelerators in the Universe. Neutrinos are not affected by the electromagnetic or strong interactions and thus point back to the source where they were produced, unlike charged cosmic rays. Gamma-rays are another cosmic probe that provides information about the acceleration mechanisms that occur in astrophysical sources. Due to this, there are several dedicated instruments, both ground or space based, performing a continuous survey of the Universe characterizing gamma-ray sources, for example, the Fermi Gamma-Ray Space Telescope and the imaging atmospheric Cherenkov telescopes HESS, MAGIC, VERITAS, and FACT.

The HAWC observatory was designed to detect and characterize the sources of high-energy gamma-rays in the energy range between 100 GeV and 100 TeV [2] and started full operations in April 2015. It is a ground based instrument that detects atmospheric showers by measuring with high precision the arrival time of the particles that compose the air showers. This is done via the Cherenkov light produced by the air shower particles as they enter the 300 water detector tanks that constitute the observatory, with a total water volume of 54 million litres. At energies above 100 TeV, gamma-rays suffer strong absorption from pair production with photons from the cosmic microwave background radiation that strongly reduces their mean free path [3]. For this reason, ultrahigh-energy neutrinos may be a better tool to study the most energetic extragalactic particle accelerators.

The Earth-skimming technique to detect ultrahigh-energy neutrinos has been proposed before; see, for instance, [4–8]. The method consists of using the interaction between a neutrino and a nucleon via the exchange of a $W^{\pm}$ boson to produce a charged lepton of the same flavour as the incoming neutrino. Since the neutrino-nucleon cross section is very small, a large-mass target is needed. The natural candidate to produce such interactions is the Earth crust, either by focusing the searches on quasi-horizontal neutrinos that travel along a chord inside the Earth or by using mountains as targets. The produced charged lepton travels essentially in the same direction as the neutrino. Thus, in the considered scheme, the charged lepton will travel either upward if it was moving through a chord inside the Earth crust or quasi-horizontally if the neutrino passed through a mountain.

From the three families of leptons, the electron neutrinos are unfavourable for this type of study because the produced high-energy electrons initiate electromagnetic showers that are easily absorbed by the target mass shortly after production. For μ neutrinos, the relatively large mean life of the produced charged lepton combined with the ultrahigh energy will produce a detectable signal only as an ultraenergetic μ. The τ neutrinos are the ones that have attracted more interest from the experimental point of view. The reason is that because of their very short mean life, even if they are very energetic the produced τ charged leptons will decay into secondary particles that would make the detection of the signal easier. However, the τ neutrinos have proven to be one of the most elusive particles of the standard model, with less than 15 detections up to now [13]. Ultrahigh-energy τ neutrinos (with searches up to 72 PeV) have eluded direct detection so far, even after the analysis of three years of IceCube data [14]. Even though τ neutrinos are disfavoured in production mechanisms at the astrophysical acceleration sites, the neutrino flavour mixing that occurs in cosmological distances is expected to produce approximately equal proportions of all neutrino flavours at the Earth.

There have been already experimental attempts to detect ultrahigh-energy neutrinos using the Earth-skimming technique. For instance, the Pierre Auger observatory in Argentina used their surface detectors to look for the electromagnetic signature of extensive air showers initiated by the decay of τ charged leptons of EeV energies that develop close to the detector [15–17], without finding candidate signals so far. There are also studies that propose to use the fluorescence detector of the Pierre Auger observatory to detect the decay in the atmosphere of τ charged leptons produced by ultrahigh-energy neutrinos [18]; however, this idea has not been implemented yet. The Ashra-1 collaboration [19] searched for neutrino emission from a GRB in the PeV–EeV energy range using the Earth-skimming technique. The Ashra-1 experiment, located on the Mauna Loa volcano and facing the Mauna Kea volcano in the Hawaii island, aimed to detect τ neutrinos that converted into τ charged leptons inside Mauna Kea. Their method consisted of measuring the Cherenkov light emitted by the particles of the atmospheric shower initiated by the decay products of the τ charged lepton. Their analysis of the GRB081203A did not find signals associated to τ neutrinos in the PeV–EeV energy range. There have been also studies about the feasibility to use the MAGIC telescopes for the detection of τ neutrinos [20, 21], by pointing their telescopes below the horizon towards the sea, or by searching for reflections of Cherenkov light by the nearby ground, the sea, or clouds [22, 23]. However, no experimental results have been published yet.

In this paper we propose to adapt the Earth-Skimming technique to use it with the HAWC gamma-ray observatory, employing the Pico de Orizaba volcano as a target for the neutrino-nucleon interactions. However, we propose not to follow the method explored so far of studying the decay products of a τ charged lepton on the atmosphere. Instead, we propose to reconstruct directly the trajectory of the charged lepton or their decay products as they travel through the HAWC detectors. In this way we do not only restrict our studies to τ neutrinos but also include the possibility for the detection of ultrahigh-energy μ neutrinos (a first suggestion of this method was mentioned in [11]). At ultrahigh energies, the produced charged leptons will have an energy approximately equal to that of the original neutrino [7, 11]. These charged leptons, or their decay products, could travel crossing several HAWC detectors depositing large amounts of Cherenkov light, well beyond the average left from both atmospheric muons ($\approx$30 photoelectrons (PEs) [24]) located far from the shower core used in the gamma/hadron discrimination algorithms of HAWC and the high PE noise, also associated to atmospheric muons, considered by the HAWC collaboration to be in the range of 10–200 PEs [25]. Here it is important to point out that the dynamical range of the HAWC electronics goes from a fraction of a PE up to thousands PEs [26].

In order to demonstrate that our proposal to search for tracks produced by τ charged leptons or their boosted decay products is feasible, we performed some GEANT4 [27] simulations of the decay of τ charged leptons with an energy of 1 PeV (that decay approximately 50 m after their creation point) and studied the shower evolution in air. We choose an energy of 1 PeV so the τ charged leptons decay quickly and are in the energy regime of our studies. After decay, the opening angle of the charged products is smaller than 0.7°, thus, after the decay products have travelled 2050 m (the approximate distance from the edge of the volcano to the HAWC array at $\approx$4090 m a.s.l. is of two kilometers), they would hit at most three columns of HAWC tanks. Moreover, over 95% of the secondary charged particles are contained within an opening angle smaller than 0.2°, therefore producing large Cherenkov signals only within a single row of HAWC tanks, producing a clearly identifiable track. This simplified exercise was done for 1 PeV charged τ's, making it easy to extrapolate the results to higher energy charged leptons, since the opening angle of the decay products is inversely proportional to the energy of the primary charged lepton [28].

Based on this information, we believe that the tracking method is possible in the search for charged leptons, produced by neutrino-nucleon interactions, in the PeV energy range. The amount of Cherenkov light that could be detected by the HAWC PMTs by such energetic particles is expected to be in the range of thousands to tens of thousands PEs. However, given the dynamic range of the current HAWC electronics one could anticipate that the collected light could be used as a proxy to at least set a lower boundary on the energy of the incoming charged lepton, in a similar manner to what IceCube does for muons that pass through the detector.

The paper is organized as follows: in Section 2 we present the calculation of the flux of charged leptons that could be produced by neutrino-nucleon interactions in the Pico de Orizaba volcano. In Section 3, we describe a method to calculate the HAWC effective area based on purely geometrical considerations and evaluate it for different trigger conditions. Then, in Section 4, we present our results for the possible detection rate and address the issue of the expected background signals. In Section 5, we discuss how the current trigger of HAWC can be useful in selecting data for these studies and discuss further a possible background rejection strategy. Finally, the conclusions are presented in Section 6.

2. Calculation of the Flux of Charged Leptons Produced by Earth-Skimming Neutrinos

In order to obtain an estimate of the number of ultrahigh-energy charged leptons that could be produced via the Earth-skimming technique, we follow the formalism developed in [7, 29]. However we use further simplifications due to the detection method that we propose in this paper and also because of the energy of the neutrinos that we plan to study. The differences between the original formalism of [7, 29] and our implementation are pointed out in the text.

The number of charged leptons (N_L) that could be detected by the observatory is given by

$$N_L = \Phi_L (A\Omega)_{\text{eff}} TD, \tag{1}$$

where Φ_L is the charged lepton flux, $(A\Omega)_{\text{eff}}$ is the effective area of the observatory, T is the live time of the experiment, and D is the duty cycle for observations, which basically describes which fraction of the time T the experiment is actually able to take data. In this section we describe the calculation of the flux of charged leptons Φ_L produced by the neutrino-nucleon interactions that occur while the neutrinos traverse the volcano. We start with the differential flux of ultrahigh-energy neutrinos produced by astrophysical sources, which we consider to be isotropic. This differential flux is given by

$$\frac{d\Phi_\nu}{dE_\nu d\cos\theta_\nu d\phi_\nu} \tag{2}$$

with $(E_\nu, \theta_\nu, \phi_\nu)$ being, respectively, the energy, polar, and azimuthal angle of the neutrinos. Since we are interested in integrating the flux on only a certain region (the one covered by the Pico de Orizaba volcano) and because we are considering an isotropic flux, we can simplify the differential neutrino flux to

$$\frac{d\Phi_\nu}{dE_\nu d\cos\theta_\nu d\phi_\nu} = \frac{1}{\Omega}\frac{d\Phi_\nu}{dE_\nu}, \tag{3}$$

where Ω is the solid angle covered by the volcano that is being used as the target for the neutrino-nucleon interactions. After the proposed isotropic neutrino flux passes through the Pico de Orizaba volcano, the produced differential charged lepton flux is given by

$$\frac{d\Phi_L}{dE_L d\cos\theta_L d\phi_L} \tag{4}$$

with (E_L, θ_L, ϕ_L) being, respectively, the energy, polar, and azimuthal angle of the produced charged leptons. The relation between the differential fluxes of incoming neutrinos and the produced charged leptons is thus given by

$$\begin{aligned} &\frac{d\Phi_L}{dE_L d\cos\theta_L d\phi_L} \\ &\quad = \int dE_\nu d\cos\theta_\nu d\phi_\nu \frac{1}{\Omega}\frac{d\Phi_\nu}{dE_\nu}\kappa\left(E_\nu, \theta_\nu, \phi_\nu; E_L, \theta_L, \phi_L\right), \end{aligned} \tag{5}$$

where κ is a function that physically represents the convolution of the probabilities of the different processes that need to take place in order that an ultrahigh-energy neutrino converts into a charged lepton inside the target material and is able to escape the mountain. Given that we are interested in studying ultrahigh-energy neutrinos ($E_\nu > 10$ PeV), the first simplification comes from the fact that the produced charged lepton will approximately follow the same direction as the original neutrino. The angle between the original neutrino and the produced charged lepton ($\theta_\nu - \theta_L$) has been estimated to be smaller than 1 arcmin for energies above 1 PeV for τ's [19]; so, we expect this angle to be negligible. This makes that the angular dependence of κ can be approximated by delta functions.

$$\begin{aligned} &\kappa\left(E_\nu, \theta_\nu, \phi_\nu; E_L, \theta_L, \phi_L\right) \\ &\quad \approx \kappa\left(E_\nu; E_L\right)\delta\left(\cos\theta_\nu - \cos\theta_L\right)\delta\left(\phi_\nu - \phi_L\right). \end{aligned} \tag{6}$$

The energy dependent part of the κ function can be written as the integral along the path of the neutrino and the corresponding charged lepton inside the volcano

$$\kappa\left(E_\nu; E_L\right) = \int P_1 P_2 P_3 (L) P_4, \tag{7}$$

where P_1 is the survival probability for a neutrino travelling a certain distance inside the volcano. This probability can be written as

$$P_1 = \exp\left[-\int_0^A \frac{dz'}{L_{\text{CC}}^\nu (E_\nu)}\right], \tag{8}$$

where L_{CC}^ν is the charged current interaction length. Since the average width of the volcano (A) is much smaller than the interaction length, almost all of the neutrinos will traverse the whole mountain. That defines the integration limit of (8). The interaction length is a function of the density of the medium that the neutrino travels through. In the case of the formalism developed in [7], where the trajectories of the neutrinos were across chords inside Earth, the authors had to consider variations of the Earth density. In this particular case, since we are interested in neutrinos that travel through volcanoes, it is reasonable to consider a constant density ρ along the path of the neutrinos. Thus, the expression for the charged current interaction length can be written as follows:

$$L_{\text{CC}}^\nu (E_\nu) = \frac{1}{\sigma_{\text{CC}}^\nu (E_\nu)\rho N_A}, \tag{9}$$

where N_A is Avogadro's constant and $\sigma_{\text{CC}}(E_\nu)$ is the charged current cross section. The second probability (P_2) that enters

TABLE 1: Numerical values of $P_3(L)$ for the different energy ranges considered in this work.

Energy bin [PeV]	$P_3(\mu)$	$P_3(\tau)$
$[10^1, 10^2]$	1	0.5
$[10^2, 10^4]$	1	0.9

the calculation of κ is that of a neutrino to convert into a charged lepton through a charged current in an infinitesimal distance dz. This probability is given by

$$P_2 = \frac{dz}{L^{\nu}_{\mathrm{CC}}(E_\nu)}. \quad (10)$$

The third process (P_3) that is taken into account in the κ function is the probability that the produced charged lepton is able to escape the volcano, taking into account the charged lepton energy losses in the medium. This process is described by a system of two coupled differential equations:

$$\frac{dE_L}{dz} = -(\alpha_L + \beta_L E_L)\rho, \quad (11)$$

$$\frac{dP_3(L)}{dz} = -\frac{m_L P_3(L)}{cT_L E_L}. \quad (12)$$

Where (11) describes the energy loss processes, with α_L describing the ionization energy loss and β_L the radiative energy loss. According to the literature, for example, [7], the effects of α_L are negligible at the energy regime of interest of this work; so, we consider in (11) that $\alpha_L \to 0$. In (12), m_L is the mass of the charged lepton, c is the speed of light, and T_L is the charged lepton lifetime. A Monte Carlo study from [11] shows that, for μ's with energies of 1 PeV, $P_3(\mu) \approx 0.98$ after 6 km water equivalent (km.w.e) (6 km.w.e. $\approx$ 2.3 km in "standard rock," the average path length inside the volcano), that is, in the extreme case where the charged lepton has to travel the average width of the volcano, and this value approaches unity as the energy of the μ charged lepton increases. Thus, we consider that, for the energy regime studied in this work, it is appropriate that for μ's we can take a value of $P_3(\mu) \approx 1.0$ (see Table 1).

The case of the survival probability for τ charged leptons is more complicated to evaluate, since most of the research has been done for energies above or at 100 PeV [30–32]. We take as a base for our calculations the results presented in [31]. For E_τ = 10 EeV, $P_3(\tau) \approx 0.99$ and, for E_τ = 1 EeV, $P_3(\tau) \approx 0.93$ after 2.3 km in "standard rock," that is, in the extreme case where τ's are produced just after entering the volcano. However, the values of $P_3(\tau)$ at lower energies cannot be obtained by a simple extrapolation since the value of $P_3(\tau)$ below 100 PeV is dominated by the mean life time of τ. By considering this fact, we obtain approximate values of $P_3(\tau)$ for the PeV energy regime. Table 1 shows the average values of $P_3(\tau)$ in the energy range of interest of this work. Finally, the factor P_4 makes sure that the charged lepton escapes the volume of the volcano with an energy E_L. Based on (11), taking $\alpha_L \approx 0$, then P_4 can be written as

$$P_4 = \delta(E_L - E_\nu \exp[-\beta_L \rho z]). \quad (13)$$

Thus, we can approximate the κ function as

$$\kappa(E_\nu; E_L) = \int P_1 P_2 P_3(L) P_4 \approx P_3(L) \int P_1 P_2 P_4. \quad (14)$$

After evaluating the integral we obtain an expression for the κ function.

$$\kappa(E_\nu; E_L) \approx \frac{P_3(L)}{L_{\mathrm{CC}}(E_\nu)} \exp\left[\frac{-A}{L_{\mathrm{CC}}(E_\nu)}\right] \frac{1}{\beta_L \rho E_L}. \quad (15)$$

Equation (15) is valid in the case where we neglect variations of the density of the volcano. Going back to (5), we can substitute (6) and (15) into it. By doing this, we obtain an equation for the differential flux with respect to the energy of the leptons

$$\frac{d\Phi_L}{dE_L} = \frac{P_3(L)}{\beta_L \rho E_L} \int dE_\nu \frac{d\Phi_\nu(E_\nu)}{dE_\nu} \frac{1}{L_{\mathrm{CC}}(E_\nu)} \cdot \exp\left[-\frac{A}{L_{\mathrm{CC}}(E_\nu)}\right]. \quad (16)$$

The next step is to define a function that describes the differential flux of neutrinos as a function of energy. For this, we use the parametrization of the measured astrophysical neutrino flux made by IceCube [12, 33, 34], which has been found to be well described by an unbroken power law

$$\frac{d\Phi_\nu(E_\nu)}{dE_\nu} = \phi \times \left(\frac{E_\nu}{100\,\mathrm{TeV}}\right)^{-\gamma}. \quad (17)$$

We choose to use the parametrization presented in [33] for the measurement of the $\nu_\mu + \bar{\nu}_\mu$ astrophysical flux. For this particular parametrization we shifted the mean values of the normalization and spectral index in order to obtain the highest flux, within the allowed range given by their statistical uncertainties. These values are ϕ = 0.3343 $\mathrm{GeV^{-1}\,Km^{-2}\,sr^{-1}\,yr^{-1}}$ and γ = 1.71. We assume for our calculations an expected equal contribution to the astrophysical flux for $\nu_\tau + \bar{\nu}_\tau$, due to the neutrino flavour mixing over cosmological distances that would produce approximately equal proportions of all flavours. Then, we extrapolate the measured flux to the energy range 10 PeV to 100 EeV. Moreover, motivated by the most energetic event found by the diffuse flux muon neutrino search (2009–2015) done by IceCube (a track event that deposited 2.6 ± 0.3 PeV in the sensible volume of the detector [35]), we also include in the flux estimations the models proposed in [9], which account for sources of multi-PeV neutrinos that are constrained by the most recent ultrahigh-energy neutrino upper limits set by IceCube [36] and Pierre Auger [15]. These models, for the sum of all neutrino flavours, have the smoothly broken power law functional forms:

$$\frac{d\Phi_\nu(E_\nu)}{dE_\nu} = \phi_i \times \left[\left(\frac{E_\nu}{E_i}\right)^{\alpha\eta} + \left(\frac{E_\nu}{E_i}\right)^{\beta\eta}\right]^{1/\eta}. \quad (18)$$

The values of the parameters are α = −1, β = −3 and η = −1. E_i take the values of 10^7 GeV for what we refer in

TABLE 2: Values of the normalization constants ϕ_i and pivot energies for the different PeV neutrino models presented in [9].

Model	E_i [GeV]	ϕ_i [GeV^{-1} Km^{-2} sr^{-1} yr^{-1}]
A	10^7	6.2545×10^{-5}
B	10^8	1.5780×10^{-6}
C	10^9	1.5798×10^{-8}

this work as Model A, 10^8 GeV for Model B, and 10^9 GeV for Model C. The values of the normalizations ϕ_i for each of these models are presented in Table 2.

According to [9], Models A and B would correspond to the spectra produced by BL Lac AGNs, and combinations of Models A and C would follow the expected shape of GZK neutrinos produced from EBL and CMB interactions.

In order to calculate the flux of charged leptons produced by neutrino-nucleon interactions we need to define the input parameters that enter the calculation of the number of produced charged leptons (see (16)). For the value of the parameter that describes the radiative energy losses of the charged leptons as they travel through the rock (β_L), one can find in the literature β_τ parametrizations that may differ by up to a factor of two. For instance, it ranges from 0.26 to 0.59×10^{-6} cm^2/g at 100 PeV in [10]. For our calculations we decided to use the β_τ values obtained with the ASW structure functions calculated in [10] and for β_μ the results obtained in [11]. For this latter case the results are available up to an energy of 1 EeV, but these values are enough for our calculations. For the numerical integrations, we used intermediate values of β_τ and β_μ in different bins of energy as shown in Table 3.

For the density of the volcano, we take that of "standard rock," $\rho = 2.65$ g/cm^3 [11]. For the calculation of the charged current interaction length we use the results from [37], with the cross section for neutrino-nucleon interactions via charged currents in the energy range 10 PeV $\le E_\nu \le$ 1 ZeV given by

$$\sigma_{CC}(\nu N) = 5.53 \times 10^{-36}\,\text{cm}^2 \left(\frac{E_\nu}{1\,\text{GeV}}\right)^{0.363}. \quad (19)$$

This parametrization is a well known result. However, as a cross check, we compared the parametrization given by [37] with more recent calculations [38–41]. The result of this is presented in Figure 1. One can notice that the result from [37] follows the general trend predicted by the work of [38] and Sarkar et al. Due to this and the fact that there is an analytical expression for the cross section of [37], we decided to use (19) in our numerical calculations. The anti neutrino-nucleon cross section is taken to be the same [37]. Using these results we get a value of $L^{\nu}_{CC}(100\,\text{PeV}) = 1413$ km, much smaller than the light-year interaction length for average neutrinos in lead [42].

For the average width of the Pico de Orizaba, we calculated the typical length of a chord that goes through a cone that follows the geometry of the volcano, as will be shown in the following section of the paper. At 4100 m a.s.l.,

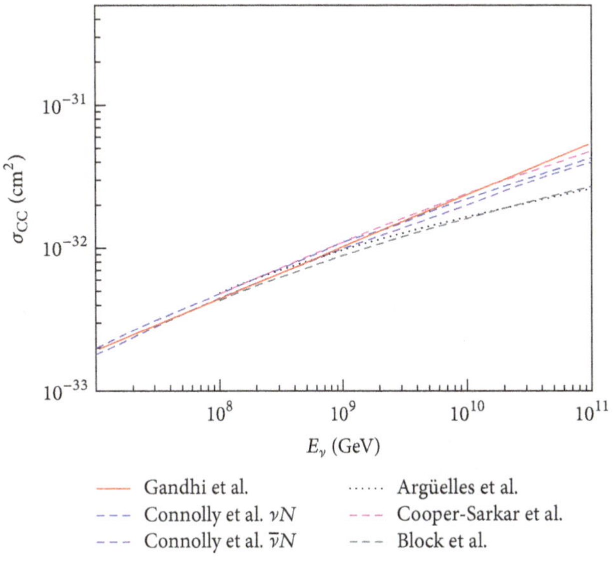

FIGURE 1: Comparison of the neutrino-nucleon cross section as a function of the neutrino energy, for the model from [37] with those from [38–41].

the volcano has a width of ≈6 km and at the summit at ≈5,500 m a.s.l., a width of ≈0.35 km (see Figures 4 and 5). The average path length for the neutrinos is of 2.33 km inside the volcano. Finally, we assume a duty cycle $D = 95\%$, which agrees with the reported value observed during actual HAWC operations [43]. The energy range for the charged leptons that exit the volcano is taken in the range ($E_{min} = (1/10)E_\nu$, $E_{max} = E_\nu$). In this way we consider an energy range that contains a good fraction of the charged leptons that exit the volcano. This is because the charged leptons escape with an energy distribution, after loosing some of their energy in the medium. For a study of the propagation of a monoenergetic beam of τ charged leptons through rock see, for instance, the work from [31].

We integrate numerically (16) using the differential neutrinos fluxes from (17) and (18) to obtain the number of charged leptons (Φ_L) that escape the mountain. The results are presented in Tables 4 and 5.

One can notice that the flux of μ charged leptons is more than an order of magnitude lower than that of τ's. The reason for this can be seen in (16), where the charged lepton flux depends inversely on the radiative energy losses of the charged leptons, which are more than an order of magnitude higher for μ's with respect to τ's (see Table 3). This does not contradict the fact that, at the highest energies, for both μ's and τ's the survival probability $P_3(L) \rightarrow 1$. $P_3(L)$ quantifies the probability that the charged leptons are able to escape the volcano, while the β_L factor in (16) appears because of the form of P_4 and is independent of $P_3(L)$. This result is consistent with the arguments developed in [44], where it is pointed out that τ charged leptons have a higher probability to escape the Earth crust compared to μ's.

In the following section we present a simple method that allows to approximate the effective area of the HAWC observatory to ultrahigh-energy charged leptons.

Table 3: Numerical values of β_L calculated with the ASW structure functions from [10] for τ's and from the results obtained by [11] for μ's. The table presents the energy at which the parameter was evaluated and the energy bin in which it is used in the numerical calculations.

β_L evaluated at [PeV]	Energy bin [PeV]	β_τ [cm^2/g]	β_μ [cm^2/g]
50	$[10^1, 10^2]$	2.496×10^{-7}	4.960×10^{-6}
500	$[10^2, 10^3]$	2.987×10^{-7}	5.143×10^{-6}
5×10^3	$[10^3, 10^4]$	3.554×10^{-7}	N/A
5×10^4	$[10^4, 10^5]$	4.184×10^{-7}	N/A

Table 4: Tau charged lepton fluxes ($\Phi_{\tau+\bar{\tau}}$) produced by neutrino-nucleon interactions in "standard rock" with average width A. The astrophysical flux is obtained from the extrapolation of the measured neutrino flux by IceCube. See the text for details.

Neutrino energy $[E_\nu]$	$\Phi_{\tau+\bar{\tau}}$ [Km^{-2} sr^{-1} yr^{-1}]			
	Astrophysical	Model A	Model B	Model C
10 PeV–100 PeV	10.7471	0.4732	0.8269	0.1008
100 PeV–10 EeV	11.0342	0.0233	0.4251	0.3690

Table 5: Muon charged lepton fluxes ($\Phi_{\mu+\bar{\mu}}$) produced by neutrino-nucleon interactions in "standard rock" with average width A. The astrophysical flux is obtained from the extrapolation of the measured neutrino flux by IceCube. See the text for details.

Neutrino energy $[E_\nu]$	$\Phi_{\mu+\bar{\mu}}$ [Km^{-2} sr^{-1} yr^{-1}]			
	Astrophysical	Model A	Model B	Model C
10 PeV–100 PeV	1.0816	0.0476	0.0832	0.0101
100 PeV–1 EeV	0.4685	0.0015	0.0267	0.0185

3. Effective Area Determination

In general, the effective area or acceptance of an observatory is calculated using a detailed Monte Carlo simulation; see, for instance, in [45, 46] studies of the sensitivity of air Cherenkov and fluorescence observatories to τ neutrinos using the Earth-skimming technique and in [22] an early proposal about using fluorescence measurements on observatories as Pierre Auger in order to search for τ charged lepton showers. The full Monte Carlo method of calculating the effective area represents a complex and time consuming task. In the early results obtained by the Pierre Auger collaboration, their effective area calculation neglected the effect of the topography that surrounds the observatory and took into account its effect in the systematic error of their observation limits [17]. However, their most recent results have incorporated the topography into their effective area calculations in two of their three analysis channels [15]. In our case we are specifically interested in the effect of the largest volcano that surrounds the HAWC observatory as a target for the neutrino-nucleon interactions. In order to reproduce the topography that surrounds HAWC, we use data from the *Instituto Nacional de Estadística y Geografía* (INEGI, México) [47]. Figure 2 shows the topography that surrounds the HAWC site (indicated by a small red rectangle). The observatory is located between two volcanoes: in the direction Northing-Easting by the Pico de Orizaba and in the opposite direction by the Sierra Negra volcano. Due to its much larger volume, we focus our attention on the Pico de Orizaba volcano as the target for the neutrino-nucleon interactions.

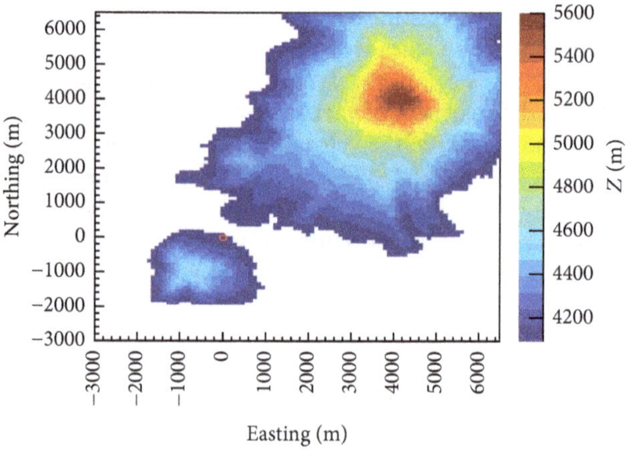

Figure 2: Topography of the HAWC site according to INEGI data. One can see two mountains, the Sierra Negra volcano at bottom left and the much larger Pico de Orizaba volcano on the top right. The small red rectangle, at the origin of the coordinate system, indicates the location and dimensions of the HAWC observatory.

Figure 3 shows the topography of the Pico de Orizaba volcano as seen in a coordinate system centered at the location of the HAWC array at ≈4090 m a.s.l. The approximate symmetry of the Pico de Orizaba volcano makes it easy to motivate a simplification in the calculation of the effective area. We can assume that the geometry of the mountain can be modelled to be conical. Figures 4 and 5 show the profile of the volcano along the Easting and Northing directions, and how we can approximate this profile using a cone with a height of 1500 m

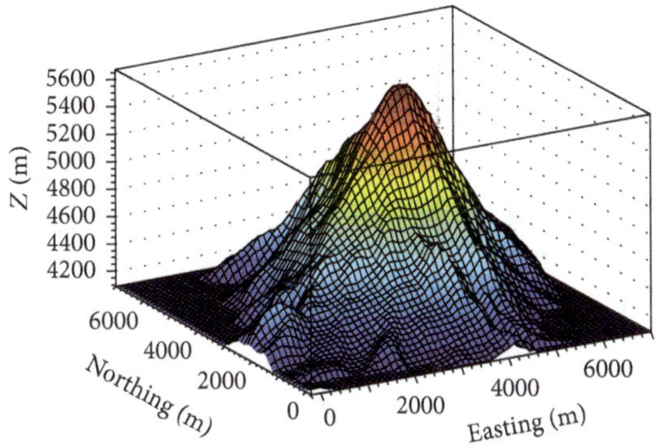

FIGURE 3: Topography of the Pico de Orizaba volcano, located in the state of Puebla in Mexico. The origin of the coordinate system is centered at the location of the HAWC array.

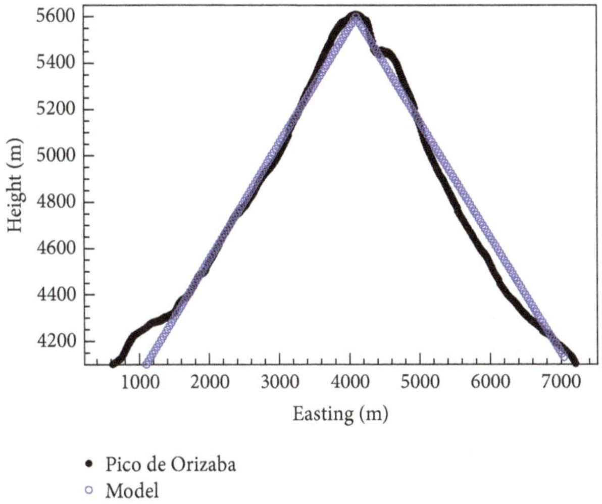

FIGURE 4: Projection of the Pico de Orizaba profile along the Easting axis. The blue open markers show the profile of the conical approximation to the volcano, while the black closed circles indicate the real profile of the volcano.

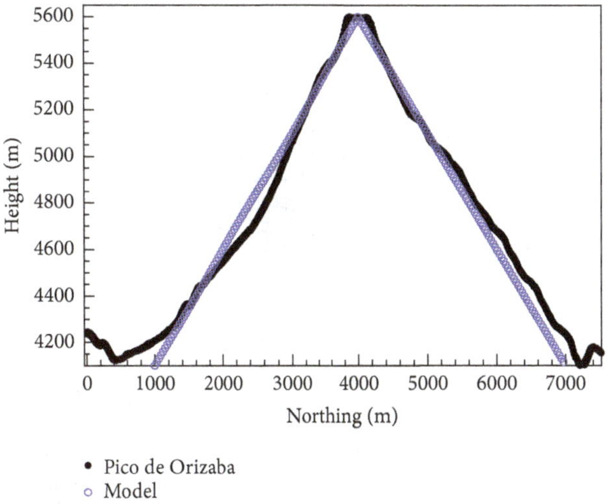

FIGURE 5: Projection of the Pico de Orizaba profile along the Northing axis. The blue open markers show the profile of the conical approximation to the volcano, while the black closed circles indicate the real profile of the volcano.

($z = 0$ taken at the altitude above sea level of the HAWC observatory) and a diameter of 6000 m.

Our method to calculate the effective area to detect ultrahigh-energy charged leptons produced by neutrino-nucleon interaction is as follows:

(1) We consider that the ultrahigh-energy charged leptons, or their highly boosted decay products, will travel following a straight trajectory along the direction of the initial neutrino.

(2) We approximate the region where the charged leptons are produced as a triangular surface, a 2D projection of the volcano, that faces the HAWC observatory. We can use this simplification since here we are only interested in studying the trajectories of the produced charged leptons. The width of the mountain is considered in the charged lepton flux calculation, as presented in Section 2. On the triangular surface we draw a grid made of lines parallel to the x- and z-axis in the coordinate system shown in Figure 6. The lines of the grid have a separation of 0.5 m in each direction. The surface is located at a distance of 5.7 km to the center of the HAWC observatory, approximately the distance between the center of the detector array and that of the volcano.

(3) The detection volume of HAWC is modelled as a rectangular prism, with dimensions (x, y, z) 140 m × 140 m × 4.5 m. This approximation is motivated by the actual configuration of the HAWC array. The observatory is made of a compact group of 300 Water Cherenkov Detectors (WCDs). Each WCD is a cylinder 7.3 m in diameter and 4.5 m in height. The array covers an area of approximately 22,000 m^2 [43].

(4) From each cell of the grid (~14 million), we generate vectors that point towards our model of HAWC, approximated as a rectangular prism. Figure 7 shows a diagram of the definition of the angles used in this work. The azimuth angle ϕ covers the range from 0 to π (with steps of 1 degree), pointing towards the $-y'$ direction in the coordinate system shown in Figure 6. This is done for the i different orientations of the polar angle θ. The polar angle is defined such that the vector orientation that lies in the x'-y' plane corresponds to 90° and increases as the orientation gets closer to the x'-z' plane (see Figure 7).

(5) For each orientation θ_i, we calculate a differential element of effective area

$$f(\theta_i) = A\Delta\phi(\theta_i) = \frac{N_{\text{trig}}}{N_{\text{gen}}} \times A_i \times \Delta\phi_i, \qquad (20)$$

where N_{trig} is the total number of vectors whose directions points towards the detection volume of HAWC with a minimum length of the trajectory

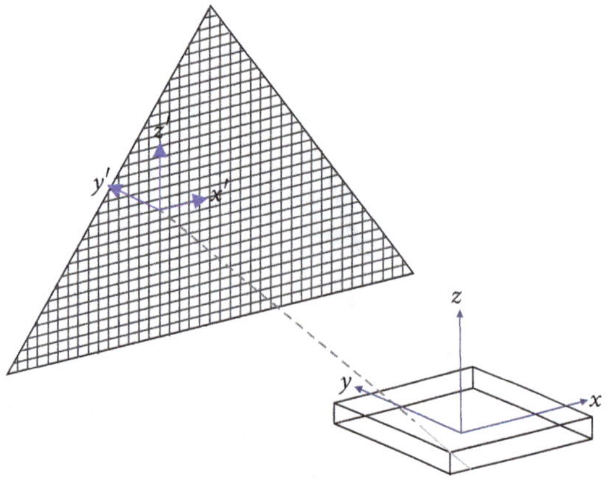

Figure 6: Schematic of the geometry and coordinate system used to simplify the problem. The Pico de Orizaba volcano is represented as a triangular surface located at 5.7 km from the center of the HAWC array. The HAWC detection volume is modelled as a rectangular prism. An example of a vector is shown with a dashed line that becomes solid as the vector passes through the detection volume. The figure is not drawn to scale.

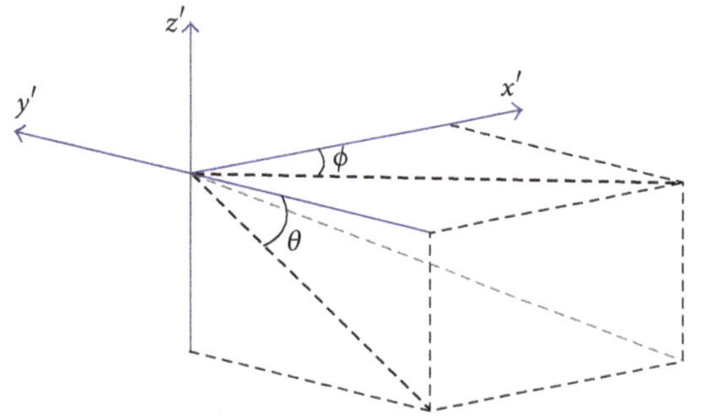

Figure 7: Definition of the angles used in the effective area calculation. The grey dashed line represents the orientation of one of the generated vectors. The polar angle θ is measured between the $-y'$ axis direction and the projection of each vector in the y'-z' plane; it has values equal to or greater than $\pi/2$. The azimuthal angle ϕ is measured between the x' axis direction and the projection of the vectors in the x'-y' plane; it ranges from 0 to π.

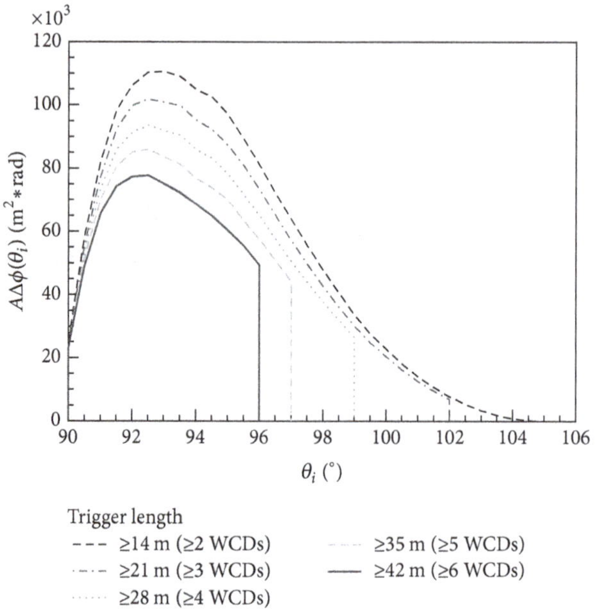

Figure 8: Values of the differential effective area as a function of the polar angle θ_i measured from zenith (the horizontal orientation, parallel to the y' axis, for the vectors is at 90°). The results are shown for different trigger conditions.

across the x-y plane shown in Figure 6 (inside the volume of the rectangular prism). We will refer to this minimum length as the trigger condition that approximately represents how many WCDs would measure Cherenkov light produced by the incoming lepton or their collimated decay products. N_{gen} is the total number of vectors that point towards the rectangular prism, regardless of the trajectory that they have inside the prism. A_i is the section of the area of the triangular surface over which the vectors that point towards the rectangular prism (N_{gen}) were generated. Finally, $\Delta\phi_i$ is the azimuthal angle covered by the area A_i, as seen from the center of the rectangular prism.

By following the procedure described above, we can obtain the differential elements of effective area as a function of the polar angle orientation. Figure 8 shows the results for different orientations of the polar angle.

One can see that the differential effective area increases with increasing values of the polar angle, up to a maximum at ≈93°, and then decreases because the area over which the vectors are generated decreases as well, as it corresponds to the upper sections of the triangular grid where the vectors are generated. The different line styles used in the plot show the results obtained for different trigger conditions. For instance, the curve with 14 m as trigger condition indicates the results obtained when the requirement of the path of the charged lepton through the detection volume corresponds to at least a trajectory in the x-y plane inside the HAWC rectangular prism of 14 m. This would correspond roughly to having the ultrahigh-energy charged lepton (or its boosted decay products) to traverse through at least two WCDs. This may not seem enough to even provide a rough idea of the direction of the incoming lepton. However, one should keep in mind that each WCD is equipped with four photomultiplier tubes (PMTs), recording the amplitude and arrival time of the Cherenkov light. By combining the information from the eight PMTs from two WCDs it may be possible to have enough information to reconstruct a trajectory, although details of the angular reconstruction of the tracks are beyond the scope of this work. In Figure 8, we present the results for trigger conditions that correspond roughly to having the charged lepton (or its collimated decay products) traversing from two up to six WCDs. As expected, the effective area that corresponds to a given trigger condition decreases as one increases the number of detectors required to have signals. We are interested in the total effective area $(A\Omega)_{\text{eff}}$ that goes

TABLE 6: Results of the effective area calculation for different trigger conditions. The trigger condition is the path length that a charged lepton or its highly boosted decay products travel along the x-y plane inside the detection volume of HAWC modelled as a rectangular prism.

Trigger [m]	$A\Omega$ [km^2 sr]
≥14	0.013867
≥21	0.012544
≥28	0.010578
≥35	0.008418
≥42	0.006782

into the calculation of the number of possible detections of (1). This total effective area is given by

$$(A\Omega)_{\text{eff}} = \int_{\theta_i}^{\theta_f} f(\theta_i)\, d\theta_i, \quad (21)$$

where the integration range in θ_i goes from 90° to 105° (the angular height of the Pico de Orizaba volcano as seen from the center of the HAWC observatory is of ≈15°). Our results of the effective area calculation as a function of the different trigger conditions are shown in Table 6.

With the results from Sections 2 and 3 we have the flux of charged leptons given a certain isotropic flux of neutrinos and the effective area of the HAWC observatory obtained with a simple trigger condition. Using this, in the next section we present our results for the flux of ultrahigh-energy charged leptons that could be detected by the HAWC observatory.

4. Results, Possible Background Signals, and Discussion

Table 7 shows the number of detectable charged τ's produced by neutrino-nucleon interactions in the Pico de Orizaba volcano, obtained using the effective area results presented in Table 6 and the ultrahigh-energy τ lepton fluxes from Table 4. We restrict the results to those from τ charged leptons since, as it was shown in Section 2, the flux from μ charged leptons is approximately an order of magnitude lower, making their detection not feasible with the method proposed in this work. The detection estimates are shown as a function of the different trigger conditions and of the different ultrahigh-energy neutrino fluxes discussed in Section 2. The eighth column presents an estimate of the background signals that are expected using this method. We estimated the background considering that the main contribution is that of ultrahigh-energy atmospheric muons, coming from the direction of the Pico de Orizaba. In order to calculate this, we used the characterization of the atmospheric muon flux above 15 TeV measured by IceCube [48], that can be modelled as an unbroken power law:

$$\frac{d\Phi_\mu}{dE_\mu} = \phi_\mu \times \left(\frac{E_\mu}{10\,\text{TeV}}\right)^{-\delta}. \quad (22)$$

In this case, we shifted the mean values (within the statistical and systematic uncertainties of the IceCube measurement) of the normalization and spectral index in order to obtain the worst case scenario of the background and then extrapolated the muon flux to the energy range of our interest. The values of the parameters that maximize the background muon flux are $\phi_\mu = 4.6673 \times 10^7\,\text{TeV}^{-1}\,\text{Km}^{-2}\,\text{sr}^{-1}\,\text{yr}^{-1}$ and a spectral index of 3.73. Thus, the background flux can be calculated using (1), where in this case the charged lepton flux corresponds to the integral of (22) within the energy bins used in this work. Of course this is an overestimation, since these atmospheric muons would have to be able to survive their travel through the volcano; however, we consider this estimation of the background appropriate because the study of the propagation of atmospheric muons through the volcano is beyond the scope of this work. Moreover, it has already been noticed that the atmospheric muon background is relevant up to energies of ≈100 TeV [49]. The last two columns of Table 7 show the estimated time to get a detection for each trigger condition, first for the case of only having as a source the neutrino flux given by the extrapolation of the measurement done by IceCube [12] and also for the case of including GZK neutrinos from [9].

From Table 7, we find that it would take around eight years of data taking in order to be able to find a track-like signal traversing three WCDs based on the extrapolation of the measured astrophysical flux, for τ neutrinos in the energy range from 10 PeV to 100 PeV. In order to find a track-like signal that propagates through four WCDs it could take more than nine years of data taking, but such a long track may allow reconstructing with better accuracy the direction of the original neutrino. We have verified the predicted detection rate by doing a back of the envelope comparison with the PeV neutrino measurements reported by IceCube [1] and found that, taking into account the different experimental conditions, our results fall in the expected order of magnitude.

The HAWC observatory is currently planned to operate for at least five years [50], but given its importance as a trigger for future instruments such as CTA, it is not unlikely that it may operate for at least 10 years. Table 7 shows also for completeness the results expected for neutrinos in the energy range from 100 PeV to 100 EeV. Charged τ leptons that do not decay before arriving or inside the HAWC array (energies of ≈100 PeV or larger) will leave almost the same signal because the values of β_L (see Table 3) change very slowly with energy [8]. This condition will make it very difficult to estimate the energy of the incoming τ. The results also show that the background from ultrahigh-energy muons at the energy regime of interest of this work is very low and would allow a clean identification of a signal coming from the direction of the Pico de Orizaba volcano.

5. Triggering of Signals and Background Rejection

The current software trigger used by the HAWC observatory consists of requiring a certain number of PMTs to be above a charge threshold (it actually consists of two thresholds used

TABLE 7: Number of detectable τ charged leptons per year (N) in HAWC for two different energy bins: 10 PeV–100 PeV and 100 PeV–100 EeV for the extrapolation of the measured IceCube (IC) flux [12] and for the ultrahigh-energy predictions (Φ_A, Φ_B, and Φ_C for the models A, B, and C discussed in Section 2) from [9]. The results are presented for different trigger conditions that correspond to how many HAWC WCDs participate in a track-like event. The addition of the flux $\Phi_A + \Phi_C$ to the extrapolation of the IC flux is motivated by their approximation to GZK neutrinos produced by interactions with the EBL and CMB [9]. The eighth column (Bckg) shows the expected number of background signals per year; see the text for details. The last two columns show the time needed (in years) for one charged τ lepton detection (ΔT), using the extrapolation of the IceCube measured flux and also considering the additional GZK flux at ultrahigh energy proposed in [9].

Neutrino energy [E_ν]	Trigger [m]	N: IC	N: Φ_A	N: Φ_B	N: Φ_C	N: IC + Φ_A + Φ_C	Bckg	ΔT IC [yr]	ΔT IC + Φ_A + Φ_C [yr]
10 PeV–100 PeV	≥14	0.1416	0.0063	0.0109	0.0014	0.1492	0.0145	7.1	6.7
	≥21	0.1281	0.0057	0.0099	0.0012	0.1349	0.0131	7.8	7.4
	≥28	0.1080	0.0048	0.0083	0.0010	0.1138	0.0111	9.3	8.8
	≥35	0.0860	0.0038	0.0066	0.0008	0.0906	0.0088	11.6	11.0
	≥42	0.0693	0.0031	0.0054	0.0007	0.0730	0.0071	14.4	13.7
100 PeV–100 EeV	≥14	0.1454	0.0003	0.0056	0.0049	0.1505	$<10^{-5}$	6.9	6.6
	≥21	0.1315	0.0003	0.0050	0.0044	0.1321	$<10^{-5}$	7.6	7.6
	≥28	0.1109	0.0003	0.0042	0.0037	0.1148	$<10^{-5}$	9.0	8.7
	≥35	0.0882	0.0002	0.0034	0.0030	0.0914	$<10^{-5}$	11.3	10.9
	≥42	0.0711	0.0002	0.0027	0.0023	0.0736	$<10^{-5}$	14.1	13.6

to estimate the charge with the Time-over-Threshold method; the thresholds are of 0.25 and 4 PEs) within a time sliding window of the order of ≈150 ns [24]. For the HAWC results presented in [43], the number of PMTs required to trigger an event was of 15. Thus, the multiplicity trigger used by the HAWC collaboration is already useful for the collection of signals that are needed for the neutrino searches proposed in this work. To show this, we can estimate the average number of PMTs that have signals during any 150 ns time window. From [51] we know that the sum of the signals from 112 HAWC's PMTs in a 60 s time window was of $\approx 2.285 \times 10^8$. Then, we can use this number to estimate the number of PMTs fired in any 150 ns triggering window, which is of ≈6. Thus, only 9 additional PMTs would be required to have signals within the specified time window in order to produce a trigger useful for the neutrino search proposed in this work. This condition can be easily fulfilled if an ultrahigh-energy lepton or its decay products pass through 3 WCDs, firing at least nine of its 12 PMTs (with a propagation time of the corresponding signals along 3 WCDs of ≈70 ns, well within the current trigger window). Of course, a specialized trigger can be developed in order to obtain a better trigger efficiency. This could be done, for instance, by selecting events with topologies consistent with tracks propagating through the WCDs. This would also have the consequence of reducing the bandwidth needed for this additional trigger.

A second point of interest to discuss is the ability of the instrument to separate the signals produced by ultrahigh-energy charged leptons from those from the background of lower energy atmospheric muons. The amount of emitted Cherenkov photons per unit length inside a HAWC WCD is given by the Frank-Tamm equation; and the total amount of light will be proportional to the the sum of the path lengths of the primary and all the produced secondary charged particles (from bremsstrahlung and pair production processes) that travel inside a WCD and have a velocity larger than the threshold for Cherenkov light production. According to the study presented in [52] this sum of path lengths increases linearly with the energy of the primary particle. Thus, the way to discriminate background signals could be based on the total light yield detected in the WCDs that belong to a given track (e.g., a 1 TeV primary electron would produce an order of magnitude more photons than those from a 100 GeV primary electron). Based on the results presented in [53], the mean momentum of nearly horizontal tracks ($\theta = 75^\circ$) is of ≈200 GeV/c, and in this work we aim to detect the tracks produced by leptons of PeV energies, so we could expect a factor larger than 10^3 in the deposited light in the WCDs for the ultrahigh-energy neutrino initiated signals relative to the average muon background tracks. Moreover, the results presented in [54] show that the integral intensity of muons of ≈1 TeV/c is larger than that of ≈200 GeV/c muons by a factor of ≈52, making it reasonable to easily handle the data rate if a specialized trigger is set to keep the data of tracks related to large deposits of Cherenkov light.

An additional experimental proof that this analysis could be performed, is the ability to observe the cosmic ray shadow produced by the volcano, as, for instance, pointed out in [22]. However, this requires the use of data and algorithms property of the HAWC collaboration that are not publicly available. Nonetheless we are confident, because of the interest on this result, that the Pico de Orizaba cosmic ray shadow will be publicly available soon.

Due to these facts, we believe that it is feasible to separate the huge atmospheric background from the signals we are looking for. We are aware that the precision that could be achieved in determining the energy of the primary tau lepton will be low; nonetheless the primary interest of this work is the detection technique and we leave the development of a possible energy estimator for future work.

6. Conclusions

We presented an estimate of the detection capabilities of the HAWC observatory to study ultrahigh-energy neutrinos interacting in the Pico de Orizaba volcano. We based our study in the analytic method developed by [7, 29] and modified it to a simpler case where the neutrino conversion takes place inside a medium of constant density. The effective area of the HAWC observatory for the detection of ultrahigh-energy charged leptons was calculated geometrically for different trigger conditions. We used the astrophysical neutrino flux measured by IceCube [12] and extrapolated it to energy range from 10 PeV up to 100 EeV and also considered models for multi-PeV neutrinos [9] that are constrained by the most recent data from both IceCube [36] and the Pierre Auger observatory [15]. With this, we found that finding a signal consistent with the propagation of an ultrahigh-energy charged lepton coming from the direction of the Pico de Orizaba volcano will require approximately nine years of data taking if the signal is required to propagate in four WCDs of HAWC. We estimated the expected background for this analysis using the measured atmospheric muon flux above 15 TeV [48] and found that it should be feasible to perform this study with a reasonable signal to background ratio during the lifetime of the HAWC observatory. We also showed that the current software trigger used by HAWC should be sufficient to acquire data for this analysis and that given the dynamic range of the HAWC electronics it is feasible to be able to discriminate the background signals produced by lower energy muons using the light yield detected by the HAWC PMTs.

As an anonymous referee pointed out to us, the detection rate should be taken with caution since there are several quantities that are uncertain to at least some degree, such as the flux of neutrinos at the ultrahigh-energy regime, the neutrino-nucleon cross section, and the parameters that describes the radiative energy loss processes for the leptons traversing the Earth crust. By selecting a different set of the input parameters for the calculations one could find a detection rate that could decrease by a factor of three or even more. However we also did not consider scenarios where the neutrino-nucleon cross section at ultrahigh energies could have enhanced values relative to the standard model predictions, due to the presence of new physics, as described, for instance, by [55]. Further work should be done using a complete Monte Carlo simulation that incorporates the detailed topography of the volcanoes that surround the

HAWC observatory and the complete detector simulation to study the detector response. However, this first step shows encouraging results to pursue more detailed studies. The detection rate is certainly low, not comparable to the one that dedicated neutrino experiment can achieve, but, as pointed out in [9], a single detection of a neutrino with an energy higher than 10 PeV would give evidence of a flux beyond what is firmly established. Our results indicate that, although extremely challenging, it is worth trying to detect ultrahigh-energy neutrinos, interacting in the Pico de Orizaba with the HAWC observatory.

Acknowledgments

This work was supported by the Programme UNAM-DGAPA-PAPIIT IA102715, Consejo Nacional de Ciencia y Tecnología (CONACyT), Mexico Grant 254964, and Coordinación de la Investigación Científica UNAM.

References

[1] M. G. Aartsen, "Evidence for high-energy extraterrestrial neutrinos at the IceCube detector," *Science*, vol. 342, no. 6161, Article ID 1242856, 2013.

[2] A. U. Abeysekara, R. Alfaro, C. Alvarez et al., "Sensitivity of the high altitude water Cherenkov detector to sources of multi-TeV gamma rays," *Astroparticle Physics*, vol. 50-52, pp. 26–32, 2013.

[3] R. J. Gould and G. Schréder, "Opacity of the universe to high-energy photons," *Physical Review Letters*, vol. 16, no. 6, pp. 252–254, 1966.

[4] D. Fargion, A. Aiello, and R. Conversano, "Horizontal tau air showers from mountains in deep vally: traces of ultrahigh neutrino tau," *International Cosmic Ray Conference*, vol. 2, 396 pages, 1999.

[5] D. Fargion, "Discovering ultra-high-energy neutrinos through horizontal and upward τ air showers: Evidence in terrestrial gamma flashes?" *Astrophysical Journal*, vol. 570, no. 2 I, pp. 909–925, 2002.

[6] A. Letessier-Selvon, "Establishing the GZK cutoff with ultra high energy tau neutrinos," in *Proceedings of the The international workshop on observing ultrahigh energy cosmic rays from space and earth*, pp. 157–171, Metepec, Puebla (Mexico).

[7] J. L. Feng, P. Fisher, F. Wilczek, and T. M. Yu, "Observability of Earth-Skimming Ultrahigh Energy Neutrinos," *Physical Review Letters*, vol. 88, no. 16, 2002.

[8] D. Fargion, P. G. De Sanctis Lucentini, M. De Santis, and M. Grossi, "TAU air showers from earth," *Astrophysical Journal*, vol. 613, no. 2 I, pp. 12853-11301, 2004.

[9] M. D. Kistler and R. Laha, Multi-PeV Signals from a New Astrophysical Neutrino Flux Beyond the Glashow Resonance. May 2016.

[10] N. Armesto, C. Merino, G. Parente, and E. Zas, "Charged current neutrino cross section and tau energy loss at ultrahigh energies," *Physical Review D - Particles, Fields, Gravitation and Cosmology*, vol. 77, no. 1, Article ID 013001, 2008.

[11] P. Lipari and T. Stanev, "Propagation of multi-TeV muons," *Physical Review D*, vol. 44, no. 11, pp. 3543–3554, 1991.

[12] M. G. Aartsen, K. Abraham, M. Ackermann et al., "The IceCube Neutrino Observatory—Contributions to ICRC 2015 Part II: Atmospheric and Astrophysical Diffuse Neutrino Searches of All Flavors: Combined Analysis of the High-Energy Cosmic Neutrino Flux at the IceCube Detector," October 2015.

[13] J. Conrad, A. De Gouvêa, S. Shalgar, and J. Spitz, "Atmospheric tau neutrinos in a multikiloton liquid argon detector," *Physical Review D - Particles, Fields, Gravitation and Cosmology*, vol. 82, no. 9, Article ID 093012, 2010.

[14] M. G. Aartsen, K. Abraham, M. Ackermann et al., "Search for astrophysical tau neutrinos in three years of IceCube data," *PhysicalReviewD*, vol. 930, no. 2, Article ID 022001, 2016 a.

[15] A. Aab, P. Abreu, M. Aglietta et al., "Improved limit to the diffuse flux of ultrahigh energy neutrinos from the pierre auger observatory," *Physical Review D*, vol. 910, no. 9, Article ID 092008, 2015.

[16] J. Abraham, P. Abreu, M. Aglietta et al., "A study of the effect of molecular and aerosol conditions in the atmosphere on air fluorescence measurements at the Pierre Auger Observatory," *Astroparticle Physics*, vol. 33, no. 2, pp. 108–129, 2010.

[17] J. Abraham, P. Abreu, M. Aglietta et al., "Erratum to "Atmospheric effects on extensive air showers observed with the surface detector of the Pierre Auger observatory" [Astroparticle Physics 32(2) (2009), 89–99]," *Astroparticle Physics*, vol. 33, no. 1, pp. 65–67, 2010.

[18] G. Miele, S. Pastor, and O. Pisanti, "The aperture for UHE tau neutrinos of the Auger fluorescence detector using a Digital Elevation Map," *Physics Letters, Section B: Nuclear, Elementary Particle and High-Energy Physics*, vol. 634, no. 2-3, pp. 137–142, 2006.

[19] Y. Aita, T. Aoki, Y. Asaoka et al., "Observational search for PeV-EeV tau neutrino from GRB081203A," *Astrophysical Journal Letters*, vol. 736, no. 1, article no. L12, 2011.

[20] M. Gaug, C. Hsu, J. K. Becker et al., "Tau neutrino search with the MAGIC telescope," in *Proceedings of the 30th International Cosmic Ray Conference, ICRC 2007*, pp. 1273–1276, July 2007.

[21] J. K. Becker, M. Gaug, C.-C. Hsu, and W. Rhode, "GRB neutrino search with MAGIC," in *Proceedings of the Santa Fe Conference on Gamma-Ray Bursts (GRB '07)*, pp. 245–248, November 2007.

[22] D. Fargion, P. Oliva, F. Massa, and G. Moreno, "Cherenkov flashes and fluorescence flares on telescopes: New lights on UHECR spectroscopy while unveiling neutrinos astronomy," *Nuclear Instruments and Methods in Physics Research, Section A: Accelerators, Spectrometers, Detectors and Associated Equipment*, vol. 588, no. 1-2, pp. 146–150, 2008.

[23] D. Fargion, M. Gaug, and P. Oliva, "Reflecting on Čerenkov reflections," *Journal of Physics: Conference Series*, vol. 110, no. 6, p. 062008, 2008.

[24] A. J. Smith, "HAWC: Design, operation, reconstruction and analysis," in *Proceedings of the 34th International Cosmic Ray Conference, ICRC 2015*, INSPIRE, The Netherlands, 2015.

[25] A. U. Abeysekara, A. Albert, R. Alfaro et al., "Observation of the crab nebula with the HAWC gamma-ray observatory," *e-prints*, 2017.

[26] H. A. Ayala Solares, M. Gerhardt, C. M. Hui et al., "HAWC Collaboration. The Calibration System of the HAWC Gamma-Ray Observatory, 2015".

[27] S. Agostinelli, J. Allison, K. Amako, J. Apostolakis, H. Araujo, and P. Arce, "Geant4—a simulation toolkit," *Nuclear Instruments and Methods in Physics Research A: Accelerators, Spectrometers, Detectors and Associated Equipment*, vol. 506, no. 3, pp. 250–303, 2003.

[28] S. Atağ and E. Gürkanlı, "Prediction for CP violation via electric dipole moment of τ lepton in $\gamma\gamma \rightarrow \tau^+\tau^-$ process at CLIC," *Journal of High Energy Physics*, vol. 2016, no. 118, 2016.

[29] J. L. Feng, P. Fisher, F. Wilczek, and T. M. Yu, "Observability of earth-skimming ultrahigh energy neutrinos," *Physical Review Letters*, vol. 88, Article ID 161102, 2002.

[30] S. I. Dutta, Y. Huang, and M. H. Reno, "Tau neutrino propagation and tau energy loss," *Physical Review D - Particles, Fields, Gravitation and Cosmology*, vol. 72, no. 1, Article ID 013005, pp. 1–10, 2005.

[31] O. Blanch Bigas, O. Deligny, K. Payet, and V. Van Elewyck, "Tau energy losses at ultrahigh energy: Continuous versus stochastic treatment," *Physical Review D - Particles, Fields, Gravitation and Cosmology*, vol. 77, no. 10, Article ID 103004, 2008.

[32] S. Iyer Dutta, M. H. Reno, I. Sarcevic, and D. Seckel, "Propagation of muons and taus at high energies," *Physical Review D*, vol. 63, no. 9, Article ID 094020, 2001.

[33] M. G. Aartsen, K. Aartsen, M. Ackermann et al., "The IceCube Neutrino Observatory - Contributions to ICRC 2015 Part II: Atmospheric and Astrophysical Diffuse Neutrino Searches of All Flavors: A measurement of the diffuse astrophysical muon neutrino flux using multiple years of IceCube data," October 2015.

[34] M. G. Aartsen, K. Abraham, M. Ackermann et al., "A Combined Maximum-likelihood Analysis of the High-energy Astrophysical Neutrino Flux Measured with IceCube," *Astrophysical Journal*, vol. 809, no. 98, 2015.

[35] S. Schoenen and L. Raedel, "Detection of a multi-PeV neutrino-induced muon event from the northern sky with IceCube," *The Astronomer's Telegram*, vol. 7856, 2015.

[36] M. G. Aartsen, K. Abraham, M. Ackermann et al., "The IceCube Neutrino Observatory - Contributions to ICRC 2015 Part II: Atmospheric and Astrophysical Diffuse Neutrino Searches of All Flavors: Observation of Astrophysical Neutrinos in Four Years of IceCube Data," October 2015.

[37] R. Gandhi, C. Quigg, M. H. Reno, and I. Sarcevic, "Neutrino interactions at ultrahigh energies," *Physical Review D*, vol. 580, no. 9, Article ID 093009, 1998.

[38] A. Connolly, R. S. Thorne, and D. Waters, "Calculation of high energy neutrino-nucleon cross sections and uncertainties using the Martin-Stirling-Thorne-Watt parton distribution functions and implications for future experiments," *Physical Review D*, vol. 83, no. 11, Article ID 113009, 13 pages, 2011.

[39] C. A. Argüelles, F. Halzen, L. Wille, M. Kroll, and M. H. Reno, "High-energy behavior of photon, neutrino, and proton cross sections," *Physical Review D - Particles, Fields, Gravitation and Cosmology*, vol. 92, no. 7, Article ID 074040, 2015.

[40] A. Cooper-Sarkar, P. Mertsch, and S. Sarkar, "The high energy neutrino cross-section in the Standard Model and its uncertainty," *Journal of High Energy Physics*, vol. 42, no. 8, 2011.

[41] M. M. Block, L. Durand, and P. Ha, "Connection of the virtual $\gamma^* p$ cross section of ep deep inelastic scattering to real γp scattering, and the implications for νN and ep total cross sections," *Physical Review D*, vol. 89, Article ID 094027, 2014.

[42] A. Cho, "Physicists snare a precious few neutrinos from the cosmos," *Science*, vol. 342, no. 6161, p. 920, 2013.

[43] A. U. Abeysekara, R. Alfaro, C. Alvarez et al., "Observation of small-scale anisotropy in the arrival direction distribution of TeV cosmic rays with HAWC," *Astrophysical Journal*, vol. 796, no. 2, article 796, 2014.

[44] A. Kusenko and T. J. Weiler, "Neutrino cross sections and future observations of ultrahigh-energy cosmic rays," *Physical Review Letters*, vol. 880, no. 16, Article ID 161101, 2002.

[45] D. Góra, E. Bernardini, and A. Kappes, "Searching for tau neutrinos with Cherenkov telescopes," *Astroparticle Physics*, vol. 61, pp. 12–16, 2015.

[46] Z. Cao, M. A. Huang, P. Sokolsky, and Y. Hu, "Ultra high energy $\nu\tau$ detection with a cosmic ray tau neutrino telescope using fluorescence/Cerenkov light technique," *Journal of Physics G: Nuclear and Particle Physics*, vol. 31, no. 7, pp. 571–582, 2005.

[47] INEGI., "Continuo de elevaciones mexicano 3.0 (cem 3.0)," 2012, http://www.inegi.org.mx/geo/contenidos/datosrelieve/continental/continuoelevaciones.aspx.

[48] M. G. Aartsen, K. Abraham, M. Ackermann et al., "Characterization of the atmospheric muon flux in IceCube," *Astroparticle Physics*, vol. 78, pp. 1–27, 2016.

[49] L. A. Anchordoqui, V. Barger, I. Cholis et al., "Cosmic neutrino pevatrons: A brand new pathway to astronomy, astrophysics, and particle physics," *Journal of High Energy Astrophysics*, vol. 1-2, pp. 1–30, 2014.

[50] A. A. Abdo, A. U. Abeysekara, R. Alfaro et al., "Milagro limits and HAWC sensitivity for the rate-density of evaporating Primordial Black Holes," *Astroparticle Physics*, vol. 64, pp. 4–12, 2015.

[51] A. U. Abeysekara, R. Alfaro, C. Alvarez et al., "Search for gamma-rays from the unusually bright GRB 130427A with the HAWC gamma-ray observatory," *The Astrophysical Journal*, vol. 800, no. 2, p. 78, 2015.

[52] L. Rädel and C. Wiebusch, "Calculation of the Cherenkov light yield from electromagnetic cascades in ice with Geant4," *Astroparticle Physics*, vol. 44, pp. 102–113, 2013.

[53] K. Olive, "Review of Particle Physics," *Chinese Physics C*, vol. 40, no. 10, Article ID 100001, 2016.

[54] H. Jokisch, K. Carstensen, W. D. Dau, H. J. Meyer, and O. C. Allkofer, "Cosmic-ray muon spectrum up to 1 TeV at 75∘zenith angle," *Physical Review D*, vol. 19, no. 5, pp. 1368–1372, 1979.

[55] I. Sarcevic, "Ultrahigh energy cosmic neutrinos and the physics beyond the Standard Model," *Journal of Physics: Conference Series*, vol. 60, no. 1, article no. 035, pp. 175–178, 2007.

PERMISSIONS

All chapters in this book were first published in AA, by Hindawi Publishing Corporation; hereby published with permission under the Creative Commons Attribution License or equivalent. Every chapter published in this book has been scrutinized by our experts. Their significance has been extensively debated. The topics covered herein carry significant findings which will fuel the growth of the discipline. They may even be implemented as practical applications or may be referred to as a beginning point for another development.

The contributors of this book come from diverse backgrounds, making this book a truly international effort. This book will bring forth new frontiers with its revolutionizing research information and detailed analysis of the nascent developments around the world.

We would like to thank all the contributing authors for lending their expertise to make the book truly unique. They have played a crucial role in the development of this book. Without their invaluable contributions this book wouldn't have been possible. They have made vital efforts to compile up to date information on the varied aspects of this subject to make this book a valuable addition to the collection of many professionals and students.

This book was conceptualized with the vision of imparting up-to-date information and advanced data in this field. To ensure the same, a matchless editorial board was set up. Every individual on the board went through rigorous rounds of assessment to prove their worth. After which they invested a large part of their time researching and compiling the most relevant data for our readers.

The editorial board has been involved in producing this book since its inception. They have spent rigorous hours researching and exploring the diverse topics which have resulted in the successful publishing of this book. They have passed on their knowledge of decades through this book. To expedite this challenging task, the publisher supported the team at every step. A small team of assistant editors was also appointed to further simplify the editing procedure and attain best results for the readers.

Apart from the editorial board, the designing team has also invested a significant amount of their time in understanding the subject and creating the most relevant covers. They scrutinized every image to scout for the most suitable representation of the subject and create an appropriate cover for the book.

The publishing team has been an ardent support to the editorial, designing and production team. Their endless efforts to recruit the best for this project, has resulted in the accomplishment of this book. They are a veteran in the field of academics and their pool of knowledge is as vast as their experience in printing. Their expertise and guidance has proved useful at every step. Their uncompromising quality standards have made this book an exceptional effort. Their encouragement from time to time has been an inspiration for everyone.

The publisher and the editorial board hope that this book will prove to be a valuable piece of knowledge for researchers, students, practitioners and scholars across the globe.

LIST OF CONTRIBUTORS

Yinhu Zhan
State Key Laboratory of Geo-Information Engineering, Xian, China

Donghan He
Zhengzhou Institute of Surveying and Mapping, Zhengzhou, China

Shaojie Chen
National Time Service Center, Chinese Academy of Sciences, Xian, China

Masaomi Tanaka
National Astronomical Observatory of Japan, Mitaka, Tokyo 181-8588, Japan

Ivan Zhelyazkov
Faculty of Physics, Sofia University, 1164 Sofia, Bulgaria

Ramesh Chandra
Department of Physics, Kumaun University, Nainital 263001, India

Abhishek K. Srivastava
Department of Physics, Indian Institute of Technology, Banaras Hindu University, Varanasi 221005, India

Ayodele Abiola Periola and Olabisi Emmanuel Falowo
Communication Research Group, Department of Electrical Engineering, University of Cape Town, Rondebosch, Cape Town, South Africa

In-Saeng Suh
Center for Astrophysics, Department of Physics and Center for Research Computing, University of Notre Dame, Notre Dame, IN 46556, USA

Grant J. Mathews and J. Reese Haywood
Center for Astrophysics, Department of Physics, University of Notre Dame, Notre Dame, IN 46556, USA

N. Q. Lan
Hanoi National University of Education, 136 XuanThuy, Hanoi, Vietnam
Joint Institute for Nuclear Astrophysics (JINA), University of Notre Dame, Notre Dame, IN 46556, USA

René Hudec
Astronomický Ústav AV ČR, Ondřejov (AS ÚAV ČR), Ond řejov, Czech Republic

Martin Jelínek, Alberto J. Castro-Tirado, Ronan Cunniffe and Javier Gorosabel
Instituto de Astrofísica de Andalucía- (IAA-) CSIC, 18008 Granada, Spain

Dolores Pérez-Ramírez
Departamento de Ingeniería de Sistemas y Automática (Unidad Asociada al CSIC), Universidad de Málaga, 29010 Málaga, Spain

Javier Gorosabel
Unidad Asociada Grupo Ciencia Planetarias UPV/EHU-IAA/CSIC, Departamento de Física Aplicada I, E.T.S. de Ingeniería, Universidad del Páis Vasco (UPV)/EHU, Alameda de Urquijo s/n, 48013 Bilbao, Spain
Ikerbasque, Basque Foundation for Science, Alameda de Urquijo 36-5, 48008 Bilbao, Spain

Stanislav Vítek, Petr Páta and René Hudec
České Vysoké Učení Technické, Fakulta Elektrotechnická (ČVUT-FEL), Praha, Czech Republic

Petr Kubánek
Fyzikální ústav AV ČR, Na Slovance 2, 182 21 Praha 8, Czech Republic

Víctor Reglero
Image Processing Laboratory, Universidad de Valencia, Burjassot, Valencia, Spain

Soomin Jeong
Institute for Science and Technology in Space, Natural Science Campus, Sungkyunkwan University, Suwon 440-746, Republic of Korea

Sebastián Castillo-Carrión
Universidad de Málaga, Campus de Teatinos, Málaga, Spain

Tomás Mateo Sanguino
Departamento de Ingeniería de Sistemas y Automática, Universidad de Huelva, E.P.S. de La R ábida, Huelva, Spain

Ovidio Rabaza
Department of Civil Engineering, University of Granada, 18071 Granada, Spain

Dolores Pérez-Ramírez
Universidad de Jaén, Campus las Lagunillas, 23071 Jaén, Spain

Rafael Fernández-Muñoz
Instituto de Hortofruticultura Subtropicaly Mediterránea "La Mayora" (IHSM-CSIC), Algarrobo, 29750 Málaga, Spain

Benito A. de la Morena Carretero
Estación de Sondeos Atmosféricos (ESAt) de El Arenosillo (CEDEA-INTA), Mazagón, Huelva, Spain

Lola Sabau-Graziati
División de Ciencias del Espacio, INTA, Torrejón de Ardoz, Madrid, Spain

Makoto Miyoshi
National Astronomical Observatory, 2-21-1 Osawa, Mitaka, Tokyo 181-8588, Japan

Takashi Kasuga
Department of Advanced Sciences, Faculty of Science and Engineering, Hosei University, 3-7-2 Kajino, Koganei, Tokyo 184-8584, Japan

Jose K. Ishitsuka Iba
Geophysical Institute of Peru, Carretera Ing. Alberto A. Giesecke M. Km 15, Huachac, Peru

Tomoharu Oka
Department of Physics, Institute of Science and Technology, Keio University, 3-14-1 Hiyoshi, Yokohama, Kanagawa 223-8522, Japan

Mamoru Sekido and Kazuhiro Takefuji
Kashima Space Technology Center, National Institute of Information and Communications Technology, 893-1 Hirai, Kashima, Ibaraki 314-8501, Japan

Masaaki Takahashi
Department of Physics and Astronomy, Aichi University of Education, Kariya 448-8542, Japan

Hiromi Saida
Department of Physics, Daido University, Minami-ku, Nagoya 457-8530, Japan

Rohta Takahashi
National Institute of Technology, Tomakomai College, 443 Nishikioka, Tomakomai, Hokkaido 059-1275, Japan

Sayantan Auddy and Shantanu Basu
Department of Physics and Astronomy, The University of Western Ontario, London, ON, Canada N6A 3K7

S. R. Valluri
King's University College,The University ofWestern Ontario, London, ON, Canada N6A 2M3

Zach Cano
Centre for Astrophysics and Cosmology, Science Institute, University of Iceland, Dunhagi 5, 107 Reykjavik, Iceland
Instituto de Astrofísica de Andalucía (IAA-CSIC), Glorieta de la Astronomía s/n, 18008 Granada, Spain

Shan-Qin Wang and Zi-Gao Dai
School of Astronomy and Space Science, Nanjing University, Nanjing 210093, China
Key Laboratory of Modern Astronomy and Astrophysics (Nanjing University), Ministry of Education, Nanjing, China

Xue-Feng Wu
Purple Mountain Observatory, Chinese Academy of Sciences, Nanjing 210008, China
Joint Center for Particle, Nuclear Physics and Cosmology, Nanjing University-Purple Mountain Observatory, Nanjing 210008, China

Poonam Chandra
National Centre for Radio Astrophysics, Tata Institute of Fundamental Research, Pune University Campus, Pune 411007, India

F. B. Gao, X. Liu and R. F. Wang
School of Mathematical Science, Yangzhou University, Yangzhou 225002, China

X. H. Zhu
Department of Mathematics, Shanghai University, Shanghai 200444, China

Takashi Ito
National Astronomical Observatory of Japan, Osawa 2-21-1, Mitaka, Tokyo 181-8588, Japan

Zhi-ming Song
University of Chinese Academy of Sciences, Beijing 100049, China

Zhong-quan Qu
Yunnan Observatories, Chinese Academy of Science, Kunming 650011, Yunnan, China

Hermes León Vargas, Andrés Sandoval, Ernesto Belmont and Rubén Alfaro
Instituto de Física, Universidad Nacional Autónoma de México, Apartado Postal 20-364, 01000 Ciudad de México, Mexico

Index

www.ingramcontent.com/pod-product-compliance
Lightning Source LLC
Chambersburg PA
CBHW080523200326
41458CB00012B/4316